STUDENT'S SOLUTIONS MANUAL

CINDY TRIMBLE & ASSOCIATES

D0209974

INTRODUCTORY ALGEBRA

FOURTH EDITION

ELAYN MARTIN-GAY

University of New Orleans

Prentice Hall
is an imprint of

PEARSON

The author and publisher of this book have used their best efforts in preparing this book. These efforts include the development, research, and testing of the theories and programs to determine their effectiveness. The author and publisher make no warranty of any kind, expressed or implied, with regard to these programs or the documentation contained in this book. The author and publisher shall not be liable in any event for incidental or consequential damages in connection with, or arising out of, the furnishing, performance, or use of these programs.

Reproduced by Prentice Hall from electronic files supplied by the author.

Copyright © 2012, 2007, 2003 Pearson Education, Inc.
Publishing as Pearson Prentice Hall, 75 Arlington Street, Boston, MA 02116.

All rights reserved. No part of this publication may be reproduced, stored in a retrieval system, or transmitted, in any form or by any means, electronic, mechanical, photocopying, recording, or otherwise, without the prior written permission of the publisher. Printed in the United States of America.

ISBN-13: 978-0-321-74532-3
ISBN-10: 0-321-74532-9

1 2 3 4 5 6 EB 15 14 13 12 11

Prentice Hall
is an imprint of

PEARSON

www.pearsonhighered.com

Contents

Chapter R	1
Chapter 1	15
Chapter 2	46
Chapter 3	107
Chapter 4	150
Chapter 5	189
Chapter 6	245
Chapter 7	304
Chapter 8	349
Chapter 9	382
Appendices	431
Practice Final Exam	435

Chapter R

Section R.1 Practice

1. $10 = 1 \cdot 10$, $10 = 2 \cdot 5$
   The factors of 10 are 1, 2, 5, and 10.

2. $18 = 1 \cdot 18$, $18 = 2 \cdot 9$, $18 = 3 \cdot 6$
   The factors of 18 are 1, 2, 3, 6, 9, and 18.

3. 5 is a prime number. Its factors are 1 and 5 only.
   16 is a composite number. Its factors are 1, 2, 4, 8, and 16.
   23 is a prime number. Its factors are 1 and 23 only.
   42 is a composite number. Its factors are 1, 2, 3, 6, 7, 14, 21, and 42.

4. $44 = 4 \cdot 11 = 2 \cdot 2 \cdot 11$
   The prime factorization of 44 is $2 \cdot 2 \cdot 11$.

5. $60 = 4 \cdot 15 = 2 \cdot 2 \cdot 3 \cdot 5$
   The prime factorization of 60 is $2 \cdot 2 \cdot 3 \cdot 5$.

6. $$\begin{array}{r} 11 \\ 3\overline{)\,33} \\ 3\overline{)\,99} \\ 3\overline{)297} \end{array}$$
   The prime factorization of 297 is $3 \cdot 3 \cdot 3 \cdot 11$.

7. $14 = 2 \cdot 7$
   $35 = 5 \cdot 7$
   $\text{LCM} = 2 \cdot 5 \cdot 7 = 70$

8. $5 = 5$
   $9 = 3 \cdot 3$
   $\text{LCM} = 3 \cdot 3 \cdot 5 = 45$

9. $4 = 2 \cdot 2$
   $15 = 3 \cdot 5$
   $10 = 2 \cdot 5$
   $\text{LCM} = 2 \cdot 2 \cdot 3 \cdot 5 = 60$

Vocabulary and Readiness Check

1. The number 40 equals $2 \cdot 2 \cdot 2 \cdot 5$. Since each factor is prime, we call $2 \cdot 2 \cdot 2 \cdot 5$ the <u>prime factorization</u> of 40.

2. A natural number, other than 1, that is not prime is called a <u>composite</u> number.

3. A natural number that has exactly two different factors, 1 and itself, is called a <u>prime</u> number.

4. The <u>least common multiple</u> of a list of numbers is the smallest number that is a multiple of all the numbers in the list.

5. To <u>factor</u> means to write as a product.

6. A <u>multiple</u> of a number is the product of that number and any natural number.

Exercise Set R.1

1. $9 = 1 \cdot 9$, $9 = 3 \cdot 3$
   The factors of 9 are 1, 3, and 9.

3. $24 = 1 \cdot 24$, $24 = 2 \cdot 12$, $24 = 3 \cdot 8$, $24 = 4 \cdot 6$
   The factors of 24 are 1, 2, 3, 4, 6, 8, 12, and 24.

5. $42 = 1 \cdot 42$, $42 = 2 \cdot 21$, $42 = 3 \cdot 14$,
   $42 = 6 \cdot 7$
   The factors of 42 are 1, 2, 3, 6, 7, 14, 21, and 42.

7. $80 = 1 \cdot 80$, $80 = 2 \cdot 40$, $80 = 4 \cdot 20$,
   $80 = 5 \cdot 16$, $80 = 8 \cdot 10$
   The factors of 80 are 1, 2, 4, 5, 8, 10, 16, 20, 40, and 80.

9. $19 = 1 \cdot 19$
   The factors of 19 are 1 and 19.

11. 13 is a prime number. Its factors are only 1 and 13.

13. 39 is a composite number. Its factors are 1, 3, 13, and 39.

15. 41 is a prime number. Its factors are only 1 and 41.

17. 201 is composite. Its factors are 1, 3, 67, and 201.

19. 2065 is a composite number. Its factors are 1, 5, 7, 35, 59, 295, 413, and 2065.

21. $18 = 2 \cdot 9 = 2 \cdot 3 \cdot 3$
    The prime factorization of 18 is $2 \cdot 3 \cdot 3$.

23. $20 = 4 \cdot 5 = 2 \cdot 2 \cdot 5$
    The prime factorization of 20 is $2 \cdot 2 \cdot 5$.

Copyright © 2012 Pearson Education, Inc. Publishing as Prentice Hall.

25. $56 = 2 \cdot 28 = 2 \cdot 2 \cdot 14 = 2 \cdot 2 \cdot 2 \cdot 7$
The prime factorization of 56 is $2 \cdot 2 \cdot 2 \cdot 7$.

27. $81 = 3 \cdot 27 = 3 \cdot 3 \cdot 9 = 3 \cdot 3 \cdot 3 \cdot 3$
The prime factorization of 81 is $3 \cdot 3 \cdot 3 \cdot 3$.

29.
$$\begin{array}{r} 5 \\ 5\overline{)25} \\ 3\overline{)75} \\ 2\overline{)150} \\ 2\overline{)300} \end{array}$$
The prime factorization of 300 is
$2 \cdot 2 \cdot 3 \cdot 5 \cdot 5$.

31.
$$\begin{array}{r} 7 \\ 7\overline{)49} \\ 3\overline{)147} \\ 2\overline{)294} \\ 2\overline{)588} \end{array}$$
The prime factorization of 588 is
$2 \cdot 2 \cdot 3 \cdot 7 \cdot 7$.

33. $48 = 1 \cdot 48$, $48 = 2 \cdot 24$, $48 = 3 \cdot 16$,
$48 = 4 \cdot 12$, $48 = 6 \cdot 8$
The factors of 48 are 1, 2, 3, 4, 6, 8, 12, 16, and 48, which is choice d.

35. $3 = 3$
$4 = 2 \cdot 2$
$\text{LCM} = 2 \cdot 2 \cdot 3 = 12$

37. $6 = 2 \cdot 3$
$14 = 2 \cdot 7$
$\text{LCM} = 2 \cdot 3 \cdot 7 = 42$

39. $20 = 2 \cdot 2 \cdot 5$
$30 = 2 \cdot 3 \cdot 5$
$\text{LCM} = 2 \cdot 2 \cdot 3 \cdot 5 = 60$

41. $5 = 5$
$7 = 7$
$\text{LCM} = 5 \cdot 7 = 35$

43. $9 = 3 \cdot 3$
$12 = 2 \cdot 2 \cdot 3$
$\text{LCM} = 2 \cdot 2 \cdot 3 \cdot 3 = 36$

45. $16 = 2 \cdot 2 \cdot 2 \cdot 2$
$20 = 2 \cdot 2 \cdot 5$
$\text{LCM} = 2 \cdot 2 \cdot 2 \cdot 2 \cdot 5 = 80$

47. $40 = 2 \cdot 2 \cdot 2 \cdot 5$
$90 = 2 \cdot 3 \cdot 3 \cdot 5$
$\text{LCM} = 2 \cdot 2 \cdot 2 \cdot 3 \cdot 3 \cdot 5 = 360$

49. $24 = 2 \cdot 2 \cdot 2 \cdot 3$
$36 = 2 \cdot 2 \cdot 3 \cdot 3$
$\text{LCM} = 2 \cdot 2 \cdot 2 \cdot 3 \cdot 3 = 72$

51. $2 = 2$
$8 = 2 \cdot 2 \cdot 2$
$15 = 3 \cdot 5$
$\text{LCM} = 2 \cdot 2 \cdot 2 \cdot 3 \cdot 5 = 120$

53. $2 = 2$
$3 = 3$
$7 = 7$
$\text{LCM} = 2 \cdot 3 \cdot 7 = 42$

55. $8 = 2 \cdot 2 \cdot 2$
$24 = 2 \cdot 2 \cdot 2 \cdot 3$
$48 = 2 \cdot 2 \cdot 2 \cdot 2 \cdot 3$
$\text{LCM} = 2 \cdot 2 \cdot 2 \cdot 2 \cdot 3 = 48$

57. $8 = 2 \cdot 2 \cdot 2$
$18 = 2 \cdot 3 \cdot 3$
$30 = 2 \cdot 3 \cdot 5$
$\text{LCM} = 2 \cdot 2 \cdot 2 \cdot 3 \cdot 3 \cdot 5 = 360$

59. a. $40 = 2 \cdot 20 = 2 \cdot 2 \cdot 10 = 2 \cdot 2 \cdot 2 \cdot 5$
The prime factorization of 40 is
$2 \cdot 2 \cdot 2 \cdot 5$.

b. $40 = 4 \cdot 10 = 2 \cdot 2 \cdot 10 = 2 \cdot 2 \cdot 2 \cdot 5$
The prime factorization of 40 is $2 \cdot 2 \cdot 2 \cdot 5$.

c. answers may vary

61. $5 = 5$
$7 = 7$
$\text{LCM} = 5 \cdot 7 = 35$
They have the same night off every 35 days.

63. $315 = 3 \cdot 3 \cdot 5 \cdot 7$
$504 = 2 \cdot 2 \cdot 2 \cdot 3 \cdot 3 \cdot 7$
$\text{LCM} = 2 \cdot 2 \cdot 2 \cdot 3 \cdot 3 \cdot 5 \cdot 7 = 2520$

Section R.2 Practice

1. $\frac{4}{4} = 1$ since $4 \div 4 = 1$.

2. $\frac{9}{3} = 3$ since $9 \div 3 = 3$.

Copyright © 2012 Pearson Education, Inc. Publishing as Prentice Hall.

3. $\frac{10}{10}=1$ since $10 \div 10 = 1$.

4. $\frac{5}{1}=5$ since $5 \div 1 = 5$.

5. $\frac{0}{11}=0$ since $0 \cdot 11 = 0$.

6. $\frac{11}{0}$ is undefined because there is no number that when multiplied by 0 gives 11.

7. $\frac{1}{4}=\frac{1}{4}\cdot\frac{5}{5}=\frac{1\cdot 5}{4\cdot 5}=\frac{5}{20}$

8. $\frac{20}{35}=\frac{2\cdot 2\cdot 5}{5\cdot 7}=\frac{4}{7}$

9. $\frac{7}{20}$ is already simplified.

10. $\frac{12}{40}=\frac{4\cdot 3}{4\cdot 10}=\frac{3}{10}$

11. $\frac{3}{4}\cdot\frac{8}{9}=\frac{3\cdot 8}{4\cdot 9}=\frac{3\cdot 4\cdot 2}{4\cdot 3\cdot 3}=\frac{2}{3}$

12. $\frac{2}{9}\div\frac{3}{4}=\frac{2}{9}\cdot\frac{4}{3}=\frac{2\cdot 4}{9\cdot 3}=\frac{8}{27}$

13. $\frac{8}{11}\div 24=\frac{8}{11}\div\frac{24}{1}=\frac{8}{11}\cdot\frac{1}{24}=\frac{8\cdot 1}{11\cdot 8\cdot 3}=\frac{1}{33}$

14. $\frac{5}{4}\div\frac{15}{8}=\frac{5}{4}\cdot\frac{8}{15}=\frac{5\cdot 4\cdot 2}{4\cdot 5\cdot 3}=\frac{2}{3}$

15. $\frac{2}{11}+\frac{5}{11}=\frac{2+5}{11}=\frac{7}{11}$

16. $\frac{1}{8}+\frac{3}{8}=\frac{1+3}{8}=\frac{4}{8}=\frac{4}{2\cdot 4}=\frac{1}{2}$

17. $\frac{7}{6}-\frac{2}{6}=\frac{7-2}{6}=\frac{5}{6}$

18. $\frac{13}{10}-\frac{3}{10}=\frac{13-3}{10}=\frac{10}{10}=1$

19. $\frac{3}{8}+\frac{1}{20}=\frac{3}{8}\cdot\frac{5}{5}+\frac{1}{20}\cdot\frac{2}{2}=\frac{15}{40}+\frac{2}{40}=\frac{17}{40}$

20. $\frac{8}{15}-\frac{1}{3}=\frac{8}{15}-\frac{1}{3}\cdot\frac{5}{5}=\frac{8}{15}-\frac{5}{15}=\frac{3}{15}=\frac{3}{3\cdot 5}=\frac{1}{5}$

21. $5\frac{1}{6}=\frac{6\cdot 5+1}{6}=\frac{31}{6}$; $4\frac{2}{5}=\frac{5\cdot 4+2}{5}=\frac{22}{5}$

$5\frac{1}{6}\cdot 4\frac{2}{5}=\frac{31}{6}\cdot\frac{22}{5}=\frac{31\cdot 2\cdot 11}{2\cdot 3\cdot 5}=\frac{341}{15}=22\frac{11}{15}$

22. $7\frac{3}{8}+6\frac{3}{4}=\frac{59}{8}+\frac{27}{4}=\frac{59}{8}+\frac{54}{8}=\frac{113}{8}=14\frac{1}{8}$

23.

$$\begin{array}{rrr} 76\frac{1}{12} & 76\frac{1}{12} & 75\frac{13}{12} \\ -\,35\frac{1}{4} & -\,35\frac{3}{12} & -\,35\frac{3}{12} \\ \hline & & 40\frac{10}{12}=40\frac{5}{6} \end{array}$$

Vocabulary and Readiness Check

1. The number $\frac{17}{31}$ is called a <u>fraction</u>. The number 31 is called its <u>denominator</u> and 17 is called its <u>numerator</u>.

2. The fraction $\frac{8}{3}$ is called an <u>improper</u> fraction, the fraction $\frac{3}{8}$ is called a <u>proper</u> fraction, and $10\frac{3}{8}$ is called a <u>mixed number</u>.

3. In $\frac{11}{48}$, since 11 and 48 have no common factors other than 1, $\frac{11}{48}$ is in <u>simplest form</u>.

4. Fractions that represent the same portion of a whole are called <u>equivalent</u> fractions.

5. To multiply two fractions, we write $\frac{a}{b}\cdot\frac{c}{d}=\underline{\frac{a\cdot c}{b\cdot d}}$.

6. Two numbers are <u>reciprocals</u> of each other if their product is 1.

Copyright © 2012 Pearson Education, Inc. Publishing as Prentice Hall.

7. To divide two fractions, we write $\frac{a}{b} \div \frac{c}{d} = \underline{\frac{a \cdot d}{b \cdot c}}$.

8. $\frac{a}{b} + \frac{c}{b} = \underline{\frac{a+c}{b}}$ and $\frac{a}{b} - \frac{c}{b} = \underline{\frac{a-c}{b}}$.

9. The smallest positive number divisible by all the denominators of a list of fractions is called the <u>least common denominator (LCD)</u>.

10. The LCD for $\frac{1}{6}$ and $\frac{5}{8}$ is <u>24</u>.

Exercise Set R.2

1. $\frac{14}{14} = 1$ since $14 \cdot 1 = 14$.

3. $\frac{20}{2} = 10$ since $2 \cdot 10 = 20$.

5. $\frac{13}{1} = 13$ since $1 \cdot 13 = 13$.

7. $\frac{0}{9} = 0$ since $9 \cdot 0 = 0$.

9. $\frac{9}{0}$ is undefined.

11. $\frac{7}{10} = \frac{7 \cdot 3}{10 \cdot 3} = \frac{21}{30}$

13. $\frac{2}{9} = \frac{2 \cdot 2}{9 \cdot 2} = \frac{4}{18}$

15. $\frac{4}{5} = \frac{4 \cdot 4}{5 \cdot 4} = \frac{16}{20}$

17. $\frac{2}{4} = \frac{2 \cdot 1}{2 \cdot 2} = \frac{1}{2}$

19. $\frac{10}{15} = \frac{2 \cdot 5}{3 \cdot 5} = \frac{2}{3}$

21. $\frac{3}{7}$ cannot be simplified further.

23. $\frac{18}{30} = \frac{3 \cdot 6}{5 \cdot 6} = \frac{3}{5}$

25. $\frac{16}{20} = \frac{4 \cdot 4}{4 \cdot 5} = \frac{4}{5}$

27. $\frac{66}{48} = \frac{6 \cdot 11}{6 \cdot 8} = \frac{11}{8}$

29. $\frac{120}{244} = \frac{4 \cdot 30}{4 \cdot 61} = \frac{30}{61}$

31. $\frac{192}{264} = \frac{8 \cdot 24}{11 \cdot 24} = \frac{8}{11}$

33. $\frac{1}{2} \cdot \frac{3}{4} = \frac{1 \cdot 3}{2 \cdot 4} = \frac{3}{8}$

35. $\frac{2}{3} \cdot \frac{3}{4} = \frac{2 \cdot 3}{3 \cdot 4} = \frac{2 \cdot 3 \cdot 1}{2 \cdot 3 \cdot 2} = \frac{1}{2}$

37. $\frac{1}{2} \div \frac{7}{12} = \frac{1}{2} \cdot \frac{12}{7} = \frac{1 \cdot 12}{2 \cdot 7} = \frac{2 \cdot 6}{2 \cdot 7} = \frac{6}{7}$

39. $\frac{3}{4} \div \frac{1}{20} = \frac{3}{4} \cdot \frac{20}{1} = \frac{3 \cdot 20}{4 \cdot 1} = \frac{3 \cdot 4 \cdot 5}{4 \cdot 1} = \frac{15}{1} = 15$

41. $5\frac{1}{9} \cdot 3\frac{2}{3} = \frac{46}{9} \cdot \frac{11}{3} = \frac{506}{27} = 18\frac{20}{27}$

43.
$$\begin{aligned} 8\frac{3}{5} \div 2\frac{9}{10} &= \frac{43}{5} \div \frac{29}{10} \\ &= \frac{43}{5} \cdot \frac{10}{29} \\ &= \frac{43 \cdot 5 \cdot 2}{5 \cdot 29} \\ &= \frac{86}{29} \\ &= 2\frac{28}{29} \end{aligned}$$

45. $\frac{4}{5} + \frac{1}{5} = \frac{4+1}{5} = \frac{5}{5} = 1$

47.
$$\begin{aligned} \frac{4}{15} - \frac{1}{12} &= \frac{4}{15} \cdot \frac{4}{4} - \frac{1}{12} \cdot \frac{5}{5} \\ &= \frac{16}{60} - \frac{5}{60} \\ &= \frac{16-5}{60} \\ &= \frac{11}{60} \end{aligned}$$

Copyright © 2012 Pearson Education, Inc. Publishing as Prentice Hall.

49. $\frac{2}{3}+\frac{3}{7}=\frac{2\cdot 7}{3\cdot 7}+\frac{3\cdot 3}{7\cdot 3}=\frac{14}{21}+\frac{9}{21}=\frac{14+9}{21}=\frac{23}{21}$

51. $\frac{10}{3}-\frac{5}{21}=\frac{10\cdot 7}{3\cdot 7}-\frac{5}{21}$
$=\frac{70}{21}-\frac{5}{21}$
$=\frac{70-5}{21}$
$=\frac{65}{21}$

53.
$$\begin{array}{r} 8\frac{1}{8} \\ -\,6\frac{3}{8} \\ \hline \end{array} \qquad \begin{array}{r} 7\frac{9}{8} \\ -\,6\frac{3}{8} \\ \hline 1\frac{6}{8} \end{array} = 1\frac{3}{4}$$

55.
$$\begin{array}{r} 1\frac{1}{2} \\ +\,3\frac{2}{3} \\ \hline \end{array} \qquad \begin{array}{r} 1\frac{3}{6} \\ +\,3\frac{4}{6} \\ \hline 4\frac{7}{6} \end{array} = 4+1\frac{1}{6}=5\frac{1}{6}$$

57. $\frac{23}{105}+\frac{4}{105}=\frac{23+4}{105}=\frac{27}{105}=\frac{3\cdot 9}{3\cdot 35}=\frac{9}{35}$

59. $\frac{17}{21}-\frac{10}{21}=\frac{17-10}{21}=\frac{7}{21}=\frac{1\cdot 7}{3\cdot 7}=\frac{1}{3}$

61. $\frac{7}{10}\cdot\frac{5}{21}=\frac{7\cdot 5}{10\cdot 21}=\frac{7\cdot 5}{5\cdot 2\cdot 7\cdot 3}=\frac{1}{2\cdot 3}=\frac{1}{6}$

63. $\frac{9}{20}\div 12=\frac{9}{20}\div\frac{12}{1}=\frac{9}{20}\cdot\frac{1}{12}=\frac{3\cdot 3\cdot 1}{20\cdot 3\cdot 4}=\frac{3}{80}$

65. $\frac{5}{22}-\frac{5}{33}=\frac{5\cdot 3}{22\cdot 3}-\frac{5\cdot 2}{33\cdot 2}$
$=\frac{15}{66}-\frac{10}{66}$
$=\frac{15-10}{66}$
$=\frac{5}{66}$

67. $17\frac{2}{5}+30\frac{2}{3}=\frac{87}{5}+\frac{92}{3}$
$=\frac{87\cdot 3}{5\cdot 3}+\frac{92\cdot 5}{3\cdot 5}$
$=\frac{261}{15}+\frac{460}{15}$
$=\frac{721}{15}$
$=48\frac{1}{15}$

69. $7\frac{2}{5}\div\frac{1}{5}=\frac{37}{5}\div\frac{1}{5}=\frac{37}{5}\cdot\frac{5}{1}=\frac{37\cdot 5}{5\cdot 1}=\frac{37}{1}=37$

71. $4\frac{2}{11}\cdot 2\frac{1}{2}=\frac{46}{11}\cdot\frac{5}{2}=\frac{2\cdot 23\cdot 5}{11\cdot 2}=\frac{115}{11}=10\frac{5}{11}$

73. $\frac{12}{5}-1=\frac{12}{5}-\frac{5}{5}=\frac{12-5}{5}=\frac{7}{5}$

75.
$$\begin{array}{r} 8\frac{11}{12} \\ -\,1\frac{5}{6} \\ \hline \end{array} \qquad \begin{array}{r} 8\frac{11}{12} \\ -\,1\frac{10}{12} \\ \hline 7\frac{1}{12} \end{array}$$

77. $\frac{2}{3}-\frac{5}{9}+\frac{5}{6}=\frac{2\cdot 6}{3\cdot 6}-\frac{5\cdot 2}{9\cdot 2}+\frac{5\cdot 3}{6\cdot 3}$
$=\frac{12}{18}-\frac{10}{18}+\frac{15}{18}$
$=\frac{12-10+15}{18}$
$=\frac{17}{18}$

79. The work is incorrect.
$\frac{12}{24}=\frac{2\cdot 2\cdot 3}{2\cdot 2\cdot 2\cdot 3}=\frac{1}{2}$

81. The work is incorrect.
$\frac{2}{7}+\frac{9}{7}=\frac{2+9}{7}=\frac{11}{7}$

83. answers may vary

Copyright © 2012 Pearson Education, Inc. Publishing as Prentice Hall.

85. $1-\frac{1}{3}-\frac{5}{12}-\frac{1}{6}=\frac{12}{12}-\frac{4}{12}-\frac{5}{12}-\frac{2}{12}$

$=\frac{12-4-5-2}{12}$

$=\frac{1}{12}$

The unknown part is $\frac{1}{12}$.

87. $1-\frac{3}{11}-\frac{2}{11}=\frac{11}{11}-\frac{3}{11}-\frac{2}{11}=\frac{11-3-2}{11}=\frac{6}{11}$

The unknown part is $\frac{6}{11}$.

89.

$$\begin{array}{r} 11\frac{1}{4} \\ -\,3\frac{3}{5} \\ \hline \end{array} \qquad \begin{array}{r} 11\frac{5}{20} \\ -\,3\frac{12}{20} \\ \hline \end{array} \qquad \begin{array}{r} 10\frac{25}{20} \\ -\,3\frac{12}{20} \\ \hline 7\frac{13}{20} \end{array}$$

On average, Tucson gets $7\frac{13}{20}$ inches more rain than Yuma.

91. a. The graph shows $\frac{21}{100}$ are miniplexes.

b. $\frac{1}{25}+\frac{23}{50}=\frac{2}{50}+\frac{23}{50}=\frac{25}{50}=\frac{1}{2}$

$\frac{1}{2}$ of U.S. movie screens are in single-screen theaters or multiplexes.

c. The largest sector of the graph corresponds to multiplexes, so multiplexes have the greatest number of screens.

d. $\frac{29}{100}+\frac{21}{100}=\frac{50}{100}=\frac{1}{2}$

$\frac{1}{2}$ of U.S. movie screens are in megaplexes or miniplexes.

93. $A=lw=\frac{2}{5}\cdot\frac{3}{11}=\frac{2\cdot 3}{5\cdot 11}=\frac{6}{55}$

The area is $\frac{6}{55}$ square meter.

Section R.3 Practice

1. $0.27=\frac{27}{100}$

2. $5.1=\frac{51}{10}$

3. $7.685=\frac{7685}{1000}$

4. a.

$$\begin{array}{r} 7.19 \\ 19.782 \\ +\,1.006 \\ \hline 27.978 \end{array}$$

b.

$$\begin{array}{r} 12. \\ 0.79 \\ +\,0.03 \\ \hline 12.82 \end{array}$$

5. a.

$$\begin{array}{r} 84.230 \\ -\,26.982 \\ \hline 57.248 \end{array}$$

b.

$$\begin{array}{r} 90.00 \\ -\,0.19 \\ \hline 89.81 \end{array}$$

6. a.

$$\begin{array}{r} 0.31 \\ \times\,4.6 \\ \hline 186 \\ 1\,24 \\ \hline 1.426 \end{array}$$

b.

$$\begin{array}{r} 1.26 \\ \times\,0.03 \\ \hline 0.0378 \end{array}$$

7. a.

$$\begin{array}{r} 43.5 \\ 0.5\overline{)\,21.75} \\ \underline{-20} \\ 1\,7 \\ \underline{-1\,5} \\ 25 \\ \underline{-25} \\ 0 \end{array}$$

Copyright © 2012 Pearson Education, Inc. Publishing as Prentice Hall.

b.
$$\begin{array}{r} 2600 \\ 0.006\overline{)\,15.600} \\ \underline{-12} \\ 36 \\ \underline{-36} \\ 0 \end{array}$$

8. 12.9187 rounded to the nearest hundredth is 12.92.

9. 245.348 rounded to the nearest tenth is 245.3.

10.
$$\begin{array}{r} 0.4 \\ 5\overline{)\,2.0} \\ \underline{-20} \\ 0 \end{array}$$

$\frac{2}{5} = 0.4$

11.
$$\begin{array}{r} 0.833 \\ 6\overline{)\,5.000} \\ \underline{-48} \\ 20 \\ \underline{-18} \\ 20 \\ \underline{-18} \\ 2 \end{array}$$

$\frac{5}{6} = 0.833... = 0.8\overline{3}$

12.
$$\begin{array}{r} 0.1111 \\ 9\overline{)1.0000} \\ \underline{-9} \\ 10 \\ \underline{-9} \\ 10 \\ \underline{-9} \\ 10 \\ \underline{-9} \\ 1 \end{array}$$

$\frac{1}{9} = 0.1111... \approx 0.111$

13. a. $20\% = 20.\% = 0.20$

b. $1.2\% = 01.2\% = 0.012$

c. $465\% = 465.\% = 4.65$

14. a. $0.42 = 0.42 = 42\%$

b. $0.003 = 0.003 = 0.3\%$

c. $2.36 = 2.36 = 236\%$

d. $0.7 = 0.70 = 70\%$

Vocabulary and Readiness Check

1. Like fractional notation, decimal notation is used to denote a part of a whole.

2. To write fractions as decimals, divide the numerator by the denominator.

3. To add or subtract decimals, write the decimals so that the decimal points line up vertically.

4. When multiplying decimals, the decimal point in the product is placed so that the number of decimal places in the product is equal to the sum of the number of decimal places in the factors.

5. Percent means "per hundred."

6. 100% = 1

7. The % symbol is read as percent.

8. To write a percent as a *decimal*, drop the % symbol and move the decimal point two places to the left.

9. To write a decimal as a *percent*, move the decimal point two places to the right and attach the % symbol.

Exercise Set R.3

1. $0.6 = \frac{6}{10}$

3. $1.86 = \frac{186}{100}$

5. $0.114 = \frac{114}{1000}$

7. $123.1 = \frac{1231}{10}$

Copyright © 2012 Pearson Education, Inc. Publishing as Prentice Hall.

9.
$$\begin{array}{r} 5.7 \\ +\,1.13 \\ \hline 6.83 \end{array}$$

11.
$$\begin{array}{r} 24.6 \\ 2.39 \\ +\,0.0678 \\ \hline 27.0578 \end{array}$$

13.
$$\begin{array}{r} 8.8 \\ -\,2.3 \\ \hline 6.5 \end{array}$$

15.
$$\begin{array}{r} 18.00 \\ -\,2.78 \\ \hline 15.22 \end{array}$$

17.
$$\begin{array}{r} 0.2 \\ \times\,0.6 \\ \hline 0.12 \end{array}$$

19.
$$\begin{array}{r} 0.063 \\ \times\quad 4.2 \\ \hline 126 \\ 252 \\ \hline 0.2646 \end{array}$$

21.
$$\begin{array}{r} 1.68 \\ 5\overline{)\,8.40} \\ \underline{-5} \\ 3\,4 \\ \underline{-3\,0} \\ 40 \\ \underline{-40} \\ 0 \end{array}$$

23.
$$\begin{array}{r} 5.8 \\ 0.82\overline{)\,4.756} \\ \underline{-4\,10} \\ 656 \\ \underline{-656} \\ 0 \end{array}$$

25.
$$\begin{array}{r} 45.02 \\ 3.006 \\ +\,8.405 \\ \hline 56.431 \end{array}$$

27.
$$\begin{array}{r} 6.75 \\ \times\,10 \\ \hline 67.5 \end{array}$$

29.
$$\begin{array}{r} 7\,0. \\ 0.6\overline{)\,42.0} \\ \underline{-42} \\ 00 \end{array}$$

31.
$$\begin{array}{r} 654.90 \\ -\,56.67 \\ \hline 598.23 \end{array}$$

33.
$$\begin{array}{r} 5.62 \\ \times\quad 7.7 \\ \hline 3\,934 \\ 39\,34 \\ \hline 43.274 \end{array}$$

35.
$$\begin{array}{r} 840. \\ 0.063\overline{)\,52.920} \\ \underline{-50\,4} \\ 252 \\ \underline{-252} \\ 00 \end{array}$$

37.
$$\begin{array}{r} 16.003 \\ \times\quad 5.31 \\ \hline 16003 \\ 4\,8009 \\ 80\,015 \\ \hline 84.97593 \end{array}$$

39. 0.57 rounded to the nearest tenth is 0.6.

41. 0.234 rounded to the nearest hundredth is 0.23.

43. 0.5945 rounded to the nearest thousandth is 0.595.

45. 98,207.23 rounded to the nearest tenth is 98,207.2.

47. 12.347 rounded to the nearest hundredth is 12.35.

49. $\frac{3}{4} = \frac{3 \cdot 25}{4 \cdot 25} = \frac{75}{100} = 0.75$

Copyright © 2012 Pearson Education, Inc. Publishing as Prentice Hall.

51.
$$\begin{array}{r} 0.333\ldots \\ 3\overline{)1.000} \\ \underline{-9} \\ 10 \\ \underline{-9} \\ 10 \\ \underline{-9} \\ 1 \end{array}$$

$\frac{1}{3} = 0.\overline{3} \approx 0.33$

53. $\frac{7}{16} = \frac{7 \cdot 625}{16 \cdot 625} = \frac{4375}{10{,}000} = 0.4375$

55.
$$\begin{array}{r} 0.5454\ldots \\ 11\overline{)6.0000} \\ \underline{-5\ 5} \\ 50 \\ \underline{-44} \\ 60 \\ \underline{-55} \\ 50 \\ \underline{-44} \\ 6 \end{array}$$

$\frac{6}{11} = 0.\overline{54} \approx 0.55$

57.
$$\begin{array}{r} 4.833\ldots \\ 6\overline{)29.000} \\ \underline{-24} \\ 5\ 0 \\ \underline{-4\ 8} \\ 20 \\ \underline{-18} \\ 20 \\ \underline{-18} \\ 2 \end{array}$$

$\frac{29}{6} = 4.8\overline{3} \approx 4.83$

59. 28% = 0.28

61. 3.1% = 0.031

63. 135% = 1.35

65. 200% = 2.00 or 2

67. 96.55% = 0.9655

69. 0.1% = 0.001

71. 15.8% = 0.158

73. 0.68 = 68%

75. 0.876 = 87.6%

77. 1 = 1.00 = 100%

79. 0.5 = 50%

81. 1.92 = 192%

83. 0.004 = 0.4%

85. 0.781 = 78.1%

87. 0.5% = 0.005

$0.5\% = \frac{0.5}{100} = \frac{5}{1000} = \frac{1}{200}$

89. 14.2% = 0.142

$14.2\% = \frac{14.2}{100} = \frac{142}{1000} = \frac{71}{500}$

91. **a.** In the number 3.659, the place value of the 6 is tenths.

b. In the number 3.659, the place value of the 9 is thousandths.

c. In the number 3.659, the place value of the 3 is ones.

93. answers may vary

95.
$$\begin{array}{r} 213.4 \\ -\ 30.8 \\ \hline 182.6 \end{array}$$

The consumption of fluid milk products is 182.6 pounds more than cheese products.

97. **a.** 52.8647% rounded to the nearest tenth percent is 52.9%.

b. 52.8647% rounded to the nearest hundredth percent is 52.86%.

99. **a.** $6.5\% = 0.065 \neq 0.65$

b. 7.8% = 0.078

c. $120\% = 1.20 \neq 0.12$

d. 0.35% = 0.0035

b and **d** are correct.

Copyright © 2012 Pearson Education, Inc. Publishing as Prentice Hall.

101. 45% + 40% + 11% = 96%
100% − 96% = 4%
4% of the U.S. population has AB blood type.

103. The longest bar corresponds to network systems and data communication analysts, so that is predicted to be the fastest growing occupation.

105. 35% = 0.35

107. answers may vary

Chapter R Vocabulary Check

1. To factor means to write as a product.
2. A multiple of a number is the product of that number and any natural number.
3. A composite number is a natural number greater than 1 that is not prime.
4. The word percent means per 100.
5. Fractions that represent the same portion of a whole are called equivalent fractions.
6. An improper fraction is a fraction whose numerator is greater than or equal to its denominator.
7. A prime number is a natural number greater than 1 whose only factors are 1 and itself.
8. A fraction is simplified when the numerator and the denominator have no factors in common other than 1.
9. A proper fraction is one whose numerator is less than its denominator.
10. A mixed number contains a whole number part and a fraction part.

Chapter R Review

1. $42 = 2 \cdot 21 = 2 \cdot 3 \cdot 7$

2. $800 = 2 \cdot 400$
$= 2 \cdot 2 \cdot 200$
$= 2 \cdot 2 \cdot 2 \cdot 100$
$= 2 \cdot 2 \cdot 2 \cdot 2 \cdot 50$
$= 2 \cdot 2 \cdot 2 \cdot 2 \cdot 2 \cdot 25$
$= 2 \cdot 2 \cdot 2 \cdot 2 \cdot 2 \cdot 5 \cdot 5$

3. $12 = 2 \cdot 2 \cdot 3$
$30 = 2 \cdot 3 \cdot 5$
LCM $= 2 \cdot 2 \cdot 3 \cdot 5 = 60$

4. $7 = 7$
$42 = 2 \cdot 3 \cdot 7$
LCM $= 2 \cdot 3 \cdot 7 = 42$

5. $4 = 2 \cdot 2$
$6 = 2 \cdot 3$
$10 = 2 \cdot 5$
LCM $= 2 \cdot 2 \cdot 3 \cdot 5 = 60$

6. $2 = 2$
$5 = 5$
$7 = 7$
LCM $= 2 \cdot 5 \cdot 7 = 70$

7. $\frac{5}{8} = \frac{5 \cdot 3}{8 \cdot 3} = \frac{15}{24}$

8. $\frac{2}{3} = \frac{2 \cdot 20}{3 \cdot 20} = \frac{40}{60}$

9. $\frac{8}{20} = \frac{2 \cdot 2 \cdot 2}{2 \cdot 2 \cdot 5} = \frac{2}{5}$

10. $\frac{15}{100} = \frac{3 \cdot 5}{2 \cdot 2 \cdot 5 \cdot 5} = \frac{3}{20}$

11. $\frac{12}{6} = \frac{2 \cdot 6}{6} = 2$

12. $\frac{8}{8} = 1$

13. $\frac{1}{7} \cdot \frac{8}{11} = \frac{1 \cdot 8}{7 \cdot 11} = \frac{8}{77}$

14. $\frac{5}{12} + \frac{2}{15} = \frac{5 \cdot 5}{12 \cdot 5} + \frac{2 \cdot 4}{15 \cdot 4}$
$= \frac{25}{60} + \frac{8}{60}$
$= \frac{25 + 8}{60}$
$= \frac{33}{60}$
$= \frac{11 \cdot 3}{20 \cdot 3}$
$= \frac{11}{20}$

Copyright © 2012 Pearson Education, Inc. Publishing as Prentice Hall.

15. $\frac{3}{10} \div 6 = \frac{3}{10} \div \frac{6}{1} = \frac{3}{10} \cdot \frac{1}{6} = \frac{3 \cdot 1}{10 \cdot 6} = \frac{3 \cdot 1}{10 \cdot 2 \cdot 3} = \frac{1}{20}$

16. $\frac{7}{9} - \frac{1}{6} = \frac{7 \cdot 2}{9 \cdot 2} - \frac{1 \cdot 3}{6 \cdot 3} = \frac{14}{18} - \frac{3}{18} = \frac{14-3}{18} = \frac{11}{18}$

17. $3\frac{3}{8} \cdot 4\frac{1}{4} = \frac{27}{8} \cdot \frac{17}{4} = \frac{27 \cdot 17}{8 \cdot 4} = \frac{459}{32} = 14\frac{11}{32}$

18. $2\frac{1}{3} - 1\frac{5}{6} = \frac{7}{3} - \frac{11}{6} = \frac{14}{6} - \frac{11}{6} = \frac{14-11}{6} = \frac{3}{6} = \frac{1}{2}$

19.
$$\begin{array}{r} 16\frac{9}{10} \\ +\ 3\frac{2}{3} \\ \hline \end{array} \qquad \begin{array}{r} 16\frac{27}{30} \\ +\ 3\frac{20}{30} \\ \hline 19\frac{47}{30} \end{array} = 19 + 1\frac{17}{30} = 20\frac{17}{30}$$

20.
$$\begin{aligned} 6\frac{2}{7} \div 2\frac{1}{5} &= \frac{44}{7} \div \frac{11}{5} \\ &= \frac{44}{7} \cdot \frac{5}{11} \\ &= \frac{44 \cdot 5}{7 \cdot 11} \\ &= \frac{4 \cdot 11 \cdot 5}{7 \cdot 11} \\ &= \frac{20}{7} \\ &= 2\frac{6}{7} \end{aligned}$$

21. $A = lw = \frac{11}{12} \cdot \frac{3}{5} = \frac{11 \cdot 3}{12 \cdot 5} = \frac{11 \cdot 3}{3 \cdot 4 \cdot 5} = \frac{11}{20}$

The area is $\frac{11}{20}$ square mile.

22. $A = \frac{1}{2}bh = \frac{1}{2} \cdot \frac{5}{4} \cdot \frac{1}{2} = \frac{1 \cdot 5 \cdot 1}{2 \cdot 4 \cdot 2} = \frac{5}{16}$

The area is $\frac{5}{16}$ square meter.

23. $1.81 = \frac{181}{100}$

24. $0.035 = \frac{35}{1000}$

25.
$$\begin{array}{r} 76.358 \\ +\ 18.76 \\ \hline 95.118 \end{array}$$

26.
$$\begin{array}{r} 35 \\ 0.02 \\ +\ 1.765 \\ \hline 36.785 \end{array}$$

27.
$$\begin{array}{r} 18.00 \\ -\ 4.62 \\ \hline 13.38 \end{array}$$

28.
$$\begin{array}{r} 804.062 \\ -\ 112.489 \\ \hline 691.573 \end{array}$$

29.
$$\begin{array}{r} 7.6 \\ \times\ 12 \\ \hline 15\ 2 \\ 76 \\ \hline 91.2 \end{array}$$

30.
$$\begin{array}{r} 14.63 \\ \times\ \ 3.2 \\ \hline 2\ 926 \\ 43\ 89 \\ \hline 46.816 \end{array}$$

31.
$$\begin{array}{r} 28.6 \\ 27\overline{)\ 772.2} \\ -54 \\ \hline 232 \\ -216 \\ \hline 16\ 2 \\ -16\ 2 \\ \hline 0 \end{array}$$

32.
$$\begin{array}{r} 230. \\ 0.06\overline{)\ 13.80} \\ -12 \\ \hline 1\ 8 \\ -1\ 8 \\ \hline 00 \end{array}$$

33. 0.7652 rounded to the nearest hundredth is 0.77.

34. 25.6293 rounded to the nearest tenth is 25.6.

Copyright © 2012 Pearson Education, Inc. Publishing as Prentice Hall.

35. $\frac{1}{2}=\frac{1\cdot 5}{2\cdot 5}=\frac{5}{10}=0.5$

36. $\frac{3}{8}=\frac{3\cdot 125}{8\cdot 125}=\frac{375}{1000}=0.375$

37.
$$\begin{array}{r} 0.3636\ldots \\ 11\overline{)\,4.0000} \\ \underline{-33} \\ 70 \\ \underline{-66} \\ 40 \\ \underline{-33} \\ 70 \\ \underline{-66} \\ 4 \end{array}$$

$\frac{4}{11}=0.\overline{36}\approx 0.364$

38.
$$\begin{array}{r} 0.833\ldots \\ 6\overline{)\,5.000} \\ \underline{-48} \\ 20 \\ \underline{-18} \\ 20 \\ \underline{-18} \\ 2 \end{array}$$

$\frac{5}{6}=0.8\overline{3}\approx 0.833$

39. 29% = 0.29

40. 1.4% = 0.014

41. 0.39 = 39%

42. 1.2 = 120%

43. 68.3% = 0.683
The decimal is 0.683.

44. 2.3% = 0.023
5 = 500%
40% = 0.4
The true statement is b.

Chapter R Test

1.
$$\begin{array}{r} 3 \\ 3\overline{)\,9} \\ 2\overline{)18} \\ 2\overline{)36} \\ 2\overline{)72} \end{array}$$

The prime factorization of 72 is $2\cdot 2\cdot 2\cdot 3\cdot 3$.

2. $5=5$
$18=2\cdot 3\cdot 3$
$20=2\cdot 2\cdot 5$
LCM $=2\cdot 2\cdot 3\cdot 3\cdot 5=180$

3. $\frac{5}{12}=\frac{5\cdot 5}{12\cdot 5}=\frac{25}{60}$

4. $\frac{15}{20}=\frac{3\cdot 5}{4\cdot 5}=\frac{3}{4}$

5. $\frac{48}{100}=\frac{4\cdot 12}{4\cdot 25}=\frac{12}{25}$

6. $1.3=1\frac{3}{10}=\frac{13}{10}$

7. $\frac{5}{8}+\frac{7}{10}=\frac{5\cdot 5}{8\cdot 5}+\frac{7\cdot 4}{10\cdot 4}=\frac{25}{40}+\frac{28}{40}=\frac{25+28}{40}=\frac{53}{40}$

8. $\frac{2}{3}\cdot\frac{27}{49}=\frac{2\cdot 27}{3\cdot 49}=\frac{2\cdot 3\cdot 9}{3\cdot 49}=\frac{18}{49}$

9.
$$\begin{aligned}\frac{9}{10}\div 18&=\frac{9}{10}\div\frac{18}{1}\\&=\frac{9}{10}\cdot\frac{1}{18}\\&=\frac{9\cdot 1}{10\cdot 18}\\&=\frac{9\cdot 1}{10\cdot 9\cdot 2}\\&=\frac{1}{20}\end{aligned}$$

10. $\frac{8}{9}-\frac{1}{12}=\frac{8\cdot 4}{9\cdot 4}-\frac{1\cdot 3}{12\cdot 3}=\frac{32}{36}-\frac{3}{36}=\frac{32-3}{36}=\frac{29}{36}$

11. $1\frac{2}{9}+3\frac{2}{3}=\frac{11}{9}+\frac{11}{3}=\frac{11}{9}+\frac{33}{9}=\frac{11+33}{9}=\frac{44}{9}=4\frac{8}{9}$

Copyright © 2012 Pearson Education, Inc. Publishing as Prentice Hall.

12.
$$\begin{array}{r} 5\frac{6}{11} \\ -\,3\frac{7}{22} \\ \hline \end{array} \qquad \begin{array}{r} 5\frac{12}{22} \\ -\,3\frac{7}{22} \\ \hline 2\frac{5}{22} \end{array}$$

13. $6\frac{7}{8} \div \frac{1}{8} = \frac{55}{8} \div \frac{1}{8} = \frac{55}{8} \cdot \frac{8}{1} = \frac{55 \cdot 8}{8 \cdot 1} = \frac{55}{1} = 55$

14. $2\frac{1}{10} \cdot 6\frac{1}{2} = \frac{21}{10} \cdot \frac{13}{2} = \frac{21 \cdot 13}{10 \cdot 2} = \frac{273}{20} = 13\frac{13}{20}$

15.
$$\begin{array}{r} 43 \\ 0.21 \\ +\,1.9 \\ \hline 45.11 \end{array}$$

16.
$$\begin{array}{r} 123.60 \\ -\,57.72 \\ \hline 65.88 \end{array}$$

17.
$$\begin{array}{r} 7.93 \\ \times\ \ 1.6 \\ \hline 4758 \\ 793 \\ \hline 12.688 \end{array}$$

18.
$$\begin{array}{r} 320. \\ 0.25\overline{)\ 80.00} \\ -75 \\ \hline 5\,0 \\ -5\,0 \\ \hline 00 \end{array}$$

19. 23.7272 rounded to the nearest hundredth is 23.73.

20. $\frac{7}{8} = \frac{7 \cdot 125}{8 \cdot 125} = \frac{875}{1000} = 0.875$

21.
$$\begin{array}{r} 0.166\ldots \\ 6\overline{)1.000} \\ -6 \\ \hline 40 \\ -36 \\ \hline 40 \\ -36 \\ \hline 4 \end{array}$$

$\frac{1}{6} = 0.1\overline{6} \approx 0.167$

22. $63.2\% = 0.632$

23. $0.09 = 9\%$

24. $\frac{3}{4} = \frac{3 \cdot 25}{4 \cdot 25} = \frac{75}{100} = 0.75 = 75\%$

25. $\frac{3}{4}$ of the fresh water is icecaps and glaciers.

26. $\frac{1}{200}$ of the fresh water is active water.

27.
$$\begin{aligned} 1 - \frac{3}{4} - \frac{1}{200} &= \frac{200}{200} - \frac{150}{200} - \frac{1}{200} \\ &= \frac{200 - 150 - 1}{200} \\ &= \frac{49}{200} \end{aligned}$$

$\frac{49}{200}$ of the fresh water is groundwater.

28. $1 - \frac{1}{200} = \frac{200}{200} - \frac{1}{200} = \frac{200 - 1}{200} = \frac{199}{200}$

$\frac{199}{200}$ of the fresh water is groundwater or icecaps and glaciers.

29.
$$\begin{aligned} \text{Area} &= \frac{1}{2}(\text{base})(\text{height}) \\ &= \frac{1}{2} \cdot \frac{3}{4} \cdot \frac{1}{3} \\ &= \frac{1 \cdot 3 \cdot 1}{2 \cdot 4 \cdot 3} \\ &= \frac{1}{8} \end{aligned}$$

The area is $\frac{1}{8}$ square foot.

Copyright © 2012 Pearson Education, Inc. Publishing as Prentice Hall.

30. $A = lw = \frac{9}{8} \cdot \frac{7}{8} = \frac{9 \cdot 7}{8 \cdot 8} = \frac{63}{64}$

The area is $\frac{63}{64}$ square centimeter.

Copyright © 2012 Pearson Education, Inc. Publishing as Prentice Hall.

Chapter 1

Section 1.2 Practice

1. Since 8 is to the right of 6 on the number line, the statement $8 < 6$ is false.

2. Since 100 is to the right of 10 on the number line, the statement $100 > 10$ is true.

3. Since $21 = 21$, the statement $21 \le 21$ is true.

4. Since $21 = 21$, the statement $21 \ge 21$ is true.

5. Since neither $0 > 5$ nor $0 = 5$ is true, the statement $0 \ge 5$ is false.

6. Since $25 > 22$, the statement $25 \ge 22$ is true.

7. a. Fourteen is greater than or equal to fourteen is written as $14 \ge 14$.

   b. Zero is less than five is written as $0 < 5$.

   c. Nine is not equal to 10 is written as $9 \ne 10$.

8. The integer -8 represents 8 feet below sea level.

9. $\frac{5}{4} = 1\frac{1}{4}$

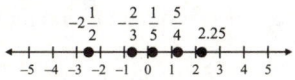

10. a. $-11 < -9$ since -11 is to the left of -9 on the number line.

    b. By comparing digits in the same places, we find that $4.511 > 4.151$, since $0.5 > 0.1$.

    c. By dividing, we find that $\frac{7}{8} = 0.875$ and $\frac{2}{3} = 0.66....$ Since $0.875 > 0.66...$, then $\frac{7}{8} > \frac{2}{3}$.

11. a. The natural numbers are 6 and 913.

    b. The whole numbers are 0, 6, and 913.

    c. The integers are -100, 0, 6, and 913.

    d. The rational numbers are -100, $-\frac{2}{5}$, 0, 6, and 913.

    e. The irrational number is π.

    f. All numbers in the given set are real numbers.

12. a. $|7| = 7$ since 7 is 7 units from 0 on the number line.

    b. $|-8| = 8$ since -8 is 8 units from 0 on the number line.

    c. $\left|\frac{2}{3}\right| = \frac{2}{3}$

    d. $|0| = 0$ since 0 is 0 units from 0 on the number line.

    e. $|-3.06| = 3.06$

13. a. $|-4| = 4$

    b. $-3 < |0|$ since $-3 < 0$.

    c. $|-2.7| > |-2|$ since $2.7 > 2$.

    d. $|-6| \le |-16|$ since $6 < 16$.

    e. $|10| < \left|-10\frac{1}{3}\right|$ since $10 < 10\frac{1}{3}$.

Vocabulary and Readiness Check

1. The <u>whole</u> numbers are $\{0, 1, 2, 3, 4, ...\}$.

2. The <u>natural</u> numbers are $\{1, 2, 3, 4, 5, ...\}$.

3. The symbols $\ne$, $\le$, and $>$ are called <u>inequality</u> symbols.

4. The <u>integers</u> are $\{..., -3, -2, -1, 0, 1, 2, 3, ...\}$.

5. The <u>real</u> numbers are {all numbers that correspond to points on the number line}.

6. The <u>rational</u> numbers are $\left\{\frac{a}{b} \middle| a \text{ and } b \text{ are integers, } b \ne 0\right\}$.

Copyright © 2012 Pearson Education, Inc. Publishing as Prentice Hall.

7. The integer 0 is neither positive nor negative.

8. The point on a number line halfway between 0 and $\frac{1}{2}$ can be represented by $\frac{1}{4}$.

9. The distance between a real number a and 0 is called the absolute value of a.

10. The absolute value of a is written in symbols as $|a|$.

Exercise Set 1.2

1. Since 4 is to the left of 10 on the number line, $4 < 10$.

3. Since 7 is to the right of 3 on the number line, $7 > 3$.

5. $6.26 = 6.26$

7. Since 0 is to the left of 7 on the number line, $0 < 7$.

9. Since 32 is to the left of 212 on the number line, $32 < 212$.

11. Since 30 is to the left of 45 on the number line, $30 \le 45$.

13. Since $11 = 11$, the statement $11 \le 11$ is true.

15. Since -11 is to the left of -10 on the number line, $-11 > -10$ is false.

17. Comparing digits with the same place value, we have $0.0 < 0.9$. Thus the statement $5.092 < 5.902$ is true.

19. Rewrite the fractions with a common denominator and compare numerators.

$$\frac{9}{10} = \frac{81}{90}; \frac{8}{9} = \frac{80}{90}$$

Since $81 > 80$, then $\frac{9}{10} \le \frac{8}{9}$ is false.

21. $25 \ge 20$ has the same meaning as $20 \le 25$.

23. $0 < 6$ has the same meaning as $6 > 0$.

25. $-10 > -12$ has the same meaning as $-12 < -10$.

27. Seven is less than eleven is written as $7 < 11$.

29. Five is greater than or equal to four is written as $5 \ge 4$.

31. Fifteen is not equal to negative two is written as $15 \ne -2$.

33. The integer 14,494 represents 14,494 feet above sea level. The integer -282 represents 282 feet below sea level.

35. The integer $-28{,}000$ represents 28,000 fewer students.

37. The integer 475 represents a \$475 deposit. The integer -195 represents a \$195 withdrawal.

39.

41.

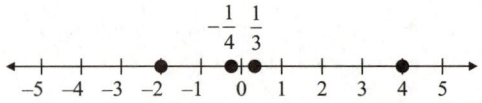

43.

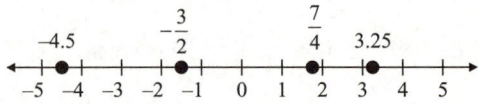

45. 0 is a whole number, an integer, a rational number, and a real number.

47. -7 is an integer, a rational number, and a real number.

49. 265 is a natural number, a whole number, an integer, a rational number, and a real number.

51. $\frac{2}{3}$ is a rational number and a real number.

53. False; the rational number $\frac{2}{3}$ is not an integer.

55. True; 0 is a real number.

57. False; the negative number $-\sqrt{2}$ is not a rational number.

59. False; the real number $\sqrt{7}$ is not a rational number.

61. $|8.9| = 8.9$ since 8.9 is 8.9 units from 0 on the number line.

Copyright © 2012 Pearson Education, Inc. Publishing as Prentice Hall.

63. $|-20| = 20$ since -20 is 20 units from 0 on the number line.

65. $\left|\frac{9}{2}\right| = \frac{9}{2}$ since $\frac{9}{2}$ is $\frac{9}{2}$ units from 0 on the number line.

67. $\left|-\frac{12}{13}\right| = \frac{12}{13}$ since $-\frac{12}{13}$ is $\frac{12}{13}$ unit from 0 on the number line.

69. $|-5| = 5$
$-4 = -4$
Since 5 is to the right of -4 on the number line, $|-5| > -4$.

71. $\left|-\frac{5}{8}\right| = \frac{5}{8}$
$\left|\frac{5}{8}\right| = \frac{5}{8}$
Since $\frac{5}{8} = \frac{5}{8}$, then $\left|-\frac{5}{8}\right| = \left|\frac{5}{8}\right|$.

73. $|-2| = 2$
$|-2.7| = 2.7$
Since 2 is to the left of 2.7 on the number line, $|-2| < |-2.7|$.

75. $|0| = 0$
$|-8| = 8$
Since 0 is to the left of 8 on the number line, $|0| < |-8|$.

77. The apple production in 2008 was 905 thousand bushels, while the apple production in 2009 was 823 thousand bushels.
905 thousand > 823 thousand, or
905,000 > 823,000

79. The apple production in 2003 was 1024 thousand bushels, while the production in 2004 was 1000 thousand bushels.
1024 – 1000 = 24
The production decreased by 24 thousand bushels, or –24 thousand bushels.

81. The 2009 cranberry production in Oregon was 49 million pounds, while the 2009 cranberry production in Washington was 16 million pounds.
49 million > 16 million, or
49,000,000 > 16,000,000

83. The 2009 cranberry production in Washington was 16 million pounds, while the 2009 cranberry production in New Jersey was 54 million pounds.
54 – 16 = 38
The production in Washington was 38 million pounds less or –38 million.

85. Since –0.04 is to the right of –26.7 on the number line, $-0.04 > -26.7$.

87. Sun: –26.7
Arcturus: –0.04
Since $-26.7 < -0.04$, the sun is brighter than Arcturus.

89. Since the brightest star corresponds to the smallest apparent magnitude, which is –26.7, the brightest star is the sun.

91. answers may vary

Section 1.3 Practice

1. a. $4^2 = 4 \cdot 4 = 16$

b. $2^2 = 2 \cdot 2 = 4$

c. $3^4 = 3 \cdot 3 \cdot 3 \cdot 3 = 81$

d. $9^1 = 9$

e. $\left(\frac{2}{5}\right)^3 = \left(\frac{2}{5}\right)\left(\frac{2}{5}\right)\left(\frac{2}{5}\right) = \frac{2 \cdot 2 \cdot 2}{5 \cdot 5 \cdot 5} = \frac{8}{125}$

f. $(0.8)^2 = (0.8)(0.8) = 0.64$

2. $3 \cdot 2 + 4^2 = 3 \cdot 2 + 16 = 6 + 16 = 22$

3. $28 \div 7 \cdot 2 = 4 \cdot 2 = 8$

4. $\frac{9}{5} \cdot \frac{1}{3} - \frac{1}{3} = \frac{9}{15} - \frac{1}{3} = \frac{9}{15} - \frac{5}{15} = \frac{4}{15}$

5.
$$\begin{aligned} 5 + 3[2(3 \cdot 4 + 1) - 20] &= 5 + 3[2(12 + 1) - 20] \\ &= 5 + 3[2(13) - 20] \\ &= 5 + 3[26 - 20] \\ &= 5 + 3[6] \\ &= 5 + 18 \\ &= 23 \end{aligned}$$

Copyright © 2012 Pearson Education, Inc. Publishing as Prentice Hall.

6. $\dfrac{1+|7-4|+3^2}{8-5}=\dfrac{1+|3|+3^2}{8-5}$
$=\dfrac{1+3+3^2}{3}$
$=\dfrac{1+3+9}{3}$
$=\dfrac{13}{3}$

7. a. Replace y with 4.
$3y^2=3\cdot(4)^2=3\cdot 16=48$

b. Replace x with 1 and y with 4.
$2y-x=2(4)-1=8-1=7$

c. Replace x with 1 and y with 4.
$\dfrac{11x}{3y}=\dfrac{11\cdot 1}{3\cdot 4}=\dfrac{11}{12}$

d. Replace x with 1 and y with 4.
$\dfrac{x}{y}+\dfrac{6}{y}=\dfrac{1}{4}+\dfrac{6}{4}=\dfrac{7}{4}$

e. Replace x with 1 and y with 4.
$y^2-x^2=4^2-1^2=16-1=15$

8. $5x-10=x+2$
$5(3)-10\stackrel{?}{=}3+2$
$15-10\stackrel{?}{=}5$
$5=5$ True
3 is a solution.

9. a. $5\cdot x$ and $5x$ are both ways to denote the product of 5 and x.

b. A number added to 7 is denoted by $7+x$.

c. A number divided by 11.2 is denoted by $x\div 11.2$ or $\dfrac{x}{11.2}$.

d. A number subtracted from 8 is denoted by $8-x$.

e. Twice a number, plus 1 is denoted by $2x+1$.

10. a. The ratio of a number and 6 is 24 is written as $\dfrac{x}{6}=24$.

b. The difference of 10 and a number is 18 is written as $10-x=18$.

c. One less than twice a number is 99 is written as $2x-1=99$.

Calculator Explorations

1. $5^3=125$

2. $7^4=2401$

3. $9^5=59{,}049$

4. $8^6=262{,}144$

5. $2(20-5)=30$

6. $3(14-7)+21=42$

7. $24(862-455)+89=9857$

8. $99+(401+962)=1462$

9. $\dfrac{4623+129}{36-34}=2376$

10. $\dfrac{956-452}{89-86}=168$

Vocabulary and Readiness Check

1. In 2^5, the 2 is called the <u>base</u> and the 5 is called the <u>exponent</u>.

2. True or false: 2^5 means 2.5. <u>false</u>

3. To simplify $8+2\cdot 6$, which operation should be performed first? <u>multiplication</u>

4. To simplify $(8+2)\cdot 6$, which operation should be performed first? <u>addition</u>

5. To simplify $9(3-2)\div 3+6$, which operation should be performed first? <u>subtraction</u>

6. To simplify $8\div 2\cdot 6$, which operation should be performed first? <u>division</u>

7. A combination of operations on letters (variables) and numbers is an <u>expression</u>.

8. A letter that represents a number is a <u>variable</u>.

Copyright © 2012 Pearson Education, Inc. Publishing as Prentice Hall.

9. $3x - 2y$ is called an <u>expression</u> and the letters x and y are <u>variables</u>.

10. Replacing a variable in an expression by a number and then finding the value of the expression is called <u>evaluating the expression</u>.

11. A statement of the form "expression = expression" is called an <u>equation</u>.

12. A value for the variable that makes the equation a true statement is called a <u>solution</u>.

Exercise Set 1.3

1. $3^5 = 3\cdot3\cdot3\cdot3\cdot3 = 243$

3. $3^3 = 3\cdot3\cdot3 = 27$

5. $1^5 = 1\cdot1\cdot1\cdot1\cdot1 = 1$

7. $5^1 = 5$

9. $7^2 = 7\cdot7 = 49$

11. $\left(\frac{2}{3}\right)^4 = \left(\frac{2}{3}\right)\left(\frac{2}{3}\right)\left(\frac{2}{3}\right)\left(\frac{2}{3}\right) = \frac{2\cdot2\cdot2\cdot2}{3\cdot3\cdot3\cdot3} = \frac{16}{81}$

13. $\left(\frac{1}{5}\right)^3 = \left(\frac{1}{5}\right)\left(\frac{1}{5}\right)\left(\frac{1}{5}\right) = \frac{1\cdot1\cdot1}{5\cdot5\cdot5} = \frac{1}{125}$

15. $(1.2)^2 = 1.2\cdot1.2 = 1.44$

17. $(0.7)^3 = 0.7\cdot0.7\cdot0.7 = 0.343$

19. $5\cdot5 = 5^2$ square meters

21. $5 + 6\cdot2 = 5 + 12 = 17$

23. $4\cdot8 - 6\cdot2 = 32 - 12 = 20$

25. $18 \div 3\cdot2 = 6\cdot2 = 12$

27. $2+(5-2)+4^2 = 2+3+4^2 = 2+3+16 = 21$

29. $5\cdot3^2 = 5\cdot9 = 45$

31. $\frac{1}{4}\cdot\frac{2}{3} - \frac{1}{6} = \frac{2}{12} - \frac{1}{6} = \frac{1}{6} - \frac{1}{6} = 0$

33. $\frac{6-4}{9-2} = \frac{2}{7}$

35. $2[5+2(8-3)] = 2[5+2(5)]$
$= 2[5+10]$
$= 2[15]$
$= 30$

37. $\frac{19-3\cdot5}{6-4} = \frac{19-15}{2} = \frac{4}{2} = 2$

39. $\frac{|6-2|+3}{8+2\cdot5} = \frac{4+3}{8+10} = \frac{7}{18}$

41. $\frac{3+3(5+3)}{3^2+1} = \frac{3+3(8)}{9+1} = \frac{3+24}{10} = \frac{27}{10}$

43. $\frac{6+|8-2|+3^2}{18-3} = \frac{6+|6|+9}{15}$
$= \frac{6+6+9}{15}$
$= \frac{21}{15}$
$= \frac{7}{5}$

45. $2+3[10(4\cdot5-16)-30]$
$= 2+3[10(20-16)-30]$
$= 2+3[10(4)-30]$
$= 2+3[40-30]$
$= 2+3[10]$
$= 2+30$
$= 32$

47. $\left(\frac{2}{3}\right)^3 + \frac{1}{9} + \frac{1}{3}\cdot\frac{4}{3} = \frac{8}{27} + \frac{1}{9} + \frac{1}{3}\cdot\frac{4}{3}$
$= \frac{8}{27} + \frac{1}{9} + \frac{4}{9}$
$= \frac{8}{27} + \frac{3}{27} + \frac{12}{27}$
$= \frac{23}{27}$

49. Replace y with 3.
$3y = 3(3) = 9$

51. Replace x with 1 and z with 5.
$\frac{z}{5x} = \frac{5}{5(1)} = \frac{5}{5} = 1$

Copyright © 2012 Pearson Education, Inc. Publishing as Prentice Hall.

53. Replace x with 1.
$3x - 2 = 3(1) - 2 = 3 - 2 = 1$

55. Replace x with 1 and y with 3.
$|2x + 3y| = |2(1) + 3(3)| = |2 + 9| = |11| = 11$

57. Replace x with 1, y with 3, and z with 5.
$xy + z = 1(3) + 5 = 3 + 5 = 8$

59. Replace y with 3.
$5y^2 = 5(3^2) = 5(9) = 45$

61. Replace x with 12, y with 8, and z with 4.
$\frac{x}{z} + 3y = \frac{12}{4} + 3(8) = 3 + 3(8) = 3 + 24 = 27$

63. Replace x with 12 and y with 8.

$$\begin{aligned} x^2 - 3y + x &= (12)^2 - 3(8) + 12 \\ &= 144 - 3(8) + 12 \\ &= 144 - 24 + 12 \\ &= 120 + 12 \\ &= 132 \end{aligned}$$

65. Replace x with 12, y with 8, and z with 4.

$$\begin{aligned} \frac{x^2 + z}{y^2 + 2z} &= \frac{(12)^2 + 4}{(8)^2 + 2(4)} \\ &= \frac{144 + 4}{64 + 2(4)} \\ &= \frac{148}{64 + 8} \\ &= \frac{148}{72} \\ &= \frac{4 \cdot 37}{4 \cdot 18} \\ &= \frac{37}{18} \end{aligned}$$

67.
$3x - 6 = 9$
$3(5) - 6 \stackrel{?}{=} 9$
$15 - 6 \stackrel{?}{=} 9$
$9 = 9$ True
Since the result is true, 5 is a solution of the given equation.

69.
$2x + 6 = 5x - 1$
$2(0) + 6 \stackrel{?}{=} 5(0) - 1$
$0 + 6 \stackrel{?}{=} 0 - 1$
$6 = -1$ False
Since the result is false, 0 is not a solution of the given equation.

71.
$2x - 5 = 5$
$2(8) - 5 \stackrel{?}{=} 5$
$16 - 5 \stackrel{?}{=} 5$
$11 = 5$ False
Since the result is false, 8 is not a solution of the given equation.

73.
$x + 6 = x + 6$
$2 + 6 \stackrel{?}{=} 2 + 6$
$8 = 8$ True
Since the result is true, 2 is a solution of the given equation.

75.
$x = 5x + 15$
$0 \stackrel{?}{=} 5(0) + 15$
$0 \stackrel{?}{=} 0 + 15$
$0 = 15$ False
Since the result is false, 0 is not a solution of the given equation.

77.
$\frac{1}{3}x = 9$
$\frac{1}{3}(27) \stackrel{?}{=} 9$
$9 = 9$ True
Since the result is true, 27 is a solution of the given equation.

79. Fifteen more than a number is written as $x + 15$.

81. Five subtracted from a number is written as $x - 5$.

83. The ratio of a number and 4 is written as $\frac{x}{4}$.

85. Three times a number, increased by 22 is written as $3x + 22$.

87. One increased by two equals the quotient of nine and three is written as $1 + 2 = 9 \div 3$.

89. Three is not equal to four divided by two is written as $3 \neq 4 \div 2$.

91. The sum of 5 and a number is 20 is written as $5 + x = 20$.

93. The product 7.6 and a number is 17 is written as $7.6x = 17$.

95. Thirteen minus three times a number is 13 is written as $13 - 3x = 13$.

Copyright © 2012 Pearson Education, Inc. Publishing as Prentice Hall.

97. no; answers may vary

99. **a.** $(6 + 2) \cdot (5 + 3) = 8 \cdot 8 = 64$

b. $(6 + 2) \cdot 5 + 3 = 8 \cdot 5 + 3 = 40 + 3 = 43$

c. $6 + 2 \cdot 5 + 3 = 6 + 10 + 3 = 19$

d. $6 + 2 \cdot (5 + 3) = 6 + 2 \cdot 8 = 6 + 16 = 22$

	Length, l	Width, w	Perimeter of Rectangle: $2l + 2w$	Area of Rectangle: lw
101.	4 in.	3 in.	$2l + 2w$ = 2(4 in.) + 2(3 in.) = 8 in. + 6 in. = 14 in.	lw = (4 in.)(3 in.) = 12 sq in.
103.	5.3 in.	1.7 in.	$2l + 2w$ = 2(5.3 in.) + 2(1.7 in.) = 10.6 in. + 3.4 in. = 14 in.	lw = (5.3 in.)(1.7 in.) = 9.01 sq in.

105. Rectangles with the same perimeter can have different areas.

107. $(20 - 4) \cdot 4 \div 2 = 16 \cdot 4 \div 2 = 64 \div 2 = 32$

109. **a.** $5x + 6$ is an expression since it does not contain the equal symbol, "="

b. $2a = 7$ is an equation since it contains the equal symbol.

c. $3a + 2 = 9$ is an equation since it contains the equal symbol.

d. $4x + 3y - 8z$ is an expression since it does not contain the equal symbol.

e. $5^2 - 2(6 - 2)$ is an expression since it does not contain the equal symbol.

111. answers may vary

113. answers may vary; for example, $-2(5) - 1$: $-2(5) - 1 = -10 - 1 = -11$

Section 1.4 Practice

1.

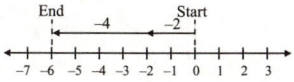

$-2 + (-4) = -6$

2.

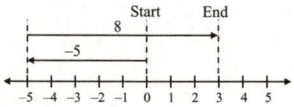

$-5 + 8 = 3$

Copyright © 2012 Pearson Education, Inc. Publishing as Prentice Hall.

3. 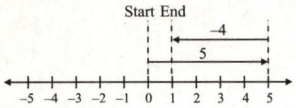

$5 + (-4) = 1$

4. $-8 + (-5) = -13$

5. $(-14) + 6 = -8$

6. $(-17) + (-10) = -27$

7. $(-4) + 12 = 8$

8. $1.5 + (-3.2) = -1.7$

9. $-\frac{5}{12}+\left(-\frac{1}{12}\right)=-\frac{6}{12}=-\frac{6\cdot 1}{6\cdot 2}=-\frac{1}{2}$

10. $12.1 + (-3.6) = 8.5$

11. $-\frac{4}{5}+\frac{2}{3}=-\frac{12}{15}+\frac{10}{15}=-\frac{2}{15}$

12. a. $16 + (-9) + (-9) = 7 + (-9) = -2$

b. $[3+(-13)]+[-4+(-7\} = [-10]+[-11]$
$= -21$

13. The opposite of -35 is 35.

14. The opposite of 12 is -12.

15. The opposite of $-\frac{3}{11}$ is $\frac{3}{11}$.

16. The opposite of 1.9 is -1.9.

17. a. $-(-22) = 22$

b. $-\left(-\frac{2}{7}\right)=\frac{2}{7}$

c. $-(-x) = x$

d. $-|-14| = -14$

e. $-|2.3| = -2.3$

18. $30 + (-30) = 0$

19. $-81 + 81 = 0$

20. $x + 3y = -6 + 3(2) = -6 + 6 = 0$

21. $x + y = -13 + (-9) = -22$

22. Temperature at 8 a.m. $= -7 + (+4) + (+7)$
$= -3 + (+7)$
$= 4$
The temperature was 4°F at 8 a.m.

Vocabulary and Readiness Check

1. If n is a number, then $-n + n =$ <u>0</u>.

2. Since $x + n = n + x$, we say that addition is <u>commutative</u>.

3. If a is a number, then $-(-a) =$ <u>a</u>.

4. Since $n + (x + a) = (n + x) + a$, we say that addition is <u>associative</u>.

Exercise Set 1.4

1. $6 + (-3) = 3$

3. $-6 + (-8) = -14$

5. $8 + (-7) = 1$

7. $-14 + 2 = -12$

9. $-2 + (-3) = -5$

11. $-9 + (-3) = -12$

13. $-7 + 3 = -4$

15. $10 + (-3) = 7$

17. $5 + (-7) = -2$

19. $-16 + 16 = 0$

21. $27 + (-46) = -19$

23. $-18 + 49 = 31$

25. $-33 + (-14) = -47$

27. $6.3 + (-8.4) = -2.1$

29. $117 + (-79) = 38$

31. $-9.6 + (-3.5) = -13.1$

33. $-\frac{3}{8}+\frac{5}{8}=\frac{2}{8}=\frac{2\cdot 1}{2\cdot 4}=\frac{1}{4}$

Copyright © 2012 Pearson Education, Inc. Publishing as Prentice Hall.

35. $-\frac{7}{16}+\frac{1}{4}=-\frac{7}{16}+\frac{4}{16}=-\frac{3}{16}$

37. $-\frac{7}{10}+\left(-\frac{3}{5}\right)=-\frac{7}{10}+\left(-\frac{6}{10}\right)=-\frac{13}{10}$

39. $|-8| + (-16) = 8 + (-16) = -8$

41. $-15 + 9 + (-2) = -6 + (-2) = -8$

43. $-21 + (-16) + (-22) = -37 + (-22) = -59$

45. $-23 + 16 + (-2) = -7 + (-2) = -9$

47. $|5 + (-10)| = |-5| = 5$

49. $6 + (-4) + 9 = 2 + 9 = 11$

51. $[-17+(-4)]+[-12+15]=[-21]+[3]=-18$

53. $|9+(-12)|+|-16|=|-3|+|-16|$
$=3+16$
$=19$

55. $-13+[5+(-3)+4]=-13+[2+4]$
$=-13+[6]$
$=-7$

57. $-38 + 12 = -26$
The sum of -38 and 12 is -26.

59. The additive inverse of 6 is -6.

61. The additive inverse of -2 is 2.

63. The additive inverse of 0 is 0.

65. Since $|-6| = 6$, the additive inverse of $|-6|$ is -6.

67. $-|-2| = -2$

69. $-(-7) = 7$

71. $-(-7.9) = 7.9$

73. $-(-5z) = 5z$

75. $\left|-\frac{2}{3}\right|=\frac{2}{3}$

77. $x + y = -20 + (-50) = -70$

79. $3x + y = 3(2) + (-3) = 6 + (-3) = 3$

81. The sum of -6 and 25 is $-6 + 25 = 19$.

83. The sum of -31, -9, and 30 is
$-31 + (-9) + 30 = -40 + 30 = -10$.

85. $0 + (-215) + (-16) = -215 + (-16) = -231$
The diver's final depth is 231 feet below the surface.

87. $-35 + 142 = 107$
The highest recorded temperature in Massachusetts was 107°F.

89. $-411 + 316 = -95$
Your elevation is -95 meters.

91. $-6+(-9)+(-4)+(-2)=-15+(-4)+(-2)$
$=-19+(-2)$
$=-21$
His total score was -21.

93. $2.5+9+(-14.2)+(-4.2)$
$=11.5+(-14.2)+(-4.2)$
$=-2.7+(-4.2)$
$=-6.9$
The total net income for fiscal year 2009 was $-\$6.9$ million.

95. The highest bar corresponds to July, so the highest low temperature was in July.

97. The shortest bar above 0 corresponds to October.

99. $\frac{-9.1+14.4+8.8}{3}=\frac{5.3+8.8}{3}=\frac{14.1}{3}=4.7$
The average for the months of April, May, and October is 4.7°F.

101. answers may vary

103. $7 + (-10) = -3$

105. $-10 + (-12) = -22$

107. True

109. False; for example, $4 + (-2) = 2 > 0$.

111. answers may vary

Section 1.5 Practice

1. a. $-20 - 6 = -20 + (-6) = -26$

b. $3 - (-5) = 3 + 5 = 8$

Copyright © 2012 Pearson Education, Inc. Publishing as Prentice Hall.

c. $7 - 17 = 7 + (-17) = -10$

d. $-4 - (-9) = -4 + 9 = 5$

2. $9.6 - (-5.7) = 15.3$

3. $-\frac{4}{9} - \frac{2}{9} = -\frac{4}{9} + \left(-\frac{2}{9}\right) = -\frac{6}{9} = -\frac{2}{3}$

4. $-\frac{1}{4} - \left(-\frac{2}{5}\right) = -\frac{1}{4} + \frac{2}{5} = -\frac{5}{20} + \frac{8}{20} = \frac{3}{20}$

5. a. $-11 - 7 = -11 + (-7) = -18$

b. $35 - (-25) = 35 + 25 = 60$

6. a.
$$\begin{aligned} -20-5+12-(-3) &= -20+(-5)+12+3 \\ &= -10 \end{aligned}$$

b.
$$\begin{aligned} 5.2-(-4.4)+(-8.8) &= 5.2+4.4+(-8.8) \\ &= 0.8 \end{aligned}$$

7. a.
$$\begin{aligned} -9+[(-4-1)-10] &= -9+[(-4+(-1))-10] \\ &= -9+[(-5)-10] \\ &= -9+[-5+(-10)] \\ &= -9+[-15] \\ &= -24 \end{aligned}$$

b.
$$\begin{aligned} 5^2-20+[-11-(-3)] &= 5^2-20+[-11+3] \\ &= 5^2-20+[-8] \\ &= 25-20+(-8) \\ &= 25+(-20)+(-8) \\ &= 5+(-8) \\ &= -3 \end{aligned}$$

8. a. Replace x with 1 and y with -4.
$$\frac{x-y}{14+x} = \frac{1-(-4)}{14+1} = \frac{1+4}{15} = \frac{5}{15} = \frac{1}{3}$$

b. Replace x with 1 and y with -4.
$$x^2 - y = (1)^2 - (-4) = 1-(-4) = 1+4 = 5$$

9.
$$\begin{aligned} -1+x &= 1 \\ -1+(-2) &\stackrel{?}{=} 1 \\ -3 &= 1 \text{ False} \end{aligned}$$
-2 is not a solution.

10. $29{,}028 - (-1312) = 29{,}028 + 1312 = 30{,}340$
Mount Everest is 30,340 feet higher than the Dead Sea.

11. a. These angles are supplementary, so their sum is 180°. This means that $m\angle x$ is $180° - 78°$.
$m\angle x = 180° - 78° = 102°$

b. These angles are complementary, so their sum is 90°. This means that $m\angle y$ is $90° - 81°$.
$m\angle y = 90° - 81° = 9°$

Vocabulary and Readiness Check

1. It is true that $a - b = \underline{a + (-b)}$. b

2. The opposite of n is $\underline{-n}$. a

3. To evaluate $x - y$ for $x = -10$ and $y = -14$, we replace x with -10 and y with -14 and evaluate $\underline{-10 - (-14)}$. d

4. The expression $-5 - 10$ equals $\underline{-5 + (-10)}$. c

Exercise Set 1.5

1. $-6 - 4 = -6 + (-4) = -10$

3. $4 - 9 = 4 + (-9) = -5$

5. $16 - (-3) = 16 + 3 = 19$

7. $7 - (-4) = 7 + 4 = 11$

9. $-26 - (-18) = -26 + 18 = -8$

11. $-6 - 5 = -6 + (-5) = -11$

13. $16 - (-21) = 16 + 21 = 37$

15. $-6 - (-11) = -6 + 11 = 5$

17. $-44 - 27 = -44 + (-27) = -71$

19. $-21 - (-21) = -21 + 21 = 0$

21. $-\frac{3}{11} - \left(-\frac{5}{11}\right) = -\frac{3}{11} + \frac{5}{11} = \frac{2}{11}$

23. $9.7 - 16.1 = 9.7 + (-16.1) = -6.4$

25. $-2.6 - (-6.7) = -2.6 + 6.7 = 4.1$

27. $\frac{1}{2} - \frac{2}{3} = \frac{1}{2} + \left(-\frac{2}{3}\right) = \frac{3}{6} + \left(-\frac{4}{6}\right) = -\frac{1}{6}$

Copyright © 2012 Pearson Education, Inc. Publishing as Prentice Hall.

29. $-\frac{1}{6}-\frac{3}{4}=-\frac{1}{6}+\left(-\frac{3}{4}\right)$
$=-\frac{2}{12}+\left(-\frac{9}{12}\right)$
$=-\frac{11}{12}$

31. $8.3-(-0.62)=8.3+0.62=8.92$

33. $0-8.92=0+(-8.92)=-8.92$

35. $8-(-5)=8+5=13$
-5 subtracted from 8 is 13.

37. $-6-(-1)=-6+1=-5$
The difference between -6 and -1 is -5.

39. $7-8=7+(-8)=-1$
8 subtracted from 7 is -1.

41. $-8-15=-8+(-15)=-23$
-8 decreased by 15 is -23.

43. $-10-(-8)+(-4)-20$
$=-10+8+(-4)+(-20)$
$=-2+(-4)+(-20)$
$=-6+(-20)$
$=-26$

45. $5-9+(-4)-8-8$
$=5+(-9)+(-4)+(-8)+(-8)$
$=-4+(-4)+(-8)+(-8)$
$=-8+(-8)+(-8)$
$=-16+(-8)$
$=-24$

47. $-6-(2-11)=-6-(-9)=-6+9=3$

49. $3^3-8\cdot 9=27-72=27+(-72)=-45$

51. $2-3(8-6)=2-3[8+(-6)]$
$=2-3(2)$
$=2-6$
$=2+(-6)$
$=-4$

53. $(3-6)+4^2=[3+(-6)]+16=[-3]+16=13$

55. $-2+[(8-11)-(-2-9)]$
$=-2+[(8+(-11))-(-2+(-9))]$
$=-2+[(-3)-(-11)]$
$=-2+[-3+11]$
$=-2+8$
$=6$

57. $|-3|+2^2+[-4-(-6)]=|-3|+4+[-4+6]$
$=3+4+2$
$=7+2$
$=9$

59. Replace x with -5 and y with 4.
$x-y=-5-4=-5+(-4)=-9$

61. Replace x with -5 and y with 4.
$\frac{9-x}{y+6}=\frac{9-(-5)}{4+6}=\frac{9+5}{10}=\frac{14}{10}=\frac{2\cdot 7}{2\cdot 5}=\frac{7}{5}$

63. Replace x with -5, y with 4, and t with 10.
$|x|+2t-8y=|-5|+2(10)-8(4)$
$=5+20-32$
$=25-32$
$=25+(-32)$
$=-7$

65. Replace x with -5 and y with 4.
$y^2-x=4^2-(-5)=16+5=21$

67. Replace x with -5 and t with 10.
$\frac{|x-(-10)|}{2t}=\frac{|-5-(-10)|}{2(10)}$
$=\frac{|-5+10|}{2(10)}$
$=\frac{|5|}{2(10)}$
$=\frac{5}{20}$
$=\frac{1\cdot 5}{4\cdot 5}$
$=\frac{1}{4}$

69. $x-9=5$
$-4-9\stackrel{?}{=}5$
$-4+(-9)\stackrel{?}{=}5$
$-13=5$ False

Since the result is false, -4 is not a solution of the given equation.

Copyright © 2012 Pearson Education, Inc. Publishing as Prentice Hall.

71.
$$\begin{aligned} -x+6 &= -x-1 \\ -(-2)+6 &\stackrel{?}{=} -(-2)-1 \\ 2+6 &\stackrel{?}{=} 2-1 \\ 8 &\stackrel{?}{=} 2+(-1) \\ 8 &= 1 \quad \text{False} \end{aligned}$$
Since the result is false, –2 is not a solution of the given equation.

73.
$$\begin{aligned} -x-13 &= -15 \\ -2-13 &\stackrel{?}{=} -15 \\ -2+(-13) &\stackrel{?}{=} -15 \\ -15 &= -15 \quad \text{True} \end{aligned}$$
Since the result is true, 2 is a solution of the given equation.

75. 136 – (–129) = 136 + 129 = 265
Therefore, 136°F is 265°F warmer than –129°F.

77. 13,796 – (–21,857) = 13,796 + 21,857 = 35,653
The difference in elevation is 35,653 feet.

79. Complementary angles sum to 90°.
$$\begin{aligned} 90^\circ - 60^\circ &= x \\ 90^\circ + (-60^\circ) &= x \\ 30^\circ &= x \end{aligned}$$

81.
$$\begin{aligned} -250+120-178 &= -250+120+(-178) \\ &= -130+(-178) \\ &= -308 \end{aligned}$$
The overall vertical change is –308 feet.

83. 19,340 – (–512) = 19,340 + 512 = 19,852
Mt. Kilimanjaro is 19,852 feet higher than Lake Assal.

85. $y = 180^\circ - 50^\circ$
$y = 180^\circ + (-50^\circ)$
$y = 130^\circ$

87. The sum of –5 and a number is $-5 + x$.

89. Subtract a number from –20 is $-20 - x$.

91.

Month	Monthly Increase or Decrease
February	–23.7 – (–19.3) = –23.7 + 19.3 = –4.4°
March	–21.1 – (–23.7) = –21.1 + 23.7 = 2.6°
April	–9.1 – (–21.1) = –9.1 + 21.1 = 12°
May	14.4 – (–9.1) = 14.4 + 9.1 = 23.5°
June	29.7 – 14.4 = 29.7 + (–14.4) = 15.3°

93. The largest positive number corresponds to May.

95. answers may vary.

Copyright © 2012 Pearson Education, Inc. Publishing as Prentice Hall.

97. $9 - (-7) = 9 + 7 = 16$

99. $10 - 30 = 10 + (-30) = -20$

101. true; answers may vary

103. false; answers may vary.

105. Since 56,875 is less than 87,262, the answer is negative.
$56{,}875 - 87{,}262 = -30{,}387$

Integrated Review

1. The opposite of a positive number is a negative number.

2. The sum of two negative numbers is a negative number.

3. The absolute value of a negative number is a positive number.

4. The absolute value of zero is 0.

5. The sum of two positive numbers is a positive number.

6. The sum of a number and its opposite is 0.

7. The absolute value of a positive number is a positive number.

8. The reciprocal of a negative number is a negative number.

9. The opposite of $\frac{1}{7}$ is $-\frac{1}{7}$.

The absolute value of $\frac{1}{7}$ is $\frac{1}{7}$.

10. The opposite of $-\frac{12}{5}$ is $\frac{12}{5}$.

The absolute value of $-\frac{12}{5}$ is $\frac{12}{5}$.

11. The number whose opposite is -3 is 3.
The absolute value of 3 is 3.

12. The number whose opposite is $\frac{9}{11}$ is $-\frac{9}{11}$.

The absolute value of $-\frac{9}{11}$ is $\frac{9}{11}$.

13. $-19 + (-23) = -42$

14. $7 - (-3) = 7 + 3 = 10$

15. $-15 + 17 = 2$

16. $-8 - 10 = -8 + (-10) = -18$

17. $18 + (-25) = -7$

18. $-2 + (-37) = -39$

19. $-14 - (-12) = -14 + 12 = -2$

20. $5 - 14 = 5 + (-14) = -9$

21. $4.5 - 7.9 = 4.5 + (-7.9) = -3.4$

22. $-8.6 - 1.2 = -8.6 + (-1.2) = -9.8$

23. $-\frac{3}{4} - \frac{1}{7} = -\frac{3}{4} + \left(-\frac{1}{7}\right) = -\frac{21}{28} + \left(-\frac{4}{28}\right) = -\frac{25}{28}$

24. $\frac{2}{3} - \frac{7}{8} = \frac{2}{3} + \left(-\frac{7}{8}\right) = \frac{16}{24} + \left(-\frac{21}{24}\right) = -\frac{5}{24}$

25.
$$\begin{aligned} -9-(-7)+4-6 &= -9+7+4+(-6) \\ &= -2+4+(-6) \\ &= 2+(-6) \\ &= -4 \end{aligned}$$

26.
$$\begin{aligned} 11-20+(-3)-12 &= 11+(-20)+(-3)+(-12) \\ &= -9+(-3)+(-12) \\ &= -12+(-12) \\ &= -24 \end{aligned}$$

27.
$$\begin{aligned} 24-6(14-11) &= 24-6[14+(-11)] \\ &= 24-6(3) \\ &= 24+(-6)(3) \\ &= 24+(-18) \\ &= 6 \end{aligned}$$

28.
$$\begin{aligned} 30-5(10-8) &= 30-5[10+(-8)] \\ &= 30-5(2) \\ &= 30+(-5)(2) \\ &= 30+(-10) \\ &= 20 \end{aligned}$$

29.
$$\begin{aligned} (7-17)+4^2 &= [7+(-17)]+4^2 \\ &= -10+4^2 \\ &= -10+16 \\ &= 6 \end{aligned}$$

Copyright © 2012 Pearson Education, Inc. Publishing as Prentice Hall.

30. $9^2+(10-30)=9^2+[10+(-30)]$
$=9^2+(-20)$
$=81+(-20)$
$=61$

31. $|-9|+3^2+(-4-20)=|-9|+3^2+[-4+(-20)]$
$=9+9+(-24)$
$=18+(-24)$
$=-6$

32. $|-4-5|+5^2+(-50)=|-4+(-5)|+5^2+(-50)$
$=|-9|+5^2+(-50)$
$=9+25+(-50)$
$=34+(-50)$
$=-16$

33. $-7+[(1-2)+(-2-9)]$
$=-7+[(1+(-2))+(-2+(-9))]$
$=-7+[(-1)+(-11)]$
$=-7+(-12)$
$=-19$

34. $-6+[(-3+7)+(4-15)]$
$=-6+[(-3+7)+(4+(-15))]$
$=-6+[4+(-11)]$
$=-6+(-7)$
$=-13$

35. $1-5=1+(-5)=-4$

36. $-3-(-2)=-3+2=-1$

37. $\frac{1}{4}-\left(-\frac{2}{5}\right)=\frac{1}{4}+\frac{2}{5}=\frac{5}{20}+\frac{8}{20}=\frac{13}{20}$

38. $-\frac{5}{8}-\frac{1}{10}=-\frac{5}{8}+\left(-\frac{1}{10}\right)=-\frac{25}{40}+\left(-\frac{4}{40}\right)=-\frac{29}{40}$

39. $2(19-17)^3-3(-7+9)^2$
$=2[19+(-17)]^3+(-3)(-7+9)^2$
$=2(2)^3+(-3)(2)^2$
$=2(8)+(-3)(4)$
$=16+(-12)$
$=4$

40. $3(10-9)^2+6(20-19)^3$
$=3[10+(-9)]^2+6[20+(-19)]^3$
$=3(1)^2+6(1)^3$
$=3(1)+6(1)$
$=3+6$
$=9$

41. Replace x with -2 and y with -1.
$x-y=-2-(-1)=-2+1=-1$

42. Replace x with -2 and y with -1.
$x+y=-2+(-1)=-3$

43. Replace y with -1 and z with 9.
$y+z=-1+9=8$

44. Replace y with -1 and z with 9.
$z-y=9-(-1)=9+1=10$

45. Replace x with -2, y with -1, and z with 9.
$\frac{|5z-x|}{y-x}=\frac{|5(9)-(-2)|}{-1-(-2)}$
$=\frac{|45+2|}{-1+2}$
$=\frac{|47|}{1}$
$=\frac{47}{1}$
$=47$

46. Replace x with -2, y with -1, and z with 9.
$\frac{|-x-y+z|}{2z}=\frac{|-(-2)-(-1)+9|}{2(9)}$
$=\frac{|2+1+9|}{18}$
$=\frac{|12|}{18}$
$=\frac{12}{18}$
$=\frac{2\cdot 6}{3\cdot 6}$
$=\frac{2}{3}$

Section 1.6 Practice

1. $-8(3)=-24$

2. $5(-30)=-150$

3. $-4(-12)=48$

Copyright © 2012 Pearson Education, Inc. Publishing as Prentice Hall.

4. $-\frac{5}{6}\cdot\frac{1}{4}=-\frac{5\cdot 1}{6\cdot 4}=-\frac{5}{24}$

5. $6(-2.3)=-13.8$

6. $-15(-2)=30$

7. a. $5(0)(-3)=0(-3)=0$

b. $(-1)(-6)(-7)=(6)(-7)=-42$

c. $(-2)(4)(-8)(-1)=(-8)(-8)(-1)$
$=64(-1)$
$=-64$

8. a. $(-2)^4=(-2)(-2)(-2)(-2)=16$

b. $-2^4=-(2\cdot 2\cdot 2\cdot 2)=-16$

c. $(-1)^5=(-1)(-1)(-1)(-1)(-1)=-1$

d. $-1^5=-(1\cdot 1\cdot 1\cdot 1\cdot 1)=-1$

e. $\left(-\frac{7}{9}\right)^2=\left(-\frac{7}{9}\right)\left(-\frac{7}{9}\right)=\frac{49}{81}$

9. a. The reciprocal of 13 is $\frac{1}{13}$ since $13\cdot\frac{1}{13}=1$.

b. The reciprocal of $\frac{7}{15}$ is $\frac{15}{7}$ since $\frac{7}{15}\cdot\frac{15}{7}=1$.

c. The reciprocal of -5 is $-\frac{1}{5}$ since $-5\cdot-\frac{1}{5}=1$.

d. The reciprocal of $-\frac{8}{11}$ is $-\frac{11}{8}$ since $-\frac{8}{11}\cdot-\frac{11}{8}=1$.

e. The reciprocal of 7.9 is $\frac{1}{7.9}$ since $7.9\cdot\frac{1}{7.9}=1$.

10. a. $-12\div 4=-12\cdot\frac{1}{4}=-3$

b. $\frac{-20}{-10}=-20\cdot-\frac{1}{10}=2$

c. $\frac{36}{-4}=36\cdot-\frac{1}{4}=-9$

11. a. $\frac{-25}{5}=-5$

b. $\frac{-48}{-6}=8$

c. $\frac{50}{-2}=-25$

d. $\frac{-72}{0.2}=-360$

12. $-\frac{5}{9}\div\frac{2}{3}=-\frac{5}{9}\cdot\frac{3}{2}=-\frac{15}{18}=-\frac{3\cdot 5}{3\cdot 6}=-\frac{5}{6}$

13. $-\frac{2}{7}\div\left(-\frac{1}{5}\right)=-\frac{2}{7}\cdot\left(-\frac{5}{1}\right)=\frac{10}{7}$

14. a. $\frac{-7}{0}$ is undefined.

b. $\frac{0}{-2}=0$

15. a. $\frac{0(-5)}{3}=\frac{0}{3}=0$

b. $-3(-9)-4(-4)=27-(-16)=27+16=43$

c. $(-3)^2+2\left[(5-15)-|-4-1|\right]$
$=(-3)^2+2\left[(-10)-|-5|\right]$
$=(-3)^2+2[-10-5]$
$=(-3)^2+2(-15)$
$=9+(-30)$
$=-21$

d. $\frac{-7(-4)+2}{-10-(-5)}=\frac{28+2}{-10+5}=\frac{30}{-5}=-6$

Copyright © 2012 Pearson Education, Inc. Publishing as Prentice Hall.

e. $\dfrac{5(-2)^3+52}{-4+1}=\dfrac{5(-8)+52}{-3}$
$=\dfrac{-40+52}{-3}$
$=\dfrac{12}{-3}$
$=-4$

16. **a.** Replace x with -1 and y with -5.
$\dfrac{3y}{45x}=\dfrac{3(-5)}{45(-1)}=\dfrac{-15}{-45}=\dfrac{15\cdot 1}{15\cdot 3}=\dfrac{1}{3}$

b. Replace x with -1 and y with -5.
$x^2-y^3=(-1)^2-(-5)^3$
$=1-(-125)$
$=1+125$
$=126$

c. Replace x with -1 and y with -5.
$\dfrac{x+y}{3x}=\dfrac{-1+(-5)}{3(-1)}=\dfrac{-6}{-3}=2$

17. $\dfrac{x}{4}-3=x+3$
$\dfrac{-8}{4}-3\stackrel{?}{=}-8+3$
$-2-3\stackrel{?}{=}-5$
$-5=-5$ True
-8 is a solution.

18. total score $=4\cdot(-13)=-52$
The card player's total score was -52.

Calculator Explorations

1. $-38(26-27)=38$

2. $-59(-8)+1726=2198$

3. $134+25(68-91)=-441$

4. $45(32)-8(218)=-304$

5. $\dfrac{-50(294)}{175-205}=490$

6. $\dfrac{-444-444.8}{-181-(-181)}$ is undefined.

7. $9^5-4550=54,499$

8. $5^8-6259=384,366$

9. $(-125)^2=15,625$

10. $-125^2=-15,625$

Vocabulary and Readiness Check

1. The product of a negative number and a positive number is a <u>negative</u> number.

2. The product of two negative numbers is a <u>positive</u> number.

3. The quotient of two negative numbers is a <u>positive</u> number.

4. The quotient of a negative number and a positive number is a <u>negative</u> number.

5. The product of a negative number and zero is <u>0</u>.

6. The reciprocal of a negative number is a <u>negative</u> number.

7. The quotient of 0 and a negative number is <u>0</u>.

8. The quotient of a negative number and 0 is <u>undefined</u>.

Exercise Set 1.6

1. $-6(4)=-24$

3. $2(-1)=-2$

5. $-5(-10)=50$

7. $-3\cdot 15=-45$

9. $-\dfrac{1}{2}\left(-\dfrac{3}{5}\right)=\dfrac{3}{10}$

11. $5(-1.4)=-7$

13. $(-1)(-3)(-5)=3(-5)=-15$

15. $(2)(-1)(-3)(0)=-2(-3)(0)=6(0)=0$

17. $(-4)^2=(-4)(-4)=16$

19. $-4^2=-4\cdot 4=-16$

Copyright © 2012 Pearson Education, Inc. Publishing as Prentice Hall.

21. $\left(-\frac{3}{4}\right)^2 = \left(-\frac{3}{4}\right)\left(-\frac{3}{4}\right) = \frac{9}{16}$

23. $-0.7^2 = -0.7 \cdot 0.7 = -0.49$

25. The reciprocal of $\frac{2}{3}$ is $\frac{3}{2}$ since $\frac{2}{3} \cdot \frac{3}{2} = 1$.

27. The reciprocal of -14 is $-\frac{1}{14}$ since $-14 \cdot \left(-\frac{1}{14}\right) = 1$.

29. The reciprocal of $-\frac{3}{11}$ is $-\frac{11}{3}$ since $-\frac{3}{11} \cdot \left(-\frac{11}{3}\right) = 1$.

31. The reciprocal of 0.2 is $\frac{1}{0.2}$ since $0.2\left(\frac{1}{0.2}\right) = 1$.

33. $\frac{18}{-2} = -9$

35. $-48 \div 12 = -48 \cdot \frac{1}{12} = -4$

37. $\frac{0}{-4} = 0$

39. $\frac{5}{0}$ is undefined.

41. $\frac{6}{7} \div \left(-\frac{1}{3}\right) = \frac{6}{7} \cdot \left(-\frac{3}{1}\right) = -\frac{18}{7}$

43. $-3.2 \div -0.02 = \frac{-3.2}{-0.02} = 160$

45. $(-8)(-8) = 64$

47. $\frac{2}{3}\left(-\frac{4}{9}\right) = -\frac{8}{27}$

49. $\frac{-12}{-4} = 3$

51. $\frac{30}{-2} = -15$

53. $(-5)^3 = (-5)(-5)(-5) = 25(-5) = -125$

55. $(-0.2)^3 = (-0.2)(-0.2)(-0.2)$
$= 0.04(-0.2)$
$= -0.008$

57. $\left(-\frac{3}{4}\right)\left(-\frac{8}{9}\right) = \frac{24}{36} = \frac{2 \cdot 12}{3 \cdot 12} = \frac{2}{3}$

59. $-\frac{5}{9} \div \left(-\frac{3}{4}\right) = -\frac{5}{9} \cdot \left(-\frac{4}{3}\right) = \frac{20}{27}$

61. $-2.1(-0.4) = 0.84$

63. $\frac{-48}{1.2} = -40$

65. $(-3)^4 = (-3)(-3)(-3)(-3)$
$= 9(-3)(-3)$
$= -27(-3)$
$= 81$

67. $-1^7 = -1 \cdot 1 \cdot 1 \cdot 1 \cdot 1 \cdot 1 \cdot 1 = -1$

69. $-11 \cdot 11 = -121$

71. $-\frac{4}{9} \div \frac{4}{9} = -\frac{4}{9} \cdot \frac{9}{4} = -\frac{36}{36} = -1$

73. $-9 - 10 = -9 + (-10) = -19$

75. $-9(-10) = 90$

77. $7(-12) = -84$

79. $7 + (-12) = -5$

81. $\frac{-9(-3)}{-6} = \frac{27}{-6} = -\frac{9 \cdot 3}{2 \cdot 3} = -\frac{9}{2}$

83. $-3(2 - 8) = -3[2 + (-8)] = -3(-6) = 18$

85. $-7(-2) - 3(-1) = 14 - (-3) = 14 + 3 = 17$

Copyright © 2012 Pearson Education, Inc. Publishing as Prentice Hall.

87. $2^2 - 3[(2-8)-(-6-8)]$
$= 2^2 - 3[(-6)-(-14)]$
$= 2^2 - 3[-6+14]$
$= 4 - 3(8)$
$= 4 - 24$
$= 4 + (-24)$
$= -20$

89. $\dfrac{-6^2+4}{-2} = \dfrac{-36+4}{-2} = \dfrac{-32}{-2} = 16$

91. $\dfrac{-3-5^2}{2(-7)} = \dfrac{-3-25}{-14}$
$= \dfrac{-3+(-25)}{-14}$
$= \dfrac{-28}{-14}$
$= \dfrac{-14\cdot 2}{-14\cdot 1}$
$= 2$

93. $\dfrac{22+(3)(-2)^2}{-5-2} = \dfrac{22+3(4)}{-5+(-2)}$
$= \dfrac{22+12}{-7}$
$= \dfrac{34}{-7}$
$= -\dfrac{34}{7}$

95. $\dfrac{(-4)^2-16}{4-12} = \dfrac{16-16}{4-12} = \dfrac{16+(-16)}{4+(-12)} = \dfrac{0}{-8} = 0$

97. $\dfrac{6-2(-3)}{4-3(-2)} = \dfrac{6-(-6)}{4-(-6)} = \dfrac{6+6}{4+6} = \dfrac{12}{10} = \dfrac{2\cdot 6}{2\cdot 5} = \dfrac{6}{5}$

99. $\dfrac{|5-9|+|10-15|}{|2(-3)|} = \dfrac{|5+(-9)|+|10+(-15)|}{|-6|}$
$= \dfrac{|-4|+|-5|}{|-6|}$
$= \dfrac{4+5}{6}$
$= \dfrac{9}{6}$
$= \dfrac{3\cdot 3}{3\cdot 2}$
$= \dfrac{3}{2}$

101. $\dfrac{-7(-1)+(-3)4}{(-2)(5)+(-6)(-8)} = \dfrac{7+(-12)}{-10+48} = \dfrac{-5}{38} = -\dfrac{5}{38}$

103. Replace x with -5 and y with -3.
$\dfrac{2x-5}{y-2} = \dfrac{2(-5)-5}{-3-2}$
$= \dfrac{-10-5}{-3-2}$
$= \dfrac{-10+(-5)}{-3+(-2)}$
$= \dfrac{-15}{-5}$
$= 3$

105. Replace x with -5 and y with -3.
$\dfrac{6-y}{x-4} = \dfrac{6-(-3)}{-5-4} = \dfrac{6+3}{-5+(-4)} = \dfrac{9}{-9} = -1$

107. Replace x with -5 and y with -3.
$\dfrac{4-2x}{y+3} = \dfrac{4-2(-5)}{-3+3}$
$= \dfrac{4-(-10)}{0}$
$= \dfrac{4+10}{0}$
$= \dfrac{14}{0}$ is undefined.

109. Replace x with -5 and y with -3.
$\dfrac{x^2+y}{3y} = \dfrac{(-5)^2+(-3)}{3(-3)}$
$= \dfrac{25+(-3)}{-9}$
$= \dfrac{22}{-9}$
$= -\dfrac{22}{9}$

111. $-3x-5=-20$
$-3(5)-5 \stackrel{?}{=} -20$
$-15-5 \stackrel{?}{=} -20$
$-15+(-5) \stackrel{?}{=} -20$
$-20=-20$ True
Since the result is true, 5 is a solution of the given equation.

Copyright © 2012 Pearson Education, Inc. Publishing as Prentice Hall.

113. $\frac{x}{5}+2=-1$

$\frac{15}{5}+2 \stackrel{?}{=} -1$

$3+2 \stackrel{?}{=} -1$

$5=-1$ False

Since the result is false, 15 is not a solution of the given equation.

115. $\frac{x-3}{7}=-2$

$\frac{-11-3}{7} \stackrel{?}{=} -2$

$\frac{-11+(-3)}{7} \stackrel{?}{=} -2$

$\frac{-14}{7} \stackrel{?}{=} -2$

$-2=-2$ True

Since the result is true, -11 is a solution of the given equation.

117. The product of -71 and a number is $-71 \cdot x$ or $-71x$.

119. Subtract a number from -16 is $-16 - x$.

121. -29 increased by a number is $-29 + x$.

123. Divide a number by -33 is $\frac{x}{-33}$ or $x \div (-33)$.

125. A loss of 4 yards is represented by -4.
$3 \cdot (-4) = -12$
The team had a total loss of 12 yards.

127. Each move of 20 feet down is represented by -20.
$5 \cdot (-20) = -100$
The diver is at a depth of 100 feet.

129. True since the product of an odd number of negative numbers is negative.

131. False since the product of an even number of negative integers is positive.

133. $2(-81) = -162$
The surface temperature of Jupiter is $-162°$F.

135. answers may vary

137. 1 and -1; answers may vary

139. $\frac{0}{5}-7=0-7=0+(-7)=-7$

141. $-8(-5) + (-1) = 40 + (-1) = 39$

Section 1.7 Practice

1. a. $7 \cdot y = y \cdot 7$

b. $4 + x = x + 4$

2. a. $5 \cdot (-3 \cdot 6) = (5 \cdot -3) \cdot 6$

b. $(-2 + 7) + 3 = -2 + (7 + 3)$

c. $(q + r) + 17 = q + (r + 17)$

d. $(ab) \cdot 21 = a \cdot (b \cdot 21)$

3. Since the order of two numbers was changed but their grouping was not, the statement is true by the commutative property of multiplication.

4. Since the grouping of the numbers was changed and their order was not, the statement is true by the associative property of addition.

5. $(-3+x)+17=-3+(x+17)$
$=-3+(17+x)$
$=(-3+17)+x$
$=14+x$

6. $4(5x) = (4 \cdot 5) \cdot x = 20x$

7. $5(x + y) = 5(x) + 5(y) = 5x + 5y$

8. $-3(2 + 7x) = -3(2) + (-3)(7x) = -6 - 21x$

9. $4(x+6y-2z)=4(x)+4(6y)-4(2z)$
$=4x+24y-8z$

10. $-1(3 - a) = (-1)(3) - (-1)(a) = -3 + a$

11. $-(8+a-b)=-1(8+a-b)$
$=(-1)(8)+(-1)(a)-(-1)(b)$
$=-8-a+b$

12. $\frac{1}{2}(2x+4)+9=\frac{1}{2}(2x)+\frac{1}{2}(4)+9$
$=1x+2+9$
$=x+11$

13. $9 \cdot 3 + 9 \cdot y = 9(3 + y)$

14. $4x + 4y = 4(x + y)$

Copyright © 2012 Pearson Education, Inc. Publishing as Prentice Hall.

15. $7(a + b) = 7 \cdot a + 7 \cdot b$ illustrates the distributive property.

16. $12 + y = y + 12$ illustrates the commutative property of addition.

17. $-4 \cdot (6 \cdot x) = (-4 \cdot 6) \cdot x$ illustrates the associative property of multiplication.

18. $6 + (z + 2) = 6 + (2 + z)$ illustrates the commutative property of addition.

19. $3\left(\frac{1}{3}\right) = 1$ illustrates the multiplicative inverse property.

20. $(x + 0) + 23 = x + 23$ illustrates the identity element for addition.

21. $(7 \cdot y) \cdot 10 = y \cdot (7 \cdot 10)$ illustrates the commutative and associative properties of multiplication.

Vocabulary and Readiness Check

1. $x + 5 = 5 + x$ is a true statement by the <u>commutative property of addition</u>.

2. $x \cdot 5 = 5 \cdot x$ is a true statement by the <u>commutative property of multiplication</u>.

3. $3(y + 6) = 3 \cdot y + 3 \cdot 6$ is true by the <u>distributive property</u>.

4. $2 \cdot (x \cdot y) = (2 \cdot x) \cdot y$ is a true statement by the <u>associative property of multiplication</u>.

5. $x + (7 + y) = (x + 7) + y$ is a true statement by the <u>associative property of addition</u>.

6. The numbers $-\frac{2}{3}$ and $-\frac{3}{2}$ are called <u>reciprocals or multiplicative inverses</u>.

7. The numbers $-\frac{2}{3}$ and $\frac{2}{3}$ are called <u>opposites or additive inverses</u>.

Exercise Set 1.7

1. $x + 16 = 16 + x$ by the commutative property of addition.

3. $-4 \cdot y = y \cdot (-4)$ by the commutative property of multiplication.

5. $xy = yx$ by the commutative property of multiplication.

7. $2x + 13 = 13 + 2x$ by the commutative property of addition.

9. $(xy) \cdot z = x \cdot (yz)$ by the associative property of multiplication.

11. $2 + (a + b) = (2 + a) + b$ by the associative property of addition.

13. $4 \cdot (ab) = 4a \cdot (b)$ by the associative property of multiplication.

15. $(a + b) + c = a + (b + c)$ by the associative property of addition.

17. $8 + (9 + b) = (8 + 9) + b = 17 + b$

19. $4(6y) = (4 \cdot 6)y = 24y$

21. $\frac{1}{5}(5y) = \left(\frac{1}{5} \cdot 5\right)y = 1y = y$

23. $(13 + a) + 13 = (a + 13) + 13$
$= a + (13 + 13)$
$= a + 26$

25. $-9(8x) = (-9 \cdot 8)x = -72x$

27. $\frac{3}{4}\left(\frac{4}{3}s\right) = \left(\frac{3}{4} \cdot \frac{4}{3}\right)s = 1s = s$

29. $-\frac{1}{2}(5x) = \left(-\frac{1}{2} \cdot 5\right)x = -\frac{5}{2}x$

31. $4(x + y) = 4(x) + 4(y) = 4x + 4y$

33. $9(x - 6) = 9(x) - 9(6) = 9x - 54$

35. $2(3x + 5) = 2(3x) + 2(5) = 6x + 10$

37. $7(4x - 3) = 7(4x) - 7(3) = 28x - 21$

39. $3(6 + x) = 3(6) + 3(x) = 18 + 3x$

41. $-2(y - z) = -2(y) - (-2)z = -2y + 2z$

43. $-\frac{1}{3}(3y + 5) = -\frac{1}{3}(3y) - \frac{1}{3}(5) = -y - \frac{5}{3}$

45. $5(x + 4m + 2) = 5(x) + 5(4m) + 5(2)$
$= 5x + 20m + 10$

Copyright © 2012 Pearson Education, Inc. Publishing as Prentice Hall.

47. $-4(1-2m+n)+4 = -4(1)-4(-2m)-4(n)+4$
$= -4+8m-4n+4$
$= (-4+4)+8m-4n$
$= 0+8m-4n$
$= 8m-4n$

49. $-(5x+2) = -1(5x+2)$
$= -1(5x)+(-1)(2)$
$= -5x-2$

51. $-(r-3-7p) = -1(r-3-7p)$
$= -1(r)-1(-3)-1(-7p)$
$= -r+3+7p$

53. $\frac{1}{2}(6x+7)+\frac{1}{2} = \frac{1}{2}(6x)+\frac{1}{2}(7)+\frac{1}{2}$
$= \left(\frac{1}{2}\cdot 6\right)x+\frac{7}{2}+\frac{1}{2}$
$= 3x+\frac{8}{2}$
$= 3x+4$

55. $-\frac{1}{3}(3x-9y) = -\frac{1}{3}(3x)-\frac{1}{3}(-9y) = -x+3y$

57. $3(2r+5)-7 = 3(2r)+3(5)-7$
$= 6r+15-7$
$= 6r+8$

59. $-9(4x+8)+2 = -9(4x)-9(8)+2$
$= -36x-72+2$
$= -36x-70$

61. $-0.4(4x+5)-0.5 = -0.4(4x)+(-0.4)(5)-0.5$
$= -1.6x-2-0.5$
$= -1.6x-2.5$

63. $4 \cdot 1 + 4 \cdot y = 4(1 + y)$

65. $11x + 11y = 11 \cdot x + 11 \cdot y = 11(x + y)$

67. $(-1) \cdot 5 + (-1) \cdot x = -1(5 + x) = -(5 + x)$

69. $30a + 30b = 30 \cdot a + 30 \cdot b = 30(a + b)$

71. $3 \cdot 5 = 5 \cdot 3$ illustrates the commutative property of multiplication.

73. $2 + (x + 5) = (2 + x) + 5$ illustrates the associative property of addition.

75. $(x + 9) + 3 = (9 + x) + 3$ illustrates the commutative property of addition.

77. $(4 \cdot y) \cdot 9 = 4 \cdot (y \cdot 9)$ illustrates the associative property of multiplication.

79. $0 + 6 = 6$ illustrates the identity property of addition.

81. $-4(y + 7) = -4 \cdot y + (-4) \cdot 7$ illustrates the distributive property.

83. $6 \cdot \frac{1}{6} = 1$ illustrates the multiplicative inverse property.

85. $-6 \cdot 1 = -6$ illustrates the identity element for multiplication.

87. The opposite of 8 is -8.
The reciprocal of 8 is $\frac{1}{8}$.

89. The opposite of x is $-x$.
The reciprocal of x is $\frac{1}{x}$.

91. The expression is the reciprocal of $\frac{1}{2x}$ or $2x$.
The opposite of $2x$ is $-2x$.

93. False; the opposite of $-\frac{a}{2}$ is $\frac{a}{2}$. $-\frac{2}{a}$ is the reciprocal of $-\frac{a}{2}$.

95. "Taking a test" and "studying for the test" are not commutative, since the order in which they are performed affects the outcome.

97. "Putting on your left shoe" and "putting on your right shoe" are commutative, since the order in which they are performed does not affect the outcome.

99. "Mowing the lawn" and "trimming the hedges" are commutative, since the order in which they are performed does not affect the outcome.

101. "Feeding the dog" and "feeding the cat" are commutative, since the order in which they are performed does not affect the outcome.

103. **a.** The property illustrated is the commutative property of addition since the order in which they are added changed.

Copyright © 2012 Pearson Education, Inc. Publishing as Prentice Hall.

b. The property illustrated is the commutative property of addition since the order in which they are added changed.

c. The property illustrated is the associative property of addition since the grouping of addition changed.

105. answers may vary

107. answers may vary

Section 1.8 Practice

1. a. The numerical coefficient of $-4x$ is -4.

b. The numerical coefficient of $15y^3$ is 15.

c. The numerical coefficient of x is 1, since $x = 1x$.

d. The numerical coefficient of $-y$ is -1, since $-y = -1y$.

e. The numerical coefficient of $\frac{z}{4}$ is $\frac{1}{4}$, since $\frac{z}{4}$ is $\frac{1}{4} \cdot z$.

2. a. $7x^2$ and $-6x^3$ are unlike terms, since the exponents on x are not the same.

b. $3x^2y^2$, $-x^2y^2$, and $4x^2y^2$ are like terms, since each variable and its exponent match.

c. $-5ab$ and $3ba$ are like terms, since $ab = ba$ by the commutative property.

d. $2x^3$ and $4y^3$ are unlike terms, since the variables are not the same.

e. $-7m^4$ and $7m^4$ are like terms, since the variable and its exponent match.

3. a. $9y - 4y = (9 - 4)y = 5y$

b. $11x^2 + x^2 = (11+1)x^2 = 12x^2$

c. $5y - 3x + 4x = 5y + (-3+4)x$
$= 5y + 1x$
$= 5y + x$

d. $14m^2 - m^2 + 3m^2 = (14-1+3)m^2 = 16m^2$

4. $7y + 2y + 6 + 10 = (7 + 2)y + (6 + 10) = 9y + 16$

5. $-2x + 4 + x - 11 = -2x + x + 4 - 11$
$= (-2+1)x + (4-11)$
$= -x - 7$

6. The terms $3z$ and $-3z^2$ cannot be combined because they are unlike terms.

7. $8.9y + 4.2y - 3 = (8.9 + 4.2)y - 3 = 13.1y - 3$

8. $3(11y + 6) = 3(11y) + 3(6) = 33y + 18$

9. $-4(x + 0.2y - 3) = -4(x) + (-4)(0.2y) - (-4)(3)$
$= -4x - 0.8y + 12$

10. $-(3x + 2y + z - 1)$
$= -1(3x + 2y + z - 1)$
$= -1(3x) + (-1)(2y) + (-1)(z) - (-1)(1)$
$= -3x - 2y - z + 1$

11. $4(4x - 6) + 20 = 16x - 24 + 20 = 16x - 4$

12. $5 - (3x + 9) + 6x = 5 - 3x - 9 + 6x$
$= -3x + 6x + 5 - 9$
$= 3x - 4$

13. $-3(7x + 1) - (4x - 2) = -21x - 3 - 4x + 2$
$= -25x - 1$

14. $8 + 11(2y - 9) = 8 + 22y - 99 = -91 + 22y$

15. $(4x - 3) - (9x - 10) = 4x - 3 - 9x + 10 = -5x + 7$

16. Three times a number subtracted from 10 is written as $10 - 3x$. This expression cannot be simplified.

17. The sum of a number and 2, divided by 5, is written as $(x + 2) \div 5$ or $\frac{x+2}{5}$. This expression cannot be simplified.

18. Three times a number, added to the sum of a number and 6, is written as $(x + 6) + 3x$.
$(x + 6) + 3x = x + 6 + 3x = 4x + 6$

19. Seven times the difference of a number and 4 is written as $7(x - 4)$.
$7(x - 4) = 7x - 28$

Copyright © 2012 Pearson Education, Inc. Publishing as Prentice Hall.

Vocabulary and Readiness Check

1. $14y^2 + 2x - 23$ is called an expression while $14y^2$, $2x$, and -23 are each called a term.
2. To multiply $3(-7x + 1)$, we use the distributive property.
3. To simplify an expression like $y + 7y$, we combine like terms.
4. The term z has an understood numerical coefficient of 1.
5. The terms $-x$ and $5x$ are like terms and the terms $5x$ and $5y$ are unlike terms.
6. For the term $-3x^2y$, -3 is called the numerical coefficient.
7. The numerical coefficient of $-7y$ is -7.
8. The numerical coefficient of $3x$ is 3.
9. The numerical coefficient of x is 1, since $x = 1x$.
10. The numerical coefficient of $-y$ is -1, since $-y = -1y$.
11. The numerical coefficient of $17x^2y$ is 17.
12. The numerical coefficient of $1.2xyz$ is 1.2.
13. $5y$ and $-y$ are like terms, since the variable and its exponent match.
14. $-2x^2y$ and $6xy$ are unlike terms, since the exponents on x are not the same.
15. $2z$ and $3z^2$ are unlike terms, since the exponents on z are not the same.
16. ab^2 and $-7ab^2$ are like terms, since each variable and its exponent match.
17. $8wz$ and $\frac{1}{7}zw$ are like terms, since $wz = zw$ by the commutative property.
18. $7.4p^3q^2$ and $6.2p^3q^2r$ are unlike terms, since the exponents on r are not the same.

Exercise Set 1.8

1. $7y + 8y = (7 + 8)y = 15y$

3. $$\begin{aligned} 8w - w + 6w &= 8w - 1w + 6w \\ &= (8 - 1 + 6)w \\ &= 13w \end{aligned}$$

5. $$\begin{aligned} 3b - 5 - 10b - 4 &= 3b - 10b - 5 - 4 \\ &= (3 - 10)b + (-5 - 4) \\ &= -7b - 9 \end{aligned}$$

7. $$\begin{aligned} m - 4m + 2m - 6 &= (1 - 4 + 2)m - 6 \\ &= -1m - 6 \\ &= -m - 6 \end{aligned}$$

9. $$\begin{aligned} 5g - 3 - 5 - 5g &= 5g - 5g - 3 - 5 \\ &= (5 - 5)g + (-3 - 5) \\ &= 0g - 8 \\ &= -8 \end{aligned}$$

11. $$\begin{aligned} 6.2x - 4 + x - 1.2 &= 6.2x + 1x - 4 - 1.2 \\ &= (6.2 + 1)x + (-4 - 1.2) \\ &= 7.2x - 5.2 \end{aligned}$$

13. $$\begin{aligned} 2k - k - 6 &= 2k - 1k - 6 \\ &= (2 - 1)k - 6 \\ &= 1k - 6 \\ &= k - 6 \end{aligned}$$

15. $$\begin{aligned} -9x + 4x + 18 - 10x &= -9x + 4x - 10x + 18 \\ &= (-9 + 4 - 10)x + 18 \\ &= -15x + 18 \end{aligned}$$

17. $$\begin{aligned} 6x - 5x + x - 3 + 2x &= 6x - 5x + x + 2x - 3 \\ &= (6 - 5 + 1 + 2)x - 3 \\ &= 4x - 3 \end{aligned}$$

19. $7x^2 + 8x^2 - 10x^2 = (7 + 8 - 10)x^2 = 5x^2$

21. $$\begin{aligned} 3.4m - 4 - 3.4m - 7 &= 3.4m - 3.4m - 4 - 7 \\ &= (3.4 - 3.4)m + (-4 - 7) \\ &= 0m - 11 \\ &= -11 \end{aligned}$$

23. $$\begin{aligned} &6x + 0.5 - 4.3x - 0.4x + 3 \\ &= 6x - 4.3x - 0.4x + 0.5 + 3 \\ &= (6 - 4.3 - 0.4)x + (0.5 + 3) \\ &= 1.3x + 3.5 \end{aligned}$$

25. $5(y + 4) = 5(y) + 5(4) = 5y + 20$

27. $-2(x + 2) = -2(x) - 2(2) = -2x - 4$

Copyright © 2012 Pearson Education, Inc. Publishing as Prentice Hall.

29. $-5(2x-3y+6)$
$=-5(2x)-(-5)(3y)+(-5)(6)$
$=-10x+15y-30$

31. $-(3x-2y+1)=-1(3x-2y+1)$
$=-1(3x)-1(-2y)-1(1)$
$=-3x+2y-1$

33. $7(d-3)+10=7d-21+10=7d-11$

35. $-4(3y-4)+12y=-12y+16+12y$
$=0y+16$
$=16$

37. $3(2x-5)-5(x-4)=6x-15-5x+20$
$=6x-5x-15+20$
$=x+5$

39. $-2(3x-4)+7x-6=-6x+8+7x-6$
$=-6x+7x+8-6$
$=1x+2$
$=x+2$

41. $5k-(3k-10)=5k-3k+10=2k+10$

43. $(3x+4)-(6x-1)=3x+4-6x+1$
$=3x-6x+4+1$
$=-3x+5$

45. $5(x+2)-(3x-4)=5x+10-3x+4$
$=5x-3x+10+4$
$=2x+14$

47. $\frac{1}{3}(7y-1)+\frac{1}{6}(4y+7)=\frac{7}{3}y-\frac{1}{3}+\frac{4}{6}y+\frac{7}{6}$
$=\frac{7}{3}y+\frac{2}{3}y-\frac{2}{6}+\frac{7}{6}$
$=\frac{9}{3}y+\frac{5}{6}$
$=3y+\frac{5}{6}$

49. $2+4(6x-6)=2+24x-24$
$=2-24+24x$
$=-22+24x$

51. $0.5(m+2)+0.4m=0.5m+1+0.4m$
$=0.5m+0.4m+1$
$=0.9m+1$

53. $10-3(2x+3y)=10-6x-9y$

55. $6(3x-6)-2(x+1)-17x$
$=18x-36-2x-2-17x$
$=18x-2x-17x-36-2$
$=-1x-38$
$=-x-38$

57. $\frac{1}{2}(12x-4)-(x+5)=6x-2-x-5$
$=6x-x-2-5$
$=5x-7$

59. $(4x-10)+(6x+7)=4x-10+6x+7$
$=4x+6x-10+7$
$=10x-3$

61. $(3x-8)-(7x+1)=3x-8-7x-1$
$=3x-7x-8-1$
$=-4x-9$

63. $(m-9)-(5m-6)=m-9-5m+6$
$=m-5m-9+6$
$=-4m-3$

65. Twice a number, decreased by four is written as $2x-4$.

67. Three-fourths of a number, increased by twelve is written as $\frac{3}{4}x+12$.

69. The sum of 5 times a number and -2, added to 7 times the number is written as
$5x+(-2)+7x=5x+7x-2=12x-2$.

71. Eight times the sum of a number and six is written as $8(x+6)=8x+48$.

73. Double a number minus the sum of the number and ten is written as
$2x-(x+10)=2x-x-10=x-10$.

75. Since 1 cone balances 1 cube and 1 cylinder balances 2 cubes, 1 cone and 1 cylinder balances $1+2=3$ cubes. The scale shown is balanced.

77. Since 1 cylinder balances 2 cubes, the left side is equivalent to 5 cubes. Since 1 cone balances 1 cube, the right side is equivalent to 5 cubes. The scale is balanced.

79. answers may vary

Copyright © 2012 Pearson Education, Inc. Publishing as Prentice Hall.

81. $2[(4x-1)+5x] = 2(4x-1+5x)$
$= 2(9x-1)$
$= 18x-2$
The perimeter is $(18x - 2)$ feet.

83. The length of the first board in inches is $12(x + 2)$.
$12(x+2)+(3x-1) = 12x+24+3x-1$
$= 12x+3x+24-1$
$= 15x+23$
The total length is $(15x + 23)$ inches.

85. answers may vary

Chapter 1 Vocabulary Check

1. The symbols $\neq$, $<$, and $>$ are called inequality symbols.

2. A mathematical statement that two expressions are equal is called and equation.

3. The absolute value of a number is the distance between that number and 0 on a number line.

4. A symbol used to represent a number is called a variable.

5. Two numbers that are the same distance from 0 but lie on opposite sides of 0 are called opposites.

6. The number in a fraction above the fraction bar is called the numerator.

7. A solution of an equation is a value for the variable that makes the equation a true statement.

8. Two numbers whose product is 1 are called reciprocals.

9. In 2^3, the 2 is called the base and the 3 is called the exponent.

10. The numerical coefficient of a term is its numerical factor.

11. The number in a fraction below the fraction bar is called the denominator.

12. Parentheses and brackets are examples of grouping symbols.

13. A term is a number or the product of a number and variables raised to powers.

14. Terms with the same variables raised to the same powers are called like terms.

15. If terms are not like terms, then they are unlike terms.

Chapter 1 Review

1. Since 8 is to the left of 10 on the number line, $8 < 10$.

2. Since 7 is to the right of 2 on the number line, $7 > 2$.

3. Since -4 is to the right of -5 on the number line, $-4 > -5$.

4. Since $\frac{12}{2} = 6$ is to the right of -8 on the number line, $\frac{12}{2} > -8$.

5. Since $|-7| = 7$ is to the left of $|-8| = 8$ on the number line, $|-7| < |-8|$.

6. Since $|-9| = 9$ is to the right of -9 on the number line, $|-9| > -9$.

7. $-|-1| = -1$

8. Since $|-14| = 14$ and $-(-14) = 14$, $|-14| = -(-14)$.

9. Since 1.2 is to the right of 1.02 on the number line, $1.2 > 1.02$.

10. Since $-\frac{3}{2} = -\frac{6}{4}$ and $-\frac{6}{4}$ is to the left of $-\frac{3}{4}$ on the number line, $-\frac{3}{2} < -\frac{3}{4}$.

11. Four is greater than or equal to negative three is written as $4 \geq -3$.

12. Six is not equal to five is written as $6 \neq 5$.

13. 0.03 is less than 0.3 is written as $0.03 < 0.3$.

14. Since 155 is to the left of 400 on the number line, $155 < 400$.

15. **a.** The natural numbers are 1, 3.

b. The whole numbers are 0, 1, 3.

Copyright © 2012 Pearson Education, Inc. Publishing as Prentice Hall.

c. The integers are –6, 0, 1, 3.

d. The rational numbers are –6, 0, 1, $1\frac{1}{2}$, 3, 9.62.

e. The irrational number is π.

f. The real numbers are all numbers in the set.

16. a. The natural numbers are 2, 5.

b. The whole numbers are 2, 5.

c. The integers are –3, 2, 5.

d. The rational numbers are –3, –1.6, 2, 5, $\frac{11}{2}$, 15.1.

e. The irrational numbers are $\sqrt{5}$, 2π.

f. The real numbers are all numbers in the set.

17. Since $-4 < -2$, the most negative number is –4 which corresponds to Friday. Thus, Friday showed the greatest loss.

18. The greatest positive number is +5 which corresponds to Wednesday. Thus, Wednesday showed the greatest gain.

19. $6 \cdot 3^2 + 2 \cdot 8 = 6 \cdot 9 + 2 \cdot 8 = 54 + 16 = 70$
The answer is c.

20. $68 - 5 \cdot 2^3 = 68 - 5 \cdot 8 = 68 - 40 = 68 + (-40) = 28$
The answer is b.

21.
$$\begin{aligned} 3(1+2\cdot 5)+4 &= 3(1+10)+4 \\ &= 3(11)+4 \\ &= 33+4 \\ &= 37 \end{aligned}$$

22.
$$\begin{aligned} 8+3(2\cdot 6-1) &= 8+3(12-1) \\ &= 8+3[12+(-1)] \\ &= 8+3(11) \\ &= 8+33 \\ &= 41 \end{aligned}$$

23.
$$\begin{aligned} \frac{4+|6-2|+8^2}{4+6\cdot 4} &= \frac{4+|6+(-2)|+8^2}{4+6\cdot 4} \\ &= \frac{4+|4|+8^2}{4+6\cdot 4} \\ &= \frac{4+4+64}{4+24} \\ &= \frac{72}{28} \\ &= \frac{18\cdot 4}{7\cdot 4} \\ &= \frac{18}{7} \end{aligned}$$

24.
$$\begin{aligned} 5[3(2+5)-5] &= 5[3(7)-5] \\ &= 5[21-5] \\ &= 5[21+(-5)] \\ &= 5(16) \\ &= 80 \end{aligned}$$

25. The difference of twenty and twelve is equal to the product of two and four is written as $20 - 12 = 2 \cdot 4$.

26. The quotient of nine and two is greater than negative five is written as $\frac{9}{2} > -5$.

27. Replace x with 6 and y with 2.
$2x+3y = 2(6)+3(2) = 12+6 = 18$

28. Replace x with 6, y with 2, and z with 8.
$x(y+2z) = 6[2+2(8)] = 6(2+16) = 6(18) = 108$

29. Replace x with 6, y with 2, and z with 8.
$\frac{x}{y}+\frac{z}{2y} = \frac{6}{2}+\frac{8}{2(2)} = \frac{6}{2}+\frac{8}{4} = 3+2 = 5$

30. Replace x with 6 and y with 2.
$$\begin{aligned} x^2-3y^2 &= 6^2-3(2)^2 \\ &= 36-3(4) \\ &= 36-12 \\ &= 36+(-12) \\ &= 24 \end{aligned}$$

31. Replace a with 37 and b with 80.
$$\begin{aligned} 180-a-b &= 180-37-80 \\ &= 180+(-37)+(-80) \\ &= 143+(-80) \\ &= 63 \end{aligned}$$
The measure of the unknown angle is 63°.

Copyright © 2012 Pearson Education, Inc. Publishing as Prentice Hall.

32. Replace a with 93, b with 80, and c with 82.

$$\begin{aligned} 360-a-b-c &= 360-93-80-82 \\ &= 360+(-93)+(-80)+(-82) \\ &= 267+(-80)+(-82) \\ &= 187+(-82) \\ &= 105 \end{aligned}$$

The measure of the unknown angle is 105°.

33.

$$\begin{aligned} 7x-3 &= 18 \\ 7(3)-3 &\stackrel{?}{=} 18 \\ 21-3 &\stackrel{?}{=} 18 \\ 21+(-3) &\stackrel{?}{=} 18 \\ 18 &= 18 \end{aligned}$$

Since the results is true, 3 is a solution of the given equation.

34.

$$\begin{aligned} 3x^2+4 &= x-1 \\ 3(1)^2+4 &\stackrel{?}{=} 1-1 \\ 3(1)+4 &\stackrel{?}{=} 0 \\ 3+4 &\stackrel{?}{=} 0 \\ 7 &= 0 \end{aligned}$$

Since the result is false, 1 is not a solution of the given equation.

35. The opposite of -9 is 9.

36. The opposite of $\frac{2}{3}$ is $-\frac{2}{3}$.

37. The opposite of $|-2| = 2$ is -2.

38. The opposite of $-|-7| = -7$ is 7.

39. $-15 + 4 = -11$

40. $-6 + (-11) = -17$

41. $\frac{1}{16}+\left(-\frac{1}{4}\right)=\frac{1}{16}+\left(-\frac{4}{16}\right)=-\frac{3}{16}$

42. $-8 + |-3| = -8 + 3 = -5$

43. $-4.6 + (-9.3) = -13.9$

44. $-2.8 + 6.7 = 3.9$

45. $6 - 20 = 6 + (-20) = -14$

46. $-3.1 - 8.4 = -3.1 + (-8.4) = -11.5$

47. $-6 - (-11) = -6 + 11 = 5$

48. $4 - 15 = 4 + (-15) = -11$

49.

$$\begin{aligned} -21-16+3(8-2) &= -21+(-16)+3[8+(-2)] \\ &= -21+(-16)+3(6) \\ &= -21+(-16)+18 \\ &= -37+18 \\ &= -19 \end{aligned}$$

50.

$$\begin{aligned} \frac{11-(-9)+6(8-2)}{2+3\cdot 4} &= \frac{11+9+6[8+(-2)]}{2+3\cdot 4} \\ &= \frac{11+9+6(6)}{2+3\cdot 4} \\ &= \frac{11+9+36}{2+12} \\ &= \frac{56}{14} \\ &= 4 \end{aligned}$$

51. Replace x with 3, y with -6, and z with -9.

$$\begin{aligned} 2x^2-y+z &= 2(3)^2-(-6)+(-9) \\ &= 2(9)+6+(-9) \\ &= 18+6+(-9) \\ &= 24+(-9) \\ &= 15 \end{aligned}$$

The answer is a.

52. Replace x with 3 and y with -6.

$$\begin{aligned} \frac{|y-4x|}{2x} &= \frac{|-6-4(3)|}{2(3)} \\ &= \frac{|-6-12|}{6} \\ &= \frac{|-6+(-12)|}{6} \\ &= \frac{|-18|}{6} \\ &= \frac{18}{6} \\ &= 3 \end{aligned}$$

The answer is a.

53.

$$\begin{aligned} &50+1+(-2)+5+1+(-4) \\ &= 51+(-2)+5+1+(-4) \\ &= 49+5+1+(-4) \\ &= 54+1+(-4) \\ &= 55+(-4) \\ &= 51 \end{aligned}$$

The price at the end of the week is \$51.

54. $50+1+(-2)+5 = 51+(-2)+5 = 49+5 = 54$

The price at the end of the day on Wednesday is \$54.

Copyright © 2012 Pearson Education, Inc. Publishing as Prentice Hall.

55. The reciprocal of -6 is $-\frac{1}{6}$.

56. The reciprocal of $\frac{3}{5}$ is $\frac{5}{3}$.

57. $6(-8) = -48$

58. $(-2)(-14) = 28$

59. $\frac{-18}{-6} = 3$

60. $\frac{42}{-3} = -14$

61. $-3(-6)(-2) = 18(-2) = -36$

62. $(-4)(-3)(0)(-6) = 12(0)(-6) = 0(-6) = 0$

63. $\frac{4(-3)+(-8)}{2+(-2)} = \frac{-12+(-8)}{2+(-2)} = \frac{-20}{0}$ is undefined.

64.
$$\begin{aligned}\frac{3(-2)^2-5}{-14} &= \frac{3(4)-5}{-14}\\ &= \frac{12-5}{-14}\\ &= \frac{12+(-5)}{-14}\\ &= \frac{7}{-14}\\ &= -\frac{1}{2}\end{aligned}$$

65. $-6 + 5 = 5 + (-6)$ illustrates the commutative property of addition.

66. $6 \cdot 1 = 6$ illustrates the multiplicative identity property.

67. $3(8 - 5) = 3 \cdot 8 - 3 \cdot 5$ illustrates the distributive property.

68. $4 + (-4) = 0$ illustrates the additive inverse property.

69. $2 + (3 + 9) = (2 + 3) + 9$ illustrates the associative property of addition.

70. $2 \cdot 8 = 8 \cdot 2$ illustrates the commutative property of multiplication.

71. $6(8 + 5) = 6 \cdot 8 + 6 \cdot 5$ illustrates the distributive property.

72. $(3 \cdot 8) \cdot 4 = 3 \cdot (8 \cdot 4)$ illustrates the associative property of multiplication.

73. $4 \cdot \frac{1}{4} = 1$ illustrates the multiplicative inverse property.

74. $8 + 0 = 8$ illustrates the additive identity property.

75. $4(8 + 3) = 4(3 + 8)$ illustrates the commutative property of addition.

76. $5(2 + 1) = 5 \cdot 2 + 5 \cdot 1$ illustrates the distributive property.

77. $5x - x + 2x = 5x - 1x + 2x = (5 - 1 + 2)x = 6x$

78. $0.2z - 4.6z - 7.4z = (0.2 - 4.6 - 7.4)z = -11.8z$

79.
$$\begin{aligned}\frac{1}{2}x+3+\frac{7}{2}x-5 &= \frac{1}{2}x+\frac{7}{2}x+3-5\\ &= \left(\frac{1}{2}+\frac{7}{2}\right)x+3+(-5)\\ &= \frac{8}{2}x+(-2)\\ &= 4x-2\end{aligned}$$

80.
$$\begin{aligned}\frac{4}{5}y+1+\frac{6}{5}y+2 &= \frac{4}{5}y+\frac{6}{5}y+1+2\\ &= \left(\frac{4}{5}+\frac{6}{5}\right)y+3\\ &= \frac{10}{5}y+3\\ &= 2y+3\end{aligned}$$

81.
$$\begin{aligned}2(n-4)+n-10 &= 2n-8+n-10\\ &= 2n+n-8-10\\ &= 3n-18\end{aligned}$$

82.
$$\begin{aligned}3(w+2)-(12-w) &= 3w+6-12+w\\ &= 3w+w+6-12\\ &= 4w-6\end{aligned}$$

83.
$$\begin{aligned}(x+5)-(7x-2) &= x+5-7x+2\\ &= x-7x+5+2\\ &= -6x+7\end{aligned}$$

Copyright © 2012 Pearson Education, Inc. Publishing as Prentice Hall.

84. $(y-0.7)-(1.4y-3) = y-0.7-1.4y+3$
$= y-1.4y-0.7+3$
$= -0.4y+2.3$

85. Three times a number decreased by 7 is written as $3x-7$.

86. Twice the sum of a number and 2.8, added to 3 times the number is written as
$2(x+2.8)+3x = 2x+5.6+3x$
$= 2x+3x+5.6$
$= 5x+5.6.$

87. $-|-11| = -11$
$|11.4| = 11.4$
So, $-|-11| < |11.4|$.

88. Since $-1\frac{1}{2}$ is to the right of $-2\frac{1}{2}$ on the number line, $-1\frac{1}{2} > -2\frac{1}{2}$.

89. $-7.2 + (-8.1) = -15.3$

90. $14 - 20 = 14 + (-20) = -6$

91. $4(-20) = -80$

92. $\frac{-20}{4} = -5$

93. $-\frac{4}{5}\left(\frac{5}{16}\right) = -\frac{20}{80} = -\frac{1\cdot 20}{4\cdot 20} = -\frac{1}{4}$

94. $-0.5(-0.3) = 0.15$

95. $8 \div 2 \cdot 4 = 4 \cdot 4 = 16$

96. $(-2)^4 = (-2)(-2)(-2)(-2)$
$= 4(-2)(-2)$
$= -8(-2)$
$= 16$

97. $\frac{-3-2(-9)}{-15-3(-4)} = \frac{-3+18}{-15+12} = \frac{15}{-3} = -5$

98. $5+2[(7-5)^2+(1-3)]$
$= 5+2[(7+(-5))^2+(1+(-3))]$
$= 5+2[(2)^2+(-2)]$
$= 5+2[4+(-2)]$
$= 5+2(2)$
$= 5+4$
$= 9$

99. $-\frac{5}{8}\div\frac{3}{4} = -\frac{5}{8}\cdot\frac{4}{3} = -\frac{20}{24} = -\frac{5\cdot 4}{6\cdot 4} = -\frac{5}{6}$

100. $\frac{-15+(-4)^2+|-9|}{10-2\cdot 5} = \frac{-15+16+9}{10-10} = \frac{10}{0}$ is undefined.

101. $7(3x-3)-5(x+4) = 21x-21-5x-20$
$= 21x-5x-21-20$
$= 16x-41$

102. $8+2(9x-10) = 8+18x-20$
$= 18x+8-20$
$= 18x-12$

Chapter 1 Test

1. The absolute value of negative seven is greater than five is written as $|-7| > 5$.

2. The sum of nine and five is greater than or equal to four is written as $9 + 5 \geq 4$.

3. $-13 + 8 = -5$

4. $-13 - (-2) = -13 + 2 = -11$

5. $6 \cdot 3 - 8 \cdot 4 = 18 - 32 = 18 + (-32) = -14$

6. $(13)(-3) = -39$

7. $(-6)(-2) = 12$

8. $\frac{|-16|}{-8} = \frac{16}{-8} = -2$

9. $\frac{-8}{0}$ is undefined.

10. $\frac{|-6|+2}{5-6} = \frac{6+2}{5-6} = \frac{8}{-1} = -8$

Copyright © 2012 Pearson Education, Inc. Publishing as Prentice Hall.

11. $\frac{1}{2}-\frac{5}{6}=\frac{3}{6}-\frac{5}{6}=-\frac{2}{6}=-\frac{1}{3}$

12. $-1\frac{1}{8}+5\frac{3}{4}=-1\frac{1}{8}+5\frac{6}{8}=4\frac{5}{8}$

13. $-\frac{3}{5}+\frac{15}{8}=-\frac{24}{40}+\frac{75}{40}=\frac{51}{40}$ or $1\frac{11}{40}$

14. $3(-4)^2-80=3(16)-80=48-80=-32$

15.
$$\begin{aligned}6[5+2(3-8)-3]&=6[5+2(3+(-8))-3]\\&=6[5+2(-5)-3]\\&=6[5+(-10)+(-3)]\\&=6[-5+(-3)]\\&=6[-8]\\&=-48\end{aligned}$$

16. $\frac{-12+3\cdot 8}{4}=\frac{-12+24}{4}=\frac{12}{4}=3$

17. $\frac{(-2)(0)(-3)}{-6}=\frac{0(-3)}{-6}=\frac{0}{-6}=0$

18. Since -3 is to the right of -7 on the number line, $-3>-7$.

19. Since 4 is to the right of -8 on the number line, $4>-8$.

20. Since $|-3|=3$ is to the right of 2 on the number line, $|-3|>2$.

21. $|-2|=2$
$-1-(-3)=-1+3=2$
Since $2=2$, $|-2|=-1-(-3)$.

22. **a.** The natural numbers are 1, 7.

b. The whole numbers are 0, 1, 7.

c. The integers are $-5, -1, 0, 1, 7$.

d. The rational numbers are $-5, -1, \frac{1}{4}, 0, 1, 7, 11.6$.

e. The irrational numbers are $\sqrt{7}, 3\pi$.

f. The real numbers are $-5, -1, \frac{1}{4}, 0, 1, 7, 11.6, \sqrt{7}, 3\pi$.

23. Replace x with 6 and y with -2.
$x^2+y^2=6^2+(-2)^2=36+4=40$

24. Replace x with 6, y with -2, and z with -3.
$x+yz=6+(-2)(-3)=6+6=12$

25. Replace x with 6 and y with -2.
$$\begin{aligned}2+3x-y&=2+3(6)-(-2)\\&=2+18+2\\&=20+2\\&=22\end{aligned}$$

26. Replace x with 6, y with -2, and z with -3.
$$\begin{aligned}\frac{y+z-1}{x}&=\frac{-2+(-3)-1}{6}\\&=\frac{-5-1}{6}\\&=\frac{-5+(-1)}{6}\\&=\frac{-6}{6}\\&=-1\end{aligned}$$

27. $8+(9+3)=(8+9)+3$ illustrates the associative property of addition.

28. $6\cdot 8=8\cdot 6$ illustrates the commutative property of multiplication.

29. $-6(2+4)=-6\cdot 2+(-6)\cdot 4$ illustrates the distributive property.

30. $\frac{1}{6}(6)=1$ illustrates the multiplicative inverse property.

31. The opposite of -9 is 9.

32. The reciprocal of $-\frac{1}{3}$ of -3.

33. Losses of yardage occurred on the second and third downs. -10 indicates a loss of 10 yards while -2 indicates a loss of 2 yards, so the greatest loss of yardage occurred on the second down.

Copyright © 2012 Pearson Education, Inc. Publishing as Prentice Hall.

34. $5+(-10)+(-2)+29=-5+(-2)+29$
$=-7+29$
$=22$
Since the team was 22 yards from the goal, a touchdown was scored.

35. $-14+31=17$
The temperature was 17° at noon.

36. $-1.5(280)=-420$
She lost $420.

37. $2y-6-y-4=2y-y-6-4$
$=1y-10$
$=y-10$

38. $2.7x+6.1+3.2x-4.9=2.7x+3.2x+6.1-4.9$
$=5.9x+1.2$

39. $4(x-2)-3(2x-6)=4x-8-6x+18$
$=4x-6x-8+18$
$=-2x+10$

40. $-5(y+1)+2(3-5y)=-5y-5+6-10y$
$=-5y-10y-5+6$
$=-15y+1$

Copyright © 2012 Pearson Education, Inc. Publishing as Prentice Hall.

Chapter 2

Section 2.1 Practice

1.
$$x-5=8$$
$$x-5+5=8+5$$
$$x=13$$
Check: $x-5=8$
$$13-5 \stackrel{?}{=} 8$$
$$8=8 \quad \text{True}$$
The solution is 13.

2.
$$y+1.7=0.3$$
$$y+1.7-1.7=0.3-1.7$$
$$y=-1.4$$
Check: $y+1.7=0.3$
$$-1.4+1.7 \stackrel{?}{=} 0.3$$
$$0.3=0.3 \quad \text{True}$$
The solution is -1.4.

3.
$$\frac{7}{8}=y-\frac{1}{3}$$
$$\frac{7}{8}+\frac{1}{3}=y-\frac{1}{3}+\frac{1}{3}$$
$$\frac{7}{8}\cdot\frac{3}{3}+\frac{1}{3}\cdot\frac{8}{8}=y$$
$$\frac{21}{24}+\frac{8}{24}=y$$
$$\frac{29}{24}=y$$
Check: $\frac{7}{8}=y-\frac{1}{3}$
$$\frac{7}{8} \stackrel{?}{=} \frac{29}{24}-\frac{1}{3}$$
$$\frac{7}{8} \stackrel{?}{=} \frac{29}{24}-\frac{8}{24}$$
$$\frac{7}{8} \stackrel{?}{=} \frac{21}{24}$$
$$\frac{7}{8}=\frac{7}{8} \quad \text{True}$$
The solution is $\frac{29}{24}$.

4.
$$3x+10=4x$$
$$3x+10-3x=4x-3x$$
$$10=x$$
Check: $3x+10=4x$
$$3(10)+10 \stackrel{?}{=} 4(10)$$
$$30+10 \stackrel{?}{=} 40$$
$$40=40 \quad \text{True}$$
The solution is 10.

5.
$$10w+3-4w+4=-2w+3+7w$$
$$6w+7=5w+3$$
$$-5w+6w+7=-5w+5w+3$$
$$w+7=3$$
$$w+7-7=3-7$$
$$w=-4$$
Check:
$$10w+3-4w+4=-2w+3+7w$$
$$10(-4)+3-4(-4)+4 \stackrel{?}{=} -2(-4)+3+7(-4)$$
$$-40+3+16+4 \stackrel{?}{=} 8+3-28$$
$$-17=-17 \quad \text{True}$$
The solution is -4.

6.
$$3(2w-5)-(5w+1)=-3$$
$$3(2w)-3(5)-1(5w)-1(1)=-3$$
$$6w-15-5w-1=-3$$
$$w-16=-3$$
$$w-16+16=-3+16$$
$$w=13$$
Check: $3(2w-5)-(5w+1)=-3$
$$3(2\cdot 13-5)-(5\cdot 13+1) \stackrel{?}{=} -3$$
$$3(26-5)-(65+1) \stackrel{?}{=} -3$$
$$3(21)-66 \stackrel{?}{=} -3$$
$$63-66 \stackrel{?}{=} -3$$
$$-3=-3 \quad \text{True}$$
The solution is 13.

7.
$$12-y=9$$
$$12-y-12=9-12$$
$$-y=-3$$
$$y=3$$
Check: $12-y=9$
$$12-3 \stackrel{?}{=} 9$$
$$9=9 \quad \text{True}$$
The solution is 3.

8. a. If the sum of two numbers is 11 and one number is 4, find the other number by subtracting 4 from 11. The other number is $11-4$, or 7.

Copyright © 2012 Pearson Education, Inc. Publishing as Prentice Hall.

b. If the sum of two numbers is 11 and one number is x, find the other number by subtracting x from 11. The other number is $11 - x$.

c. If the sum of two numbers is 56 and one number is a, find the other number by subtracting a from 56. The other number is $56 - a$.

9. Mike received 100,445 more votes than Zane, who received n votes. So, Mike received $(n + 100{,}445)$ votes.

Vocabulary and Readiness Check

1. A combination of operations on variables and numbers is called an <u>expression</u>.

2. A statement of the form "expression = expression" is called an <u>equation</u>.

3. An <u>equation</u> contains an equal sign (=).

4. An <u>expression</u> does not contain an equal sign (=).

5. An <u>expression</u> may be simplified and evaluated while an <u>equation</u> may be solved.

6. A <u>solution</u> of an equation is a number that when substituted for a variable makes the equation a true statement.

7. <u>Equivalent</u> equations have the same solution.

8. By the <u>addition</u> property of equality, the same number may be added to or subtracted from both sides of an equation without changing the solution of the equation.

9. $x+4=6$
$x=2$

10. $x+7=17$
$x=10$

11. $n+18=30$
$n=12$

12. $z+22=40$
$z=18$

13. $b-11=6$
$b=17$

14. $d-16=5$
$d=21$

Exercise Set 2.1

1. $x+7=10$
$x+7-7=10-7$
$x=3$
Check: $x+7=10$
$3+7 \stackrel{?}{=} 10$
$10=10$ True
The solution is 3.

3. $x-2=-4$
$x-2+2=-4+2$
$x=-2$
Check: $x-2=-4$
$-2-2 \stackrel{?}{=} -4$
$-4=-4$ True
The solution is -2.

5. $-11=3+x$
$-11-3=3+x-3$
$-14=x$
Check: $-11=3+x$
$-11 \stackrel{?}{=} 3+(-14)$
$-11=-11$ True
The solution is -14.

7. $r-8.6=-8.1$
$r-8.6+8.6=-8.1+8.6$
$r=0.5$
Check: $r-8.6=-8.1$
$0.5-8.6 \stackrel{?}{=} -8.1$
$-8.1=-8.1$ True
The solution is 0.5.

9. $x-\frac{2}{5}=-\frac{3}{20}$
$x-\frac{2}{5}+\frac{2}{5}=-\frac{3}{20}+\frac{2}{5}$
$x=-\frac{3}{20}+\frac{8}{20}$
$x=\frac{5}{20}$
$x=\frac{1}{4}$

Copyright © 2012 Pearson Education, Inc. Publishing as Prentice Hall.

Check: $x-\frac{2}{5}=-\frac{3}{20}$

$\frac{1}{4}-\frac{2}{5}\stackrel{?}{=}-\frac{3}{20}$

$\frac{5}{20}-\frac{8}{20}\stackrel{?}{=}-\frac{3}{20}$

$-\frac{3}{20}=-\frac{3}{20}$ True

The solution is $\frac{1}{4}$.

11. $\frac{1}{3}+f=\frac{3}{4}$

$-\frac{1}{3}+\frac{1}{3}+f=-\frac{1}{3}+\frac{3}{4}$

$f=-\frac{4}{12}+\frac{9}{12}$

$f=\frac{5}{12}$

Check: $\frac{1}{3}+f=\frac{3}{4}$

$\frac{1}{3}+\frac{5}{12}\stackrel{?}{=}\frac{3}{4}$

$\frac{4}{12}+\frac{5}{12}\stackrel{?}{=}\frac{3}{4}$

$\frac{9}{12}\stackrel{?}{=}\frac{3}{4}$

$\frac{3}{4}=\frac{3}{4}$ True

The solution is $\frac{5}{12}$.

13. $7x+2x=8x-3$

$9x=8x-3$

$9x-8x=8x-3-8x$

$x=-3$

Check: $7x+2x=8x-3$

$7(-3)+2(-3)\stackrel{?}{=}8(-3)-3$

$-21-6\stackrel{?}{=}-24-3$

$-27=-27$ True

The solution is -3.

15. $\frac{5}{6}x+\frac{1}{6}x=-9$

$\frac{6}{6}x=-9$

$x=-9$

Check: $\frac{5}{6}x+\frac{1}{6}x=-9$

$\frac{5}{6}(-9)+\frac{1}{6}(-9)\stackrel{?}{=}-9$

$-\frac{45}{6}-\frac{9}{6}\stackrel{?}{=}-9$

$-\frac{54}{6}\stackrel{?}{=}-9$

$-9=-9$ True

The solution is -9.

17. $2y+10=5y-4y$

$2y+10=y$

$2y+10-2y=y-2y$

$10=-y$

$-10=y$

Check: $2y+10=5y-4y$

$2(-10)+10\stackrel{?}{=}5(-10)-4(-10)$

$-20+10\stackrel{?}{=}-50+40$

$-10=-10$ True

The solution is -10.

19. $-5(n-2)=8-4n$

$-5n+10=8-4n$

$5n-5n+10=5n+8-4n$

$10=n+8$

$10-8=n+8-8$

$2=n$

Check: $-5(n-2)=8-4n$

$-5(2-2)\stackrel{?}{=}8-4(2)$

$-5(0)\stackrel{?}{=}8-8$

$0=0$ True

The solution is 2.

21. $\frac{3}{7}x+2=-\frac{4}{7}x-5$

$\frac{3}{7}x+2+\frac{4}{7}x=-\frac{4}{7}x-5+\frac{4}{7}x$

$x+2=-5$

$x+2-2=-5-2$

$x=-7$

Check: $\frac{3}{7}x+2=-\frac{4}{7}x-5$

$\frac{3}{7}(-7)+2\stackrel{?}{=}-\frac{4}{7}(-7)-5$

$-3+2\stackrel{?}{=}4-5$

$-1=-1$ True

The solution is -7.

Copyright © 2012 Pearson Education, Inc. Publishing as Prentice Hall.

23.
$$\begin{aligned} 5x-6&=6x-5\\ -5x+5x-6&=-5x+6x-5\\ -6&=x-5\\ -6+5&=x-5+5\\ -1&=x \end{aligned}$$
Check:
$$\begin{aligned} 5x-6&=6x-5\\ 5(-1)-6&\stackrel{?}{=}6(-1)-5\\ -5-6&\stackrel{?}{=}-6-5\\ -11&=-11 \quad \text{True} \end{aligned}$$
The solution is -1.

25.
$$\begin{aligned} 8y+2-6y&=3+y-10\\ 2y+2&=y-7\\ 2y+2-y&=y-7-y\\ y+2&=-7\\ y+2-2&=-7-2\\ y&=-9 \end{aligned}$$
Check:
$$\begin{aligned} 8y+2-6y&=3+y-10\\ 8(-9)+2-6(-9)&\stackrel{?}{=}3+(-9)-10\\ -72+2+54&\stackrel{?}{=}3-9-10\\ -16&=-16 \quad \text{True} \end{aligned}$$
The solution is -9.

27.
$$\begin{aligned} -3(x-4)&=-4x\\ -3x+12&=-4x\\ 3x-3x+12&=3x-4x\\ 12&=-x\\ -12&=x \end{aligned}$$
Check:
$$\begin{aligned} -3(x-4)&=-4x\\ -3(-12-4)&\stackrel{?}{=}-4(-12)\\ -3(-16)&\stackrel{?}{=}48\\ 48&=48 \quad \text{True} \end{aligned}$$
The solution is -12.

29.
$$\begin{aligned} \frac{3}{8}x-\frac{1}{6}&=-\frac{5}{8}x-\frac{2}{3}\\ \frac{3}{8}x-\frac{1}{6}+\frac{5}{8}x&=-\frac{5}{8}x-\frac{2}{3}+\frac{5}{8}x\\ x-\frac{1}{6}&=-\frac{2}{3}\\ x-\frac{1}{6}+\frac{1}{6}&=-\frac{2}{3}+\frac{1}{6}\\ x&=-\frac{4}{6}+\frac{1}{6}\\ x&=-\frac{3}{6}\\ x&=-\frac{1}{2} \end{aligned}$$
Check:
$$\begin{aligned} \frac{3}{8}x-\frac{1}{6}&=-\frac{5}{8}x-\frac{2}{3}\\ \frac{3}{8}\left(-\frac{1}{2}\right)-\frac{1}{6}&\stackrel{?}{=}-\frac{5}{8}\left(-\frac{1}{2}\right)-\frac{2}{3}\\ -\frac{3}{16}-\frac{1}{6}&\stackrel{?}{=}\frac{5}{16}-\frac{2}{3}\\ -\frac{9}{48}-\frac{8}{48}&\stackrel{?}{=}\frac{15}{48}-\frac{32}{48}\\ -\frac{17}{48}&=-\frac{17}{48} \quad \text{True} \end{aligned}$$
The solution is $-\frac{1}{2}$.

31.
$$\begin{aligned} 2(x-4)&=x+3\\ 2x-8&=x+3\\ -x+2x-8&=-x+x+3\\ x-8&=3\\ x-8+8&=3+8\\ x&=11 \end{aligned}$$
Check:
$$\begin{aligned} 2(x-4)&=x+3\\ 2(11-4)&\stackrel{?}{=}11+3\\ 2(7)&\stackrel{?}{=}14\\ 14&=14 \quad \text{True} \end{aligned}$$
The solution is 7.

33.
$$\begin{aligned} 3(n-5)-(6-2n)&=4n\\ 3n-15-6+2n&=4n\\ 5n-21&=4n\\ 5n-21-5n&=4n-5n\\ -21&=-n\\ 21&=n \end{aligned}$$
Check:
$$\begin{aligned} 3(n-5)-(6-2n)&=4n\\ 3(21-5)-(6-2\cdot 21)&\stackrel{?}{=}4(21)\\ 3(21-5)-(6-42)&\stackrel{?}{=}84\\ 3(16)-(-36)&\stackrel{?}{=}84\\ 48+36&\stackrel{?}{=}84\\ 84&=84 \quad \text{True} \end{aligned}$$
The solution is 21.

35.
$$\begin{aligned} -2(x+6)+3(2x-5)&=3(x-4)+10\\ -2x-12+6x-15&=3x-12+10\\ 4x-27&=3x-2\\ -3x+4x-27&=-3x+3x-2\\ x-27&=-2\\ x-27+27&=-2+27\\ x&=25 \end{aligned}$$

Copyright © 2012 Pearson Education, Inc. Publishing as Prentice Hall.

Check:
$$\begin{aligned} -2(x+6)+3(2x-5) &= 3(x-4)+10 \\ -2(25+6)+3(2\cdot 25-5) &\stackrel{?}{=} 3(25-4)+10 \\ -2(31)+3(50-5) &\stackrel{?}{=} 3(21)+10 \\ -62+3(45) &\stackrel{?}{=} 63+10 \\ -62+135 &\stackrel{?}{=} 73 \\ 73 &= 73 \quad \text{True} \end{aligned}$$

The solution is 25.

37.
$$\begin{aligned} 13x-3 &= 14x \\ 13x-3-13x &= 14x-13x \\ -3 &= x \end{aligned}$$

39.
$$\begin{aligned} 5b-0.7 &= 6b \\ 5b-0.7-5b &= 6b-5b \\ -0.7 &= b \end{aligned}$$

41.
$$\begin{aligned} 3x-6 &= 2x+5 \\ 3x-6+6 &= 2x+5+6 \\ 3x &= 2x+11 \\ 3x-2x &= 2x+11-2x \\ x &= 11 \end{aligned}$$

43.
$$\begin{aligned} 13x-9+2x-5 &= 12x-1+2x \\ 15x-14 &= 14x-1 \\ 15x-14-14x &= 14x-1-14x \\ x-14 &= -1 \\ x-14+14 &= -1+14 \\ x &= 13 \end{aligned}$$

45.
$$\begin{aligned} 7(6+w) &= 6(2+w) \\ 42+7w &= 12+6w \\ 42+7w-6w &= 12+6w-6w \\ 42+w &= 12 \\ 42+w-42 &= 12-42 \\ w &= -30 \end{aligned}$$

47.
$$\begin{aligned} n+4 &= 3.6 \\ n+4-4 &= 3.6-4 \\ n &= -0.4 \end{aligned}$$

49.
$$\begin{aligned} 10-(2x-4) &= 7-3x \\ 10-2x+4 &= 7-3x \\ 14-2x &= 7-3x \\ 14-2x+3x &= 7-3x+3x \\ 14+x &= 7 \\ 14+x-14 &= 7-14 \\ x &= -7 \end{aligned}$$

51.
$$\begin{aligned} \frac{1}{3} &= x+\frac{2}{3} \\ \frac{1}{3}-\frac{2}{3} &= x+\frac{2}{3}-\frac{2}{3} \\ -\frac{1}{3} &= x \end{aligned}$$

53.
$$\begin{aligned} -6.5-4x-1.6-3x &= -6x+9.8 \\ -8.1-7x &= -6x+9.8 \\ -8.1-7x+7x &= -6x+9.8+7x \\ -8.1 &= x+9.8 \\ -8.1-9.8 &= x+9.8-9.8 \\ -17.9 &= x \end{aligned}$$

55. If the sum of the lengths of the two pieces is 10 feet and one piece is x feet, then the other piece has a length of $(10-x)$ feet.

57. If the sum of the measures of two angles is 180° and one angle measures $x°$, then the other angle measures $(180-x)°$.

59. If the number of undergraduate students is n, and the number of graduate students is 28,000 fewer than n, then the number of graduate students is $n-28{,}000$.

61. If the area of the Gobi Desert is x square miles and the area of the Sahara Desert is 7 times the area of the Gobi Desert, then the area of the Sahara Desert is $7x$ square miles.

63. The multiplicative inverse of $\frac{5}{8}$ is $\frac{8}{5}$, since $\frac{5}{8}\cdot\frac{8}{5}=1$.

65. The multiplicative inverse of 2 is $\frac{1}{2}$, since $2\cdot\frac{1}{2}=1$.

67. The multiplicative inverse of $-\frac{1}{9}$ is -9, since $-\frac{1}{9}\cdot(-9)=1$.

69. $\frac{3x}{3}=\frac{3\cdot x}{3\cdot 1}=\frac{x}{1}=x$

Copyright © 2012 Pearson Education, Inc. Publishing as Prentice Hall.

71. $-5\left(-\frac{1}{5}y\right)=\left[-5\cdot\left(-\frac{1}{5}\right)\right]y=1y=y$

73. $\frac{3}{5}\left(\frac{5}{3}x\right)=\left(\frac{3}{5}\cdot\frac{5}{3}\right)x=1x=x$

75. answers may vary

77.
$$\begin{aligned} x-4&=-9\\ x-4+4&=-9+4\\ x&=-5 \end{aligned}$$

79. answers may vary

81.
$$\begin{aligned} 180-x-(2x+7)&=180-x-2x-7\\ &=173-3x \end{aligned}$$
The measure of the third angle is $(173-3x)°$.

83. answers may vary

85.
$$\begin{aligned} 36.766+x&=-108.712\\ 36.766+x-36.766&=-108.712-36.766\\ x&=-145.478 \end{aligned}$$

Section 2.2 Practice

1.
$$\begin{aligned} \frac{3}{7}x&=9\\ \frac{7}{3}\cdot\left(\frac{3}{7}x\right)&=\frac{7}{3}\cdot 9\\ \left(\frac{7}{3}\cdot\frac{3}{7}\right)x&=\frac{7}{3}\cdot 9\\ 1x&=21\\ x&=21 \end{aligned}$$

Check:
$$\begin{aligned} \frac{3}{7}x&=9\\ \frac{3}{7}(21)&\stackrel{?}{=}9\\ 9&=9 \quad \text{True} \end{aligned}$$
The solution is 21.

2.
$$\begin{aligned} 7x&=42\\ \frac{7x}{7}&=\frac{42}{7}\\ 1\cdot x&=6\\ x&=6 \end{aligned}$$
Check:
$$\begin{aligned} 7x&=42\\ 7\cdot 6&\stackrel{?}{=}42\\ 42&=42 \quad \text{True} \end{aligned}$$
The solution is 6.

3.
$$\begin{aligned} -4x&=52\\ \frac{-4x}{-4}&=\frac{52}{-4}\\ 1x&=-13\\ x&=-13 \end{aligned}$$
Check:
$$\begin{aligned} -4x&=52\\ -4(-13)&\stackrel{?}{=}52\\ 52&=52 \quad \text{True} \end{aligned}$$
The solution is -13.

4.
$$\begin{aligned} \frac{y}{5}&=13\\ \frac{1}{5}y&=13\\ 5\cdot\frac{1}{5}y&=5\cdot 13\\ 1y&=65\\ y&=65 \end{aligned}$$
Check:
$$\begin{aligned} \frac{y}{5}&=13\\ \frac{65}{5}&\stackrel{?}{=}13\\ 13&=13 \quad \text{True} \end{aligned}$$
The solution is 65.

5.
$$\begin{aligned} 2.6x&=13.52\\ \frac{2.6x}{2.6}&=\frac{13.52}{2.6}\\ x&=5.2 \end{aligned}$$
Check:
$$\begin{aligned} 2.6x&=13.52\\ 2.6(5.2)&\stackrel{?}{=}13.52\\ 13.52&=13.52 \quad \text{True} \end{aligned}$$
The solution is 5.2.

6.
$$\begin{aligned} -\frac{5}{6}y&=-\frac{3}{5}\\ -\frac{6}{5}\cdot-\frac{5}{6}y&=-\frac{6}{5}\cdot-\frac{3}{5}\\ y&=\frac{18}{25} \end{aligned}$$
Check:
$$\begin{aligned} -\frac{5}{6}y&=-\frac{3}{5}\\ -\frac{5}{6}\left(\frac{18}{25}\right)&\stackrel{?}{=}-\frac{3}{5}\\ -\frac{3}{5}&=-\frac{3}{5} \quad \text{True} \end{aligned}$$
The solution is $\frac{18}{25}$.

Copyright © 2012 Pearson Education, Inc. Publishing as Prentice Hall.

7.
$$\begin{aligned} -x+7&=-12\\ -x+7-7&=-12-7\\ -x&=-19\\ \frac{-x}{-1}&=\frac{-19}{-1}\\ 1x&=19\\ x&=19 \end{aligned}$$
Check: $-x+7=-12$
$-19+7 \stackrel{?}{=} -12$
$-12=-12$ True
The solution is 19.

8.
$$\begin{aligned} -7x+2x+3-20&=-2\\ -5x-17&=-2\\ -5x-17+17&=-2+17\\ -5x&=15\\ \frac{-5x}{-5}&=\frac{15}{-5}\\ x&=-3 \end{aligned}$$
Check: $-7x+2x+3-20=-2$
$-7(-3)+2(-3)+3-20 \stackrel{?}{=} -2$
$21-6+3-20 \stackrel{?}{=} -2$
$-2=-2$ True
The solution is -3.

9.
$$\begin{aligned} 10x-4&=7x+14\\ 10x-4-7x&=7x+14-7x\\ 3x-4&=14\\ 3x-4+4&=14+4\\ 3x&=18\\ \frac{3x}{3}&=\frac{18}{3}\\ x&=6 \end{aligned}$$
Check: $10x-4=7x+14$
$10(6)-4 \stackrel{?}{=} 7(6)+14$
$60-4 \stackrel{?}{=} 42+14$
$56=56$ True
The solution is 6.

10.
$$\begin{aligned} 4(3x-2)&=-1+4\\ 4(3x)-4(2)&=-1+4\\ 12x-8&=3\\ 12x-8+8&=3+8\\ 12x&=11\\ \frac{12x}{12}&=\frac{11}{12}\\ x&=\frac{11}{12} \end{aligned}$$
Check: $4(3x-2)=-1+4$
$4\left(3\cdot\frac{11}{12}-2\right) \stackrel{?}{=} -1+4$
$4\left(\frac{11}{4}-2\right) \stackrel{?}{=} -1+4$
$11-8 \stackrel{?}{=} 3$
$3=3$ True
The solution is $\frac{11}{12}$.

11. a. If x is the first integer, then $x+1$ is the second integer.
Their sum is $x+(x+1)=x+x+1=2x+1$.

b. If x is the first odd integer, then $x+2$ is the second consecutive odd integer.
Their sum is $x+(x+2)=x+x+2=2x+2$.

Vocabulary and Readiness Check

1. By the <u>multiplication</u> property of equality, both sides of an equation may be multiplied or divided by the same nonzero number without changing the solution of the equation.

2. By the <u>addition</u> property of equality, the same number may be added to or subtracted from both sides of an equation without changing the solution of the equation.

3. An <u>equation</u> may be solved while an <u>expression</u> may be simplified and evaluated.

4. An <u>equation</u> contains an equal sign (=) while an <u>expression</u> does not.

5. <u>Equivalent</u> equations have the same solution.

6. A <u>solution</u> of an equation is a number that when substituted for a variable makes the equation a true statement.

7. $3a=27$
$a=9$

8. $9c=54$
$c=6$

9. $5b=10$
$b=2$

10. $7t=14$
$t=2$

Copyright © 2012 Pearson Education, Inc. Publishing as Prentice Hall.

11. $6x = -30$
$x = -5$

12. $8r = -64$
$r = -8$

Exercise Set 2.2

1. $-5x = -20$
$\frac{-5x}{-5} = \frac{-20}{-5}$
$x = 4$
Check: $-5x = -20$
$-5(4) \stackrel{?}{=} -20$
$-20 = -20$ True
The solution is 4.

3. $3x = 0$
$\frac{3x}{3} = \frac{0}{3}$
$x = 0$
Check: $3x = 0$
$3 \cdot 0 \stackrel{?}{=} 0$
$0 = 0$ True
The solution is 0.

5. $-x = -12$
$\frac{-x}{-1} = \frac{-12}{-1}$
$x = 12$
Check: $-x = -12$
$-12 = -12$ True
The solution is 12.

7. $\frac{2}{3}x = -8$
$\frac{3}{2} \cdot \frac{2}{3}x = \frac{3}{2} \cdot (-8)$
$x = -12$
Check: $\frac{2}{3}x = -8$
$\frac{2}{3}(-12) \stackrel{?}{=} -8$
$-8 = -8$ True
The solution is -12.

9. $\frac{1}{6}d = \frac{1}{2}$
$6 \cdot \frac{1}{6}d = 6 \cdot \frac{1}{2}$
$d = 3$
Check: $\frac{1}{6}d = \frac{1}{2}$
$\frac{1}{6}(3) \stackrel{?}{=} \frac{1}{2}$
$\frac{1}{2} = \frac{1}{2}$ True
The solution is 3.

11. $\frac{a}{2} = 1$
$2 \cdot \frac{a}{2} = 2 \cdot 1$
$a = 2$
Check: $\frac{a}{2} = 1$
$\frac{2}{2} \stackrel{?}{=} 1$
$1 = 1$ True
The solution is 2.

13. $\frac{k}{-7} = 0$
$-7\left(\frac{k}{-7}\right) = -7(0)$
$k = 0$
Check: $\frac{k}{-7} = 0$
$\frac{0}{-7} \stackrel{?}{=} 0$
$0 = 0$ True
The solution is 0.

15. $1.7x = 10.71$
$\frac{1.7x}{1.7} = \frac{10.71}{1.7}$
$x = 6.3$
Check: $1.7x = 10.71$
$1.7(6.3) \stackrel{?}{=} 10.71$
$10.71 = 10.71$ True
The solution is 6.3.

17. $2x - 4 = 16$
$2x - 4 + 4 = 16 + 4$
$2x = 20$
$\frac{2x}{2} = \frac{20}{2}$
$x = 10$

Copyright © 2012 Pearson Education, Inc. Publishing as Prentice Hall.

Check: $2x-4=16$
$2(10)-4 \stackrel{?}{=} 16$
$20-4 \stackrel{?}{=} 16$
$16=16$ True

The solution is 10.

19. $-x+2=22$
$-x+2-2=22-2$
$-x=20$
$\frac{-x}{-1}=\frac{20}{-1}$
$x=-20$

Check: $-x+2=22$
$-(-20)+2 \stackrel{?}{=} 22$
$20+2 \stackrel{?}{=} 22$
$22=22$ True

The solution is -20.

21. $6a+3=3$
$6a+3-3=3-3$
$6a=0$
$\frac{6a}{6}=\frac{0}{6}$
$a=0$

Check: $6a+3=3$
$6(0)+3 \stackrel{?}{=} 3$
$0+3 \stackrel{?}{=} 3$
$3=3$ True

The solution is 0.

23. $\frac{x}{3}-2=-5$
$\frac{x}{3}-2+2=-5+2$
$\frac{x}{3}=-3$
$3\cdot\frac{x}{3}=3\cdot(-3)$
$x=-9$

Check: $\frac{x}{3}-2=-5$
$\frac{-9}{3}-2 \stackrel{?}{=} -5$
$-3-2 \stackrel{?}{=} -5$
$-5=-5$ True

The solution is -9.

25. $6z-8-z+3=0$
$5z-5=0$
$5z-5+5=0+5$
$5z=5$
$\frac{5z}{5}=\frac{5}{5}$
$z=1$

Check: $6z-8-z+3=0$
$6(1)-8-1+3 \stackrel{?}{=} 0$
$6-8-1+3 \stackrel{?}{=} 0$
$0=0$ True

The solution is 1.

27. $1=0.4x-0.6x-5$
$1=-0.2x-5$
$1+5=-0.2x-5+5$
$6=-0.2x$
$\frac{6}{-0.2}=\frac{-0.2x}{-0.2}$
$-30=x$

Check: $1=0.4x-0.6x-5$
$1 \stackrel{?}{=} 0.4(-30)-0.6(-30)-5$
$1 \stackrel{?}{=} -12+18-5$
$1=1$ True

The solution is -30.

29. $\frac{2}{3}y-11=-9$
$\frac{2}{3}y-11+11=-9+11$
$\frac{2}{3}y=2$
$\frac{3}{2}\left(\frac{2}{3}y\right)=\frac{3}{2}(2)$
$y=3$

Check: $\frac{2}{3}y-11=-9$
$\frac{2}{3}(3)-11 \stackrel{?}{=} -9$
$2-11 \stackrel{?}{=} -9$
$-9=-9$ True

The solution is 3.

Copyright © 2012 Pearson Education, Inc. Publishing as Prentice Hall.

31.
$$\frac{3}{4}t-\frac{1}{2}=\frac{1}{3}$$
$$\frac{3}{4}t-\frac{1}{2}+\frac{1}{2}=\frac{1}{3}+\frac{1}{2}$$
$$\frac{3}{4}t=\frac{2}{6}+\frac{3}{6}$$
$$\frac{3}{4}t=\frac{5}{6}$$
$$\frac{4}{3}\cdot\frac{3}{4}t=\frac{4}{3}\cdot\frac{5}{6}$$
$$t=\frac{10}{9}$$

Check:
$$\frac{3}{4}t-\frac{1}{2}=\frac{1}{3}$$
$$\frac{3}{4}\cdot\frac{10}{9}-\frac{1}{2}\stackrel{?}{=}\frac{1}{3}$$
$$\frac{5}{6}-\frac{3}{6}\stackrel{?}{=}\frac{1}{3}$$
$$\frac{2}{6}\stackrel{?}{=}\frac{1}{3}$$
$$\frac{1}{3}=\frac{1}{3}\quad\text{True}$$

The solution is $\frac{10}{9}$.

33.
$$8x+20=6x+18$$
$$8x+20-6x=6x+18-6x$$
$$2x+20=18$$
$$2x+20-20=18-20$$
$$2x=-2$$
$$\frac{2x}{2}=\frac{-2}{2}$$
$$x=-1$$

35.
$$3(2x+5)=-18+9$$
$$6x+15=-18+9$$
$$6x+15=-9$$
$$6x+15-15=-9-15$$
$$6x=-24$$
$$\frac{6x}{6}=\frac{-24}{6}$$
$$x=-4$$

37.
$$2x-5=20x+4$$
$$2x-5-20x=20x+4-20x$$
$$-18x-5=4$$
$$-18x-5+5=4+5$$
$$-18x=9$$
$$\frac{-18x}{-18}=\frac{9}{-18}$$
$$x=-\frac{1}{2}$$

39.
$$2+14=-4(3x-4)$$
$$2+14=-12x+16$$
$$16=-12x+16$$
$$16-16=-12x+16-16$$
$$0=-12x$$
$$\frac{0}{-12}=\frac{-12x}{-12}$$
$$0=x$$

41.
$$-6y-3=-5y-7$$
$$-6y-3+6y=-5y-7+6y$$
$$-3=y-7$$
$$-3+7=y-7+7$$
$$4=y$$

43.
$$\frac{1}{2}(2x-1)=-\frac{1}{7}-\frac{3}{7}$$
$$x-\frac{1}{2}=-\frac{4}{7}$$
$$x-\frac{1}{2}+\frac{1}{2}=-\frac{4}{7}+\frac{1}{2}$$
$$x=-\frac{8}{14}+\frac{7}{14}$$
$$x=-\frac{1}{14}$$

45.
$$-10z-0.5=-20z+1.6$$
$$-10z-0.5+20z=-20z+1.6+20z$$
$$10z-0.5=1.6$$
$$10z-0.5+0.5=1.6+0.5$$
$$10z=2.1$$
$$\frac{10z}{10}=\frac{2.1}{10}$$
$$z=0.21$$

Copyright © 2012 Pearson Education, Inc. Publishing as Prentice Hall.

47.
$$\begin{aligned} -4x+20 &= 4x-20 \\ -4x-4x+20 &= -4x+4x-20 \\ -8x+20 &= -20 \\ -8x+20-20 &= -20-20 \\ -8x &= -40 \\ \frac{-8x}{-8} &= \frac{-40}{-8} \\ x &= 5 \end{aligned}$$

49.
$$\begin{aligned} 42 &= 7x \\ \frac{42}{7} &= \frac{7x}{7} \\ 6 &= x \end{aligned}$$

51.
$$\begin{aligned} 4.4 &= -0.8x \\ \frac{4.4}{-0.8} &= \frac{-0.8x}{-0.8} \\ -5.5 &= x \end{aligned}$$

53.
$$\begin{aligned} 6x+10 &= -20 \\ 6x+10-10 &= -20-10 \\ 6x &= -30 \\ \frac{6x}{6} &= \frac{-30}{6} \\ x &= -5 \end{aligned}$$

55.
$$\begin{aligned} 5-0.3k &= 5 \\ -5+5-0.3k &= -5+5 \\ -0.3k &= 0 \\ \frac{-0.3k}{-0.3} &= \frac{0}{-0.3} \\ k &= 0 \end{aligned}$$

57.
$$\begin{aligned} 13x-5 &= 11x-11 \\ 13x-5+5 &= 11x-11+5 \\ 13x &= 11x-6 \\ 13x-11x &= 11x-11x-6 \\ 2x &= -6 \\ \frac{2x}{2} &= \frac{-6}{2} \\ x &= -3 \end{aligned}$$

59.
$$\begin{aligned} 9(3x+1) &= 4x-5x \\ 27x+9 &= -x \\ -27x+27x+9 &= -27x-x \\ 9 &= -28x \\ \frac{9}{-28} &= \frac{-28x}{-28} \\ -\frac{9}{28} &= x \end{aligned}$$

61.
$$\begin{aligned} -\frac{3}{7}p &= -2 \\ -\frac{7}{3}\left(-\frac{3}{7}p\right) &= -\frac{7}{3}(-2) \\ p &= \frac{14}{3} \end{aligned}$$

63.
$$\begin{aligned} -\frac{4}{3}x &= 12 \\ -\frac{3}{4}\cdot\left(-\frac{4}{3}x\right) &= -\frac{3}{4}\cdot 12 \\ x &= -9 \end{aligned}$$

65.
$$\begin{aligned} -2x-\frac{1}{2} &= \frac{7}{2} \\ -2x-\frac{1}{2}+\frac{1}{2} &= \frac{7}{2}+\frac{1}{2} \\ -2x &= \frac{8}{2} \\ -2x &= 4 \\ \frac{-2x}{-2} &= \frac{4}{-2} \\ x &= -2 \end{aligned}$$

67.
$$\begin{aligned} 10 &= 2x-1 \\ 10+1 &= 2x-1+1 \\ 11 &= 2x \\ \frac{11}{2} &= \frac{2x}{2} \\ \frac{11}{2} &= x \end{aligned}$$

69.
$$\begin{aligned} 10-3x-6-9x &= 7 \\ 4-12x &= 7 \\ 4-12x-4 &= 7-4 \\ -12x &= 3 \\ \frac{-12x}{-12} &= \frac{3}{-12} \\ x &= -\frac{1}{4} \end{aligned}$$

71.
$$\begin{aligned} z-5z &= 7z-9-z \\ -4z &= 6z-9 \\ -4z-6z &= 6z-9-6z \\ -10z &= -9 \\ \frac{-10z}{-10} &= \frac{-9}{-10} \\ z &= \frac{9}{10} \end{aligned}$$

Copyright © 2012 Pearson Education, Inc. Publishing as Prentice Hall.

73.
$$\begin{aligned} -x-\frac{4}{5} &= x+\frac{1}{2}+\frac{2}{5} \\ -x-\frac{4}{5} &= x+\frac{5}{10}+\frac{4}{10} \\ -x-\frac{4}{5} &= x+\frac{9}{10} \\ -x-\frac{4}{5}+x &= x+\frac{9}{10}+x \\ -\frac{4}{5} &= 2x+\frac{9}{10} \\ -\frac{4}{5}-\frac{9}{10} &= 2x \\ -\frac{8}{10}-\frac{9}{10} &= 2x \\ -\frac{17}{10} &= 2x \\ \frac{1}{2}\left(-\frac{17}{10}\right) &= \frac{1}{2}(2x) \\ -\frac{17}{20} &= x \end{aligned}$$

75.
$$\begin{aligned} -15+37 &= -2(x+5) \\ 22 &= -2x-10 \\ 22+10 &= -2x-10+10 \\ 32 &= -2x \\ \frac{32}{-2} &= \frac{-2x}{-2} \\ -16 &= x \end{aligned}$$

77. If x represents the first of two consecutive odd integers, then $x + 2$ represents the second. Thus, the sum is represented by $x + x + 2 = 2x + 2$.

79. If x represents the first integer, then $x + 1$, $x + 2$, and $x + 3$ represent the second, third, and fourth integers, respectively. The sum of the first and third integers is represented by $x + x + 2 = 2x + 2$.

81. If x represents the number on the first door, then the next four door numbers are represented by $x + 2$, $x + 4$, $x + 6$, and $x + 8$.
The sum of the numbers is
$x + x + 2 + x + 4 + x + 6 + x + 8 = 5x + 20$.

83.
$$\begin{aligned} 5x+2(x-6) &= 5x+2\cdot x+2\cdot(-6) \\ &= 5x+2x-12 \\ &= 7x-12 \end{aligned}$$

85.
$$\begin{aligned} 6(2z+4)+20 &= 6\cdot 2z+6\cdot 4+20 \\ &= 12z+24+20 \\ &= 12z+44 \end{aligned}$$

87.
$$\begin{aligned} -(x-1)+x &= -x+1+x \\ &= -x+x+1 \\ &= 0+1 \\ &= 1 \end{aligned}$$

89. If the solution is -8, then replacing x by -8 results in a true statement.
$6x = 6(-8) = -48$
The missing number is -48.

91. answers may vary

93. answers may vary

95.
$$\begin{aligned} 0.07x-5.06 &= -4.92 \\ 0.07x-5.06+5.06 &= -4.92+5.06 \\ 0.07x &= 0.14 \\ \frac{0.07x}{0.07} &= \frac{0.14}{0.07} \\ x &= 2 \end{aligned}$$

Section 2.3 Practice

1.
$$\begin{aligned} 5(3x-1)+2 &= 12x+6 \\ 15x-5+2 &= 12x+6 \\ 15x-3 &= 12x+6 \\ 15x-3-12x &= 12x+6-12x \\ 3x-3 &= 6 \\ 3x-3+3 &= 6+3 \\ 3x &= 9 \\ \frac{3x}{3} &= \frac{9}{3} \\ x &= 3 \end{aligned}$$

Check:
$$\begin{aligned} 5(3x-1)+2 &= 12x+6 \\ 5[3(3)-1]+2 &\stackrel{?}{=} 12(3)+6 \\ 5(9-1)+2 &\stackrel{?}{=} 36+6 \\ 5(8)+2 &\stackrel{?}{=} 42 \\ 40+2 &\stackrel{?}{=} 42 \\ 42 &= 42 \quad \text{True} \end{aligned}$$

The solution is 3.

2.
$$\begin{aligned} 9(5-x) &= -3x \\ 45-9x &= -3x \\ 45-9x+9x &= -3x+9x \\ 45 &= 6x \\ \frac{45}{6} &= \frac{6x}{6} \\ \frac{15}{2} &= x \end{aligned}$$

Copyright © 2012 Pearson Education, Inc. Publishing as Prentice Hall.

Check: $9(5-x)=-3x$

$$9\left(5-\frac{15}{2}\right)\stackrel{?}{=}-3\left(\frac{15}{2}\right)$$
$$9\left(\frac{10}{2}-\frac{15}{2}\right)\stackrel{?}{=}-\frac{45}{2}$$
$$9\left(-\frac{5}{2}\right)\stackrel{?}{=}-\frac{45}{2}$$
$$-\frac{45}{2}=-\frac{45}{2}\quad\text{True}$$

The solution is $\frac{15}{2}$.

3.
$$\frac{5}{2}x-1=\frac{3}{2}x-4$$
$$2\left(\frac{5}{2}x-1\right)=2\left(\frac{3}{2}x-4\right)$$
$$5x-2=3x-8$$
$$5x-2-3x=3x-8-3x$$
$$2x-2=-8$$
$$2x-2+2=-8+2$$
$$2x=-6$$
$$\frac{2x}{2}=\frac{-6}{2}$$
$$x=-3$$

Check: $\frac{5}{2}x-1=\frac{3}{2}x-4$

$$\frac{5}{2}(-3)-1\stackrel{?}{=}\frac{3}{2}(-3)-4$$
$$-\frac{15}{2}-1\stackrel{?}{=}-\frac{9}{2}-4$$
$$-\frac{15}{2}-\frac{2}{2}\stackrel{?}{=}-\frac{9}{2}-\frac{8}{2}$$
$$-\frac{17}{2}=-\frac{17}{2}\quad\text{True}$$

The solution is -3.

4.
$$\frac{3(x-2)}{5}=3x+6$$
$$5\cdot\frac{3(x-2)}{5}=5(3x+6)$$
$$3(x-2)=5(3x+6)$$
$$3x-6=15x+30$$
$$3x-6-3x=15x+30-3x$$
$$-6=12x+30$$
$$-6-30=12x+30-30$$
$$-36=12x$$
$$\frac{-36}{12}=\frac{12x}{12}$$
$$-3=x$$

Check: $\frac{3(x-2)}{5}=3x+6$

$$\frac{3(-3-2)}{5}\stackrel{?}{=}3(-3)+6$$
$$\frac{3(-5)}{5}\stackrel{?}{=}-9+6$$
$$\frac{-15}{5}\stackrel{?}{=}-3$$
$$-3=-3$$

The solution is -3.

5.
$$0.06x-0.10(x-2)=-0.16$$
$$100[0.06x-0.10(x-2)]=100[-0.16]$$
$$6x-10(x-2)=-16$$
$$6x-10x+20=-16$$
$$-4x+20=-16$$
$$-4x+20-20=-16-20$$
$$-4x=-36$$
$$\frac{-4x}{-4}=\frac{-36}{-4}$$
$$x=9$$

To check, replace x with 9 in the original equation. The solution is 9.

6.
$$5(2-x)+8x=3(x-6)$$
$$10-5x+8x=3x-18$$
$$10+3x=3x-18$$
$$10+3x-3x=3x-18-3x$$
$$10=-18$$

Since the statement $10=-18$ is false, the equation has no solution.

7.
$$-6(2x+1)-14=-10(x+2)-2x$$
$$-12x-6-14=-10x-20-2x$$
$$-12x-20=-12x-20$$
$$12x-12x-20=12x-12x-20$$
$$-20=-20$$

Since $-20=-20$ is a true statement, every real number is a solution.

Calculator Explorations

1. $2x = 48 + 6x$

$\boxed{2}\boxed{\times}\boxed{-12}\boxed{=}$ Display: $\boxed{-24}$

$\boxed{48}\boxed{+}\boxed{6}\boxed{\times}\boxed{-12}\boxed{=}$ Display: $\boxed{-24}$

Since the left side equals the right side, $x=-12$ is a solution.

2. $-3x - 7 = 3x - 1$

$\boxed{-3}\boxed{\times}\boxed{-1}\boxed{-}\boxed{7}\boxed{=}$ Display: $\boxed{-4}$

$\boxed{3}\boxed{\times}\boxed{-1}\boxed{-}\boxed{1}\boxed{=}$ Display: $\boxed{-4}$

Since the left side equals the right side, $x=-1$ is a solution.

Copyright © 2012 Pearson Education, Inc. Publishing as Prentice Hall.

3. $5x - 2.6 = 2(x + 0.8)$

$\boxed{5}\boxed{\times}\boxed{4.4}\boxed{-}\boxed{2.6}\boxed{=}$ Display: $\boxed{19.4}$

$\boxed{2}\boxed{(}\boxed{4.4}\boxed{+}\boxed{0.8}\boxed{)}\boxed{=}$ Display: $\boxed{10.4}$

Since the left side does not equal the right side, $x = 4.4$ is not a solution.

4. $-1.6x - 3.9 = -6.9x - 25.6$

$\boxed{-1.6}\boxed{\times}\boxed{5}\boxed{-}\boxed{3.9}\boxed{=}$ Display: $\boxed{-11.9}$

$\boxed{-6.9}\boxed{\times}\boxed{5}\boxed{-}\boxed{25.6}\boxed{=}$ Display: $\boxed{-60.1}$

Since the left side does not equal the right side, $x = 5$ is not a solution.

5. $\dfrac{564x}{4} = 200x - 11(649)$

$\boxed{(}\boxed{564}\boxed{\times}\boxed{121}\boxed{)}\boxed{\div}\boxed{4}\boxed{=}$ Display: $\boxed{17061}$

$\boxed{200}\boxed{\times}\boxed{121}\boxed{-}\boxed{11}\boxed{\times}\boxed{649}\boxed{=}$ Display: $\boxed{17061}$

Since the left side equals the right side, $x = 121$ is a solution.

6. $20(x - 39) = 5x - 432$

$\boxed{20}\boxed{(}\boxed{23.2}\boxed{-}\boxed{39}\boxed{)}\boxed{=}$ Display: $\boxed{-316}$

$\boxed{5}\boxed{\times}\boxed{23.2}\boxed{-}\boxed{432}\boxed{=}$ Display: $\boxed{-316}$

Since the left side equals the right side, $x = 23.2$ is a solution.

Vocabulary and Readiness Check

1. $x = -7$ is an <u>equation</u>.

2. $x - 7$ is an <u>expression</u>.

3. $4y - 6 + 9y + 1$ is an <u>expression</u>.

4. $4y - 6 = 9y + 1$ is an <u>equation</u>.

5. $\dfrac{1}{x} - \dfrac{x-1}{8}$ is an <u>expression</u>.

6. $\dfrac{1}{x} - \dfrac{x-1}{8} = 6$ is an <u>equation</u>.

7. $0.1x + 9 = 0.2x$ is an <u>equation</u>.

8. $0.1x^2 + 9y - 0.2x^2$ is an <u>expression</u>.

Exercise Set 2.3

1.
$$\begin{aligned} -4y+10 &= -2(3y+1) \\ -4y+10 &= -6y-2 \\ -4y+10-10 &= -6y-2-10 \\ -4y &= -6y-12 \\ -4y+6y &= -6y-12+6y \\ 2y &= -12 \\ \frac{2y}{2} &= \frac{-12}{2} \\ y &= -6 \end{aligned}$$

3.
$$\begin{aligned} 15x-8 &= 10+9x \\ 15x-8-9x &= 10+9x-9x \\ 6x-8 &= 10 \\ 6x-8+8 &= 10+8 \\ 6x &= 18 \\ \frac{6x}{6} &= \frac{18}{6} \\ x &= 3 \end{aligned}$$

5.
$$\begin{aligned} -2(3x-4) &= 2x \\ -6x+8 &= 2x \\ -6x+8+6x &= 2x+6x \\ 8 &= 8x \\ \frac{8}{8} &= \frac{8x}{8} \\ 1 &= x \end{aligned}$$

7.
$$\begin{aligned} 5(2x-1)-2(3x) &= 1 \\ 10x-5-6x &= 1 \\ -5+4x &= 1 \\ 5-5+4x &= 5+1 \\ 4x &= 6 \\ \frac{4x}{4} &= \frac{6}{4} \\ x &= \frac{3}{2} \end{aligned}$$

9.
$$\begin{aligned} -6(x-3)-26 &= -8 \\ -6x+18-26 &= -8 \\ -6x-8 &= -8 \\ -6x-8+8 &= -8+8 \\ -6x &= 0 \\ \frac{-6x}{-6} &= \frac{0}{-6} \\ x &= 0 \end{aligned}$$

Copyright © 2012 Pearson Education, Inc. Publishing as Prentice Hall.

11.
$$\begin{aligned} 8-2(a+1) &= 9+a \\ 8-2a-2 &= 9+a \\ -2a+6 &= 9+a \\ -2a+6-a &= 9+a-a \\ -3a+6 &= 9 \\ -3a+6-6 &= 9-6 \\ -3a &= 3 \\ \frac{-3a}{-3} &= \frac{3}{-3} \\ a &= -1 \end{aligned}$$

13.
$$\begin{aligned} 4x+3 &= -3+2x+14 \\ 4x+3 &= 11+2x \\ 4x+3-2x &= 11+2x-2x \\ 2x+3 &= 11 \\ 2x+3-3 &= 11-3 \\ 2x &= 8 \\ \frac{2x}{2} &= \frac{8}{2} \\ x &= 4 \end{aligned}$$

15.
$$\begin{aligned} -2y-10 &= 5y+18 \\ -2y-10+10 &= 5y+18+10 \\ -2y &= 5y+28 \\ -2y-5y &= 5y+28-5y \\ -7y &= 28 \\ \frac{-7y}{-7} &= \frac{28}{-7} \\ y &= -4 \end{aligned}$$

17.
$$\begin{aligned} \frac{2}{3}x+\frac{4}{3} &= -\frac{2}{3} \\ 3\left(\frac{2}{3}x+\frac{4}{3}\right) &= 3\left(-\frac{2}{3}\right) \\ 2x+4 &= -2 \\ 2x+4-4 &= -2-4 \\ 2x &= -6 \\ \frac{2x}{2} &= \frac{-6}{2} \\ x &= -3 \end{aligned}$$

19.
$$\begin{aligned} \frac{3}{4}x-\frac{1}{2} &= 1 \\ 4\left(\frac{3}{4}x-\frac{1}{2}\right) &= 4(1) \\ 3x-2 &= 4 \\ 3x-2+2 &= 4+2 \\ 3x &= 6 \\ \frac{3x}{3} &= \frac{6}{3} \\ x &= 2 \end{aligned}$$

21.
$$\begin{aligned} 0.50x+0.15(70) &= 35.5 \\ 50x+15(70) &= 3550 \\ 50x+1050 &= 3550 \\ 50x+1050-1050 &= 3550-1050 \\ 50x &= 2500 \\ \frac{50x}{50} &= \frac{2500}{50} \\ x &= 50 \end{aligned}$$

23.
$$\begin{aligned} \frac{2(x+1)}{4} &= 3x-2 \\ 4\left[\frac{2(x+1)}{4}\right] &= 4(3x-2) \\ 2(x+1) &= 4(3x-2) \\ 2x+2 &= 12x-8 \\ 2x+2+8 &= 12x-8+8 \\ 2x+10 &= 12x \\ 2x+10-2x &= 12x-2x \\ 10 &= 10x \\ \frac{10}{10} &= \frac{10x}{10} \\ 1 &= x \end{aligned}$$

25.
$$\begin{aligned} x+\frac{7}{6} &= 2x-\frac{7}{6} \\ 6\left(x+\frac{7}{6}\right) &= 6\left(2x-\frac{7}{6}\right) \\ 6x+7 &= 12x-7 \\ 6x+7+7 &= 12x-7+7 \\ 6x+14 &= 12x \\ 6x+14-6x &= 12x-6x \\ 14 &= 6x \\ \frac{14}{6} &= \frac{6x}{6} \\ \frac{7}{3} &= x \end{aligned}$$

Copyright © 2012 Pearson Education, Inc. Publishing as Prentice Hall.

27. $0.12(y-6)+0.06y=0.08y-0.7$
$12(y-6)+6y=8y-70$
$12y-72+6y=8y-70$
$18y-72=8y-70$
$18y-72-8y=8y-70-8y$
$10y-72=-70$
$10y-72+72=-70+72$
$10y=2$
$\frac{10y}{10}=\frac{2}{10}$
$y=0.2$

29. $4(3x+2)=12x+8$
$12x+8=12x+8$
Since both sides of the equation are identical, the equation is an identity and every real number is a solution.

31. $\frac{x}{4}+1=\frac{x}{4}$
$\frac{x}{4}+1-\frac{x}{4}=\frac{x}{4}-\frac{x}{4}$
$1=0$
Since the statement 1 = 0 is false, the equation has no solution.

33. $3x-7=3(x+1)$
$3x-7=3x+3$
$3x-7-3x=3x+3-3x$
$-7=3$
Since the statement −7 = 3 is false, the equation has no solution.

35. $-2(6x-5)+4=-12x+14$
$-12x+10+4=-12x+14$
$-12x+14=-12x+14$
Since both sides of the equation are identical, the equation is an identity and every real number is a solution.

37. $\frac{6(3-z)}{5}=-z$
$5\cdot\frac{6(3-z)}{5}=5(-z)$
$6(3-z)=-5z$
$18-6z=-5z$
$18-6z+6z=-5z+6z$
$18=z$

39. $-3(2t-5)+2t=5t-4$
$-6t+15+2t=5t-4$
$-4t+15=5t-4$
$-4t+15+4t=5t-4+4t$
$15=9t-4$
$15+4=9t-4+4$
$19=9t$
$\frac{19}{9}=\frac{9t}{9}$
$\frac{19}{9}=t$

41. $5y+2(y-6)=4(y+1)-2$
$5y+2y-12=4y+4-2$
$7y-12=4y+2$
$7y-12+12=4y+2+12$
$7y=4y+14$
$7y-4y=4y+14-4y$
$3y=14$
$\frac{3y}{3}=\frac{14}{3}$
$y=\frac{14}{3}$

43. $\frac{3(x-5)}{2}=\frac{2(x+5)}{3}$
$6\left[\frac{3(x-5)}{2}\right]=6\left[\frac{2(x+5)}{3}\right]$
$9(x-5)=4(x+5)$
$9x-45=4x+20$
$9x-45+45=4x+20+45$
$9x=4x+65$
$9x-4x=4x+65-4x$
$5x=65$
$\frac{5x}{5}=\frac{65}{5}$
$x=13$

45. $0.7x-2.3=0.5$
$7x-23=5$
$7x-23+23=5+23$
$7x=28$
$\frac{7x}{7}=\frac{28}{7}$
$x=4$

47. $5x-5=2(x+1)+3x-7$
$5x-5=2x+2+3x-7$
$5x-5=5x-5$
Since both sides of the equation are identical, the equation is an identity and every real number is a solution.

Copyright © 2012 Pearson Education, Inc. Publishing as Prentice Hall.

49.
$$\begin{aligned} 4(2n+1) &= 3(6n+3)+1 \\ 8n+4 &= 18n+9+1 \\ 8n+4 &= 18n+10 \\ 8n+4-10 &= 18n+10-10 \\ 8n-6 &= 18n \\ 8n-6-8n &= 18n-8n \\ -6 &= 10n \\ \frac{-6}{10} &= \frac{10n}{10} \\ -\frac{3}{5} &= n \end{aligned}$$

51.
$$\begin{aligned} x+\frac{5}{4} &= \frac{3}{4}x \\ 4\left(x+\frac{5}{4}\right) &= 4\left(\frac{3}{4}x\right) \\ 4x+5 &= 3x \\ 4x+5-4x &= 3x-4x \\ 5 &= -x \\ \frac{5}{-1} &= \frac{-x}{-1} \\ -5 &= x \end{aligned}$$

53.
$$\begin{aligned} \frac{x}{2}-1 &= \frac{x}{5}+2 \\ 10\left(\frac{x}{2}-1\right) &= 10\left(\frac{x}{5}+2\right) \\ 5x-10 &= 2x+20 \\ 5x-10+10 &= 2x+20+10 \\ 5x &= 2x+30 \\ 5x-2x &= 2x+30-2x \\ 3x &= 30 \\ \frac{3x}{3} &= \frac{30}{3} \\ x &= 10 \end{aligned}$$

55.
$$\begin{aligned} 2(x+3)-5 &= 5x-3(1+x) \\ 2x+6-5 &= 5x-3-3x \\ 2x+1 &= 2x-3 \\ 2x+1-2x &= 2x-3-2x \\ 1 &= -3 \end{aligned}$$

Since the statement $1 = -3$ is false, the equation has no solution.

57.
$$\begin{aligned} 0.06-0.01(x+1) &= -0.02(2-x) \\ 6-1(x+1) &= -2(2-x) \\ 6-x-1 &= -4+2x \\ 5-x &= -4+2x \\ 5-x+x &= -4+2x+x \\ 5 &= -4+3x \\ 5+4 &= -4+3x+4 \\ 9 &= 3x \\ \frac{9}{3} &= \frac{3x}{3} \\ 3 &= x \end{aligned}$$

59.
$$\begin{aligned} \frac{9}{2}+\frac{5}{2}y &= 2y-4 \\ 2\left(\frac{9}{2}+\frac{5}{2}y\right) &= 2(2y-4) \\ 9+5y &= 4y-8 \\ 9+5y-4y &= 4y-8-4y \\ 9+y &= -8 \\ 9+y-9 &= -8-9 \\ y &= -17 \end{aligned}$$

61.
$$\begin{aligned} \frac{3}{4}x-1+\frac{1}{2}x &= \frac{5}{12}x+\frac{1}{6} \\ 12\left(\frac{3}{4}x-1+\frac{1}{2}x\right) &= 12\left(\frac{5}{12}x+\frac{1}{6}\right) \\ 9x-12+6x &= 5x+2 \\ 15x-12 &= 5x+2 \\ 15x-12-5x &= 5x+2-5x \\ 10x-12 &= 2 \\ 10x-12+12 &= 2+12 \\ 10x &= 14 \\ \frac{10x}{10} &= \frac{14}{10} \\ x &= \frac{7}{5} \end{aligned}$$

Copyright © 2012 Pearson Education, Inc. Publishing as Prentice Hall.

63.
$$\begin{aligned} 3x+\frac{5}{16} &= \frac{3}{4}-\frac{1}{8}x-\frac{1}{2} \\ 16\left(3x+\frac{5}{16}\right) &= 16\left(\frac{3}{4}-\frac{1}{8}x-\frac{1}{2}\right) \\ 48x+5 &= 12-2x-8 \\ 48x+5 &= 4-2x \\ 48x+5+2x &= 4-2x+2x \\ 50x+5 &= 4 \\ 50x+5-5 &= 4-5 \\ 50x &= -1 \\ \frac{50x}{50} &= \frac{-1}{50} \\ x &= -\frac{1}{50} \end{aligned}$$

65. The perimeter is the sum of the lengths of the sides.
$$\begin{aligned} x+(2x-3)+(3x-5) &= x+2x-3+3x-5 \\ &= 6x-8 \end{aligned}$$
The perimeter is $(6x - 8)$ meters.

67. A number subtracted from -8 is $-8 - x$.

69. The sum of -3 and twice a number is $-3 + 2x$.

71. The product of 9 and the sum of a number and 20 is $9(x + 20)$.

73. **a.** Since both sides of the equation are identical, the equation is an identity and every real number is a solution.

b. answers may vary

c. answers may vary

75. $5x + 1 = 5x + 1$
Since both sides of the equation are identical, the equation is an identity and every real number is a solution. The choice is a.

77.
$$\begin{aligned} 2x-6x-10 &= -4x+3-10 \\ -4x-10 &= -4x-7 \\ -4x-10+4x &= -4x-7+4x \\ -10 &= -7 \end{aligned}$$
Since the statement $-10 = -7$ is false, the equation has no solution. The choice is b.

79.
$$\begin{aligned} 9x-20 &= 8x-20 \\ 9x-20-8x &= 8x-20-8x \\ x-20 &= -20 \\ x-20+20 &= -20+20 \\ x &= 0 \end{aligned}$$
The choice is c.

81. answers may vary

83. **a.** The perimeter is the sum of the lengths of the sides.
$x + x + x + 2x + 2x = 28$

Copyright © 2012 Pearson Education, Inc. Publishing as Prentice Hall.

b.
$$\begin{aligned} x+x+x+2x+2x &= 28 \\ 7x &= 28 \\ \frac{7x}{7} &= \frac{28}{7} \\ x &= 4 \end{aligned}$$

c. The sides of length x cm are 4 cm and the sides of length $2x$ cm are 2(4 cm) = 8 cm.

85. answers may vary

87.
$$\begin{aligned} 1000(7x-10) &= 50(412+100x) \\ 7000x-10{,}000 &= 20{,}600+5000x \\ 7000x-10{,}000-5000x &= 20{,}600+5000x-5000x \\ 2000x-10{,}000 &= 20{,}600 \\ 2000x-10{,}000+10{,}000 &= 20{,}600+10{,}000 \\ 2000x &= 30{,}600 \\ \frac{2000x}{2000} &= \frac{30{,}600}{2000} \\ x &= 15.3 \end{aligned}$$

89.
$$\begin{aligned} 0.035x+5.112 &= 0.010x+5.107 \\ 35x+5112 &= 10x+5107 \\ 35x+5112-10x &= 10x+5107-10x \\ 25x+5112 &= 5107 \\ 25x+5112-5112 &= 5107-5112 \\ 25x &= -5 \\ \frac{25x}{25} &= \frac{-5}{25} \\ x &= -\frac{1}{5} \\ x &= -0.2 \end{aligned}$$

Integrated Review

1.
$$\begin{aligned} x-10 &= -4 \\ x-10+10 &= -4+10 \\ x &= 6 \end{aligned}$$

2.
$$\begin{aligned} y+14 &= -3 \\ y+14-14 &= -3-14 \\ y &= -17 \end{aligned}$$

3.
$$\begin{aligned} 9y &= 108 \\ \frac{9y}{9} &= \frac{108}{9} \\ y &= 12 \end{aligned}$$

4.
$$\begin{aligned} -3x &= 78 \\ \frac{-3x}{-3} &= \frac{78}{-3} \\ x &= -26 \end{aligned}$$

Copyright © 2012 Pearson Education, Inc. Publishing as Prentice Hall.

5.
$$\begin{aligned} -6x+7&=25\\ -6x+7-7&=25-7\\ -6x&=18\\ \frac{-6x}{-6}&=\frac{18}{-6}\\ x&=-3 \end{aligned}$$

6.
$$\begin{aligned} 5y-42&=-47\\ 5y-42+42&=-47+42\\ 5y&=-5\\ \frac{5y}{5}&=\frac{-5}{5}\\ y&=-1 \end{aligned}$$

7.
$$\begin{aligned} \frac{2}{3}x&=9\\ \frac{3}{2}\cdot\frac{2}{3}x&=\frac{3}{2}\cdot 9\\ x&=\frac{27}{2} \end{aligned}$$

8.
$$\begin{aligned} \frac{4}{5}z&=10\\ \frac{5}{4}\cdot\frac{4}{5}z&=\frac{5}{4}\cdot 10\\ z&=\frac{50}{4}\\ z&=\frac{25}{2} \end{aligned}$$

9.
$$\begin{aligned} \frac{r}{-4}&=-2\\ -4\cdot\frac{r}{-4}&=-4\cdot(-2)\\ r&=8 \end{aligned}$$

10.
$$\begin{aligned} \frac{y}{-8}&=8\\ -8\cdot\frac{y}{-8}&=-8\cdot 8\\ y&=-64 \end{aligned}$$

11.
$$\begin{aligned} 6-2x+8&=10\\ -2x+14&=10\\ -2x+14-14&=10-14\\ -2x&=-4\\ \frac{-2x}{-2}&=\frac{-4}{-2}\\ x&=2 \end{aligned}$$

12.
$$\begin{aligned} -5-6y+6&=19\\ -6y+1&=19\\ -6y+1-1&=19-1\\ -6y&=18\\ \frac{-6y}{-6}&=\frac{18}{-6}\\ y&=-3 \end{aligned}$$

13.
$$\begin{aligned} 2x-7&=6x-27\\ 2x-7+7&=6x-27+7\\ 2x&=6x-20\\ 2x-6x&=6x-20-6x\\ -4x&=-20\\ \frac{-4x}{-4}&=\frac{-20}{-4}\\ x&=5 \end{aligned}$$

14.
$$\begin{aligned} 3+8y&=3y-2\\ 3+8y-3y&=3y-2-3y\\ 3+5y&=-2\\ -3+3+5y&=-3-2\\ 5y&=-5\\ \frac{5y}{5}&=\frac{-5}{5}\\ y&=-1 \end{aligned}$$

15.
$$\begin{aligned} 9(3x-1)&=-4+49\\ 27x-9&=45\\ 27x-9+9&=45+9\\ 27x&=54\\ \frac{27x}{27}&=\frac{54}{27}\\ x&=2 \end{aligned}$$

16.
$$\begin{aligned} 12(2x+1)&=-6+66\\ 24x+12&=60\\ 24x+12-12&=60-12\\ 24x&=48\\ \frac{24x}{24}&=\frac{48}{24}\\ x&=2 \end{aligned}$$

17.
$$\begin{aligned} -3a+6+5a&=7a-8a\\ 6+2a&=-a\\ 6+2a-2a&=-a-2a\\ 6&=-3a\\ \frac{6}{-3}&=\frac{-3a}{-3}\\ -2&=a \end{aligned}$$

Copyright © 2012 Pearson Education, Inc. Publishing as Prentice Hall.

18.
$$\begin{aligned} 4b-8-b &= 10b-3b \\ 3b-8 &= 7b \\ -3b+3b-8 &= -3b+7b \\ -8 &= 4b \\ \frac{-8}{4} &= \frac{4b}{4} \\ -2 &= b \end{aligned}$$

19.
$$\begin{aligned} -\frac{2}{3}x &= \frac{5}{9} \\ -\frac{3}{2}\cdot\left(-\frac{2}{3}x\right) &= -\frac{3}{2}\cdot\frac{5}{9} \\ x &= -\frac{15}{18} \\ x &= -\frac{5}{6} \end{aligned}$$

20.
$$\begin{aligned} -\frac{3}{8}y &= -\frac{1}{16} \\ -\frac{8}{3}\cdot\left(-\frac{3}{8}y\right) &= -\frac{8}{3}\cdot\left(-\frac{1}{16}\right) \\ y &= \frac{1}{6} \end{aligned}$$

21.
$$\begin{aligned} 10 &= -6n+16 \\ 10-16 &= -6n+16-16 \\ -6 &= -6n \\ \frac{-6}{-6} &= \frac{-6n}{-6} \\ 1 &= n \end{aligned}$$

22.
$$\begin{aligned} -5 &= -2m+7 \\ -5-7 &= -2m+7-7 \\ -12 &= -2m \\ \frac{-12}{-2} &= \frac{-2m}{-2} \\ 6 &= m \end{aligned}$$

23.
$$\begin{aligned} 3(5c-1)-2 &= 13c+3 \\ 15c-3-2 &= 13c+3 \\ 15c-5 &= 13c+3 \\ 15c-5+5 &= 13c+3+5 \\ 15c &= 13c+8 \\ 15c-13c &= 13c+8-13c \\ 2c &= 8 \\ \frac{2c}{2} &= \frac{8}{2} \\ c &= 4 \end{aligned}$$

24.
$$\begin{aligned} 4(3t+4)-20 &= 3+5t \\ 12t+16-20 &= 3+5t \\ 12t-4 &= 3+5t \\ 12t-4-5t &= 3+5t-5t \\ 7t-4 &= 3 \\ 7t-4+4 &= 3+4 \\ 7t &= 7 \\ \frac{7t}{7} &= \frac{7}{7} \\ t &= 1 \end{aligned}$$

25.
$$\begin{aligned} \frac{2(z+3)}{3} &= 5-z \\ 3\left[\frac{2(z+3)}{3}\right] &= 3(5-z) \\ 2(z+3) &= 3(5-z) \\ 2z+6 &= 15-3z \\ 2z+6+3z &= 15-3z+3z \\ 6+5z &= 15 \\ 6+5z-6 &= 15-6 \\ 5z &= 9 \\ \frac{5z}{5} &= \frac{9}{5} \\ z &= \frac{9}{5} \end{aligned}$$

26.
$$\begin{aligned} \frac{3(w+2)}{4} &= 2w+3 \\ 4\left[\frac{3(w+2)}{4}\right] &= 4(2w+3) \\ 3(w+2) &= 4(2w+3) \\ 3w+6 &= 8w+12 \\ 3w+6-6 &= 8w+12-6 \\ 3w &= 8w+6 \\ 3w-8w &= 8w+6-8w \\ -5w &= 6 \\ \frac{-5w}{-5} &= \frac{6}{-5} \\ w &= -\frac{6}{5} \end{aligned}$$

27. $-2(2x-5) = -3x+7-x+3$
$-4x+10 = -4x+10$
Since both sides of the equation are identical, the equation is an identity and every real number is a solution.

28. $-4(5x-2) = -12x+4-8x+4$
$-20x+8 = -20x+8$
Since both sides of the equation are identical, the equation is an identity and every real number is a solution.

Copyright © 2012 Pearson Education, Inc. Publishing as Prentice Hall.

29.
$$\begin{aligned}0.02(6t-3)&=0.04(t-2)+0.02\\2(6t-3)&=4(t-2)+2\\12t-6&=4t-8+2\\12t-6&=4t-6\\12t-6-4t&=4t-6-4t\\8t-6&=-6\\8t-6+6&=-6+6\\8t&=0\\\frac{8t}{8}&=\frac{0}{8}\\t&=0\end{aligned}$$

30.
$$\begin{aligned}0.03(m+7)&=0.02(5-m)+0.03\\3(m+7)&=2(5-m)+3\\3m+21&=10-2m+3\\3m+21&=13-2m\\3m+21+2m&=13-2m+2m\\5m+21&=13\\5m+21-21&=13-21\\5m&=-8\\\frac{5m}{5}&=\frac{-8}{5}\\m&=-1.6\end{aligned}$$

31.
$$\begin{aligned}-3y&=\frac{4(y-1)}{5}\\5(-3y)&=5\left[\frac{4(y-1)}{5}\right]\\-15y&=4(y-1)\\-15y&=4y-4\\-15y-4y&=4y-4-4y\\-19y&=-4\\\frac{-19y}{-19}&=\frac{-4}{-19}\\y&=\frac{4}{19}\end{aligned}$$

32.
$$\begin{aligned}-4x&=\frac{5(1-x)}{6}\\6(-4x)&=6\cdot\frac{5(1-x)}{6}\\-24x&=5(1-x)\\-24x&=5-5x\\-24x+5x&=5-5x+5x\\-19x&=5\\\frac{-19x}{-19}&=\frac{5}{-19}\\x&=-\frac{5}{19}\end{aligned}$$

33.
$$\begin{aligned}\frac{5}{3}x-\frac{7}{3}&=x\\3\left(\frac{5}{3}x-\frac{7}{3}\right)&=3x\\5x-7&=3x\\-5x+5x-7&=-5x+3x\\-7&=-2x\\\frac{-7}{-2}&=\frac{-2x}{-2}\\\frac{7}{2}&=x\end{aligned}$$

34.
$$\begin{aligned}\frac{7}{5}n+\frac{3}{5}&=-n\\5\left(\frac{7}{5}n+\frac{3}{5}\right)&=5(-n)\\7n+3&=-5n\\-7n+7n+3&=-7n-5n\\3&=-12n\\\frac{3}{-12}&=\frac{-12n}{-12}\\-\frac{1}{4}&=n\end{aligned}$$

35.
$$\begin{aligned}\frac{1}{10}(3x-7)&=\frac{3}{10}x+5\\\frac{3}{10}x-\frac{7}{10}&=\frac{3}{10}x+5\\-\frac{3}{10}x+\frac{3}{10}x-\frac{7}{10}&=-\frac{3}{10}x+\frac{3}{10}x+5\\-\frac{7}{10}&=5\end{aligned}$$

Since the statement $-\frac{7}{10}=5$ is false, the equation has no solution.

36.
$$\begin{aligned}\frac{1}{7}(2x-5)&=\frac{2}{7}x+1\\7\cdot\frac{1}{7}(2x-5)&=7\left(\frac{2}{7}x+1\right)\\2x-5&=2x+7\\2x-5-2x&=2x+7-2x\\-5&=7\end{aligned}$$

Since the statement $-5 = 7$ is false, the equation has no solution.

Copyright © 2012 Pearson Education, Inc. Publishing as Prentice Hall.

37. $5+2(3x-6)=-4(6x-7)$
$5+6x-12=-24x+28$
$6x-7=-24x+28$
$24x+6x-7=24x-24x+28$
$30x-7=28$
$30x-7+7=28+7$
$30x=35$
$\frac{30x}{30}=\frac{35}{30}$
$x=\frac{7}{6}$

38. $3+5(2x-4)=-7(5x+2)$
$3+10x-20=-35x-14$
$10x-17=-35x-14$
$10x-17+35x=-35x-14+35x$
$45x-17=-14$
$45x-17+17=-14+17$
$45x=3$
$\frac{45x}{45}=\frac{3}{45}$
$x=\frac{1}{15}$

Section 2.4 Practice

1. Let x represent the number.
$3x-6=2x+3$
$3x-6-2x=2x+3-2x$
$x-6=3$
$x-6+6=3+6$
$x=9$
The number is 9.

2. Let x represent the number.
$3(x-5)=2x-3$
$3x-15=2x-3$
$3x-15-2x=2x-3-2x$
$x-15=-3$
$x-15+15=-3+15$
$x=12$
The number is 12.

3. Let x represent the length of the shorter piece. Then $5x$ represents the length of the longer piece. Their sum is 18 feet.
$x+5x=18$
$6x=18$
$\frac{6x}{6}=\frac{18}{6}$
$x=3$
The shorter piece is 3 feet and the longer piece is $5(3) = 15$ feet.

4. Let x represent the number of votes for Texas. Then $x + 21$ represents the number of votes for California. Their sum is 89.
$x+x+21=89$
$2x+21=89$
$2x+21-21=89-21$
$2x=68$
$\frac{2x}{2}=\frac{68}{2}$
$x=34$
Texas has 34 electoral votes and California has $34 + 21 = 55$ electoral votes.

5. Let x represent the number of miles driven. The cost for x miles is $0.15x$. The daily cost is \$28.
$0.15x+28=52$
$0.15x+28-28=52-28$
$0.15x=24$
$\frac{0.15x}{0.15}=\frac{24}{0.15}$
$x=160$
You drove 160 miles.

6. Let x represent the measure of the smallest angle. Then $2x$ represents the measure of the second angle and $3x$ represents the measure of the third angle. The sum of the measures of the angles of a triangle equals 180.
$x+2x+3x=180$
$6x=180$
$\frac{6x}{6}=\frac{180}{6}$
$x=30$
If $x = 30$, then $2x = 2(30) = 60$ and $3x = 3(30) = 90$.
The smallest is 30°, second is 60°, and third is 90°.

7. If x is the first even integer, then $x + 2$ and $x + 4$ are the next two even integers.
$x+x+2+x+4=144$
$3x+6=144$
$3x+6-6=144-6$
$3x=138$
$\frac{3x}{3}=\frac{138}{3}$
$x=46$
If $x = 46$, then $x + 2 = 48$ and $x + 4 = 50$. The integers are 46, 48, 50.

Copyright © 2012 Pearson Education, Inc. Publishing as Prentice Hall.

Vocabulary and Readiness Check

1. If x is the number, then "double the number" is $2x$, and "double the number, decreased by 31" is $2x - 31$.

2. If x is the number, then "three times the number" is $3x$, and "three times the number, increased by 17" is $3x + 17$.

3. If x is the number, then "the sum of the number and 5" is $x + 5$, and "twice the sum of the number and 5" is $2(x + 5)$.

4. If x is the number, then "the difference of the number and 11" is $x - 11$, and "seven times the difference of the number and 11" is $7(x - 11)$.

5. If y is the number, then "the difference of 20 and the number" is $20 - y$, and "the difference of 20 and the number, divided by 3" is $\frac{20-y}{3}$ or $(20 - y) \div 3$.

6. If y is the number, then "the sum of -10 and the number" is $-10 + y$, and "the sum of -10 and the number, divided by 9" is $\frac{(-10+y)}{9}$ or $(-10 + y) \div 9$.

Exercise Set 2.4

1.
$$\begin{aligned} 2x+7 &= x+6 \\ 2x+7-x &= x+6-x \\ x+7 &= 6 \\ x+7-7 &= 6-7 \\ x &= -1 \end{aligned}$$
The number is -1.

3.
$$\begin{aligned} 3x-6 &= 2x+8 \\ 3x-6-2x &= 2x+8-2x \\ x-6 &= 8 \\ x-6+6 &= 8+6 \\ x &= 14 \end{aligned}$$
The number is 14.

5.
$$\begin{aligned} 2(x-8) &= 3(x+3) \\ 2x-16 &= 3x+9 \\ 2x-16-2x &= 3x+9-2x \\ -16 &= x+9 \\ -16-9 &= x+9-9 \\ -25 &= x \end{aligned}$$
The number is -25.

7.
$$\begin{aligned} 2x(3) &= 5x-\frac{3}{4} \\ 6x &= 5x-\frac{3}{4} \\ 6x-5x &= 5x-\frac{3}{4}-5x \\ x &= -\frac{3}{4} \end{aligned}$$
The number is $-\frac{3}{4}$.

9. The sum of the three lengths is 25 inches.
$$\begin{aligned} x+2x+1+5x &= 25 \\ 1+8x &= 25 \\ 1+8x-1 &= 25-1 \\ 8x &= 24 \\ \frac{8x}{8} &= \frac{24}{8} \\ x &= 3 \end{aligned}$$
$2x = 2(3) = 6$
$1 + 5x = 1 + 5(3) = 1 + 15 = 16$
The lengths are 3 inches, 6 inches, and 16 inches.

11. Let x be the length of the first piece. Then the second piece is $2x$ and the third piece is $5x$. The sum of the lengths is 40 inches.
$$\begin{aligned} x+2x+5x &= 40 \\ 8x &= 40 \\ \frac{8x}{8} &= \frac{40}{8} \\ x &= 5 \end{aligned}$$
$2x = 2(5) = 10$
$5x = 5(5) = 25$
The 1st piece is 5 inches, 2nd piece is 10 inches, and 3rd piece is 25 inches.

Copyright © 2012 Pearson Education, Inc. Publishing as Prentice Hall.

13. Let x represent the number of millions of pounds of pecans produced in Texas. Then $x + 15$ represents the number of millions of pounds of pecans produced in New Mexico.

$$\begin{aligned} x+x+15 &= 75 \\ 2x+15 &= 75 \\ 2x+15-15 &= 75-15 \\ 2x &= 60 \\ \frac{2x}{2} &= \frac{60}{2} \\ x &= 30 \end{aligned}$$

$x + 15 = 30 + 15 = 45$

Texas produced 30 million pounds of pecans and New Mexico produced 45 million pounds.

15. Let x be the number of miles. Then the cost for x miles is $0.29x$. Each day costs \$24.95.

$$\begin{aligned} 0.29x+2(24.95) &= 100 \\ 0.29x+49.9 &= 100 \\ 0.29x+49.9-49.9 &= 100-49.9 \\ 0.29x &= 50.1 \\ \frac{0.29x}{0.29} &= \frac{50.1}{0.29} \\ x &\approx 172.8 \end{aligned}$$

You can drive 172 whole miles on a \$100 budget.

17. Let x be the number of miles. Then the total fare is $3 + 0.8x + 4.5$.

$$\begin{aligned} 3+0.8x+4.5 &= 27.5 \\ 30+8x+45 &= 275 \\ 8x+75 &= 275 \\ 8x+75-75 &= 275-75 \\ 8x &= 200 \\ \frac{8x}{8} &= \frac{200}{8} \\ x &= 25 \end{aligned}$$

You can travel 25 miles from the airport by taxi for \$27.50.

19. Let x be the measure of each of the two equal angles. Then $2x + 30$ is the measure of the third angle. Their sum is 180°.

$$\begin{aligned} x+x+2x+30 &= 180 \\ 4x+30 &= 180 \\ 4x+30-30 &= 180-30 \\ 4x &= 150 \\ \frac{4x}{4} &= \frac{150}{4} \\ x &= 37.5 \end{aligned}$$

$2x + 30 = 2(37.5) + 30 = 75 + 30 = 105$

The 1st angle measures 37.5°, the 2nd angle measures 37.5°, and the 3rd angle measures 105°.

21. Angles A and D both measure $x°$, while angles C and B both measure $(2x)°$. The sum of the angle measures is 360°.

$$\begin{aligned} x+2x+x+2x &= 360 \\ 6x &= 360 \\ \frac{6x}{6} &= \frac{360}{6} \\ x &= 60 \end{aligned}$$

$2x = 2(60) = 120$

Angles A and D measure 60°; angles B and C measure 120°.

Copyright © 2012 Pearson Education, Inc. Publishing as Prentice Hall.

	First Integer	Next Integers			Indicated Sum
23.	x	$x+1$	$x+2$		$x+(x+1)+(x+2)=3x+3$
25.	x	$x+2$	$x+4$		$x+(x+4)=2x+4$
27.	x	$x+1$	$x+2$	$x+3$	$x+(x+1)+(x+2)+(x+3)=4x+6$
29.	x	$x+2$	$x+4$		$(x+2)+(x+4)=2x+6$

31. If x is the first integer, the next consecutive integer is $x + 1$.

$$\begin{aligned} x+x+1&=469\\ 2x+1&=469\\ 2x+1-1&=469-1\\ 2x&=468\\ \frac{2x}{2}&=\frac{468}{2}\\ x&=234 \end{aligned}$$

The page numbers are 234 and 234 + 1 = 235.

33. If x is the first integer, the next two consecutive integers are $x + 1$ and $x + 2$.

$$\begin{aligned} x+x+1+x+2&=99\\ 3x+3&=99\\ 3x+3-3&=99-3\\ 3x&=96\\ \frac{3x}{3}&=\frac{96}{3}\\ x&=32 \end{aligned}$$

The code for Belgium is 32, France is 32 + 1 = 33, and Spain is 32 + 2 = 34.

35. Let x be the length of the shorter piece. Then $2x + 2$ is the length of the longer piece. The measures sum to 17 feet.

$$\begin{aligned} x+2x+2&=17\\ 3x+2&=17\\ 3x+2-2&=17-2\\ 3x&=5\\ \frac{3x}{3}&=\frac{15}{3}\\ x&=5 \end{aligned}$$

$2x + 2 = 2(5) + 2 = 10 + 2 = 12$

The pieces measure 5 feet and 12 feet.

37. Let x represent the speed of the TGV. Then the speed of the Maglev is $x + 3.8$.

$$\begin{aligned} x+x+3.8&=718.2\\ 2x+3.8&=718.2\\ 2x+3.8-3.8&=718.2-3.8\\ 2x&=714.4\\ \frac{2x}{2}&=\frac{714.4}{2}\\ x&=357.2 \end{aligned}$$

$x + 3.8 = 357.2 + 3.8 = 361$

The speed of the TGV is 357.2 miles per hour and the speed of the Maglev is 361 miles per hour.

Copyright © 2012 Pearson Education, Inc. Publishing as Prentice Hall.

39. Let x be the measure of the smaller angle. Then the larger angle measures $3x + 8$. Their sum is 180°.

$$x+3x+8=180$$
$$4x+8=180$$
$$4x+8-8=180-8$$
$$4x=172$$
$$\frac{4x}{4}=\frac{172}{4}$$
$$x=43$$

$3x + 8 = 3(43) + 8 = 129 + 8 = 137$

The angles measure 43° and 137°.

41. Let x be the first even integer. Then the next two consecutive even integers are $x + 2$ and $x + 4$. The sum of the measures of the angles of a triangle is 180°.

$$x+x+2+x+4=180$$
$$3x+6=180$$
$$3x+6-6=180-6$$
$$3x=174$$
$$\frac{3x}{3}=\frac{174}{3}$$
$$x=58$$

$x + 2 = 58 + 2 = 60$

$x + 4 = 58 + 4 = 62$

The angles measure 58°, 60°, and 62°.

43.

$$\frac{1}{5}+2x=3x-\frac{4}{5}$$
$$\frac{1}{5}+2x-2x=3x-\frac{4}{5}-2x$$
$$\frac{1}{5}=x-\frac{4}{5}$$
$$\frac{1}{5}+\frac{4}{5}=x-\frac{4}{5}+\frac{4}{5}$$
$$\frac{5}{5}=x$$
$$1=x$$

The number is 1.

45. Let x be the number of miles. Then the charge for driving x miles in one day is $39 + 0.2x$.

$$39+0.2x=95$$
$$390+2x=950$$
$$390+2x-390=950-390$$
$$2x=560$$
$$\frac{2x}{2}=\frac{560}{2}$$
$$x=280$$

You drove 280 miles.

47. Let x represent the number of points scored by Penn State. Then USC scored $x + 14$ points. Their combined scores totaled 62 points.

$$x+x+14=62$$
$$2x+14=62$$
$$2x+14-14=62-14$$
$$2x=48$$
$$\frac{2x}{2}=\frac{48}{2}$$
$$x=24$$

$x + 14 = 24 + 14 = 38$

Penn State scored 24 points and USC scored 38 points.

49. Let x represent the number of counties in Montana. Then $x + 2$ represents the number of counties in California.

$$x+x+2=114$$
$$2x+2=114$$
$$2x+2-2=114-2$$
$$2x=112$$
$$\frac{2x}{2}=\frac{112}{2}$$
$$x=56$$

$x + 2 = 56 + 2 = 58$

Montana has 56 counties and California has 58 counties.

51. Let x represent the number of moons for Neptune. Then $x + 13$ represents the number of moons for Uranus and $2x + 2$ represents the number of moons for Saturn. The total is 47.

$$x+x+13+2x+2=47$$
$$4x+15=47$$
$$4x+15-15=47-15$$
$$4x=32$$
$$\frac{4x}{4}=\frac{32}{4}$$
$$x=8$$

$x + 13 = 8 + 13 = 21$

$2x + 2 = 2(8) + 2 = 16 + 2 = 18$

Neptune has 8 moons, Uranus has 21 moons, and Saturn has 18 moons.

53.

$$3(x+5)=2x-1$$
$$3x+15=2x-1$$
$$3x+15-2x=2x-1-2x$$
$$x+15=-1$$
$$x+15-15=-1-15$$
$$x=-16$$

The number is −16.

Copyright © 2012 Pearson Education, Inc. Publishing as Prentice Hall.

55. Let x represent the area of the Gobi Desert, in square miles. Then $7x$ represents the area of the Sahara Desert.

$$x+7x=4{,}000{,}000$$
$$8x=4{,}000{,}000$$
$$\frac{8x}{8}=\frac{4{,}000{,}000}{8}$$
$$x=500{,}000$$

$7x = 7(500{,}000) = 3{,}500{,}000$

The Gobi Desert's area is 500,000 square miles and the Sahara Desert's area is 3,500,000 square miles.

57. Let x represent the number of gold medals won by Australia. Then Germany won $x + 1$ gold medals and Korea won $x + 2$ gold medals.

$$x+x+1+x+2=21$$
$$3x+3=21$$
$$3x+3-3=21-3$$
$$3x=18$$
$$\frac{3x}{3}=\frac{18}{3}$$
$$x=6$$

$x + 1 = 6 + 1 = 7$

$x + 2 = 6 + 2 = 8$

Australia won 6 gold medals, Germany won 7, and Korea won 8.

59. Let x be the number of votes for Jim Martin. Then $x + 315{,}217$ is the number of votes for Saxby Chambliss.

$$x+x+315{,}217=2{,}126{,}491$$
$$2x+315{,}217=2{,}126{,}491$$
$$2x+315{,}217-315{,}217=2{,}126{,}491-315{,}217$$
$$2x=1{,}811{,}274$$
$$\frac{2x}{2}=\frac{1{,}811{,}274}{2}$$
$$x=905{,}637$$

$x + 315{,}217 = 905{,}637 + 315{,}217 = 1{,}220{,}854$

Martin received 905,637 votes and Chambliss received 1,220,854 votes.

61. Let x be the measure of the two equal angles. Then $x + 76.5$ is the measure of the third angle.

$$x+x+x+76.5=180$$
$$3x+76.5=180$$
$$3x+76.5-76.5=180-76.5$$
$$3x=103.5$$
$$\frac{3x}{3}=\frac{103.5}{3}$$
$$x=34.5$$

$x + 76.5 = 34.5 + 76.5 = 111$

The three angles measure 34.5°, 34.5°, and 111°.

63. The longest bar represents the album Eagles: *Their Greatest Hits*, so Eagles: *Their Greatest Hits* is the best selling album of all time.

65. Let x represent sales of *The Wall*. Then $x + 4$ is the sales of *Thriller*.

$$x+x+4=50$$
$$2x+4=50$$
$$2x+4-4=50-4$$
$$2x=46$$
$$\frac{2x}{2}=\frac{46}{2}$$
$$x=23$$

$x + 4 = 27$

Thriller brought in \$27 million and *The Wall* brought in \$23 million.

67. answers may vary

69. Replace W by 7 and L by 10.

$2W + 2L = 2(7) + 2(10) = 14 + 20 = 34$

71. Replace r by 15.

$\pi r^2 = \pi(15)^2 = \pi(225) = 225\pi$

73. Let x represent the width. Then $1.6x$ represents the length. The perimeter is

$2 \cdot \text{length} + 2 \cdot \text{width}$.

$$2(1.6x)+2x=78$$
$$3.2x+2x=78$$
$$5.2x=78$$
$$\frac{5.2x}{5.2}=\frac{78}{5.2}$$
$$x=15$$

$1.6x = 1.6(15) = 24$

The dimensions of the garden are 15 feet by 24 feet.

75. 90 chirps every minute is $\frac{90 \text{ chirps}}{1 \text{ min}}$. There are 60 minutes in one hour.

$$\frac{90 \text{ chirps}}{1 \text{ min}} \cdot 60 \text{ min} = 5400 \text{ chirps}$$

At this rate, there are 5400 chirps each hour.

$24 \cdot 5400 = 129{,}600$

There are 129,600 chirps in one 24-hour day.

$365 \cdot 129{,}600 = 47{,}304{,}000$

There are 47,304,000 chirps in one year.

77. answers may vary

79. answers may vary

Copyright © 2012 Pearson Education, Inc. Publishing as Prentice Hall.

81. Measurements may vary. Rectangle (c) best approximates the shape of the golden rectangle.

Section 2.5 Practice

1. Use $d = rt$ when $d = 1180$ and $r = 50$.

$$d = rt$$
$$1180 = 50t$$
$$\frac{1180}{50} = \frac{50t}{50}$$
$$23.6 = t$$

They will spend 23.6 hours driving.

2. Use $A = lw$ when $w = 18$.

$$A = lw$$
$$450 = l \cdot 18$$
$$\frac{450}{18} = \frac{18l}{18}$$
$$25 = l$$

The length of the deck is 25 feet.

3. Use $F = \frac{9}{5}C + 32$ with $C = 5$.

$$F = \frac{9}{5}C + 32$$
$$F = \frac{9}{5} \cdot 5 + 32$$
$$F = 9 + 32$$
$$F = 41$$

Thus, 5°C is equivalent to 41°F.

4. Let x be the width. Then $4x + 1$ is the length. The perimeter is 52 meters.

$$P = 2l + 2w$$
$$52 = 2(4x+1) + 2x$$
$$52 = 8x + 2 + 2x$$
$$52 = 10x + 2$$
$$52 - 2 = 10x + 2 - 2$$
$$50 = 10x$$
$$\frac{50}{10} = \frac{10x}{10}$$
$$5 = x$$

$4x + 1 = 4(5) + 1 = 20 + 1 = 21$

The width is 5 meters and the length is 21 meters.

5.

$$C = 2\pi r$$
$$\frac{C}{2\pi} = \frac{2\pi r}{2\pi}$$
$$\frac{C}{2\pi} = r \text{ or } r = \frac{C}{2\pi}$$

6.

$$P = 2l + 2w$$
$$P - 2w = 2l + 2w - 2w$$
$$P - 2w = 2l$$
$$\frac{P-2w}{2} = \frac{2l}{2}$$
$$\frac{P-2w}{2} = l \text{ or } l = \frac{P-2w}{2}$$

7.

$$P = 2a + b - c$$
$$P + c = 2a + b - c + c$$
$$P + c = 2a + b$$
$$P + c - b = 2a + b - b$$
$$P + c - b = 2a$$
$$\frac{P+c-b}{2} = a \text{ or } a = \frac{P-b-c}{2}$$

8.

$$A = \frac{a+b}{2}$$
$$2A = 2 \cdot \frac{a+b}{2}$$
$$2A = a + b$$
$$2A - a = a + b - a$$
$$2A - a = b \text{ or } b = 2A - a$$

Exercise Set 2.5

1. Use $A = bh$ when $A = 45$ and $b = 15$.

$$A = bh$$
$$45 = 15 \cdot h$$
$$\frac{45}{15} = \frac{15h}{15}$$
$$3 = h$$

3. Use $S = 4lw + 2wh$ when $S = 102$, $l = 7$, and $w = 3$.

$$S = 4lw + 2wh$$
$$102 = 4 \cdot 7 \cdot 3 + 2 \cdot 3 \cdot h$$
$$102 = 84 + 6h$$
$$102 - 84 = 84 + 6h - 84$$
$$18 = 6h$$
$$\frac{18}{6} = \frac{6h}{6}$$
$$3 = h$$

Copyright © 2012 Pearson Education, Inc. Publishing as Prentice Hall.

5. Use $A = \frac{1}{2}h(B+b)$ when $A = 180$, $B = 11$, and $b = 7$.

$$A = \frac{1}{2}h(B+b)$$
$$180 = \frac{1}{2}h(11+7)$$
$$180 = \frac{1}{2}h(18)$$
$$180 = 9h$$
$$\frac{180}{9} = \frac{9h}{9}$$
$$20 = h$$

7. Use $P = a + b + c$ when $P = 30$, $a = 8$, and $b = 10$.

$$P = a+b+c$$
$$30 = 8+10+c$$
$$30 = 18+c$$
$$30-18 = 18+c-18$$
$$12 = c$$

9. Use $C = 2\pi r$ when $C = 15.7$ and 3.14 is used as an approximation for π.

$$C = 2\pi r$$
$$15.7 = 2(3.14)r$$
$$15.7 = 6.28r$$
$$\frac{15.7}{6.28} = \frac{6.28r}{6.28}$$
$$2.5 = r$$

11.
$$f = 5gh$$
$$\frac{f}{5g} = \frac{5gh}{5g}$$
$$\frac{f}{5g} = h$$

13.
$$V = lwh$$
$$\frac{V}{lh} = \frac{lwh}{lh}$$
$$\frac{V}{lh} = w$$

15.
$$3x + y = 7$$
$$3x + y - 3x = 7 - 3x$$
$$y = 7 - 3x$$

17.
$$A = P + PRT$$
$$A - P = P + PRT - P$$
$$A - P = PRT$$
$$\frac{A-P}{PT} = \frac{PRT}{PT}$$
$$\frac{A-P}{PT} = R$$

19.
$$V = \frac{1}{3}Ah$$
$$3V = 3\cdot\frac{1}{3}Ah$$
$$3V = Ah$$
$$\frac{3V}{h} = \frac{Ah}{h}$$
$$\frac{3V}{h} = A$$

21.
$$P = a+b+c$$
$$P-b-c = a+b+c-b-c$$
$$P-b-c = a$$

23.
$$S = 2\pi rh + 2\pi r^2$$
$$S - 2\pi r^2 = 2\pi rh + 2\pi r^2 - 2\pi r^2$$
$$S - 2\pi r^2 = 2\pi rh$$
$$\frac{S-2\pi r^2}{2\pi r} = \frac{2\pi rh}{2\pi r}$$
$$\frac{S-2\pi r^2}{2\pi r} = h$$

25. Use $A = lw$ when $A = 10{,}080$ and $w = 84$.

$$A = lw$$
$$10{,}080 = l(84)$$
$$\frac{10{,}080}{84} = \frac{84l}{84}$$
$$120 = l$$

The length (height) of the sign is 120 feet.

27. a.
$$\text{Area} = \frac{1}{2}h(B+b)$$
$$= \frac{1}{2}\cdot 12(56+24)$$
$$= 6(80)$$
$$= 480$$

Perimeter = 24 + 20 + 56 + 20 = 120
The area is 480 square inches and the perimeter is 120 inches.

b. The frame goes around the edges of the picture, so it involves perimeter. The glass covers the picture, so it involves area.

Copyright © 2012 Pearson Education, Inc. Publishing as Prentice Hall.

29. a. Area $= l \cdot w = (11.5)(9) = 103.5$

$$\begin{aligned}\text{Perimeter} &= 2l + 2w \\ &= 2(11.5) + 2(9) \\ &= 23 + 18 \\ &= 41\end{aligned}$$

The area is 103.5 square feet and the perimeter is 41 feet.

b. The baseboard goes around the edges of the room, so it involves the perimeter. The carpet covers the floor of the room, so it involves area.

31. Use $F = \frac{9}{5}C + 32$ when $F = 14$.

$$\begin{aligned}F &= \frac{9}{5}C + 32 \\ 14 &= \frac{9}{5}C + 32 \\ 14 - 32 &= \frac{9}{5}C + 32 - 32 \\ -18 &= \frac{9}{5}C \\ \frac{5}{9}\cdot(-18) &= \frac{5}{9}\cdot\frac{9}{5}C \\ -10 &= C\end{aligned}$$

Thus, 14°F is equivalent to −10°C.

33. Use $d = rt$ when $d = 25{,}000$ and $r = 4000$.

$$\begin{aligned}d &= rt \\ 25{,}000 &= 4000t \\ \frac{25{,}000}{4000} &= \frac{4000t}{4000} \\ 6.25 &= t\end{aligned}$$

It will take the X-30 6.25 hours to travel around the Earth.

35. Let x be the length. Then $\frac{2}{3}x$ is the width. Use $P = 2 \cdot \text{length} + 2 \cdot \text{width}$ when $P = 260$.

$$\begin{aligned}P &= 2\cdot\text{length} + 2\cdot\text{width} \\ 260 &= 2x + 2\cdot\frac{2}{3}x \\ 260 &= 2x + \frac{4}{3}x \\ 260 &= \frac{6}{3}x + \frac{4}{3}x \\ 260 &= \frac{10}{3}x \\ \frac{3}{10}\cdot 260 &= \frac{3}{10}\cdot\frac{10}{3}x \\ 78 &= x\end{aligned}$$

The length is 78 feet and the width is $\frac{2}{3}\cdot 78 = 52$ feet.

37. Let x represent the length of the shortest side. Then the second side has length $2x$ and the third side has length $30 + x$. The perimeter is the sum of the lengths of the sides.

$$\begin{aligned}x + 2x + 30 + x &= 102 \\ 4x + 30 &= 102 \\ 4x + 30 - 30 &= 102 - 30 \\ 4x &= 72 \\ \frac{4x}{4} &= \frac{72}{4} \\ x &= 18\end{aligned}$$

$2x = 2(18) = 36$

$30 + x = 30 + 18 = 48$

The flower bed has sides of length 18 feet, 36 feet, and 48 feet.

39. Use $d = rt$ when $r = 55$ and $t = 2\frac{1}{2}$.

$$\begin{aligned}d &= rt \\ d &= 55\cdot 2\frac{1}{2} \\ d &= 55\cdot 2.5 \\ d &= 137.5\end{aligned}$$

The distance between Bar Harbor and Yarmouth is 137.5 miles.

41. Use $N = 86$.

$$\begin{aligned}T &= 50 + \frac{N-40}{4} \\ T &= 50 + \frac{86-40}{4} \\ T &= 50 + \frac{46}{4} \\ T &= 50 + 11.5 \\ T &= 61.5\end{aligned}$$

The temperature is 61.5° Fahrenheit.

Copyright © 2012 Pearson Education, Inc. Publishing as Prentice Hall.

43. Use $T = 55$.

$$T = 50 + \frac{N-40}{4}$$
$$55 = 50 + \frac{N-40}{4}$$
$$55 - 50 = 50 + \frac{N-40}{4} - 50$$
$$5 = \frac{N-40}{4}$$
$$4 \cdot 5 = 4 \cdot \frac{N-40}{4}$$
$$20 = N - 40$$
$$20 + 40 = N - 40 + 40$$
$$60 = N$$

There are 60 chirps per minute.

45. As the number of cricket chirps per minute increases, the air temperature of their environment increases.

47. To find the amount of water in the tank, use $V = lwh$ with $l = 8$, $w = 3$, and $h = 6$.

$V = lwh = 8 \cdot 3 \cdot 6 = 144$

The tank holds 144 cubic feet of water. Let x represent the number of piranhas the tank could hold. Then $1.5x = 144$.

$$1.5x = 144$$
$$\frac{1.5x}{1.5} = \frac{144}{1.5}$$
$$x = 96$$

The tank could hold 96 piranhas.

49. Use $A = \frac{1}{2}h(B+b)$ to find the area of the lawn.

$$A = \frac{1}{2}h(B+b)$$
$$A = \frac{1}{2}(60)(130+70) = 30(200) = 6000$$

Let x be the number of bags of fertilizer.

$$4000x = 6000$$
$$\frac{4000x}{4000} = \frac{6000}{4000}$$
$$x = 1.5$$

Since $\frac{1}{2}$ bag cannot be purchased, 2 bags must be purchased to cover the lawn.

51. Use $A = \pi r^2$ to find the area of a pizza.

For the 16-inch pizza, $r = \frac{16}{2} = 8$.

$$A = \pi r^2 = \pi(8)^2 = 64\pi$$

For a 10-inch pizza, $r = \frac{10}{2} = 5$.

$$A = \pi r^2 = \pi(5)^2 = 25\pi$$

Two 10-inch pizzas have an area of $2 \cdot 25\pi = 50\pi$ square inches. Since $50\pi < 64\pi$, you get more pizza by buying the 16-inch pizza.

53. Use $d = rt$ when $r = 552$ and $d = 42.8$.

$$d = rt$$
$$42.8 = 552t$$
$$\frac{42.8}{552} = \frac{552t}{552}$$
$$0.0775 = t$$

It would last 0.0775 hour or $0.0775(60) \approx 4.65$ minutes.

55. Let s represent the length of one side of the square. Then the perimeter of the square is $4s$. A side of the triangle is $s + 5$ and the triangle's perimeter is $3(s + 5)$.

$$3(s+5) = 4s+7$$
$$3s+15 = 4s+7$$
$$3s+15-3s = 4s+7-3s$$
$$15 = s+7$$
$$15-7 = s+7-7$$
$$8 = s$$

$s + 5 = 8 + 5 = 13$

Each side of the triangle has length 13 inches.

57. Use $d = rt$ when $d = 135$ an $r = 60$.

$$d = rt$$
$$135 = 60t$$
$$\frac{135}{60} = \frac{60t}{60}$$
$$2.25 = t$$

It will take 2.25 hours.

59. Use $A = lw$ when $A = 1{,}813{,}500$ and $w = 150$.

$$A = lw$$
$$1{,}813{,}500 = l(150)$$
$$\frac{1{,}813{,}500}{150} = \frac{150l}{150}$$
$$12{,}090 = l$$

The length of the runway is 12,090 feet (more than 2 miles!).

Copyright © 2012 Pearson Education, Inc. Publishing as Prentice Hall.

61. Use $F = \frac{9}{5}C + 32$ when $F = 122$.

$$122 = \frac{9}{5}C + 32$$
$$122 - 32 = \frac{9}{5}C + 32 - 32$$
$$90 = \frac{9}{5}C$$
$$\frac{5}{9} \cdot 90 = \frac{5}{9} \cdot \frac{9}{5}C$$
$$50 = C$$

Thus, 122°F is equivalent to 50°C.

63. Use $V = lwh$ when $l = 199$, $w = 78.5$, and $h = 33$.
$V = lwh = 199(78.5)(33) = 515{,}509.5$
The smallest possible shipping crate has a volume of 515,509.5 cubic inches.

65. Use $V = \frac{4}{3}\pi r^3$ when $r = \frac{9.5}{2} = 4.75$ and $\pi = 3.14$.

$$V = \frac{4}{3}\pi r^3 = \frac{4}{3}(3.14)(4.75)^3 \approx 449$$

The volume of the sphere is 449 cubic inches.

67. Use $F = \frac{9}{5}C + 32$ when $C = 167$.

$$\begin{aligned} F &= \frac{9}{5}C + 32 \\ &= \frac{9}{5}(167) + 32 \\ &= 300.6 + 32 \\ &= 332.6 \\ &\approx 333 \end{aligned}$$

The average temperature on the planet Mercury is 333°F.

69. 32% = 0.32

71. 200% = 2.00 or 2

73. 0.17 = 0.17(100%) = 17%

75. 7.2 = 7.2(100%) = 720%

77.
$$N = R + \frac{V}{G}$$
$$N - R = R + \frac{V}{G} - R$$
$$N - R = \frac{V}{G}$$
$$G(N - R) = G \cdot \frac{V}{G}$$
$$G(N - R) = V$$

79. Use $V = lwh$. If the length is doubled, the new length is $2l$. If the width and height are doubled, the new width and height are $2w$ and $2h$, respectively.
$V = (2l)(2w)(2h) = 2 \cdot 2 \cdot 2lwh = 8lwh$
The volume of the box is multiplied by 8.

81. Replace T with N and solve for N.

$$T = 50 + \frac{N - 40}{4}$$
$$N = 50 + \frac{N - 40}{4}$$
$$N - 50 = 50 + \frac{N - 40}{4} - 50$$
$$N - 50 = \frac{N - 40}{4}$$
$$4(N - 50) = 4 \cdot \frac{N - 40}{4}$$
$$4N - 200 = N - 40$$
$$4N - 200 - N = N - 40 - N$$
$$3N - 200 = -40$$
$$3N - 200 + 200 = -40 + 200$$
$$3N = 160$$
$$\frac{3N}{3} = \frac{160}{3}$$
$$N = 53\frac{1}{3}$$

They are the same when the number of cricket chirps per minute is $53\frac{1}{3}$.

83.

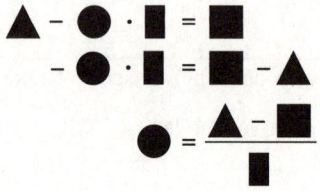

Copyright © 2012 Pearson Education, Inc. Publishing as Prentice Hall.

85. $\frac{20 \text{ miles}}{1 \text{ hour}} \cdot \frac{5280 \text{ feet}}{1 \text{ mile}} \cdot \frac{1 \text{ hour}}{60 \text{ minutes}} \cdot \frac{1 \text{ minute}}{60 \text{ seconds}}$

$= \frac{20 \cdot 5280 \text{ feet}}{60 \cdot 60 \text{ seconds}}$

≈ 29.3 feet/second

Use $d = rt$ when $d = 1300$ and $r = 29.3$.

$d = rt$

$1300 = 29.3t$

$\frac{1300}{29.3} = \frac{29.3t}{29.3}$

$44.3 \approx t$

It took 44.3 seconds to travel that distance.

87. Use $I = PRT$ when $I = 1{,}056{,}000$, $R = 0.055$, and $T = 6$.

$I = PRT$

$1{,}056{,}000 = P(0.055)(6)$

$1{,}056{,}000 = 0.33P$

$\frac{1{,}056{,}000}{0.33} = \frac{0.33P}{0.33}$

$3{,}200{,}000 = P$

89. Use $V = \frac{4}{3}\pi r^3$ when $r = 3$.

$V = \frac{4}{3}\pi \cdot 3^3$

$V \approx 113.1$

Section 2.6 Practice

1. Let x be the unknown percent.

$22 = x \cdot 40$

$22 = 40x$

$\frac{22}{40} = \frac{40x}{40}$

$0.55 = x$

$55\% = x$

The number 22 is 55% of 40.

2. Let x be the unknown number.

$150 = 40\% \cdot x$

$150 = 0.4x$

$\frac{150}{0.4} = \frac{0.4x}{0.4}$

$375 = x$

The number 150 is 40% of 375.

3. a. From the graph, we see 66% are for solely pleasure.

b. From the graph, 66% are for pleasure and 4% are for combined business/pleasure. The sum is 66% + 4% = 70%.

c. Find 66% of 250.
$0.66(250) = 165$
We expect 165 people to be traveling solely for pleasure.

4. discount = percent · original price

$= 40\% \cdot \$400$

$= 0.40 \cdot \$400$

$= \$160$

new price = original price − discount

$= \$400 - \$160$

$= \$240$

The discount in price is \$160 and the new price is \$240.

5. increase = new − old = 200 − 120 = 80

Let x be the percent increase.

$80 = x \cdot 120$

$\frac{80}{120} = \frac{120x}{120}$

$0.667 \approx x$

$66.7\% \approx x$

The percent increase is 66.7%.

6. Let x be the original price.

$x - 0.20x = 46$

$0.8x = 46$

$\frac{0.8x}{0.8} = \frac{46}{0.8}$

$x = 57.5$

The original price is \$57.50.

Copyright © 2012 Pearson Education, Inc. Publishing as Prentice Hall.

7. Let x represent the liters of 20% solution.

	Number of Liters	Dye Strength	Amount
20% solution	x	20%	$0.2x$
50% solution	$6-x$	50%	$0.5(6-x)$
40% solution	6	40%	$0.4(6)$

$$\begin{aligned} 0.2x+0.5(6-x) &= 0.4(6) \\ 0.2x+3-0.5x &= 2.4 \\ -0.3x+3 &= 2.4 \\ -0.3x+3-3 &= 2.4-3 \\ -0.3x &= -0.6 \\ \frac{-0.3x}{-0.3} &= \frac{-0.6}{-0.3} \\ x &= 2 \end{aligned}$$

$6-x=6-2=4$

If 2 liters of 20% solution are mixed with 4 liters of 50% solution, the result is 6 liters of 40% solution.

Vocabulary and Readiness Check

1. no; 25% + 25% + 40% ≠ 100%

2. no; 30% + 30% + 30% ≠ 100%

3. yes; 25% + 25% + 25% + 25% = 100%

4. yes; 40% + 50% + 10% = 100%

Exercise Set 2.6

1. Let x be the unknown number.

$x = 16\% \cdot 70$

$x = 0.16 \cdot 70$

$x = 11.2$

11.2 is 16% of 70.

3. Let x be the unknown percent.

$28.6 = x \cdot 52$

$\frac{28.6}{52} = \frac{52x}{52}$

$0.55 = x$

$55\% = x$

The number 28.6 is 55% of 52.

5. Let x be the unknown number.

$45 = 25\% \cdot x$

$45 = 0.25 \cdot x$

$\frac{45}{0.25} = \frac{0.25x}{0.25}$

$180 = x$

45 is 25% of 180.

Copyright © 2012 Pearson Education, Inc. Publishing as Prentice Hall.

7. From the graph, 4% of adults spend more than 121 minutes on the phone each day.

9. 37% of adults talk 16–60 minutes on the phone each day.
$37\% \cdot 27{,}000 = 0.37 \cdot 27{,}000 = 9990$
You would expect 9990 of the adults in Florence to talk 16–60 minutes each day.

11.
$$\begin{aligned}\text{discount} &= \text{percent} \cdot \text{original price}\\ &= 8\% \cdot \$18{,}500\\ &= 0.08 \cdot \$18{,}500\\ &= \$1480\end{aligned}$$
$$\begin{aligned}\text{new price} &= \text{original price} - \text{discount}\\ &= \$18{,}500 - \$1480\\ &= \$17{,}020\end{aligned}$$
The discount is \$1480 and the new price is \$17,020.

13. $15\% \cdot 40.50 = 0.15 \cdot 40.5 = 6.075$
The tip is \$6.08.
$40.5 + 6.08 = 46.58$
The total cost of the meal is \$46.58.

15.
$$\begin{aligned}\text{percent decrease} &= \frac{\text{amount of decrease}}{\text{original amount}}\\ &= \frac{148{,}600 - 73{,}900}{148{,}600}\\ &= \frac{74{,}700}{148{,}600}\\ &\approx 0.50\end{aligned}$$
The number of complaints decreased by 50%.

17.
$$\begin{aligned}\text{percent decrease} &= \frac{\text{amount of decrease}}{\text{original amount}}\\ &= \frac{40-28}{40}\\ &= \frac{12}{40}\\ &= 0.3\end{aligned}$$
The area decreased by 30%.

19. Let x represent the original price.
$$\begin{aligned}x - 25\% \cdot x &= 78\\ x - 0.25x &= 78\\ 0.75x &= 78\\ \frac{0.75x}{0.75} &= \frac{78}{0.75}\\ x &= 104\end{aligned}$$
The original price of the shoes was \$104.

Copyright © 2012 Pearson Education, Inc. Publishing as Prentice Hall.

21. Let x represent last year's salary.

$$x + 4\% \cdot x = 44,200$$
$$x + 0.04x = 44,200$$
$$1.04x = 44,200$$
$$\frac{1.04x}{1.04} = \frac{44,200}{1.04}$$
$$x = 42,500$$

Last year's salary was \$42,500.

23. Let x represent the number of gallons of pure acid.

	Number of Gallons	· Acid Strength	= Amount of Acid
Pure Acid	x	100%	$1x$
40% Acid Solution	2	40%	$0.4x$
70% Acid Solution Needed	$x + 2$	70%	$0.7(x + 2)$

The amount of acid being combined must be the same as that in the mixture.

$$x + 0.4x = 0.7(x + 2)$$
$$1.4x = 0.7x + 1.4$$
$$1.4x - 0.7x = 0.7x + 1.4 - 0.7x$$
$$0.7x = 1.4$$
$$\frac{0.7x}{0.7} = \frac{1.4}{0.7}$$
$$x = 2$$

Thus, 2 gallons of pure acid should be used.

25. Let x represent the number of pounds of coffee worth \$7 a pound.

	Number of pounds	· Cost per pound	= Value
\$7/lb coffee	x	7	$7x$
\$4/lb coffee	14	4	$4 \cdot 14 = 56$
\$5/lb coffee wanted	$x + 14$	5	$5(x + 14)$

The value of the coffee being combined must be the same as the value of the mixture.

$$7x + 56 = 5(x + 14)$$
$$7x + 56 = 5x + 70$$
$$7x + 56 - 5x = 5x + 70 - 5x$$
$$2x + 56 = 70$$
$$2x + 56 - 56 = 70 - 56$$
$$2x = 14$$
$$\frac{2x}{2} = \frac{14}{2}$$
$$x = 7$$

7 pounds of the \$4 a pound coffee should be used.

27. $23\% \cdot 20 = 0.23 \cdot 20 = 4.6$

Copyright © 2012 Pearson Education, Inc. Publishing as Prentice Hall.

29. Let x represent the unknown number.

$$40 = 80\% \cdot x$$
$$40 = 0.80 \cdot x$$
$$\frac{40}{0.8} = \frac{0.8x}{0.8}$$
$$50 = x$$

40 is 80% of 50.

31. Let x represent the unknown percent.

$$144 = x \cdot 480$$
$$\frac{144}{480} = \frac{480x}{480}$$
$$0.3 = x$$
$$30\% = x$$

144 is 30% of 480.

33. From the graph, it appears that 71% of the population of Fairbanks, Alaska shops by catalog.

35. 65% of 278,700 = 0.65 · 278,700 = 181,155
We predict 181,155 catalog shoppers live in Anchorage.

37.

Top Cranberry-Producing States in 2008 (in millions of pounds)		
	Millions of Pounds	Percent of Total (rounded to nearest percent)
Wisconsin	385	$\frac{385}{689} \approx 0.559 \approx 56\%$
Oregon	50	$\frac{50}{689} \approx 0.073 \approx 7\%$
Massachusetts	190	$\frac{190}{689} \approx 0.276 \approx 28\%$
Washington	15	$\frac{15}{689} \approx 0.022 \approx 2\%$
New Jersey	49	$\frac{49}{689} \approx 7\%$
Total	689	

39.
$$\text{percent increase} = \frac{\text{amount of increase}}{\text{original amount}}$$
$$= \frac{70-40}{40}$$
$$= \frac{30}{40}$$
$$= 0.75$$

The price increased by 75%.

Copyright © 2012 Pearson Education, Inc. Publishing as Prentice Hall.

41. Let x represent the amount Charles paid for the car.

$$\begin{aligned} x + 20\% \cdot x &= 4680 \\ x + 0.20x &= 4680 \\ 1.2x &= 4680 \\ \frac{1.2x}{1.2} &= \frac{4680}{1.2} \\ x &= 3900 \end{aligned}$$

Charles paid \$3900 for the car.

43.

$$\begin{aligned} \text{percent increase} &= \frac{\text{amount of increase}}{\text{original amount}} \\ &= \frac{144-36}{36} \\ &= \frac{108}{36} \\ &= 3 \end{aligned}$$

The area increased by 300%.

45. Markup = $5\% \cdot 2.20 = 0.05 \cdot 2.2 = 0.11$
New price = $2.20 + 0.11 = 2.31$
The markup is \$0.11 and the new price is \$2.31.

47. Let x be the ounces of alloy that is 20% copper.

	ounces	concentration	amount
20% copper	x	20%	$0.2x$
50% copper	200	50%	$0.5(200)$
30% copper	$200 + x$	30%	$0.3(200 + x)$

The amount of copper being combined must be the same as that in the mixture.

$$\begin{aligned} 0.2x + 0.5(200) &= 0.3(200 + x) \\ 0.2x + 100 &= 60 + 0.3x \\ 0.2x + 100 - 0.2x &= 60 + 0.3x - 0.2x \\ 100 &= 60 + 0.1x \\ 100 - 60 &= 60 + 0.1x - 60 \\ 40 &= 0.1x \\ \frac{40}{0.1} &= \frac{0.1x}{0.1} \\ 400 &= x \end{aligned}$$

Thus 400 ounces should be used.

49.

$$\begin{aligned} \text{percent decrease} &= \frac{\text{amount of decrease}}{\text{original amount}} \\ &= \frac{151-72}{151} \\ &= \frac{79}{151} \\ &\approx 0.523 \end{aligned}$$

The number of decisions by the Supreme Court decreased by 52.3%.

Copyright © 2012 Pearson Education, Inc. Publishing as Prentice Hall.

51. Let x be the prior number of employees.

$$x - 0.35x = 78$$
$$0.65x = 78$$
$$\frac{0.65x}{0.65} = \frac{78}{0.65}$$
$$x = 120$$

There were 120 employees prior to the layoffs.

53. decrease = 25% · 256 = 0.25 · 256 = 64
256 – 64 = 192
The price of the coat decreased by \$64. The sale price was \$192.

55. increase = 48% · 577 = 0.48 · 577 = 276.96
577 + 276.96 = 853.96
The Naga Jolokia pepper measures 854 thousand Scoville units.

57. 42% · 860 = 0.42 · 860 = 361.2
You would expect 361 students to rank flexible hours as their top priority.

59. Let x be the ounces of self-tanning lotion.

	ounces	cost (\$)	value
self-tanning	x	3	$3x$
everyday	800	0.30	0.3(800)
experimental	$800 + x$	1.20	$1.2(800 + x)$

The value of those being combined must be the same as the value of the mixture.

$$3x + 0.3(800) = 1.2(800 + x)$$
$$3x + 240 = 960 + 1.2x$$
$$3x + 240 - 1.2x = 960 + 1.2x - 1.2x$$
$$1.8x + 240 = 960$$
$$1.8x + 240 - 240 = 960 - 240$$
$$1.8x = 720$$
$$\frac{1.8x}{1.8} = \frac{720}{1.8}$$
$$x = 400$$

Therefore, 400 ounces of the self-tanning lotion should be used.

61. $-5 > -7$ since -5 is to the right of -7 on a number line.

63. $|-5| = 5$
$-(-5) = 5$
$|-5| = -(-5)$ since $5 = 5$.

65. $(-3)^2 = (-3)(-3) = 9$
$-3^2 = -(3 \cdot 3) = -9$
Since $9 > -9$, $(-3)^2 > -3^2$.

67. no; answers may vary

Copyright © 2012 Pearson Education, Inc. Publishing as Prentice Hall.

69. 230 mg is what percent of 2400 mg?
Let x represent the unknown percent.

$$x \cdot 2400 = 230$$
$$\frac{2400x}{2400} = \frac{230}{2400}$$
$$x = 0.0958\overline{3}$$

This food contains 9.6% of the daily value of sodium in one serving.

71. 35 is what percent of 130? Let x be the unknown percent.

$$35 = x \cdot 130$$
$$\frac{35}{130} = \frac{130x}{130}$$
$$0.269 \approx x$$

The percent calories from fat is 26.9%. Yes, this food satisfies the recommendation since $26.9\% \le 30\%$.

73. 12 g · 4 calories/gram = 48 calories
48 of the 280 calories come from protein.

$$\frac{48}{280} \approx 0.171$$

17.1% of the calories in this food come from protein.

Section 2.7 Practice

1. $x \ge -2$

2. $5 > x$ or $x < 5$

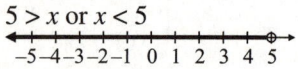

3. $-3 \le x < 1$

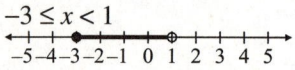

4.
$$x - 6 \ge -11$$
$$x - 6 + 6 \ge -11 + 6$$
$$x \ge -5$$

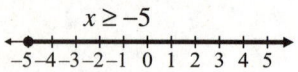

5.
$$-3x \le 12$$
$$\frac{-3x}{-3} \ge \frac{12}{-3}$$
$$x \ge -4$$

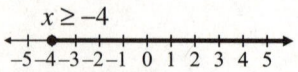

6.
$$5x > -20$$
$$\frac{5x}{5} > \frac{-20}{5}$$
$$x > -4$$

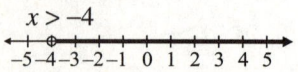

7.
$$-3x + 11 \le -13$$
$$-3x + 11 - 11 \le -13 - 11$$
$$-3x \le -24$$
$$\frac{-3x}{-3} \ge \frac{-24}{-3}$$
$$x \ge 8$$

$\{x | x \ge 8\}$

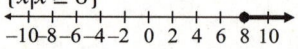

8.
$$2x - 3 > 4(x - 1)$$
$$2x - 3 > 4x - 4$$
$$2x - 3 - 4x > 4x - 4 - 4x$$
$$-2x - 3 > -4$$
$$-2x - 3 + 3 > -4 + 3$$
$$-2x > -1$$
$$\frac{-2x}{-2} < \frac{-1}{-6}$$
$$x < \frac{1}{2}$$

$\left\{x \middle| x < \frac{1}{2}\right\}$

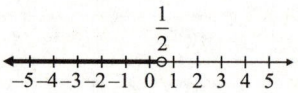

9.
$$3(x + 5) - 1 \ge 5(x - 1) + 7$$
$$3x + 15 - 1 \ge 5x - 5 + 7$$
$$3x + 14 \ge 5x + 2$$
$$3x + 14 - 5x \ge 5x + 2 - 5x$$
$$-2x + 14 \ge 2$$
$$-2x + 14 - 14 \ge 2 - 14$$
$$-2x \ge -12$$
$$\frac{-2x}{-2} \le \frac{-12}{-2}$$
$$x \le 6$$

$\{x | x \le 6\}$

10. Let x be the unknown number.

$$35 - 2x > 15$$
$$35 - 2x - 35 > 15 - 35$$
$$-2x > -20$$
$$\frac{-2x}{-2} < \frac{-20}{-2}$$
$$x < 10$$

All numbers less than 10 make the statement true.

Copyright © 2012 Pearson Education, Inc. Publishing as Prentice Hall.

11. Let x represent the minimum sales.

$$600 + 0.04x \geq 3000$$
$$0.04x \geq 2400$$
$$x \geq 60{,}000$$

Alex must have minimum sales of \$60,000.

Vocabulary and Readiness Check

1. $6x - 7(x + 9)$ is an <u>expression</u>.

2. $6x = 7(x + 9)$ is an <u>equation</u>.

3. $6x < 7(x + 9)$ is an <u>inequality</u>.

4. $5y - 2 \geq -38$ is an <u>inequality</u>.

5. $\frac{9}{7} = \frac{x+2}{14}$ is an <u>equation</u>.

6. $\frac{9}{7} - \frac{x+2}{14}$ is an <u>expression</u>.

7. $x \geq -3$
-5 is not a solution.

8. $x < 6$
$|-6| = 6$ is not a solution.

9. $x < 4.01$
4.1 is not a solution.

10. $x \geq -3$
-4 is not a solution.

Exercise Set 2.7

1.
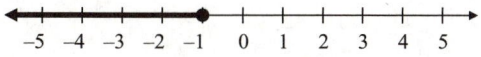

3.
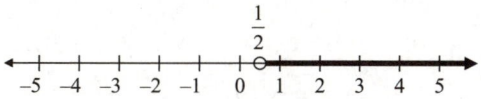

5.
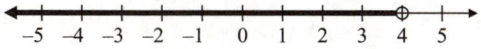

7.
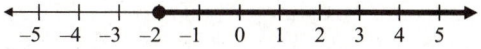

9.
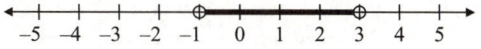

11.

13.
$$x - 2 \geq -7$$
$$x - 2 + 2 \geq -7 + 2$$
$$x \geq -5$$
$\{x|x \geq -5\}$

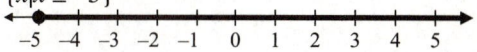

15.
$$-9 + y < 0$$
$$9 - 9 + y < 9 + 0$$
$$y < 9$$
$\{y|y < 9\}$

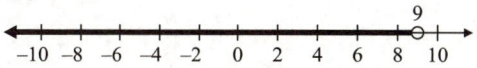

17.
$$3x - 5 > 2x - 8$$
$$3x - 5 - 2x > 2x - 8 - 2x$$
$$x - 5 > -8$$
$$x - 5 + 5 > -8 + 5$$
$$x > -3$$
$\{x|x > -3\}$

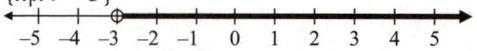

19.
$$4x - 1 \leq 5x - 2x$$
$$4x - 1 \leq 3x$$
$$4x - 1 - 4x \leq 3x - 4x$$
$$-1 \leq -x$$
$$\frac{-1}{-1} \geq \frac{-x}{-1}$$
$$1 \geq x \text{ or } x \leq 1$$
$\{x|x \leq 1\}$

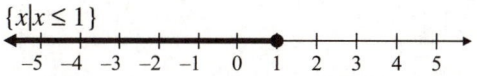

21.
$$2x < -6$$
$$\frac{2x}{2} < \frac{-6}{2}$$
$$x < -3$$
$\{x|x < -3\}$

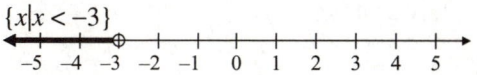

23.
$$-8x \leq 16$$
$$\frac{-8x}{-8} \geq \frac{16}{-8}$$
$$x \geq -2$$
$\{x|x \geq -2\}$

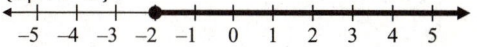

Copyright © 2012 Pearson Education, Inc. Publishing as Prentice Hall.

25.
$$-x > 0$$
$$(-1)(-x) < (-1)(0)$$
$$x < 0$$
$$\{x|x < 0\}$$

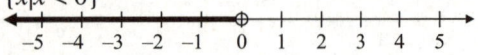

27.
$$\frac{3}{4}y \geq -2$$
$$\frac{4}{3}\cdot\frac{3}{4}y \geq \frac{4}{3}\cdot(-2)$$
$$y \geq -\frac{8}{3}$$
$$\left\{y \middle| y \geq -\frac{8}{3}\right\}$$

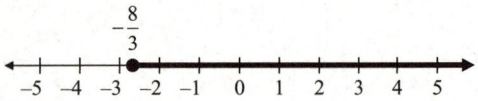

29.
$$-0.6y < -1.8$$
$$\frac{-0.6y}{-0.6} > \frac{-1.8}{-0.6}$$
$$y > 3$$
$$\{y|y > 3\}$$

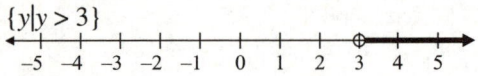

31.
$$-8 < x+7$$
$$-8-7 < x+7-7$$
$$-15 < x$$
$$\{x|x > -15\}$$

33.
$$7(x+1)-6x \geq -4$$
$$7x+7-6x \geq -4$$
$$x+7 \geq -4$$
$$x+7-7 \geq -4-7$$
$$x \geq -11$$
$$\{x|x \geq -11\}$$

35.
$$4x > 1$$
$$\frac{4x}{4} > \frac{1}{4}$$
$$x > \frac{1}{4}$$
$$\left\{x \middle| x > \frac{1}{4}\right\}$$

37.
$$-\frac{2}{3}y \leq 8$$
$$-\frac{3}{2}\left(-\frac{2}{3}y\right) \geq -\frac{3}{2}(8)$$
$$y \geq -12$$
$$\{y|y \geq -12\}$$

39.
$$4(2z+1) < 4$$
$$8z+4 < 4$$
$$8z+4-4 < 4-4$$
$$8z < 0$$
$$\frac{8z}{8} < \frac{0}{8}$$
$$z < 0$$
$$\{z|z < 0\}$$

41.
$$3x-7 < 6x+2$$
$$3x-7-3x < 6x+2-3x$$
$$-7 < 3x+2$$
$$-7-2 < 3x+2-2$$
$$-9 < 3x$$
$$\frac{-9}{3} < \frac{3x}{3}$$
$$-3 < x$$
$$\{x|x > -3\}$$

43.
$$5x-7x \leq x+2$$
$$-2x \leq x+2$$
$$-2x-x \leq x+2-x$$
$$-3x \leq 2$$
$$\frac{-3x}{-3} \geq \frac{2}{-3}$$
$$x \geq -\frac{2}{3}$$
$$\left\{x \middle| x \geq -\frac{2}{3}\right\}$$

45.
$$-6x+2 \geq 2(5-x)$$
$$-6x+2 \geq 10-2x$$
$$-6x+2+6x \geq 10-2x+6x$$
$$2 \geq 10+4x$$
$$2-10 \geq 10+4x-10$$
$$-8 \geq 4x$$
$$\frac{-8}{4} \geq \frac{4x}{4}$$
$$-2 \geq x$$
$$\{x|x \leq -2\}$$

Copyright © 2012 Pearson Education, Inc. Publishing as Prentice Hall.

47.
$$\begin{aligned} 3(x-5) &< 2(2x-1) \\ 3x-15 &< 4x-2 \\ 3x-15-3x &< 4x-2-3x \\ -15 &< x-2 \\ -15+2 &< x-2+2 \\ -13 &< x \end{aligned}$$
$\{x|x > -13\}$

49.
$$\begin{aligned} 4(3x-1) &\le 5(2x-4) \\ 12x-4 &\le 10x-20 \\ 12x-4-10x &\le 10x-20-10x \\ 2x-4 &\le -20 \\ 2x-4+4 &\le -20+4 \\ 2x &\le -16 \\ \frac{2x}{2} &\le \frac{-16}{2} \\ x &\le -8 \end{aligned}$$
$\{x|x \le -8\}$

51.
$$\begin{aligned} 3(x+2)-6 &> -2(x-3)+14 \\ 3x+6-6 &> -2x+6+14 \\ 3x &> -2x+20 \\ 3x+2x &> -2x+20+2x \\ 5x &> 20 \\ \frac{5x}{5} &> \frac{20}{5} \\ x &> 4 \end{aligned}$$
$\{x|x > 4\}$

53.
$$\begin{aligned} -5(1-x)+x &\le -(6-2x)+6 \\ -5+5x+x &\le -6+2x+6 \\ -5+6x &\le 2x \\ -5+6x-6x &\le 2x-6x \\ -5 &\le -4x \\ \frac{-5}{-4} &\ge \frac{-4x}{-4} \\ \frac{5}{4} &\ge x \end{aligned}$$
$\left\{x \middle| x \le \frac{5}{4}\right\}$

55.
$$\begin{aligned} \frac{1}{4}(x+4) &< \frac{1}{5}(2x+3) \\ 20\cdot\frac{1}{4}(x+4) &< 20\cdot\frac{1}{5}(2x+3) \\ 5(x+4) &< 4(2x+3) \\ 5x+20 &< 8x+12 \\ 5x+20-5x &< 8x+12-5x \\ 20 &< 3x+12 \\ 20-12 &< 3x+12-12 \\ 8 &< 3x \\ \frac{8}{3} &< \frac{3x}{3} \\ \frac{8}{3} &< x \end{aligned}$$
$\left\{x \middle| x > \frac{8}{3}\right\}$

57.
$$\begin{aligned} -5x+4 &\le -4(x-1) \\ -5x+4 &\le -4x+4 \\ -5x+4+4x &\le -4x+4+4x \\ -x+4 &\le 4 \\ -x+4-4 &\le 4-4 \\ -x &\le 0 \\ -1(-x) &\ge -1(0) \\ x &\ge 0 \end{aligned}$$
$\{x|x \ge 0\}$

59. Let x be the number.
$$\begin{aligned} 2x+6 &> -14 \\ 2x+6-6 &> -14-6 \\ 2x &> -20 \\ \frac{2x}{2} &> \frac{-20}{2} \\ x &> -10 \end{aligned}$$
All numbers greater than -10 make this statement true.

61. Use $P = 2l + 2w$ when $w = 15$ and $P \le 100$.
$$\begin{aligned} 2l+2(15) &\le 100 \\ 2l+30 &\le 100 \\ 2l+30-30 &\le 100-30 \\ 2l &\le 70 \\ \frac{2l}{2} &\le \frac{70}{2} \\ l &\le 35 \end{aligned}$$
The maximum length of the rectangle is 35 cm.

Copyright © 2012 Pearson Education, Inc. Publishing as Prentice Hall.

63. Let x be the score in his third game.

$$\frac{146+201+x}{3} \geq 180$$
$$\frac{347+x}{3} \geq 180$$
$$3 \cdot \frac{347+x}{3} \geq 3 \cdot 180$$
$$347+x \geq 540$$
$$347+x-347 \geq 540-347$$
$$x \geq 193$$

He must bowl at least 193 on the third game.

65. Let x represent the number of people. Then the cost is $50 + 34x$.

$$50+34x \leq 3000$$
$$50+34x-50 \leq 3000-50$$
$$34x \leq 2950$$
$$\frac{34x}{34} \leq \frac{2950}{34}$$
$$x \leq \frac{2950}{34} \approx 86.76$$

They can invite at most 86 people.

67. Let x represent the number of minutes.

$$5.8x \geq 200$$
$$\frac{5.8x}{5.8} \geq \frac{200}{5.8}$$
$$x \geq \frac{200}{5.8} \approx 35$$

The person must walk at least 35 minutes.

69. $3^4 = 3 \cdot 3 \cdot 3 \cdot 3 = 81$

71. $1^8 = 1 \cdot 1 \cdot 1 \cdot 1 \cdot 1 \cdot 1 \cdot 1 \cdot 1 = 1$

73. $\left(\frac{7}{8}\right)^2 = \left(\frac{7}{8}\right)\left(\frac{7}{8}\right) = \frac{49}{64}$

75. There were about 3200 Starbucks locations in 2002.

77. The greatest increase occurred between 2006 and 2007.

79. The number of Starbucks locations rose above 5000 in 2005.

81. Since $3 < 5$, $3(-4) > 5(-4)$.

83. If $m \leq n$, then $-2m \geq -2n$.

85. Reverse the direction of the inequality symbol when multiplying or dividing by a negative number.

87. Let x be the score on his final exam. Since the final counts as two tests, his final course average is $\frac{75+83+85+2x}{5}$.

$$\frac{75+83+85+2x}{5} \geq 80$$
$$\frac{243+2x}{5} \geq 80$$
$$5\left(\frac{243+2x}{5}\right) \geq 5(80)$$
$$243+2x \geq 400$$
$$243+2x-243 \geq 400-243$$
$$2x \geq 157$$
$$\frac{2x}{2} \geq \frac{157}{2}$$
$$x \geq 78.5$$

His final exam score must be at least 78.5 for him to get a B.

Chapter 2 Vocabulary Check

1. A <u>linear equation in one variable</u> can be written in the form $ax + b = c$.
2. Equations that have the same solution are called <u>equivalent equations</u>.
3. An equation that describes a known relationship among quantities is called a <u>formula</u>.
4. A <u>linear inequality in one variable</u> can be written in the form $ax + b < c$, (or $>$, $\leq$, $\geq$).
5. The solutions to the equation $x + 5 = x + 5$ are <u>all real numbers</u>.
6. The solution to the equation $x + 5 = x + 4$ is <u>no solution</u>.
7. If both sides of an inequality are multiplied or divided by the same positive number, the direction of the inequality symbol is <u>the same</u>.
8. If both sides of an inequality are multiplied by the same negative number, the direction of the inequality symbol is <u>reversed</u>.

Copyright © 2012 Pearson Education, Inc. Publishing as Prentice Hall.

Chapter 2 Review

1.
$$\begin{aligned} 8x+4&=9x \\ 8x+4-8x&=9x-8x \\ 4&=x \end{aligned}$$

2.
$$\begin{aligned} 5y-3&=6y \\ 5y-3-5y&=6y-5y \\ -3&=y \end{aligned}$$

3.
$$\begin{aligned} \frac{2}{7}x+\frac{5}{7}x&=6 \\ \frac{7}{7}x&=6 \\ 1x&=6 \\ x&=6 \end{aligned}$$

4.
$$\begin{aligned} 3x-5&=4x+1 \\ 3x-5-3x&=4x+1-3x \\ -5&=x+1 \\ -5-1&=x+1-1 \\ -6&=x \end{aligned}$$

5.
$$\begin{aligned} 2x-6&=x-6 \\ 2x-6-x&=x-6-x \\ x-6&=-6 \\ x-6+6&=-6+6 \\ x&=0 \end{aligned}$$

6.
$$\begin{aligned} 4(x+3)&=3(1+x) \\ 4x+12&=3+3x \\ 4x+12-3x&=3+3x-3x \\ 12+x&=3 \\ -12+12+x&=-12+3 \\ x&=-9 \end{aligned}$$

7.
$$\begin{aligned} 6(3+n)&=5(n-1) \\ 18+6n&=5n-5 \\ 18+6n-5n&=5n-5-5n \\ 18+n&=-5 \\ -18+18+n&=-18-5 \\ n&=-23 \end{aligned}$$

8.
$$\begin{aligned} 5(2+x)-3(3x+2)&=-5(x-6)+2 \\ 10+5x-9x-6&=-5x+30+2 \\ -4x+4&=-5x+32 \\ 5x-4x+4&=5x-5x+32 \\ x+4&=32 \\ x+4-4&=32-4 \\ x&=28 \end{aligned}$$

9. If the sum is 10 and one number is x, then the other number is $10-x$. The choice is b.

10. Since Mandy is 5 inches taller than Melissa, and x represents Mandy's height, then $x-5$ represents Melissa's height. The choice is a.

11. Complementary angles sum to 90°. The complement of angle x is $90-x$. The choice is b.

12. Supplementary angles sum to 180°. The supplement to $(x+5)°$ is
$180-(x+5)=180-x-5=175-x$.
The choice is c.

13.
$$\begin{aligned} \frac{3}{4}x&=-9 \\ \frac{4}{3}\cdot\frac{3}{4}x&=\frac{4}{3}\cdot(-9) \\ x&=-12 \end{aligned}$$

14.
$$\begin{aligned} \frac{x}{6}&=\frac{2}{3} \\ 6\cdot\frac{x}{6}&=6\cdot\frac{2}{3} \\ x&=4 \end{aligned}$$

15.
$$\begin{aligned} -5x&=0 \\ \frac{-5x}{-5}&=\frac{0}{-5} \\ x&=0 \end{aligned}$$

16.
$$\begin{aligned} -y&=7 \\ \frac{-y}{-1}&=\frac{7}{-1} \\ y&=-7 \end{aligned}$$

17.
$$\begin{aligned} 0.2x&=0.15 \\ 20x&=15 \\ \frac{20x}{20}&=\frac{15}{20} \\ x&=0.75 \end{aligned}$$

18.
$$\begin{aligned} \frac{-x}{3}&=1 \\ -3\left(\frac{-x}{3}\right)&=-3(1) \\ x&=-3 \end{aligned}$$

Copyright © 2012 Pearson Education, Inc. Publishing as Prentice Hall.

19.
$$\begin{aligned} -3x+1&=19\\ -3x+1-1&=19-1\\ -3x&=18\\ \frac{-3x}{-3}&=\frac{18}{-3}\\ x&=-6 \end{aligned}$$

20.
$$\begin{aligned} 5x+25&=20\\ 5x+25-25&=20-25\\ 5x&=-5\\ \frac{5x}{5}&=\frac{-5}{5}\\ x&=-1 \end{aligned}$$

21.
$$\begin{aligned} 7(x-1)+9&=5x\\ 7x-7+9&=5x\\ 7x+2&=5x\\ -7x+7x+2&=-7x+5x\\ 2&=-2x\\ \frac{2}{-2}&=\frac{-2x}{-2}\\ -1&=x \end{aligned}$$

22.
$$\begin{aligned} 7x-6&=5x-3\\ 7x-6-5x&=5x-3-5x\\ 2x-6&=-3\\ 2x-6+6&=-3+6\\ 2x&=3\\ \frac{2x}{2}&=\frac{3}{2}\\ x&=\frac{3}{2}\text{ or }1\frac{1}{2} \end{aligned}$$

23.
$$\begin{aligned} -5x+\frac{3}{7}&=\frac{10}{7}\\ 7\left(-5x+\frac{3}{7}\right)&=7\cdot\frac{10}{7}\\ -35x+3&=10\\ -35x+3-3&=10-3\\ -35x&=7\\ \frac{-35x}{-35}&=\frac{7}{-35}\\ x&=-\frac{1}{5} \end{aligned}$$

24.
$$\begin{aligned} 5x+x&=9+4x-1+6\\ 6x&=4x+14\\ 6x-4x&=4x+14-4x\\ 2x&=14\\ \frac{2x}{2}&=\frac{14}{2}\\ x&=7 \end{aligned}$$

25. Let x be the first integer. Then $x + 1$ and $x + 2$ are the next two consecutive integers. Their sum is $x + x + 1 + x + 2 = 3x + 3$.

26. Let x be the first even integer. Then $x + 2$, $x + 4$, and $x + 6$ are the 2nd, 3rd, and 4th consecutive even integers. The sum of the first and fourth is $x + x + 6 = 2x + 6$.

27.
$$\begin{aligned} \frac{5}{3}x+4&=\frac{2}{3}x\\ 3\left(\frac{5}{3}x+4\right)&=3\left(\frac{2}{3}x\right)\\ 5x+12&=2x\\ 5x+12-5x&=2x-5x\\ 12&=-3x\\ \frac{12}{-3}&=\frac{-3x}{-3}\\ -4&=x \end{aligned}$$

28.
$$\begin{aligned} \frac{7}{8}x+1&=\frac{5}{8}x\\ 8\left(\frac{7}{8}x+1\right)&=8\left(\frac{5}{8}x\right)\\ 7x+8&=5x\\ 7x+8-7x&=5x-7x\\ 8&=-2x\\ \frac{8}{-2}&=\frac{-2x}{-2}\\ -4&=x \end{aligned}$$

29.
$$\begin{aligned} -(5x+1)&=-7x+3\\ -5x-1&=-7x+3\\ -5x-1+7x&=-7x+3+7x\\ 2x-1&=3\\ 2x-1+1&=3+1\\ 2x&=4\\ \frac{2x}{2}&=\frac{4}{2}\\ x&=2 \end{aligned}$$

Copyright © 2012 Pearson Education, Inc. Publishing as Prentice Hall.

30.
$$\begin{aligned} -4(2x+1) &= -5x+5 \\ -8x-4 &= -5x+5 \\ -8x-4+8x &= -5x+5+8x \\ -4 &= 3x+5 \\ -4-5 &= 3x+5-5 \\ -9 &= 3x \\ \frac{-9}{3} &= \frac{3x}{3} \\ -3 &= x \end{aligned}$$

31.
$$\begin{aligned} -6(2x-5) &= -3(9+4x) \\ -12x+30 &= -27-12x \\ 12x-12x+30 &= 12x-27-12x \\ 30 &= -27 \end{aligned}$$
Since the statement 30 = –27 is false, the equation has no solution.

32.
$$\begin{aligned} 3(8y-1) &= 6(5+4y) \\ 24y-3 &= 30+24y \\ 24y-3-24y &= 30+24y-24y \\ -3 &= 30 \end{aligned}$$
Since the statement –3 = 30 is false, the equation has no solution.

33.
$$\begin{aligned} \frac{3(2-z)}{5} &= z \\ 5\left[\frac{3(2-z)}{5}\right] &= 5\cdot z \\ 3(2-z) &= 5z \\ 6-3z &= 5z \\ 6-3z+3z &= 5z+3z \\ 6 &= 8z \\ \frac{6}{8} &= \frac{8z}{8} \\ \frac{3}{4} &= z \end{aligned}$$

34.
$$\begin{aligned} \frac{4(n+2)}{5} &= -n \\ 5\left[\frac{4(n+2)}{5}\right] &= 5(-n) \\ 4(n+2) &= -5n \\ 4n+8 &= -5n \\ 4n+8-4n &= -5n-4n \\ 8 &= -9n \\ \frac{8}{-9} &= \frac{-9n}{-9} \\ -\frac{8}{9} &= n \end{aligned}$$

35.
$$\begin{aligned} 0.5(2n-3)-0.1 &= 0.4(6+2n) \\ 5(2n-3)-1 &= 4(6+2n) \\ 10n-15-1 &= 24+8n \\ 10n-16 &= 24+8n \\ 10n-16-8n &= 24+8n-8n \\ 2n-16 &= 24 \\ 2n-16+16 &= 24+16 \\ 2n &= 40 \\ \frac{2n}{2} &= \frac{40}{2} \\ n &= 20 \end{aligned}$$

36.
$$\begin{aligned} -9-5a &= 3(6a-1) \\ -9-5a &= 18a-3 \\ 9-5a+5a &= 18a-3+5a \\ -9 &= 23a-3 \\ -9+3 &= 23a-3+3 \\ -6 &= 23a \\ \frac{-6}{23} &= \frac{23a}{23} \\ -\frac{6}{23} &= a \end{aligned}$$

37.
$$\begin{aligned} \frac{5(c+1)}{6} &= 2c-3 \\ 6\left[\frac{5(c+1)}{6}\right] &= 6(2c-3) \\ 5(c+1) &= 6(2c-3) \\ 5c+5 &= 12c-18 \\ 5c+5-5c &= 12c-18-5c \\ 5 &= 7c-18 \\ 5+18 &= 7c-18+18 \\ 23 &= 7c \\ \frac{23}{7} &= \frac{7c}{7} \\ \frac{23}{7} &= c \end{aligned}$$

Copyright © 2012 Pearson Education, Inc. Publishing as Prentice Hall.

38.
$$\frac{2(8-a)}{3}=4-4a$$
$$3\left[\frac{2(8-a)}{3}\right]=3(4-4a)$$
$$2(8-a)=3(4-4a)$$
$$16-2a=12-12a$$
$$16-2a+12a=12-12a+12a$$
$$16+10a=12$$
$$16+10a-16=12-16$$
$$10a=-4$$
$$\frac{10a}{10}=\frac{-4}{10}$$
$$a=-\frac{2}{5}$$

39.
$$200(70x-3560)=-179(150x-19,300)$$
$$14,000x-712,000=-26,850x+3,454,700$$
$$14,000x-712,000+26,850x=-26,850x+3,454,700+26,850x$$
$$40,850x-712,000=3,454,700$$
$$40,850x-712,000+712,000=3,454,700+712,000$$
$$40,850x=4,166,700$$
$$\frac{40,850x}{40,850}=\frac{4,166,700}{40,850}$$
$$x=102$$

40.
$$1.72y-0.04y=0.42$$
$$172y-4y=42$$
$$168y=42$$
$$\frac{168y}{168}=\frac{42}{168}$$
$$y=0.25$$

41. Let x be the length of the side of the square base. Then the height is $10x + 50.5$. The sum is 7327.
$$x+10x+50.5=7327$$
$$11x+50.5=7327$$
$$11x+50.5-50.5=7327-50.5$$
$$11x=7276.5$$
$$\frac{11x}{11}=\frac{7276.5}{11}$$
$$x=661.5$$
$$10x+50.5=10(661.5)+50.5$$
$$=6615+50.5$$
$$=6665.5$$
The height is 6665.5 inches.

Copyright © 2012 Pearson Education, Inc. Publishing as Prentice Hall.

42. Let x be the length of the short piece. Then $2x$ is the length of the long piece. The lengths sum to 12.

$$x+2x=12$$
$$3x=12$$
$$\frac{3x}{3}=\frac{12}{3}$$
$$x=4$$

$2x = 2(4) = 8$

The short piece is 4 feet and the long piece is 8 feet.

43. Let x be the number of Cornell library sites. Then $2x + 2$ is the number of Harvard library sites. The total number of library sites is 119.

$$x+2x+2=119$$
$$3x+2=119$$
$$3x+2-2=119-2$$
$$3x=117$$
$$\frac{3x}{3}=\frac{117}{3}$$
$$x=39$$

$2x + 2 = 2(39) + 2 = 78 + 2 = 80$

Cornell has 39 library sites and Harvard has 80 library sites.

44. Let x be the first integer. Then $x + 1$ and $x + 2$ are the next two consecutive integers. Their sum is -114.

$$x+x+1+x+2=-114$$
$$3x+3=-114$$
$$3x+3-3=-114-3$$
$$3x=-117$$
$$\frac{3x}{3}=\frac{-117}{3}$$
$$x=-39$$

$x + 1 = -39 + 1 = -38$

$x + 2 = -39 + 2 = -37$

The integers are -39, -38, and -37.

45.

$$\frac{x}{3}=x-2$$
$$3\cdot\frac{x}{3}=3(x-2)$$
$$x=3x-6$$
$$x-3x=3x-6-3x$$
$$-2x=-6$$
$$\frac{-2x}{-2}=\frac{-6}{-2}$$
$$x=3$$

The number is 3.

46.

$$2(x+6)=-x$$
$$2x+12=-x$$
$$-2x+2x+12=-2x-x$$
$$12=-3x$$
$$\frac{12}{-3}=\frac{-3x}{-3}$$
$$-4=x$$

The number is -4.

47. Use $P = 2l + 2w$ when $P = 46$ and $l = 14$.

$$P=2l+2w$$
$$46=2(14)+2w$$
$$46=28+2w$$
$$46-28=28+2w-28$$
$$18=2w$$
$$\frac{18}{2}=\frac{2w}{2}$$
$$9=w$$

48. Use $V = lwh$ when $V = 192$, $l = 8$, and $w = 6$.

$$V=lwh$$
$$192=8\cdot 6\cdot h$$
$$192=48h$$
$$\frac{192}{48}=\frac{48h}{48}$$
$$4=h$$

49.

$$y=mx+b$$
$$y-b=mx+b-b$$
$$y-b=mx$$
$$\frac{y-b}{x}=\frac{mx}{x}$$
$$\frac{y-b}{x}=m$$

50.

$$r=vst-5$$
$$r+5=vst-5+5$$
$$r+5=vst$$
$$\frac{r+5}{vt}=\frac{vst}{vt}$$
$$\frac{r+5}{vt}=s$$

51.

$$2y-5x=7$$
$$-2y+2y-5x=-2y+7$$
$$-5x=-2y+7$$
$$\frac{-5x}{-5}=\frac{-2y+7}{-5}$$
$$x=\frac{2y-7}{5}$$

Copyright © 2012 Pearson Education, Inc. Publishing as Prentice Hall.

52.
$$\begin{aligned} 3x-6y &= -2 \\ -3x+3x-6y &= -3x-2 \\ -6y &= -3x-2 \\ \frac{-6y}{-6} &= \frac{-3x-2}{-6} \\ y &= \frac{3x+2}{6} \end{aligned}$$

53.
$$\begin{aligned} C &= \pi D \\ \frac{C}{D} &= \frac{\pi D}{D} \\ \frac{C}{D} &= \pi \end{aligned}$$

54.
$$\begin{aligned} C &= 2\pi r \\ \frac{C}{2r} &= \frac{2\pi r}{2r} \\ \frac{C}{2r} &= \pi \end{aligned}$$

55. Use $V = lwh$ when $V = 900$, $l = 20$ and $h = 3$.
$$\begin{aligned} V &= lwh \\ 900 &= 20 \cdot w \cdot 3 \\ 900 &= 60w \\ \frac{900}{60} &= \frac{60w}{60} \\ 15 &= w \end{aligned}$$
The width is 15 meters.

56. Let x be the width. Then the length is $x + 6$. Use $P = 2 \cdot \text{length} + 2 \cdot \text{width}$ when $P = 60$.
$$\begin{aligned} P &= 2 \cdot \text{length} + 2 \cdot \text{width} \\ 60 &= 2(x+6)+2x \\ 60 &= 2x+12+2x \\ 60 &= 4x+12 \\ 60-12 &= 4x+12-12 \\ 48 &= 4x \\ \frac{48}{4} &= \frac{4x}{4} \\ 12 &= x \end{aligned}$$
$x + 6 = 12 + 6 = 18$
The dimensions of the billboard are 12 feet by 18 feet.

57. Use $d = rt$ when $d = 10\text{K}$ or 10,000 m and $r = 125$.
$$\begin{aligned} d &= rt \\ 10{,}000 &= 125t \\ \frac{10{,}000}{125} &= \frac{125t}{125} \\ 80 &= t \end{aligned}$$
The time is 80 minutes or $\frac{80}{60} = 1\frac{1}{3}$ hours or 1 hour and 20 minutes.

58. Use $F = \frac{9}{5}C + 32$ when $F = 104$.
$$\begin{aligned} F &= \frac{9}{5}C+32 \\ 104 &= \frac{9}{5}C+32 \\ 104-32 &= \frac{9}{5}C+32-32 \\ 72 &= \frac{9}{5}C \\ \frac{5}{9}\cdot 72 &= \frac{5}{9}\cdot\frac{9}{5}C \\ 40 &= C \end{aligned}$$
Thus, 104°F is equivalent to 40°C.

59. Let x be the unknown percent.
$$\begin{aligned} 9 &= x \cdot 45 \\ \frac{9}{45} &= \frac{45x}{45} \\ 0.2 &= x \\ 20\% &= x \end{aligned}$$
9 is 20% of 45.

60. Let x be the unknown percent.
$$\begin{aligned} 59.5 &= x \cdot 85 \\ \frac{59.5}{85} &= \frac{85x}{85} \\ 0.7 &= x \\ 70\% &= x \end{aligned}$$
59.5 is 70% of 85.

61. Let x be the unknown number.
$$\begin{aligned} 137.5 &= 125\% \cdot x \\ 137.5 &= 1.25x \\ \frac{137.5}{1.25} &= \frac{1.25x}{1.25} \\ 110 &= x \end{aligned}$$
137.5 is 125% of 110.

62. Let x be the unknown number.
$$\begin{aligned} 768 &= 60\% \cdot x \\ 768 &= 0.6x \\ \frac{768}{0.6} &= \frac{0.6x}{0.6} \\ 1280 &= x \end{aligned}$$
768 is 60% of 1280.

Copyright © 2012 Pearson Education, Inc. Publishing as Prentice Hall.

63. increase = 11% · 1900 = 0.11 · 1900 = 209
new price = 1900 + 209 = 2109
The mark-up is \$209 and the new price is \$2109.

64. Find 66.9% of 76,000.
66.9% · 76,000 = 0.669 · 76,000 = 50,844
You would expect 50,844 to use the Internet.

65. Let x be the number of gallons of 40% solution. Then $30 - x$ is the number of gallons of 10% solution.

	gallons	concentration	amount
40% solution	x	40%	$0.4x$
10% solution	$30 - x$	10%	$0.1(30 - x)$
20% solution	30	20%	$0.2(30)$

The amount of acid in the combined solutions must be the same as in the mixture.

$$\begin{aligned} 0.4x + 0.1(30 - x) &= 0.2(30) \\ 0.4x + 3 - 0.1x &= 6 \\ 3 + 0.3x &= 6 \\ 3 + 0.3x - 3 &= 6 - 3 \\ 0.3x &= 3 \\ \frac{0.3x}{0.3} &= \frac{3}{0.3} \\ x &= 10 \end{aligned}$$

$30 - x = 30 - 10 = 20$
Mix 10 gallons of 40% solution with 20 gallons of 10% solution.

66.
$$\begin{aligned} \text{percent increase} &= \frac{\text{amount of increase}}{\text{original amount}} \\ &= \frac{21.1 - 20.7}{20.7} \\ &= \frac{0.4}{20.7} \\ &\approx 0.0193 \end{aligned}$$

The percent increase is 1.9%.

67. From the graph, 18% of motorists who use a cell phone while driving have almost hit another car.

68. The tallest bar represents the most common effect. Therefore, swerving is the most common effect of cell phone use on driving.

69. 21% of drivers cut off someone. Find 21% of 4600.
21% · 4600 = 0.21 · 4600 = 966
You expect 966 customers to cut someone off while driving and talking on their cell phones.

70. 46% + 41% + 21% + 18% = 126%
No, the percents do not sum to 100%.
Answers may vary.

71.
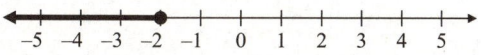

Copyright © 2012 Pearson Education, Inc. Publishing as Prentice Hall.

72. 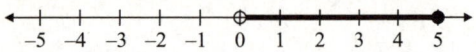

73.
$$x-5\le -4$$
$$x-5+5\le -4+5$$
$$x\le 1$$
$$\{x|x\le 1\}$$

74.
$$x+7>2$$
$$x+7-7>2-7$$
$$x>-5$$
$$\{x|x>-5\}$$

75.
$$-2x\ge -20$$
$$\frac{-2x}{-2}\le\frac{-20}{-2}$$
$$x\le 10$$
$$\{x|x\le 10\}$$

76.
$$-3x>12$$
$$\frac{-3x}{-3}<\frac{12}{-3}$$
$$x<-4$$
$$\{x|x<-4\}$$

77.
$$5x-7>8x+5$$
$$5x-7-8x>8x+5-8x$$
$$-3x-7>5$$
$$-3x-7+7>5+7$$
$$-3x>12$$
$$\frac{-3x}{-3}<\frac{12}{-3}$$
$$x<-4$$
$$\{x|x<-4\}$$

78.
$$x+4\ge 6x-16$$
$$x+4-6x\ge 6x-16-6x$$
$$-5x+4\ge -16$$
$$-5x+4-4\ge -16-4$$
$$-5x\ge -20$$
$$\frac{-5x}{-5}\le\frac{-20}{-5}$$
$$x\le 4$$
$$\{x|x\le 4\}$$

79.
$$\frac{2}{3}y>6$$
$$\frac{3}{2}\cdot\frac{2}{3}y>\frac{3}{2}\cdot 6$$
$$y>9$$
$$\{y|y>9\}$$

80.
$$-0.5y\le 7.5$$
$$\frac{-0.5y}{-0.5}\ge\frac{7.5}{-0.5}$$
$$y\ge -15$$
$$\{y|y\ge -15\}$$

81.
$$-2(x-5)>2(3x-2)$$
$$-2x+10>6x-4$$
$$-2x+10-6x>6x-4-6x$$
$$-8x+10>-4$$
$$-8x+10-10>-4-10$$
$$-8x>-14$$
$$\frac{-8x}{-8}<\frac{-14}{-8}$$
$$x<\frac{7}{4}$$
$$\left\{x\middle|x<\frac{7}{4}\right\}$$

82.
$$4(2x-5)\le 5x-1$$
$$8x-20\le 5x-1$$
$$8x-20-5x\le 5x-1-5x$$
$$3x-20\le -1$$
$$3x-20+20\le -1+20$$
$$3x\le 19$$
$$\frac{3x}{3}\le\frac{19}{3}$$
$$x\le\frac{19}{3}$$
$$\left\{x\middle|x\le\frac{19}{3}\right\}$$

83. Let x be the sales. Her weekly earnings are $175+0.05x$.
$$175+0.05x\ge 300$$
$$175+0.05x-175\ge 300-175$$
$$0.05x\ge 125$$
$$\frac{0.05x}{0.05}\ge\frac{125}{0.05}$$
$$x\ge 2500$$
She must have weekly sales of at least \$2500.

Copyright © 2012 Pearson Education, Inc. Publishing as Prentice Hall.

84. Let x be his score on the fourth round.

$$\frac{76+82+79+x}{4}<80$$
$$\frac{237+x}{4}<80$$
$$4\cdot\frac{237+x}{4}<4\cdot 80$$
$$237+x<320$$
$$237+x-237<320-237$$
$$x<83$$

His score must be less than 83.

85.
$$6x+2x-1=5x+11$$
$$8x-1=5x+11$$
$$8x-1-5x=5x+11-5x$$
$$3x-1=11$$
$$3x-1+1=11+1$$
$$3x=12$$
$$\frac{3x}{3}=\frac{12}{3}$$
$$x=4$$

86.
$$2(3y-4)=6+7y$$
$$6y-8=6+7y$$
$$6y-8-6y=6+7y-6y$$
$$-8=6+y$$
$$-8-6=6+y-6$$
$$-14=y$$

87.
$$4(3-a)-(6a+9)=-12a$$
$$12-4a-6a-9=-12a$$
$$3-10a=-12a$$
$$3-10a+10a=-12a+10a$$
$$3=-2a$$
$$\frac{3}{-2}=\frac{-2a}{-2}$$
$$-\frac{3}{2}=a$$

88.
$$\frac{x}{3}-2=5$$
$$\frac{x}{3}-2+2=5+2$$
$$\frac{x}{3}=7$$
$$3\cdot\frac{x}{3}=3\cdot 7$$
$$x=21$$

89.
$$2(y+5)=2y+10$$
$$2y+10=2y+10$$

Since both sides of the equation are identical, the equation is an identity and every real number is a solution.

90.
$$7x-3x+2=2(2x-1)$$
$$4x+2=4x-2$$
$$4x+2-4x=4x-2-4x$$
$$2=-2$$

Since the statement $2=-2$ is false, there is no solution.

91. Let x be the number.

$$6+2x=x-7$$
$$6+2x-x=x-7-x$$
$$6+x=-7$$
$$6+x-6=-7-6$$
$$x=-13$$

The number is -13.

92. Let x be the length of the shorter piece. Then $4x+3$ is the length of the longer piece. The lengths sum to 23.

$$x+4x+3=23$$
$$5x+3=23$$
$$5x+3-3=23-3$$
$$5x=20$$
$$\frac{5x}{5}=\frac{20}{5}$$
$$x=4$$

$4x+3=4(4)+3=16+3=19$

The shorter piece is 4 inches and the longer piece is 19 inches.

93.
$$V=\frac{1}{3}Ah$$
$$3V=3\cdot\frac{1}{3}Ah$$
$$3V=Ah$$
$$\frac{3V}{A}=\frac{Ah}{A}$$
$$\frac{3V}{A}=h$$

94. Let x be the number.

$x=26\%\cdot 85$

$x=0.26\cdot 85$

$x=22.1$

22.1 is 26% of 85.

Copyright © 2012 Pearson Education, Inc. Publishing as Prentice Hall.

95. Let x be the unknown number.

$$72 = 45\% \cdot x$$
$$72 = 0.45x$$
$$\frac{72}{0.45} = \frac{0.45x}{0.45}$$
$$160 = x$$

72 is 45% of 160.

96. $$\text{percent increase} = \frac{\text{amount of increase}}{\text{original amount}} = \frac{282-235}{235} = \frac{47}{235} = 0.2$$

The percent increase is 20%.

97.
$$4x-7>3x+2$$
$$4x-7-3x>3x+2-3x$$
$$x-7>2$$
$$x-7+7>2+7$$
$$x>9$$

$\{x|x>9\}$

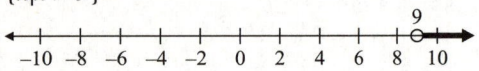

98.
$$-5x<20$$
$$\frac{-5x}{5}>\frac{20}{-5}$$
$$x>-4$$

$\{x|x>-4\}$

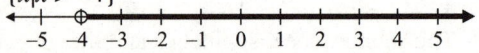

99.
$$-3(1+2x)+x\ge-(3-x)$$
$$-3-6x+x\ge-3+x$$
$$-3-5x\ge-3+x$$
$$-3-5x-x\ge-3+x-x$$
$$-3-6x\ge-3$$
$$-3-6x+3\ge-3+3$$
$$-6x\ge0$$
$$\frac{-6x}{-6}\le\frac{0}{-6}$$
$$x\le0$$

$\{x|x\le0\}$

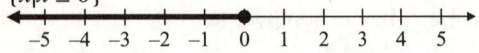

Chapter 2 Test

1.
$$-\frac{4}{5}x=4$$
$$-\frac{5}{4}\left(-\frac{4}{5}x\right)=-\frac{5}{4}(4)$$
$$x=-5$$

2.
$$4(n-5)=-(4-2n)$$
$$4n-20=-4+2n$$
$$4n-20-2n=-4+2n-2n$$
$$2n-20=-4$$
$$2n-20+20=-4+20$$
$$2n=16$$
$$\frac{2n}{2}=\frac{16}{2}$$
$$n=8$$

3.
$$5y-7+y=-(y+3y)$$
$$6y-7=-y-3y$$
$$6y-7=-4y$$
$$6y-7-6y=-4y-6y$$
$$-7=-10y$$
$$\frac{-7}{-10}=\frac{-10y}{-10}$$
$$\frac{7}{10}=y$$

4.
$$4z+1-z=1+z$$
$$3z+1=1+z$$
$$3z+1-z=1+z-z$$
$$2z+1=1$$
$$2z+1-1=1-1$$
$$2z=0$$
$$\frac{2z}{2}=\frac{0}{2}$$
$$z=0$$

5.
$$\frac{2(x+6)}{3}=x-5$$
$$3\left(\frac{2(x+6)}{3}\right)=3(x-5)$$
$$2(x+6)=3(x-5)$$
$$2x+12=3x-15$$
$$2x+12-2x=3x-15-2x$$
$$12=x-15$$
$$12+15=x-15+15$$
$$27=x$$

Copyright © 2012 Pearson Education, Inc. Publishing as Prentice Hall.

6.
$$\frac{4(y-1)}{5}=2y+3$$
$$5\left[\frac{4(y-1)}{5}\right]=5(2y+3)$$
$$4(y-1)=5(2y+3)$$
$$4y-4=10y+15$$
$$4y-4-10y=10y+15-10y$$
$$-6y-4=15$$
$$-6y-4+4=15+4$$
$$-6y=19$$
$$\frac{-6y}{-6}=\frac{19}{-6}$$
$$y=-\frac{19}{6}$$

7.
$$\frac{1}{2}-x+\frac{3}{2}=x-4$$
$$-x+\frac{4}{2}=x-4$$
$$-x+2=x-4$$
$$-x+2+x=x-4+x$$
$$2=2x-4$$
$$2+4=2x-4+4$$
$$6=2x$$
$$\frac{6}{2}=\frac{2x}{2}$$
$$3=x$$

8.
$$\frac{1}{3}(y+3)=4y$$
$$3\cdot\frac{1}{3}(y+3)=3\cdot 4y$$
$$y+3=12y$$
$$y+3-y=12y-y$$
$$3=11y$$
$$\frac{3}{11}=\frac{11y}{11}$$
$$\frac{3}{11}=y$$

9.
$$-0.3(x-4)+x=0.5(3-x)$$
$$-0.3(x-4)+1.0x=0.5(3-x)$$
$$-3(x-4)+10x=5(3-x)$$
$$-3x+12+10x=15-5x$$
$$7x+12=15-5x$$
$$7x+12+5x=15-5x+5x$$
$$12x+12=15$$
$$12x+12-12=15-12$$
$$12x=3$$
$$\frac{12x}{12}=\frac{3}{12}$$
$$x=\frac{1}{4}=0.25$$

10.
$$-4(a+1)-3a=-7(2a-3)$$
$$-4a-4-3a=-14a+21$$
$$-4-7a=-14a+21$$
$$-4-7a+14a=-14a+21+14a$$
$$-4+7a=21$$
$$-4+7a+4=21+4$$
$$7a=25$$
$$\frac{7a}{7}=\frac{25}{7}$$
$$a=\frac{25}{7}$$

11.
$$-2(x-3)=x+5-3x$$
$$-2x+6=-2x+5$$
$$2x-2x+6=2x-2x+5$$
$$6=5$$
Since the statement 6 = 5 is false, there is no solution.

12. Let x be the number.
$$x+\frac{2}{3}x=35$$
$$\frac{3}{3}x+\frac{2}{3}x=35$$
$$\frac{5}{3}x=35$$
$$\frac{3}{5}\cdot\frac{5}{3}x=\frac{3}{5}\cdot 35$$
$$x=21$$
The number is 21.

13. $A = lw = (35)(20) = 700$
The area of the deck is 700 square feet. To paint two coats of water seal means covering $2 \cdot 700 = 1400$ square feet.

Copyright © 2012 Pearson Education, Inc. Publishing as Prentice Hall.

$$1400 \text{ sq ft} \cdot \frac{1 \text{ gal}}{200 \text{ sq ft}} = 7 \text{ gal}$$

7 gallons of water seal are needed.

14. Use $y = mx + b$ when $y = -14$, $m = -2$, and $b = -2$.

$$\begin{aligned} y &= mx + b \\ -14 &= -2x + (-2) \\ -14 + 2 &= -2x + (-2) + 2 \\ -12 &= -2x \\ \frac{-12}{-2} &= \frac{-2x}{-2} \\ 6 &= x \end{aligned}$$

15.

$$\begin{aligned} V &= \pi r^2 h \\ \frac{V}{\pi r^2} &= \frac{\pi r^2 h}{\pi r^2} \\ \frac{V}{\pi r^2} &= h \end{aligned}$$

16.

$$\begin{aligned} 3x - 4y &= 10 \\ 3x - 4y - 3x &= 10 - 3x \\ -4y &= 10 - 3x \\ \frac{-4y}{-4} &= \frac{10 - 3x}{-4} \\ y &= \frac{3x - 10}{4} \end{aligned}$$

17.

$$\begin{aligned} 3x - 5 &\geq 7x + 3 \\ 3x - 5 - 3x &\geq 7x + 3 - 3x \\ -5 &\geq 4x + 3 \\ -5 - 3 &\geq 4x + 3 - 3 \\ -8 &\geq 4x \\ \frac{-8}{4} &\geq \frac{4x}{4} \\ -2 &\geq x \end{aligned}$$

$\{x | x \leq -2\}$

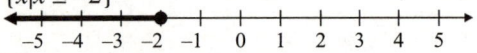

18.

$$\begin{aligned} x + 6 &> 4x - 6 \\ x + 6 - 4x &> 4x - 6 - 4x \\ -3x + 6 &> -6 \\ -3x + 6 - 6 &> -6 - 6 \\ -3x &> -12 \\ \frac{-3x}{-3} &< \frac{-12}{-3} \\ x &< 4 \end{aligned}$$

$\{x | x < 4\}$

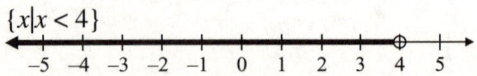

19.

$$\begin{aligned} -0.3x &\geq 2.4 \\ \frac{-0.3x}{-0.3} &\leq \frac{2.4}{-0.3} \\ x &\leq -8 \end{aligned}$$

$\{x | x \leq -8\}$

20.

$$\begin{aligned} -5(x-1) + 6 &\leq -3(x+4) + 1 \\ -5x + 5 + 6 &\leq -3x - 12 + 1 \\ -5x + 11 &\leq -3x - 11 \\ -5x + 11 + 3x &\leq -3x - 11 + 3x \\ -2x + 11 &\leq -11 \\ -2x + 11 - 11 &\leq -11 - 11 \\ -2x &\leq -22 \\ \frac{-2x}{-2} &\geq \frac{-22}{-2} \\ x &\geq 11 \end{aligned}$$

$\{x | x \geq 11\}$

21.

$$\begin{aligned} \frac{2(5x+1)}{3} &> 2 \\ 3 \cdot \frac{2(5x+1)}{3} &> 3(2) \\ 2(5x+1) &> 6 \\ 10x + 2 &> 6 \\ 10x + 2 - 2 &> 6 - 2 \\ 10x &> 4 \\ \frac{10x}{10} &> \frac{4}{10} \\ x &> \frac{2}{5} \end{aligned}$$

$\left\{x \middle| x > \frac{2}{5}\right\}$

22. From the graph, 69% are classified as weak. Find 69% of 800.

$69\% \cdot 800 = 0.69 \cdot 800 = 552$

You would expect 552 of the 800 to be classified as weak.

23. Let x be the unknown percent.

$$\begin{aligned} 72 &= x \cdot 180 \\ \frac{72}{180} &= \frac{180x}{180} \\ 0.4 &= x \end{aligned}$$

72 is 40% of 180.

24. Let x = area code 1, then $2x$ = area code 2.

$$\begin{aligned} x + 2x &= 1203 \\ 3x &= 1203 \\ \frac{3x}{3} &= \frac{1203}{3} \\ x &= 401 \end{aligned}$$

$2x = 2(401) = 802$

The area codes are 401 and 802.

Copyright © 2012 Pearson Education, Inc. Publishing as Prentice Hall.

25. Let x represent the number of public libraries in Georgia. Then there are $x + 696$ public libraries in New York.

$$\begin{aligned} x+x+696 &= 812 \\ 2x+696 &= 812 \\ 2x+696-696 &= 812-696 \\ 2x &= 116 \\ \frac{2x}{2} &= \frac{116}{2} \\ x &= 58 \end{aligned}$$

$x + 696 = 58 + 696 = 754$

Georgia has 58 public libraries and New York has 754.

Cumulative Review Chapters 1–2

1. Since $8 = 8$, the statement $8 \geq 8$ is true.

2. Since -4 is to the right of -6 on the number line, the statement $-4 < -6$ is false.

3. Since $8 = 8$, the statement $8 \leq 8$ is true.

4. Since 3 is to the right of -3 on the number line, the statement $3 > -3$ is true.

5. Since neither $23 < 0$ nor $23 = 0$ is true, the statement $23 \leq 0$ is false.

6. Since $-8 = -8$, the statement $-8 \geq -8$ is true.

7. Since $0 < 23$ is true, the statement $0 \leq 23$ is true.

8. Since $-8 = -8$, the statement $-8 \leq -8$ is true.

9. a. $|0| < 2$ since $|0| = 0$ and $0 < 2$.

b. $|-5| = 5$

c. $|-3| > |-2|$ since $3 > 2$.

d. $|-9| < |-9.7|$ since $9 < 9.7$.

e. $\left|-7\frac{1}{6}\right| > |7|$ since $7\frac{1}{6} > 7$.

10. a. $|5| = 5$ since 5 is 5 units from 0 on the number line.

b. $|-8| = 8$ since -8 is 8 units from 0 on the number line.

c. $\left|-\frac{2}{3}\right| = \frac{2}{3}$ since $-\frac{2}{3}$ is $\frac{2}{3}$ unit from 0 on the number line.

11.
$$\begin{aligned} \frac{3+|4-3|+2^2}{6-3} &= \frac{3+|1|+2^2}{6-3} \\ &= \frac{3+1+2^2}{3} \\ &= \frac{3+1+4}{3} \\ &= \frac{8}{3} \end{aligned}$$

12.
$$\begin{aligned} 1+2(9-7)^3+4^2 &= 1+2(2)^3+4^2 \\ &= 1+2(8)+16 \\ &= 1+16+16 \\ &= 33 \end{aligned}$$

13. $(-8) + (-11) = -19$

14. $-2 + (-8) = -10$

15. $(-2) + 10 = 8$

16. $-10 + 20 = 10$

17. $0.2 + (-0.5) = -0.3$

18. $1.2 + (-1.2) = 0$

19. a.
$$\begin{aligned} -3+[(-2-5)-2] &= -3+[(-2+(-5))-2] \\ &= -3+[(-7)-2] \\ &= -3+[-7+(-2)] \\ &= -3+[-9] \\ &= -12 \end{aligned}$$

b.
$$\begin{aligned} 2^3-10+[-6-(-5)] &= 2^3-10+[-6+5] \\ &= 2^3-10+[-1] \\ &= 8-10+(-1) \\ &= 8+(-10)+(-1) \\ &= -2+(-1) \\ &= -3 \end{aligned}$$

20. a. $-(-5) = 5$

b. $-\left(-\frac{2}{3}\right) = \frac{2}{3}$

c. $-(-a) = a$

d. $-|-3| = -3$

Copyright © 2012 Pearson Education, Inc. Publishing as Prentice Hall.

21. a. $7(0)(-6) = 0(-6) = 0$

b. $(-2)(-3)(-4) = (6)(-4) = -24$

c. $(-1)(-5)(-9)(-2) = 5(-9)(-2)$
$= (-45)(-2)$
$= 90$

22. a. $-2.7 - 8.4 = -2.7 + (-8.4) = -11.1$

b. $-\frac{4}{5} - \left(-\frac{3}{5}\right) = -\frac{4}{5} + \frac{3}{5} = -\frac{1}{5}$

c. $\frac{1}{4} - \left(-\frac{1}{2}\right) = \frac{1}{4} + \frac{1}{2} = \frac{1}{4} + \frac{2}{4} = \frac{3}{4}$

23. a. $-18 \div 3 = -18 \cdot \frac{1}{3} = -6$

b. $\frac{-14}{-2} = -14 \cdot -\frac{1}{2} = 7$

c. $\frac{20}{-4} = 20 \cdot -\frac{1}{4} = -5$

24. a. $(4.5)(-0.08) = -0.36$

b. $-\frac{3}{4} \cdot -\frac{8}{17} = \frac{3 \cdot 8}{4 \cdot 17} = \frac{6}{17}$

25. $-5(-3+2z) = -5(-3) + (-5)(2z) = 15 - 10z$

26. $2(x^2 - 3x + 4) = 2(x^2) - 2(3x) + 2(4)$
$= 2x^2 - 6x + 8$

27. $\frac{1}{2}(6x+14) + 10 = \frac{1}{2}(6x) + \frac{1}{2}(14) + 10$
$= 3x + 7 + 10$
$= 3x + 17$

28. $-(x+4) + 3(x+4) = -1(x+4) + 3(x+4)$
$= -1 \cdot x + (-1)(4) + 3 \cdot x + 3 \cdot 4$
$= -x - 4 + 3x + 12$
$= -x + 3x - 4 + 12$
$= 2x + 8$

29. a. $2x$ and $3x^2$ are unlike terms, since the exponents on x are not the same.

b. $4x^2y$, xy^2, and $-2x^2y$ are like terms, since each variable and its exponent match.

c. $-2yz$ and $-3zy$ are like terms, since $zy = yz$ by the commutative property.

d. $-x^4$ and x^4 are like terms. The variable and its exponent match.

e. $-8a^5$ and $8a^5$ are like terms. The variable and its exponent match.

30. a. $\frac{-32}{8} = -4$

b. $\frac{-108}{-12} = 9$

c. $\frac{-5}{7} \div \left(\frac{-9}{2}\right) = \frac{-5}{7} \cdot \left(\frac{2}{-9}\right) = \frac{5 \cdot 2}{7 \cdot 9} = \frac{10}{63}$

31. $(2x - 3) - (4x - 2) = 2x - 3 - 4x + 2 = -2x - 1$

32. $(-5x+1) - (10x+3) = -5x + 1 - 10x - 3$
$= -15x - 2$

33.
$$\begin{aligned} x - 7 &= 10 \\ x - 7 + 7 &= 10 + 7 \\ x &= 17 \end{aligned}$$

34.
$$\begin{aligned} \frac{5}{6} + x &= \frac{2}{3} \\ \frac{5}{6} + x - \frac{5}{6} &= \frac{2}{3} - \frac{5}{6} \\ x &= \frac{4}{6} - \frac{5}{6} \\ x &= -\frac{1}{6} \end{aligned}$$

35.
$$\begin{aligned} -z - 4 &= 6 \\ -z - 4 + 4 &= 6 + 4 \\ -z &= 10 \\ \frac{-z}{-1} &= \frac{10}{-1} \\ z &= -10 \end{aligned}$$

36.
$$\begin{aligned} -3x + 1 - (-4x - 6) &= 10 \\ -3x + 1 + 4x + 6 &= 10 \\ x + 7 &= 10 \\ x + 7 - 7 &= 10 - 7 \\ x &= 3 \end{aligned}$$

Copyright © 2012 Pearson Education, Inc. Publishing as Prentice Hall.

37.
$$\frac{2(a+3)}{3}=6a+2$$
$$3\cdot\frac{2(a+3)}{3}=3(6a+2)$$
$$2(a+3)=3(6a+2)$$
$$2a+6=18a+6$$
$$2a+6-18a=18a+6-18a$$
$$-16a+6=6$$
$$-16a+6-6=6-6$$
$$-16a=0$$
$$\frac{-16a}{-16}=\frac{0}{-16}$$
$$a=0$$

38.
$$\frac{x}{4}=18$$
$$4\cdot\frac{x}{4}=4\cdot 18$$
$$x=72$$

39. Let x be the number of Republicans. Then $x + 78$ is the number of Democrats. The total number is 434.
$$x+x+78=434$$
$$2x+78=434$$
$$2x+78-78=434-78$$
$$2x=356$$
$$\frac{2x}{2}=\frac{356}{2}$$
$$x=178$$
$x + 78 = 128 + 78 = 256$
There were 256 Democrats and 178 Republicans.

40.
$$6x+5=4(x+4)-1$$
$$6x+5=4x+16-1$$
$$6x+5=4x+15$$
$$6x+5-4x=4x+15-4x$$
$$2x+5=15$$
$$2x+5-5=15-5$$
$$2x=10$$
$$\frac{2x}{2}=\frac{10}{2}$$
$$x=5$$

41. Use $d = rt$ when $d = 31{,}680$ and $r = 400$.
$$d=rt$$
$$31{,}680=400t$$
$$\frac{31{,}680}{400}=\frac{400t}{400}$$
$$79.2=t$$
It will take the ice 79.2 years to reach the lake.

42.
$$x+4=3x-8$$
$$x+4-3x=3x-8-3x$$
$$-2x+4=-8$$
$$-2x+4-4=-8-4$$
$$-2x=-12$$
$$\frac{-2x}{-2}=\frac{-12}{-2}$$
$$x=6$$
The number is 6.

43. Let x be the unknown percent.
$$63=x\cdot 72$$
$$\frac{63}{72}=\frac{72x}{72}$$
$$0.875=x$$
$$87.5\%=x$$
63 is 87.5% of 72.

44.
$$C=2\pi r$$
$$\frac{C}{2\pi}=\frac{2\pi r}{2\pi}$$
$$\frac{C}{2\pi}=r \text{ or } r=\frac{C}{2\pi}$$

45.
$$5(2x+3)=-1+7$$
$$5(2x)+5(3)=-1+7$$
$$10x+15=6$$
$$10x+15-15=6-15$$
$$10x=-9$$
$$\frac{10x}{10}=\frac{-9}{10}$$
$$x=-\frac{9}{10}$$

46.
$$x-3>2$$
$$x-3+3>2+3$$
$$x>5$$
$\{x|x>5\}$

47. $-1 > x$ or $x < -1$

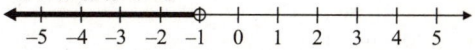

48.
$$3x-4\le 2x-14$$
$$3x-4-2x\le 2x-14-2x$$
$$x-4\le -14$$
$$x-4+4\le -14+4$$
$$x\le -10$$
$\{x|x\le -10\}$

Copyright © 2012 Pearson Education, Inc. Publishing as Prentice Hall.

49. $2(x-3)-5 \le 3(x+2)-18$

$2x-6-5 \le 3x+6-18$

$2x-11 \le 3x-12$

$-x-11 \le -12$

$-x \le -1$

$\frac{-x}{-1} \ge \frac{-1}{-1}$

$x \ge 1$

$\{x|x \ge 1\}$

50. $-3x \ge 9$

$\frac{-3x}{-3} \le \frac{9}{-3}$

$x \le -3$

$\{x|x \le -3\}$

Copyright © 2012 Pearson Education, Inc. Publishing as Prentice Hall.

Chapter 3

Section 3.1 Practice

1. $3^4 = 3 \cdot 3 \cdot 3 \cdot 3 = 81$

2. $7^1 = 7$

3. $(-2)^3 = (-2)(-2)(-2) = -8$

4. $-2^3 = -(2 \cdot 2 \cdot 2) = -8$

5. $\left(\frac{2}{3}\right)^2 = \frac{2}{3} \cdot \frac{2}{3} = \frac{4}{9}$

6. $5 \cdot 6^2 = 5 \cdot 36 = 180$

7. a. When x is 4, $3x^2 = 3 \cdot 4^2$
$= 3 \cdot (4 \cdot 4)$
$= 3 \cdot 16$
$= 48.$

b. When x is -2, $\frac{x^4}{-8} = \frac{(-2)^4}{-8}$
$= \frac{(-2)(-2)(-2)(-2)}{-8}$
$= \frac{16}{-8}$
$= -2.$

8. $7^3 \cdot 7^2 = 7^{3+2} = 7^5$

9. $x^4 \cdot x^9 = x^{4+9} = x^{13}$

10. $r^5 \cdot r = r^{5+1} = r^6$

11. $s^6 \cdot s^2 \cdot s^3 = s^{6+2+3} = s^{11}$

12. $(-3)^9 \cdot (-3) = (-3)^{9+1} = (-3)^{10}$

13. $(6x^3)(-2x^9) = (6 \cdot x^3) \cdot (-2 \cdot x^9)$
$= (6 \cdot -2) \cdot (x^3 \cdot x^9)$
$= -12x^{12}$

14. $(m^5 n^{10})(mn^8) = (m^5 \cdot m) \cdot (n^{10} \cdot n^8)$
$= m^6 \cdot n^{18}$
$= m^6 n^{18}$

15. $(-x^9 y)(4x^2 y^{11}) = (-1 \cdot 4) \cdot (x^9 \cdot x^2) \cdot (y \cdot y^{11})$
$= -4x^{11} y^{12}$

16. $(9^4)^{10} = 9^{4 \cdot 10} = 9^{40}$

17. $(z^6)^3 = z^{6 \cdot 3} = z^{18}$

18. $(xy)^7 = x^7 \cdot y^7 = x^7 y^7$

19. $(3y)^4 = 3^4 \cdot y^4 = 81y^4$

20. $(-2p^4 q^2 r)^3 = (-2)^3 \cdot (p^4)^3 \cdot (q^2)^3 \cdot (r^1)^3$
$= -8p^{12} q^6 r^3$

21. $(-a^4 b)^7 = (-1a^4 b)^7$
$= (-1)^7 \cdot (a^4)^7 \cdot (b^1)^7$
$= -1a^{28} b^7$
$= -a^{28} b^7$

22. $\left(\frac{r}{s}\right)^6 = \frac{r^6}{s^6}, \ s \neq 0$

23. $\left(\frac{5x^6}{9y^3}\right)^2 = \frac{5^2 \cdot (x^6)^2}{9^2 \cdot (y^3)^2} = \frac{25x^{12}}{81y^6}, \ y \neq 0$

24. $\frac{y^7}{y^3} = y^{7-3} = y^4$

25. $\frac{5^9}{5^6} = 5^{9-6} = 5^3 = 125$

26. $\frac{(-2)^{14}}{(-2)^{10}} = (-2)^{14-10} = (-2)^4 = 16$

27. $\frac{7a^4 b^{11}}{ab} = 7 \cdot \frac{a^4}{a^1} \cdot \frac{b^{11}}{b^1}$
$= 7 \cdot (a^{4-1}) \cdot (b^{11-1})$
$= 7a^3 b^{10}$

28. $8^0 = 1$

Copyright © 2012 Pearson Education, Inc. Publishing as Prentice Hall.

29. $(2r^2s)^0 = 1$

30. $(-7)^0 = 1$

31. $-7^0 = -1 \cdot 7^0 = -1 \cdot 1 = -1$

32. $7y^0 = 7 \cdot y^0 = 7 \cdot 1 = 7$

33. a. $\dfrac{x^7}{x^4} = x^{7-4} = x^3$

b. $(3y^4)^4 = 3^4 \cdot (y^4)^4 = 81y^{16}$

c. $\left(\dfrac{x}{4}\right)^3 = \dfrac{x^3}{4^3} = \dfrac{x^3}{64}$

34. a. $2^3 - 2^0 = 8 - 1 = 7$

b. $(y^0)^7 + (5^0)^3 = 1^7 + 1^3 = 1 + 1 = 2$

c. $\left(\dfrac{7x^9}{14y^6}\right) = \dfrac{7^2(x^9)^2}{14^2(y^6)^2} = \dfrac{49 \cdot x^{18}}{196 \cdot y^{12}} = \dfrac{x^{18}}{4y^{12}}$

d.
$$\begin{aligned}\frac{(3a^2b^5)^3}{-27a^6b^4} &= \frac{3^3(a^2)^3(b^5)^3}{-27a^6b^4}\\ &= \frac{27a^6b^{15}}{-27a^6b^4}\\ &= -1 \cdot (a^{6-6}) \cdot (b^{15-4})\\ &= -1 \cdot a^0 \cdot b^{11}\\ &= -1 \cdot 1 \cdot b^{11}\\ &= -b^{11}\end{aligned}$$

Vocabulary and Readiness Check

1. Repeated multiplication of the same factor can be written using an <u>exponent</u>.

2. In 5^2, the 2 is called the <u>exponent</u> and the 5 is called the <u>base</u>.

3. To simplify $x^2 \cdot x^7$, keep the base and <u>add</u> the exponents.

4. To simplify $(x^3)^6$, keep the base and <u>multiply</u> the exponents.

5. The understood exponent on the term y is <u>1</u>.

6. If $x^{\square} = 1$, the exponent is <u>0</u>.

7. For the exponent 2, the base is 3.

8. For the exponent 6, the base is -3.

9. For the exponent 2, the base is 4.

10. For the exponent 4, the base is 3.

11. For the exponent 2, the base is x.

12. For the exponent 2, the base is $5x$.

Exercise Set 3.1

1. $7^2 = 7 \cdot 7 = 49$

3. $(-5)^1 = -5$

5. $-2^4 = -(2 \cdot 2 \cdot 2 \cdot 2) = -16$

7. $(-2)^4 = (-2)(-2)(-2)(-2) = 16$

9. $\left(\dfrac{1}{3}\right)^3 = \dfrac{1}{3} \cdot \dfrac{1}{3} \cdot \dfrac{1}{3} = \dfrac{1}{27}$

11. $7 \cdot 2^4 = 7(2 \cdot 2 \cdot 2 \cdot 2) = 7 \cdot 16 = 112$

13. When x is -2, $x^2 = (-2)^2 = (-2)(-2) = 4$.

15. When x is 3,
$5x^3 = 5(3)^3 = 5(3)(3)(3) = 5(27) = 135$.

17. When $x = 3$ and $y = -5$,
$$\begin{aligned}2xy^2 &= 2(3)(-5)^2\\ &= 2(3)(-5)(-5)\\ &= 2(3)(25)\\ &= 150.\end{aligned}$$

Copyright © 2012 Pearson Education, Inc. Publishing as Prentice Hall.

19. When z is -2,

$$\begin{aligned}\frac{2z^4}{5} &= \frac{2(-2)^4}{5}\\ &= \frac{2(-2)(-2)(-2)(-2)}{5}\\ &= \frac{2(16)}{5}\\ &= \frac{32}{5}.\end{aligned}$$

21. $x^2 \cdot x^5 = x^{2+5} = x^7$

23. $(-3)^3 \cdot (-3)^9 = (-3)^{3+9} = (-3)^{12}$

25. $(5y^4)(3y) = (5 \cdot 3)(y^4 \cdot y) = 15y^{4+1} = 15y^5$

27. $$\begin{aligned}(x^9y)(x^{10}y^5) &= (x^9 \cdot x^{10})(y \cdot y^5)\\ &= (x^{9+10})(y^{1+5})\\ &= x^{19}y^6\end{aligned}$$

29. $$\begin{aligned}(-8mn^6)(9m^2n^2) &= (-8 \cdot 9)(m \cdot m^2)(n^6 \cdot n^2)\\ &= -72m^{1+2}n^{6+2}\\ &= -72m^3n^8\end{aligned}$$

31. $$\begin{aligned}(4z^{10})(-6z^7)(z^3) &= (4 \cdot -6 \cdot 1)(z^{10} \cdot z^7 \cdot z^3)\\ &= -24z^{10+7+3}\\ &= -24z^{20}\end{aligned}$$

33. $$\begin{aligned}\text{Area} &= (\text{length})(\text{width})\\ &= (5x^3 \text{ feet})(4x^2 \text{ feet})\\ &= (5 \cdot 4)(x^3 \cdot x^2) \text{ square feet}\\ &= 20x^{3+2} \text{ square feet}\\ &= 20x^5 \text{ square feet}\end{aligned}$$

35. $(x^9)^4 = x^{9 \cdot 4} = x^{36}$

37. $(pq)^8 = p^8 \cdot q^8 = p^8q^8$

39. $(2a^5)^3 = (2)^3(a^5)^3 = 2^3a^{15} = 8a^{15}$

41. $(x^2y^3)^5 = (x^2)^5(y^3)^5 = x^{10}y^{15}$

43. $$\begin{aligned}(-7a^2b^5c)^2 &= (-7)^2(a^2)^2(b^5)^2(c)^2\\ &= 49a^4b^{10}c^2\end{aligned}$$

45. $\left(\frac{r}{s}\right)^9 = \frac{r^9}{s^9}$

47. $\left(\frac{mp}{n}\right)^9 = \frac{m^9p^9}{n^9}$

49. $$\begin{aligned}\left(\frac{-2xz}{y^5}\right)^2 &= \frac{(-2)^2(x)^2(z)^2}{(y^5)^2}\\ &= \frac{4x^2z^2}{y^{5 \cdot 2}}\\ &= \frac{4x^2z^2}{y^{10}}\end{aligned}$$

51. $$\begin{aligned}\text{Area} &= (\text{length})(\text{length})\\ &= (8z^5 \text{ decimeters})(8z^5 \text{ decimeters})\\ &= (8 \cdot 8)(z^5 \cdot z^5) \text{ square decimeters}\\ &= 64z^{5+5} \text{ square decimeters}\\ &= 64z^{10} \text{ square decimeters}\end{aligned}$$

53. $$\begin{aligned}\text{Volume} &= (\text{length})(\text{width})(\text{height})\\ &= (3y^4 \text{ feet})(3y^4 \text{ feet})(3y^4 \text{ feet})\\ &= (3)^3(y^4)^3 \text{ cubic feet}\\ &= 27y^{4 \cdot 3} \text{ cubic feet}\\ &= 27y^{12} \text{ cubic feet}\end{aligned}$$

55. $\frac{x^3}{x} = x^{3-1} = x^2$

57. $\frac{(-4)^6}{(-4)^3} = (-4)^{6-3} = (-4)^3 = -64$

59. $\frac{p^7q^{20}}{pq^{15}} = \frac{p^7}{p} \cdot \frac{q^{20}}{q^{15}} = p^{7-1} \cdot q^{20-15} = p^6q^5$

61. $$\begin{aligned}\frac{7x^2y^6}{14x^2y^3} &= \frac{7}{14} \cdot \frac{x^2}{x^2} \cdot \frac{y^6}{y^3}\\ &= \frac{1}{2} \cdot x^{2-2} \cdot y^{6-3}\\ &= \frac{1}{2}x^0y^3\\ &= \frac{y^3}{2}\end{aligned}$$

63. $7^0 = 1$

Copyright © 2012 Pearson Education, Inc. Publishing as Prentice Hall.

65. $(2x)^0 = 1$

67. $-7x^0 = -7 \cdot x^0 = -7 \cdot 1 = -7$

69. $5^0 + y^0 = 1 + 1 = 2$

71. $-9^2 = -(9)^2 = -(9 \cdot 9) = -81$

73. $\left(\frac{1}{4}\right)^3 = \frac{1}{4} \cdot \frac{1}{4} \cdot \frac{1}{4} = \frac{1}{64}$

75. $b^4 b^2 = b^{4+2} = b^6$

77. $a^2 a^3 a^4 = a^{2+3+4} = a^9$

79. $(2x^3)(-8x^4) = (2 \cdot -8)(x^3 \cdot x^4)$
$= -16x^{3+4}$
$= -16x^7$

81. $(a^7 b^{12})(a^4 b^8) = a^7 a^4 \cdot b^{12} b^8$
$= a^{7+4} b^{12+8}$
$= a^{11} b^{20}$

83. $(-2mn^6)(-13m^8 n) = (-2)(-13)(m \cdot m^8)(n^6 \cdot n)$
$= 26m^{1+8} n^{6+1}$
$= 26m^9 n^7$

85. $(z^4)^{10} = z^{4 \cdot 10} = z^{40}$

87. $(4ab)^3 = (4)^3 a^3 b^3 = 64a^3 b^3$

89. $(-6xyz^3)^2 = (-6)^2 x^2 y^2 (z^3)^2$
$= 36x^2 y^2 z^{3 \cdot 2}$
$= 36x^2 y^2 z^6$

91. $\frac{3x^5}{x^4} = 3 \cdot \frac{x^5}{x^4} = 3x^{5-4} = 3x^1 = 3x$

93. $(9xy)^2 = 9^2 \cdot x^2 y^2 = 81x^2 y^2$

95. $2^3 + 2^0 = (2 \cdot 2 \cdot 2) + 1 = 8 + 1 = 9$

97. $\left(\frac{3y^5}{6x^4}\right)^3 = \left(\frac{y^5}{2x^4}\right)^3 = \frac{(y^5)^3}{2^3 (x^4)^3} = \frac{y^{5 \cdot 3}}{8x^{4 \cdot 3}} = \frac{y^{15}}{8x^{12}}$

99. $\frac{2x^3 y^2 z}{xyz} = 2 \cdot \frac{x^3}{x} \cdot \frac{y^2}{y} \cdot \frac{z}{z}$
$= 2x^{3-1} y^{2-1} z^{1-1}$
$= 2x^2 y^1 z^0$
$= 2x^2 y$

101. $(5^0)^3 + (y^0)^7 = 1^3 + 1^7 = 1 + 1 = 2$

103. $\left(\frac{5x^9}{10y^{11}}\right)^2 = \frac{5^2 (x^9)^2}{10^2 (y^{11})^2} = \frac{25x^{18}}{100y^{22}} = \frac{x^{18}}{4y^{22}}$

105. $\frac{(2a^5 b^3)^4}{-16a^{20} b^7} = \frac{2^4 (a^5)^4 (b^3)^4}{-16a^{20} b^7}$
$= \frac{16a^{20} b^{12}}{-16a^{20} b^7}$
$= -1 \cdot a^{20-20} \cdot b^{12-7}$
$= -1 \cdot a^0 \cdot b^5$
$= -1 \cdot 1 \cdot b^5$
$= -b^5$

107. $5 - 7 = 5 + (-7) = -2$

109. $3 - (-2) = 3 + 2 = 5$

111. $-11 - (-4) = -11 + 4 = -7$

113. The expression $(x^{14})^{23}$ can be simplified by multiplying the exponents; c.

115. The expression $x^{14} + x^{23}$ cannot be simplified by adding subtracting, multiplying, or dividing the exponents; e.

117. answers may vary

119. answers may vary

121. $V = x^3$
$= (7 \text{ meters})^3$
$= 7^3$ cubic meters
$= 343$ cubic meters

123. The volume of a cube measures the amount of material that the cube can hold, so to find the amount of water that a swimming pool can hold, the formula for volume should be used.

125. answers may vary

Copyright © 2012 Pearson Education, Inc. Publishing as Prentice Hall.

127. answers may vary

129. $x^{5a}x^{4a} = x^{5a+4a} = x^{9a}$

131. $(a^b)^5 = a^{b\cdot 5} = a^{5b}$

133. $\dfrac{x^{9a}}{x^{4a}} = x^{9a-4a} = x^{5a}$

Section 3.2 Practice

1. $5^{-3} = \dfrac{1}{5^3} = \dfrac{1}{125}$

2. $7x^{-4} = 7^1 \cdot \dfrac{1}{x^4} = \dfrac{7^1}{x^4}$ or $\dfrac{7}{x^4}$

3. $5^{-1} + 3^{-1} = \dfrac{1}{5} + \dfrac{1}{3} = \dfrac{3}{15} + \dfrac{5}{15} = \dfrac{8}{15}$

4. $(-3)^{-4} = \dfrac{1}{(-3)^4} = \dfrac{1}{(-3)(-3)(-3)(-3)} = \dfrac{1}{81}$

5. $\left(\dfrac{6}{7}\right)^{-2} = \dfrac{6^{-2}}{7^{-2}} = \dfrac{6^{-2}}{1} \cdot \dfrac{1}{7^{-2}} = \dfrac{1}{6^2} \cdot \dfrac{7^2}{1} = \dfrac{7^2}{6^2} = \dfrac{49}{36}$

6. $\dfrac{x}{x^{-4}} = \dfrac{x^1}{x^{-4}} = x^{1-(-4)} = x^5$

7. $\dfrac{y^{-9}}{z^{-5}} = y^{-9} \cdot \dfrac{1}{z^{-5}} = \dfrac{1}{y^9} \cdot z^5 = \dfrac{z^5}{y^9}$

8. $\dfrac{y^{-4}}{y^6} = y^{-4-6} = y^{-10} = \dfrac{1}{y^{10}}$

9.
$$\begin{aligned}\frac{(3x^5)^3 x}{x^4} &= \frac{3^3 \cdot x^{15} \cdot x}{x^4}\\ &= \frac{27 \cdot x^{15+1}}{x^4}\\ &= \frac{27 \cdot x^{16}}{x^4}\\ &= 27x^{16-4}\\ &= 27x^{12}\end{aligned}$$

10.
$$\begin{aligned}\left(\frac{9x^3}{y}\right)^{-2} &= \frac{9^{-2}(x^3)^{-2}}{y^{-2}}\\ &= \frac{9^{-2}x^{-6}}{y^{-2}}\\ &= \frac{y^2}{9^2x^6}\\ &= \frac{y^2}{81x^6}\end{aligned}$$

11. $(a^{-4}b^7)^{-5} = (a^{-4})^{-5}(b^7)^{-5} = a^{20}b^{-35} = \dfrac{a^{20}}{b^{35}}$

12. $\dfrac{y^{-10}}{(y^5)^4} = \dfrac{y^{-10}}{y^{20}} = y^{-10-20} = y^{-30} = \dfrac{1}{y^{30}}$

13. $(4a^2)^{-3} = 4^{-3}(a^2)^{-3} = 4^{-3}a^{-6} = \dfrac{1}{4^3a^6} = \dfrac{1}{64a^6}$

14.
$$\begin{aligned}\frac{5x^7y^3z^0}{15xy^8z^3} &= \frac{5}{15} \cdot x^{7-1} \cdot y^{3-8} \cdot z^{0-3}\\ &= \frac{1}{3} \cdot x^6y^{-5}z^{-3}\\ &= \frac{x^6}{3y^5z^3}\end{aligned}$$

15.
$$\begin{aligned}-\frac{32x^{-3}y^{-6}}{8x^{-5}y^{-2}} &= -\frac{32}{8} \cdot x^{-3-(-5)}y^{-6-(-2)}\\ &= -4x^2y^{-4}\\ &= -\frac{4x^2}{y^4}\end{aligned}$$

Copyright © 2012 Pearson Education, Inc. Publishing as Prentice Hall.

16. $\dfrac{(3x^{-2}y)^{-2}}{(2x^7y)^3} = \dfrac{3^{-2}(x^{-2})^{-2}y^{-2}}{2^3(x^7)^3y^3}$

$= \dfrac{3^{-2}x^4y^{-2}}{2^3x^{21}y^3}$

$= \dfrac{3^{-2}}{2^3}\cdot x^{4-21}y^{-2-3}$

$= \dfrac{3^{-2}}{2^3}x^{-17}y^{-5}$

$= \dfrac{1}{2^3 3^2 x^{17}y^5}$

$= \dfrac{1}{72x^{17}y^5}$

17. a. $420{,}000 = 4.2\times10^5$

b. $0.00017 = 1.7\times10^{-4}$

c. $9{,}060{,}000{,}000 = 9.06\times10^9$

d. $0.000007 = 7.0\times10^{-6}$

18. a. $3.062\times10^{-4} = 0.0003062$

b. $5.21\times10^4 = 52{,}100$

c. $9.6\times10^{-5} = 0.000096$

d. $6.002\times10^6 = 6{,}002{,}000$

19. a. $(9\times10^7)(4\times10^{-9}) = 9\cdot4\cdot10^7\cdot10^{-9}$

$= 36\times10^{-2}$

$= 0.36$

b. $\dfrac{8\times10^4}{2\times10^{-3}} = \dfrac{8}{2}\times10^{4-(-3)}$

$= 4\times10^7$

$= 40{,}000{,}000$

Calculator Explorations

1. 5.31×10^3 5.31 EE 3

2. -4.8×10^{14} –4.8 EE 14

3. 6.6×10^{-9} 6.6 EE –9

4. -9.9811×10^{-2} –9.9811 EE –2

5. $3{,}000{,}000\times5{,}000{,}000 = 1.5\times10^{13}$

6. $230{,}000\times1000 = 2.3\times10^8$

7. $(3.26\times10^6)(2.5\times10^{13}) = 8.15\times10^{19}$

8. $(8.76\times10^{-4})(1.237\times10^9) = 1.083612\times10^6$

Vocabulary and Readiness Check

1. The expression x^{-3} equals $\underline{\dfrac{1}{x^3}}$.

2. The expression 5^{-4} equals $\underline{\dfrac{1}{625}}$.

3. The number 3.021×10^{-3} is written in <u>scientific notation</u>.

4. The number 0.0261 is written in <u>standard form</u>.

5. $5x^{-2} = 5\cdot\dfrac{1}{x^2} = \dfrac{5}{x^2}$

6. $3x^{-3} = 3\cdot\dfrac{1}{x^3} = \dfrac{3}{x^3}$

7. $\dfrac{1}{y^{-6}} = y^6$

8. $\dfrac{1}{x^{-3}} = x^3$

9. $\dfrac{4}{y^{-3}} = 4\cdot\dfrac{1}{y^{-3}} = 4y^3$

10. $\dfrac{16}{y^{-7}} = 16\cdot\dfrac{1}{y^{-7}} = 16y^7$

Exercise Set 3.2

1. $4^{-3} = \dfrac{1}{4^3} = \dfrac{1}{64}$

3. $7x^{-3} = 7\cdot\dfrac{1}{x^3} = \dfrac{7}{x^3}$

Copyright © 2012 Pearson Education, Inc. Publishing as Prentice Hall.

5. $\left(-\frac{1}{4}\right)^{-3} = \frac{(-1)^{-3}}{4^{-3}}$
$= (-1)^{-3} \cdot \frac{1}{4^{-3}}$
$= \frac{1}{(-1)^3} \cdot 4^3$
$= \frac{1}{-1} \cdot 64$
$= -64$

7. $3^{-1} + 2^{-1} = \frac{1}{3} + \frac{1}{2} = \frac{2}{6} + \frac{3}{6} = \frac{5}{6}$

9. $\frac{1}{p^{-3}} = p^3$

11. $\frac{p^{-5}}{q^{-4}} = \frac{1}{p^5} \cdot \frac{q^4}{1} = \frac{q^4}{p^5}$

13. $\frac{x^{-2}}{x} = \frac{x^{-2}}{x^1} = x^{-2-1} = x^{-3} = \frac{1}{x^3}$

15. $\frac{z^{-4}}{z^{-7}} = z^{-4-(-7)} = z^{-4+7} = z^3$

17. $3^{-2} + 3^{-1} = \frac{1}{3^2} + \frac{1}{3} = \frac{1}{9} + \frac{1}{3} = \frac{4}{9}$

19. $(-3)^{-2} = \frac{1}{(-3)^2} = \frac{1}{9}$

21. $\frac{-1}{p^{-4}} = -1 \cdot \frac{1}{p^{-4}} = -1 \cdot p^4 = -p^4$

23. $-2^0 - 3^0 = -(2^0) - (3^0) = -1 - 1 = -2$

25. $\frac{x^2 x^5}{x^3} = \frac{x^{2+5}}{x^3} = \frac{x^7}{x^3} = x^{7-3} = x^4$

27. $\frac{p^2 p}{p^{-1}} = \frac{p^{2+1}}{p^{-1}} = \frac{p^3}{p^{-1}} = p^{3-(-1)} = p^{3+1} = p^4$

29. $\frac{(m^5)^4 m}{m^{10}} = \frac{m^{5 \cdot 4} m^1}{m^{10}}$
$= \frac{m^{20} m^1}{m^{10}}$
$= \frac{m^{20+1}}{m^{10}}$
$= \frac{m^{21}}{m^{10}}$
$= m^{21-10}$
$= m^{11}$

31. $\frac{r}{r^{-3} r^{-2}} = \frac{r}{r^{(-3)+(-2)}} = \frac{r}{r^{-5}} = r^{1-(-5)} = r^{1+5} = r^6$

33. $(x^5 y^3)^{-3} = (x^5)^{-3} (y^3)^{-3} = x^{-15} y^{-9} = \frac{1}{x^{15} y^9}$

35. $\frac{(x^2)^3}{x^{10}} = \frac{x^6}{x^{10}} = x^{6-10} = x^{-4} = \frac{1}{x^4}$

37. $\frac{(a^5)^2}{(a^3)^4} = \frac{a^{10}}{a^{12}} = a^{10-12} = a^{-2} = \frac{1}{a^2}$

39. $\frac{8k^4}{2k} = \frac{8}{2} \cdot \frac{k^4}{k} = 4 \cdot k^{4-1} = 4k^3$

41. $\frac{-6m^4}{-2m^3} = \frac{-6}{-2} \cdot \frac{m^4}{m^3} = 3 \cdot m^{4-3} = 3m^1 = 3m$

43. $\frac{-24a^6 b}{6ab^2} = \frac{-24}{6} \cdot \frac{a^6}{a} \cdot \frac{b}{b^2}$
$= -4a^{6-1} b^{1-2}$
$= -4a^5 b^{-1}$
$= -\frac{4a^5}{b}$

Copyright © 2012 Pearson Education, Inc. Publishing as Prentice Hall.

45. $$\begin{aligned}\frac{6x^2y^3z^0}{-7x^2y^5z^5} &= \frac{6}{-7}\cdot\frac{x^2}{x^2}\cdot\frac{y^3}{y^5}\cdot\frac{z^0}{z^5}\\ &= -\frac{6}{7}\cdot x^{2-2}\cdot y^{3-5}\cdot z^{0-5}\\ &= -\frac{6}{7}x^0\cdot y^{-2}\cdot z^{-5}\\ &= -\frac{6}{7}\cdot 1\cdot\frac{1}{y^2}\cdot\frac{1}{z^5}\\ &= -\frac{6}{7y^2z^5}\end{aligned}$$

47. $(3a^2b^{-4})^3 = 3^3a^{2\cdot 3}b^{-4\cdot 3} = 27a^6b^{-12} = \dfrac{27a^6}{b^{12}}$

49. $(a^{-5}b^2)^{-6} = (a^{-5})^{-6}(b^2)^{-6} = a^{30}b^{-12} = \dfrac{a^{30}}{b^{12}}$

51. $$\begin{aligned}\left(\frac{x^{-2}y^4z^0}{x^3y^7}\right)^2 &= \frac{(x^{-2})^2(y^4)^2(z^0)^2}{(x^3)^2(y^7)^2}\\ &= \frac{x^{-4}y^8z^0}{x^6y^{14}}\\ &= x^{-4-6}\cdot y^{8-14}\cdot z^0\\ &= x^{-10}\cdot y^{-6}\cdot 1\\ &= \frac{1}{x^{10}}\cdot\frac{1}{y^6}\\ &= \frac{1}{x^{10}y^6}\end{aligned}$$

53. $$\begin{aligned}\frac{4^2z^{-3}}{4^3z^{-5}} &= \frac{4^2}{4^3}\cdot\frac{z^{-3}}{z^{-5}}\\ &= 4^{2-3}z^{-3-(-5)}\\ &= 4^{-1}z^{-3+5}\\ &= \frac{1}{4}\cdot z^2\\ &= \frac{z^2}{4}\end{aligned}$$

55. $$\begin{aligned}\frac{3^{-1}x^4}{3^3x^{-7}} &= \frac{3^{-1}}{3^3}\cdot\frac{x^4}{x^{-7}}\\ &= 3^{-1-3}x^{4-(-7)}\\ &= 3^{-4}x^{4+7}\\ &= 3^{-4}x^{11}\\ &= \frac{x^{11}}{3^4}\\ &= \frac{x^{11}}{81}\end{aligned}$$

57. $$\begin{aligned}\frac{7ab^{-4}}{7^{-1}a^{-3}b^2} &= \frac{7}{7^{-1}}\cdot\frac{a^1}{a^{-3}}\cdot\frac{b^{-4}}{b^2}\\ &= 7^{1-(-1)}a^{1-(-3)}b^{-4-2}\\ &= 7^{1+1}a^{1+3}b^{-6}\\ &= 7^2a^4b^{-6}\\ &= \frac{49a^4}{b^6}\end{aligned}$$

59. $$\begin{aligned}\frac{-12m^5n^{-7}}{4m^{-2}n^{-3}} &= \frac{-12}{4}\cdot\frac{m^5}{m^{-2}}\cdot\frac{n^{-7}}{n^{-3}}\\ &= -3m^{5-(-2)}n^{-7-(-3)}\\ &= -3m^{5+2}n^{-7+3}\\ &= -3m^7n^{-4}\\ &= -\frac{3m^7}{n^4}\end{aligned}$$

61. $$\begin{aligned}\left(\frac{a^{-5}b}{ab^3}\right)^{-4} &= \frac{(a^{-5})^{-4}b^{-4}}{a^{-4}(b^3)^{-4}}\\ &= \frac{a^{20}b^{-4}}{a^{-4}b^{-12}}\\ &= a^{20-(-4)}b^{-4-(-12)}\\ &= a^{20+4}b^{-4+12}\\ &= a^{24}b^8\end{aligned}$$

63. $(5^2)(8)(2^0) = (25)(8)(1) = 200$

Copyright © 2012 Pearson Education, Inc. Publishing as Prentice Hall.

65. $\dfrac{(xy^3)^5}{(xy)^{-4}} = \dfrac{x^5(y^3)^5}{x^{-4}y^{-4}}$
$= \dfrac{x^5y^{15}}{x^{-4}y^{-4}}$
$= x^{5-(-4)}y^{15-(-4)}$
$= x^{5+4}y^{15+4}$
$= x^9y^{19}$

67. $\dfrac{(-2xy^{-3})^{-3}}{(xy^{-1})^{-1}} = \dfrac{-2^{-3}x^{-3}(y^{-3})^{-3}}{x^{-1}(y^{-1})^{-1}}$
$= \dfrac{-2^{-3}x^{-3}y^9}{x^{-1}y}$
$= -2^{-3}x^{-3-(-1)}y^{9-1}$
$= -2^{-3}x^{-3+1}y^8$
$= -2^{-3}x^{-2}y^8$
$= -\dfrac{y^8}{2^3x^2}$
$= -\dfrac{y^8}{8x^2}$

69. $\dfrac{(a^4b^{-7})^{-5}}{(5a^2b^{-1})^{-2}} = \dfrac{(a^4)^{-5}(b^{-7})^{-5}}{5^{-2}(a^2)^{-2}(b^{-1})^{-2}}$
$= \dfrac{a^{-20}b^{35}}{5^{-2}a^{-4}b^2}$
$= 5^2a^{-20-(-4)}b^{35-2}$
$= 25a^{-20+4}b^{33}$
$= 25a^{-16}b^{33}$
$= \dfrac{25b^{33}}{a^{16}}$

71. $V = s^3$
$= \left(\dfrac{3x^{-2}}{z}\text{ inches}\right)^3$
$= \left(\dfrac{3^3(x^{-2})^3}{z^3}\right)\text{ cubic inches}$
$= \dfrac{27x^{-6}}{z^3}\text{ cubic inches}$
$= \dfrac{27}{z^3x^6}\text{ cubic inches}$

73. $78{,}000 = 7.8\times10^4$

75. $0.00000167 = 1.67\times10^{-6}$

77. $0.00635 = 6.35\times10^{-3}$

79. $1{,}160{,}000 = 1.16\times10^6$

81. $2400 = 2.4\times10^3$

83. $8.673\times10^{-10} = 0.0000000008673$

85. $3.3\times10^{-2} = 0.033$

87. $2.032\times10^4 = 20{,}320$

89. $7.0\times10^8 = 700{,}000{,}000$

91. $184{,}000{,}000{,}000 = 1.84\times10^{11}$

93. $1.55\times10^{11} = 155{,}000{,}000{,}000$

95. $3.5\times10^4 = 35{,}000$

97. $(1.2\times10^{-3})(3\times10^{-2}) = 1.2\cdot3\cdot10^{-3}\cdot10^{-2}$
$= 3.6\times10^{-5}$
$= 0.000036$

99. $(4\times10^{-10})(7\times10^{-9}) = (4\cdot7)(10^{-10}\cdot10^{-9})$
$= 28\cdot10^{-19}$
$= 0.0000000000000000028$

101. $\dfrac{8\times10^{-1}}{16\times10^5} = \dfrac{8}{16}\times10^{-1-5}$
$= 0.5\times10^{-6}$
$= 5.0\times10^{-7}$
$= 0.0000005$

103. $\dfrac{1.4\times10^{-2}}{7\times10^{-8}} = \dfrac{1.4}{7}\cdot\dfrac{10^{-2}}{10^{-8}}$
$= 0.2\cdot10^{-2-(-8)}$
$= 0.2\cdot10^{-2+8}$
$= 0.2\cdot10^6$
$= 200{,}000$

Copyright © 2012 Pearson Education, Inc. Publishing as Prentice Hall.

105. $7.5\times10^5\cdot3600 = 7.5\times10^5\cdot3.6\times10^3$
$= 7.5\cdot3.6\cdot10^5\cdot10^3$
$= 27\times10^8$
$= 2.7\times10^9$

107. $3x - 5x + 2 = -2x + 7$

109. $y - 10 + y = y + y - 10 = 2y - 10$

111. $7x + 2 - 8x - 6 = 7x - 8x + 2 - 6 = -x - 4$

113. 90 million $= 90{,}000{,}000 = 9\times10^7$

115. 1 billion $= 1{,}000{,}000{,}000 = 1\times10^9$

117. 0.44 billion $= 440{,}000{,}000 = 4.4\times10^8$

119. no; answers may vary

121. $(2a^3)^3a^4 + a^5a^8 = 2^3(a^3)^3a^4 + a^5a^8$
$= 8a^9a^4 + a^5a^8$
$= 8a^{13} + a^{13}$
$= 9a^{13}$

123. If $x^{\square} = \dfrac{1}{x^5}$, the exponent is -5.

125. answers may vary

127. **a.** 9.7×10^{-2} or $1.3\times10^1 \Rightarrow 1.3\times10^1$

b. 8.6×10^5 or $4.4\times10^7 \Rightarrow 4.4\times10^7$

c. 6.1×10^{-2} or $5.6\times10^{-4} \Rightarrow 6.1\times10^{-2}$

129. answers may vary

131. $(x^{-3s})^3 = x^{-9s} = \dfrac{1}{x^{9s}}$

133. $a^{4m+1}\cdot a^4 = a^{4m+1+4} = a^{4m+5}$

Section 3.3 Practice

1. $-6x^6 + 4x^5 + 7x^3 - 9x^2 - 1$

Term	Coefficient
$7x^3$	7
$-9x^2$	-9
$-6x^6$	-6
$4x^5$	4
-1	-1

2. $-15x^3 + 2x^2 - 5$
The term $-15x^3$ has degree 3.
The term $2x^2$ has degree 2.
The term -5 has degree 0 since -5 is $-5x^0$.

3. **a.** The degree of the binomial $-6x + 14$ is 1.

b. The degree of the polynomial $9x - 3x^6 + 5x^4 + 2$ is 6. The polynomial is neither a monomial, binomial, or trinomial.

c. The degree of the trinomial $10x^2 - 6x - 6$ is 2.

4. **a.** When $x = -1$,
$-2x + 10 = -2(-1) + 10 = 2 + 10 = 12$.

b. When $x = -1$,
$6x^2 + 11x - 20 = 6(-1)^2 + 11(-1) - 20$
$= 6 - 11 - 20$
$= -25$.

5. When $t = 2$ seconds,
$-16t^2 + 592.1 = -16(2)^2 + 592.1$
$= -16(4) + 592.1$
$= -64 + 592.1$
$= 528.1$ feet.

When $t = 4$ seconds,
$-16t^2 + 592.1 = -16(4)^2 + 592.1$
$= -16(16) + 592.1$
$= -256 + 592.1$
$= 336.1$ feet.

6. $-6y + 8y = (-6 + 8)y = 2y$

Copyright © 2012 Pearson Education, Inc. Publishing as Prentice Hall.

7. $14y^2+3-10y^2-9=14y^2-10y^2+3-9$
$=4y^2-6$

8. $7x^3+x^3=7x^3+1x^3=8x^3$

9. $23x^2-6x-x-15=23x^2-7x-15$

10. $\frac{2}{7}x^3-\frac{1}{4}x+2-\frac{1}{2}x^3+\frac{3}{8}x$
$=\frac{2}{7}x^3-\frac{1}{2}x^3-\frac{1}{4}x+\frac{3}{8}x+2$
$=\frac{4}{14}x^3-\frac{7}{14}x^3-\frac{2}{8}x+\frac{3}{8}x+2$
$=-\frac{3}{14}x^3+\frac{1}{8}x+2$

11. Area $=5\cdot x+x\cdot x+4\cdot 5+x\cdot x+8\cdot x$
$=5x+x^2+20+x^2+8x$
$=x^2+x^2+5x+8x+20$
$=2x^2+13x+20$

12.

Terms of Polynomial	Degree of Term
$-2x^3y^2$	3 + 2 or 5
4	0
$-8xy$	1 + 1 or 2
$3x^3y$	3 + 1 or 4
$5xy^2$	1 + 2 or 3

The degree of the polynomial is 5.

13. $11ab-6a^2-ba+8b^2=(11-1)ab-6a^2+8b^2$
$=10ab-6a^2+8b^2$

14. $7x^2y^2+2y^2-4y^2x^2+x^2-y^2+5x^2$
$=7x^2y^2-4x^2y^2+2y^2-y^2+x^2+5x^2$
$=3x^2y^2+y^2+6x^2$

15. a. $x^2+9=x^2+0x^1+9$ or x^2+0x+9

b. $9m^3+m^2-5=9m^3+m^2+0m^1-5$
$=9m^3+m^2+0m-5$

c. $-3a^3+a^4=a^4-3a^3+0a^2+0a^1+0a^0$
$=a^4-3a^3+0a^2+0a+0a^0$

Vocabulary and Readiness Check

1. A binomial is a polynomial with exactly two terms.

2. A monomial is a polynomial with exactly one term.

3. A trinomial is a polynomial with exactly three terms.

4. The numerical factor of a term is called the coefficient.

5. A number term is also called a constant.

6. The degree of a polynomial is the greatest degree of any term of the polynomial.

Exercise Set 3.3

1.

Term	Coefficient
x^2	1
$-3x$	-3
5	5

3.

Term	Coefficient
$-5x^4$	-5
$3.2x^2$	3.2
x	1
-5	-5

5. $x+2=x^1+2$
This is a binomial of degree 1.

7. $9m^3-5m^2+4m-8$
This is a polynomial of degree 3. None of these.

9. $12x^4-x^6-12x^2=-x^6+12x^4-12x^2$
This is a trinomial of degree 6.

11. $3z-5z^4=-5z^4+3z$
This is a binomial of degree 4.

13. a. $5x-6=5(0)-6=0-6=-6$

Copyright © 2012 Pearson Education, Inc. Publishing as Prentice Hall.

b. $5x - 6 = 5(-1) - 6 = -5 - 6 = -11$

15. a. $x^2 - 5x - 2 = (0)^2 - 5(0) - 2 = 0 - 0 - 2 = -2$

b. $x^2 - 5x - 2 = (-1)^2 - 5(-1) - 2 = 1 + 5 - 2 = 4$

17. a. $-x^3 + 4x^2 - 15 = -(0)^3 + 4(0)^2 - 15$
$= 0 + 0 - 15$
$= -15$

b. $-x^3 + 4x^2 - 15 = -(-1)^3 + 4(-1)^2 - 15$
$= -(-1) + 4(1) - 15$
$= 1 + 4 - 15$
$= -10$

19. $-16t^2 + 200t = -16(1)^2 + 200(1)$
$= -16 + 200$
$= 184$
After 1 second, the height of the rocket is 184 feet.

21. $-16t^2 + 200t = -16(7.6)^2 + 200(7.6)$
$= -16(57.76) + 1520$
$= -924.16 + 1520$
$= 595.84$
After 7.6 seconds, the height of the rocket is 595.84 feet.

23. $-7.5x^2 + 93x - 100 = -7.5(8)^2 + 93(8) - 100$
$= -7.5(64) + 93(8) - 100$
$= -480 + 744 - 100$
$= 164$
There were 164 thousand or 164,000 visitors in 2008.

25. $0.52x^2 + 11.4x + 27.87$
$= 0.52(17)^2 + 11.4(17) + 27.87$
$= 0.52(289) + 11.4(17) + 27.87$
$= 150.28 + 193.8 + 27.87$
$= 371.95$
There will be 371.95 million or 371,950,000 subscribers in 2012.

27. $9x - 20x = (9 - 20)\,x = -11x$

29. $14x^3 + 9x^3 = (14 + 9)x^3 = 23x^3$

31. $7x^2 + 3 + 9x^2 - 10 = 7x^2 + 9x^2 + 3 - 10$
$= (7 + 9)x^2 + 3 - 10$
$= 16x^2 - 7$

33. $15x^2 - 3x^2 - 13 = (15 - 3)x^2 - 13 = 12x^2 - 13$

35. $8s - 5s + 4s = (8 - 5 + 4)s = 7s$

37. $0.1y^2 - 1.2y^2 + 6.7 - 1.9$
$= (0.1 - 1.2)\,y^2 + 6.7 - 1.9$
$= -1.1y^2 + 4.8$

39. $\frac{2}{3}x^4 + 12x^3 + \frac{1}{6}x^4 - 19x^3 - 19$
$= \frac{2}{3}x^4 + \frac{1}{6}x^4 + 12x^3 - 19x^3 - 19$
$= \left(\frac{4}{6} + \frac{1}{6}\right)x^4 + (12 - 19)x^3 - 19$
$= \frac{5}{6}x^4 - 7x^3 - 19$

41. $\frac{3}{20}x^3 + \frac{1}{10} - \frac{3}{10}x - \frac{1}{5} - \frac{7}{20}x + 6x^2$
$= \frac{3}{20}x^3 + 6x^2 - \frac{3}{10}x - \frac{7}{20}x + \frac{1}{10} - \frac{1}{5}$
$= \frac{3}{20}x^3 + 6x^2 + \left(-\frac{6}{20} - \frac{7}{20}\right)x + \frac{1}{10} - \frac{2}{10}$
$= \frac{3}{20}x^3 + 6x^2 - \frac{13}{20}x - \frac{1}{10}$

43. $4x^2 + 7x + x^2 + 5x = 4x^2 + x^2 + 7x + 5x$
$= (4 + 1)x^2 + (7 + 5)x$
$= 5x^2 + 12x$

45. $5x + 3 + 4x + 3 + 2x + 6 + 3x + 7x$
$= 5x + 4x + 2x + 3x + 7x + 3 + 3 + 6$
$= (5 + 4 + 2 + 3 + 7)x + 12$
$= 21x + 12$

47. $9ab = 9a^1b^1$ has degree $1 + 1 = 2$.
$-6a = -6a^1$ has degree 1.
$5b = 5b^1$ has degree 1.
$-3 = -3a^0b^0$ has degree 0.
$9ab - 6a + 5b - 3$ is a polynomial of degree 2.

Copyright © 2012 Pearson Education, Inc. Publishing as Prentice Hall.

49. $x^3y = x^3y^1$ has degree 3 + 1 = 4.

$-6 = -6x^0y^0$ has degree 0.

$2x^2y^2$ has degree 2 + 2 = 4.

$5y^3$ has degree 3.

$x^3y - 6 + 2x^2y^2 + 5y^3$ is a polynomial of degree 4.

51. $3ab - 4a + 6ab - 7a = 3ab + 6ab - 4a - 7a$
$= (3+6)ab - (4+7)a$
$= 9ab - 11a$

53. $4x^2 - 6xy + 3y^2 - xy = 4x^2 - 6xy - xy + 3y^2$
$= 4x^2 + (-6-1)xy + 3y^2$
$= 4x^2 - 7xy + 3y^2$

55. $5x^2y + 6xy^2 - 5yx^2 + 4 - 9y^2x$
$= 5x^2y - 5x^2y + 6xy^2 - 9xy^2 + 4$
$= (5-5)x^2y + (6-9)xy^2 + 4$
$= 0x^2y - 3xy^2 + 4$
$= -3xy^2 + 4$

57. $14y^3 - 9 + 3a^2b^2 - 10 - 19b^2a^2$
$= 14y^3 - 9 - 10 + 3a^2b^2 - 19a^2b^2$
$= 14y^3 + (-9-10) + (3-19)a^2b^2$
$= 14y^3 - 19 - 16a^2b^2$

59. $7x^2 + 3 = 7x^2 + 0x + 3$

61. $x^3 - 64 = x^3 + 0x^2 + 0x - 64$

63. $5y^3 + 2y - 10 = 5y^3 + 0y^2 + 2y - 10$

65. $8y + 2y^4 = 2y^4 + 0y^3 + 0y^2 + 8y + 0$

67. $6x^5 + x^3 - 3x + 15$
$= 6x^5 + 0x^4 + x^3 + 0x^2 - 3x + 15$

69. $4 + 5(2x + 3) = 4 + 10x + 15 = 10x + 19$

71. $2(x-5) + 3(5-x) = 2x - 10 + 15 - 3x$
$= 2x - 3x - 10 + 15$
$= (2-3)x + 5$
$= -x + 5$

73. answers may vary

75. answers may vary

77. $x^4 \cdot x^9 = x^{4+9} = x^{13}$

79. $a \cdot b^3 \cdot a^2 \cdot b^7 = a^1 \cdot a^2 \cdot b^3 \cdot b^7$
$= a^{1+2}b^{3+7}$
$= a^3b^{10}$

81. $(y^5)^4 + (y^2)^{10} = y^{20} + y^{20} = 2y^{20}$

83. answers may vary

85. answers may vary

87. $1.85x^2 - 3.76x + 9.25x^2 + 10.76 - 4.21x$
$= 1.85x^2 + 9.25x^2 - 3.76x - 4.21x + 10.76$
$= (1.85 + 9.25)x^2 - (3.76 + 4.21)x + 10.76$
$= 11.1x^2 - 7.97x + 10.76$

Section 3.4 Practice

1. $(3x^5 - 7x^3 + 2x - 1) + (3x^3 - 2x)$
$= 3x^5 - 7x^3 + 2x - 1 + 3x^3 - 2x$
$= 3x^5 + (-7x^3 + 3x^3) + (2x - 2x) - 1$
$= 3x^5 - 4x^3 - 1$

2. $(5x^2 - 2x + 1) + (-6x^2 + x - 1)$
$= 5x^2 - 2x + 1 - 6x^2 + x - 1$
$= (5x^2 - 6x^2) + (-2x + x) + (1 - 1)$
$= -x^2 - x$

3.
$$\begin{array}{r} 9y^2 - 6y + 5 \\ 4y + 3 \\ \hline 9y^2 - 2y + 8 \end{array}$$

4. $(9x+5) - (4x-3) = (9x+5) + [-(4x-3)]$
$= (9x+5) + (-4x+3)$
$= 9x + 5 - 4x + 3$
$= 5x + 8$

5. $(4x^3 - 10x^2 + 1) - (-4x^3 + x^2 - 11)$
$= (4x^3 - 10x^2 + 1) + (4x^3 - x^2 + 11)$
$= 4x^3 - 10x^2 + 1 + 4x^3 - x^2 + 11$
$= 4x^3 + 4x^3 - 10x^2 - x^2 + 1 + 11$
$= 8x^3 - 11x^2 + 12$

Copyright © 2012 Pearson Education, Inc. Publishing as Prentice Hall.

6. $$\begin{array}{r} 2y^2-2y+7 \\ \underline{-(6y^2-3y+2)} \\ \end{array} \Rightarrow \begin{array}{r} 2y^2-2y+7 \\ \underline{-6y^2+3y-2} \\ -4y^2+\ \ y+5 \end{array}$$

7. $$\begin{aligned}[(4x-3)+(12x-5)]-(3x+1) &= 4x-3+12x-5-3x-1 \\ &= 4x+12x-3x-3-5-1 \\ &= 13x-9\end{aligned}$$

8. $$\begin{aligned}(2a^2-ab+6b^2)+(-3a^2+ab-7b^2) &= 2a^2-ab+6b^2-3a^2+ab-7b^2 \\ &= -a^2-b^2\end{aligned}$$

9. $$\begin{aligned}(5x^2y^2+3-9x^2y+y^2)-(-x^2y^2+7-8xy^2+2y^2) &= 5x^2y^2+3-9x^2y+y^2+x^2y^2-7+8xy^2-2y^2 \\ &= 6x^2y^2-4-9x^2y+8xy^2-y^2\end{aligned}$$

Vocabulary and Readiness Check

1. $-9y-5y=-14y$

2. $6m^5+7m^5=13m^5$

3. $x+6x=7x$

4. $7z-z=6z$

5. $5m^2+2m=5m^2+2m$

6. $8p^3+3p^2=8p^3+3p^2$

Exercise Set 3.4

1. $$\begin{aligned}(3x+7)+(9x+5) &= 3x+7+9x+5 \\ &= 3x+9x+7+5 \\ &= 12x+12\end{aligned}$$

3. $$\begin{aligned}(-7x+5)+(-3x^2+7x+5) &= -7x+5-3x^2+7x+5 \\ &= -3x^2-7x+7x+5+5 \\ &= -3x^2+10\end{aligned}$$

5. $$\begin{aligned}(-5x^2+3)+(2x^2+1) &= -5x^2+3+2x^2+1 \\ &= -5x^2+2x^2+3+1 \\ &= -3x^2+4\end{aligned}$$

7. $$\begin{aligned}(-3y^2-4y)+(2y^2+y-1) &= -3y^2-4y+2y^2+y-1 \\ &= -3y^2+2y^2-4y+y-1 \\ &= -y^2-3y-1\end{aligned}$$

Copyright © 2012 Pearson Education, Inc. Publishing as Prentice Hall.

9. $(1.2x^3-3.4x+7.9)+(6.7x^3+4.4x^2-10.9)$
$=1.2x^3-3.4x+7.9+6.7x^3+4.4x^2-10.9$
$=1.2x^3+6.7x^3+4.4x^2-3.4x+7.9-10.9$
$=7.9x^3+4.4x^2-3.4x-3$

11. $\left(\frac{3}{4}m^2-\frac{2}{5}m+\frac{1}{8}\right)+\left(-\frac{1}{4}m^2-\frac{3}{10}m+\frac{11}{16}\right)$
$=\frac{3}{4}m^2-\frac{2}{5}m+\frac{1}{8}-\frac{1}{4}m^2-\frac{3}{10}m+\frac{11}{16}$
$=\frac{3}{4}m^2-\frac{1}{4}m^2-\frac{2}{5}m-\frac{3}{10}m+\frac{1}{8}+\frac{11}{16}$
$=\frac{3}{4}m^2-\frac{1}{4}m^2-\frac{4}{10}m-\frac{3}{10}m+\frac{2}{16}+\frac{11}{16}$
$=\frac{2}{4}m^2-\frac{7}{10}m+\frac{13}{16}$
$=\frac{1}{2}m^2-\frac{7}{10}m+\frac{13}{16}$

13.
$$\begin{array}{r} 3t^2+4 \\ \underline{5t^2-8} \\ 8t^2-4 \end{array}$$

15.
$$\begin{array}{r} 10a^3-8a^2+4a+9 \\ \underline{5a^3+9a^2-7a+7} \\ 15a^3+a^2-3a+16 \end{array}$$

17. $(2x+5)-(3x-9)=(2x+5)+(-3x+9)$
$=2x+5-3x+9$
$=2x-3x+5+9$
$=-x+14$

19. $(5x^2+4)-(-2y^2+4)=(5x^2+4)+(2y^2-4)$
$=5x^2+4+2y^2-4$
$=5x^2+2y^2+4-4$
$=5x^2+2y^2$

21. $3x-(5x-9)=3x+(-5x+9)$
$=3x-5x+9$
$=-2x+9$

23. $(2x^2+3x-9)-(-4x+7)$
$=(2x^2+3x-9)+(4x-7)$
$=2x^2+3x-9+4x-7$
$=2x^2+3x+4x-9-7$
$=2x^2+7x-16$

25. $(5x+8)-(-2x^2-6x+8)$
$=(5x+8)+(2x^2+6x-8)$
$=5x+8+2x^2+6x-8$
$=2x^2+5x+6x+8-8$
$=2x^2+11x$

27. $(0.7x^2+0.2x-0.8)-(0.9x^2+1.4)$
$=(0.7x^2+0.2x-0.8)+(-0.9x^2-1.4)$
$=0.7x^2+0.2x-0.8-0.9x^2-1.4$
$=0.7x^2-0.9x^2+0.2x-0.8-1.4$
$=-0.2x^2+0.2x-2.2$

29. $\left(\frac{1}{4}z^2-\frac{1}{5}z\right)-\left(-\frac{3}{20}z^2+\frac{1}{10}z-\frac{7}{20}\right)$
$=\left(\frac{1}{4}z^2-\frac{1}{5}z\right)+\left(\frac{3}{20}z^2-\frac{1}{10}z+\frac{7}{20}\right)$
$=\frac{1}{4}z^2-\frac{1}{5}z+\frac{3}{20}z^2-\frac{1}{10}z+\frac{7}{20}$
$=\frac{5}{20}z^2+\frac{3}{20}z^2-\frac{2}{10}z-\frac{1}{10}z+\frac{7}{20}$
$=\frac{8}{20}z^2-\frac{3}{10}z+\frac{7}{20}$
$=\frac{2}{5}z^2-\frac{3}{10}z+\frac{7}{20}$

31.
$$\begin{array}{r} 4z^2-8z+3 \\ \underline{-(6z^2+8z-3)} \end{array} \Rightarrow \begin{array}{r} 4z^2-8z+3 \\ \underline{-6z^2-8z+3} \\ -2z^2-16z+6 \end{array}$$

33.
$$\begin{array}{r} 5u^5-4u^2+3u-7 \\ \underline{-(3u^5+6u^2-8u+2)} \end{array} \quad \begin{array}{r} 5u^5-4u^2+3u-7 \\ \underline{-3u^5-6u^2+8u-2} \\ 2u^5-10u^2+11u-9 \end{array}$$

35. $(3x+5)+(2x-14)=3x+5+2x-14$
$=3x+2x+5-14$
$=5x-9$

37. $(9x-1)-(5x+2)=(9x-1)+(-5x-2)$
$=9x-1-5x-2$
$=4x-3$

39. $(14y+12)+(-3y-5)=14y+12-3y-5$
$=11y+7$

Copyright © 2012 Pearson Education, Inc. Publishing as Prentice Hall.

41. $(x^2+2x+1)-(3x^2-6x+2)=(x^2+2x+1)+(-3x^2+6x-2)$
$=x^2+2x+1-3x^2+6x-2$
$=-2x^2+8x-1$

43. $(3x^2+5x-8)+(5x^2+9x+12)-(8x^2-14)=(3x^2+5x-8)+(5x^2+9x+12)+(-8x^2+14)$
$=3x^2+5x-8+5x^2+9x+12-8x^2+14$
$=14x+18$

45. $(-a^2+1)-(a^2-3)+(5a^2-6a+7)=(-a^2+1)+(-a^2+3)+(5a^2-6a+7)$
$=-a^2+1-a^2+3+5a^2-6a+7$
$=3a^2-6a+11$

47. $(7x-3)-4x=7x-3-4x=3x-3$

49. $(4x^2-6x+1)+(3x^2+2x+1)=4x^2-6x+1+3x^2+2x+1$
$=7x^2-4x+2$

51. $(7x^2+3x+9)-(5x+7)=(7x^2+3x+9)+(-5x-7)$
$=7x^2+3x+9-5x-7$
$=7x^2-2x+2$

53. $(8y^2+7)+(6y+9)-(4y^2-6y-3)=(8y^2+7)+(6y+9)+(-4y^2+6y+3)$
$=8y^2+7+6y+9-4y^2+6y+3$
$=4y^2+12y+19$

55. $(x^2-9x+2)+(2x^2-6x+1)-(3x^2-4)=(x^2-9x+2)+(2x^2-6x+1)+(-3x^2+4)$
$=x^2-9x+2+2x^2-6x+1-3x^2+4$
$=-15x+7$

57. $(9a+6b-5)+(-11a-7b+6)=9a+6b-5-11a-7b+6$
$=-2a-b+1$

59. $(4x^2+y^2+3)-(x^2+y^2-2)=4x^2+y^2+3-x^2-y^2+2$
$=3x^2+5$

61. $(x^2+2xy-y^2)+(5x^2-4xy+20y^2)=x^2+2xy-y^2+5x^2-4xy+20y^2$
$=6x^2-2xy+19y^2$

63. $(11r^2s+16rs-3-2r^2s^2)-(3sr^2+5-9r^2s^2)=11r^2s+16rs-3-2r^2s^2-3r^2s-5+9r^2s^2$
$=8r^2s+16rs-8+7r^2s^2$

65. $(2x^2+5)+(4x-1)+(-x^2+3x)=2x^2+5+4x-1-x^2+3x$
$=x^2+7x+4$

The perimeter is (x^2+7x+4) feet.

Copyright © 2012 Pearson Education, Inc. Publishing as Prentice Hall.

67. $(2x-3)+\left(\frac{4}{5}x\right)+\left(\frac{7}{10}x-1\right)+(2x-2)+(x+4)+(3x+5)=2x-3+\frac{4}{5}x+\frac{7}{10}x-1+2x-2+x+4+3x+5$

$$=\frac{19}{2}x+3$$

The perimeter is $\left(\frac{19}{2}x+3\right)$ units.

69. $(4y^2+4y+1)-(y^2-10)=(4y^2+4y+1)+(-y^2+10)$

$$=4y^2+4y+1-y^2+10$$
$$=3y^2+4y+11$$

The remaining piece is $(3y^2+4y+11)$ meters long.

71. $[(1.2x^2-3x+9.1)-(7.8x^2-3.1+8)]+(1.2x-6)=(1.2x^2-3x+9.1)+(-7.8x^2+3.1-8)+(1.2x-6)$

$$=1.2x^2-3x+9.1-7.8x^2+3.1-8+1.2x-6$$
$$=-6.6x^2-1.8x-1.8$$

73. $3x(2x)=(3\cdot 2)(x\cdot x)=6x^2$

75. $(12x^3)(-x^5)=(12\cdot -1)(x^3\cdot x^5)=-12x^8$

77. $10x^2(20xy^2)=10\cdot 20\cdot(x^2\cdot x)(y^2)=200x^3y^2$

79. Since $3+4=7$, $3x^2+4x^2=7x^2$ is a true statement.

81. Since $2+4=6$ and $3-5=-2$, $2x^4+3x^3-5x^3+4x^4=6x^4-2x^3$ is a true statement.

83. $10y-6y^2-y=9y-6y^2$; b

85. $(5x-3)+(5x-3)=5x-3+5x-3=10x-6$; e

87. **a.** $z+3z=1z+3z=4z$

b. $z\cdot 3z=z^1\cdot 3z^1=3z^{1+1}=3z^2$

c. $-z-3z=-1z-3z=-4z$

d. $(-z)(-3z)=(-z^1)(-3z^1)=3z^{1+1}=3z^2$; answers may vary

89. **a.** $m\cdot m\cdot m=m^1\cdot m^1\cdot m^1=m^{1+1+1}=m^3$

b. $m+m+m=1m+1m+1m=(1+1+1)m=3m$

c. $(-m)(-m)(-m)=(-1\cdot m^1)(-1\cdot m^1)(-1\cdot m^1)$

$$=(-1)(-1)(-1)(m\cdot m\cdot m)$$
$$=-1m^3$$
$$=-m^3$$

Copyright © 2012 Pearson Education, Inc. Publishing as Prentice Hall.

d. $-m-m-m=-1m-1m-1m$
$=(-1-1-1)m$
$=-3m$; answers may vary

91. $(-20x^2+156x+14,437)+(894x^2-90x+10,939)=-20x^2+156x+14,437+894x^2-90x+10,939$
$=874x^2+66x+25,376$

Section 3.5 Practice

1. $10x\cdot 9x=(10\cdot 9)(x\cdot x)=90x^2$

2. $8x^3(-11x^7)=(8\cdot -11)(x^3\cdot x^7)=-88x^{10}$

3. $(-5x^4)(-x)=(-5\cdot -1)(x^4\cdot x)=5x^5$

4. $4x(x^2+4x+3)=4x(x^2)+4x(4x)+4x(3)$
$=4x^3+16x^2+12x$

5. $8x(7x^4+1)=8x(7x^4)+8x(1)=56x^5+8x$

6. $-2x^3(3x^2-x+2)=-2x^3(3x^2)-2x^3(-x)-2x^3(2)$
$=-6x^5+2x^4-4x^3$

7. a. $(x+5)(x+10)=x(x+10)+5(x+10)$
$=x\cdot x+x\cdot 10+5\cdot x+5\cdot 10$
$=x^2+10x+5x+50$
$=x^2+15x+50$

b. $(4x+5)(3x-4)=4x(3x-4)+5(3x-4)$
$=4x(3x)+4x(-4)+5(3x)+5(-4)$
$=12x^2-16x+15x-20$
$=12x^2-x-20$

8. $(3x-2y)^2=(3x-2y)(3x-2y)$
$=3x(3x)+3x(-2y)+(-2y)(3x)+(-2y)(-2y)$
$=9x^2-6xy-6xy+4y^2$
$=9x^2-12xy+4y^2$

9. $(x+3)(2x^2-5x+4)=x(2x^2)+x(-5x)+x(4)+3(2x^2)+3(-5x)+3(4)$
$=2x^3-5x^2+4x+6x^2-15x+12$
$=2x^3+x^2-11x+12$

Copyright © 2012 Pearson Education, Inc. Publishing as Prentice Hall.

10.

$$\begin{array}{r} y^2-4y+5 \\ 3y^2+1 \\ \hline y^2-4y+5 \\ 3y^4-12y^3+15y^2 \qquad\qquad \\ \hline 3y^4-12y^3+16y^2-4y+5 \end{array}$$

11.

$$\begin{array}{r} 4x^2 \quad -x-1 \\ 3x^2+6x-2 \\ \hline -8x^2+2x+2 \\ 24x^3-\ 6x^2-6x \qquad \\ 12x^4-\ 3x^3-\ 3x^2 \qquad\qquad \\ \hline 12x^4+21x^3-17x^2-4x+2 \end{array}$$

Vocabulary and Readiness Check

1. The expression $5x(3x + 2)$ equals $5x \cdot 3x + 5x \cdot 2$ by the <u>distributive</u> property.

2. The expression $(x + 4)(7x - 1)$ equals $x(7x - 1) + 4(7x - 1)$ by the <u>distributive</u> property.

3. The expression $(5y-1)^2$ equals <u>$(5y - 1)(5y - 1)$</u>.

4. The expression $9x \cdot 3x$ equals <u>$27x^2$</u>.

5. $x^3 \cdot x^5 = x^{3+5} = x^8$

6. $x^2 \cdot x^6 = x^{2+6} = x^8$

7. $x^3 + x^5$ cannot be simplified.

8. $x^2 + x^6$ cannot be simplified.

9. $x^7 \cdot x^7 = x^{7+7} = x^{14}$

10. $x^{11} \cdot x^{11} = x^{11+11} = x^{22}$

11. $x^7 + x^7 = 1x^7 + 1x^7 = (1+1)x^7 = 2x^7$

12. $x^{11} + x^{11} = 1x^{11} + 1x^{11} = (1+1)x^{11} = 2x^{11}$

13. $9y^2 \cdot 11y^2 = 9 \cdot 11 \cdot y^2 \cdot y^2 = 99y^{2+2} = 99y^4$

14. $6z^3 \cdot 7z^3 = 6 \cdot 7 \cdot z^3 \cdot z^3 = 42z^{3+3} = 42z^6$

15. $9y^2 + 11y^2 = (9+11)y^2 = 20y^2$

16. $6z^3 + 7z^3 = (6+7)z^3 = 13z^3$

Exercise Set 3.5

1. $8x^2 \cdot 3x = (8 \cdot 3)(x^2 \cdot x) = 24x^3$

3. $(-x^3)(-x) = (-1 \cdot -1)(x^3 \cdot x) = x^4$

5. $-4n^3 \cdot 7n^7 = (-4 \cdot 7)(n^3 \cdot n^7) = -28n^{10}$

7. $(-3.1x^3)(4x^9) = (-3.1 \cdot 4)(x^3 \cdot x^9) = -12.4x^{12}$

9. $\left(-\frac{1}{3}y^2\right)\left(\frac{2}{5}y\right) = \left(-\frac{1}{3}\right)\left(\frac{2}{5}\right)(y^2 \cdot y) = -\frac{2}{15}y^3$

11. $(2x)(-3x^2)(4x^5) = (2 \cdot -3 \cdot 4)(x \cdot x^2 \cdot x^5)$
$= -24x^8$

13. $3x(2x+5) = 3x(2x) + 3x(5) = 6x^2 + 15x$

15. $7x(x^2 + 2x - 1) = 7x(x^2) + 7x(2x) + 7x(-1)$
$= 7x^3 + 14x^2 - 7x$

17. $-2a(a+4) = -2a(a) + (-2a)(4) = -2a^2 - 8a$

19. $3x(2x^2 - 3x + 4) = 3x(2x^2) + 3x(-3x) + 3x(4)$
$= 6x^3 - 9x^2 + 12x$

21. $3a^2(4a^3 + 15) = 3a^2(4a^3) + 3a^2(15)$
$= 12a^5 + 45a^2$

23. $-2a^2(3a^2 - 2a + 3)$
$= -2a^2(3a^2) - 2a^2(-2a) - 2a^2(3)$
$= -6a^4 + 4a^3 - 6a^2$

25. $3x^2y(2x^3 - x^2y^2 + 8y^3)$
$= 3x^2y(2x^3) + 3x^2y(-x^2y^2) + 3x^2y(8y^3)$
$= 6x^5y - 3x^4y^3 + 24x^2y^4$

27. $-y(4x^3 - 7x^2y + xy^2 + 3y^3)$
$= -y(4x^3) - y(-7x^2y) - y(xy^2) - y(3y^3)$
$= -4x^3y + 7x^2y^2 - xy^3 - 3y^4$

Copyright © 2012 Pearson Education, Inc. Publishing as Prentice Hall.

29. $\frac{1}{2}x^2(8x^2-6x+1)=\frac{1}{2}x^2(8x^2)+\frac{1}{2}x^2(-6x)+\frac{1}{2}x^2(1)$
$=4x^4-3x^3+\frac{1}{2}x^2$

31. $(x+4)(x+3)=x(x+3)+4(x+3)$
$=x(x)+x(3)+4(x)+4(3)$
$=x^2+3x+4x+12$
$=x^2+7x+12$

33. $(a+7)(a-2)=a(a-2)+7(a-2)$
$=a(a)+a(-2)+7(a)+7(-2)$
$=a^2-2a+7a-14$
$=a^2+5a-14$

35. $\left(x+\frac{2}{3}\right)\left(x-\frac{1}{3}\right)=x\left(x-\frac{1}{3}\right)+\frac{2}{3}\left(x-\frac{1}{3}\right)$
$=x(x)+x\left(-\frac{1}{3}\right)+\frac{2}{3}(x)+\frac{2}{3}\left(-\frac{1}{3}\right)$
$=x^2-\frac{1}{3}x+\frac{2}{3}x-\frac{2}{9}$
$=x^2+\frac{1}{3}x-\frac{2}{9}$

37. $(3x^2+1)(4x^2+7)=3x^2(4x^2+7)+1(4x^2+7)$
$=3x^2(4x^2)+3x^2(7)+1(4x^2)+1(7)$
$=12x^4+21x^2+4x^2+7$
$=12x^4+25x^2+7$

39. $(4x-3)(3x-5)=4x(3x-5)+(-3)(3x-5)$
$=4x(3x)+4x(-5)+(-3)(3x)+(-3)(-5)$
$=12x^2-20x-9x+15$
$=12x^2-29x+15$

41. $(1-3a)(1-4a)=1(1-4a)+(-3a)(1-4a)$
$=1(1)+1(-4a)+(-3a)(1)+(-3a)(-4a)$
$=1-4a-3a+12a^2$
$=1-7a+12a^2$

43. $(2y-4)^2=(2y-4)(2y-4)$
$=2y(2y-4)+(-4)(2y-4)$
$=2y(2y)+2y(-4)+(-4)(2y)+(-4)(-4)$
$=4y^2-8y-8y+16$
$=4y^2-16y+16$

Copyright © 2012 Pearson Education, Inc. Publishing as Prentice Hall.

45. $(x-2)(x^2-3x+7) = x(x^2-3x+7)+(-2)(x^2-3x+7)$
$= x(x^2)+x(-3x)+x(7)+(-2)(x^2)+(-2)(-3x)+(-2)(7)$
$= x^3-3x^2+7x-2x^2+6x-14$
$= x^3-5x^2+13x-14$

47. $(x+5)(x^3-3x+4) = x(x^3-3x+4)+5(x^3-3x+4)$
$= x(x^3)+x(-3x)+x(4)+5(x^3)+5(-3x)+5(4)$
$= x^4-3x^2+4x+5x^3-15x+20$
$= x^4+5x^3-3x^2-11x+20$

49. $(2a-3)(5a^2-6a+4) = 2a(5a^2-6a+4)+(-3)(5a^2-6a+4)$
$= 2a(5a^2)+2a(-6a)+2a(4)+(-3)(5a^2)+(-3)(-6a)+(-3)(4)$
$= 10a^3-12a^2+8a-15a^2+18a-12$
$= 10a^3-27a^2+26a-12$

51. $(7xy-y)^2 = (7xy-y)(7xy-y)$
$= 7xy(7xy-y)+(-y)(7xy-y)$
$= 7xy(7xy)+7xy(-y)+(-y)(7xy)+(-y)(-y)$
$= 49x^2y^2-7xy^2-7xy^2+y^2$
$= 49x^2y^2-14xy^2+y^2$

53.

$$\begin{array}{r} 2x-11 \\ 6x+\ 1 \\ \hline 2x-11 \\ 12x^2-66x \\ \hline 12x^2-64x-11 \end{array}$$

55.

$$\begin{array}{r} 2x^2+4x-1 \\ x+3 \\ \hline 6x^2+12x-3 \\ 2x^3+\ 4x^2-\ x \\ \hline 2x^3+10x^2+11x-3 \end{array}$$

57.

$$\begin{array}{r} x^2\ +5x\ -7 \\ 2x^2\ -7x\ -9 \\ \hline -9x^2-45x+63 \\ -7x^3-35x^2+49x \\ 2x^4+10x^3-14x^2 \\ \hline 2x^4\ +3x^3-58x^2\ +4x+63 \end{array}$$

59. $-1.2y(-7y^6) = (-1.2\cdot -7)(y\cdot y^6) = 8.4y^7$

61. $-3x(x^2+2x-8) = -3x(x^2)+(-3x)(2x)+(-3x)(-8)$
$= -3x^3-6x^2+24x$

Copyright © 2012 Pearson Education, Inc. Publishing as Prentice Hall.

63. $(x+19)(2x+1) = x(2x+1)+19(2x+1)$
$= x(2x)+x(1)+19(2x)+19(1)$
$= 2x^2+x+38x+19$
$= 2x^2+39x+19$

65. $\left(x+\frac{1}{7}\right)\left(x-\frac{3}{7}\right)$
$= x\left(x-\frac{3}{7}\right)+\frac{1}{7}\left(x-\frac{3}{7}\right)$
$= x(x)+x\left(-\frac{3}{7}\right)+\frac{1}{7}(x)+\frac{1}{7}\left(-\frac{3}{7}\right)$
$= x^2-\frac{3}{7}x+\frac{1}{7}x-\frac{3}{49}$
$= x^2-\frac{2}{7}x-\frac{3}{49}$

67. $(3y+5)^2 = (3y+5)(3y+5)$
$= 3y(3y+5)+5(3y+5)$
$= 3y(3y)+3y(5)+5(3y)+5(5)$
$= 9y^2+15y+15y+25$
$= 9y^2+30y+25$

69. $(a+4)(a^2-6a+6)$
$= a(a^2-6a+6)+4(a^2-6a+6)$
$= a(a^2)+a(-6a)+a(6)+4(a^2)+4(-6a)+4(6)$
$= a^3-6a^2+6a+4a^2-24a+24$
$= a^3-2a^2-18a+24$

71. $(2x+5)(2x-5)$
$= 2x(2x-5)+5(2x-5)$
$= 2x(2x)+2x(-5)+5(2x)+5(-5)$
$= 4x^2-10x+10x-25$
$= 4x^2-25$
The area is $(4x^2-25)$ square yards.

73. Area $= \frac{1}{2}$(base)(height)
$= \frac{1}{2}(3x-2)(4x)$
$= 2x(3x-2)$
$= 2x(3x)+2x(-2)$
$= 6x^2-4x$
The area is $(6x^2-4x)$ square inches.

75. Add: $5a + 15a = (5 + 15)a = 20a$
Subtract: $5a - 15a = (5 - 15)a = -10a$
Multiply: $5a \cdot 15a = 5 \cdot 15 \cdot a^1 \cdot a^1 = 75a^{1+1} = 75a^2$
Divide: $\frac{5a}{15a} = \frac{1}{3}a^{1-1} = \frac{1}{3}a^0 = \frac{1}{3}\cdot 1 = \frac{1}{3}$

77. Add: $-3y^5+9y^4$ cannot be simplified.
Subtract: $-3y^5-9y^4$ cannot be simplified.
Multiply: $-3y^5 \cdot 9y^4 = -3 \cdot 9 \cdot y^5 \cdot y^4$
$= -27y^{5+4}$
$= -27y^9$
Divide: $\frac{-3y^5}{9y^4} = -\frac{3}{9}\cdot y^{5-4} = -\frac{1}{3}y^1 = -\frac{y}{3}$

79. a. $(3x+5)+(3x+7) = 3x+5+3x+7$
$= 6x+12$

b. $(3x+5)(3x+7)$
$= 3x(3x+7)+5(3x+7)$
$= 3x(3x)+3x(7)+5(3x)+5(7)$
$= 9x^2+21x+15x+35$
$= 9x^2+36x+35$
answers may vary

81. $(3x - 1) + (10x - 6) = 3x - 1 + 10x - 6 = 13x - 7$

83. $(3x-1)(10x-6)$
$= 3x(10x-6)+(-1)(10x-6)$
$= 3x(10x)+3x(-6)+(-1)(10x)+(-1)(-6)$
$= 30x^2-18x-10x+6$
$= 30x^2-28x+6$

85. $(3x-1)-(10x-6) = (3x-1)+(-10x+6)$
$= 3x-1-10x+6$
$= -7x+5$

87. The areas of the smaller rectangles are:
$x \cdot x = x^2$
$x \cdot 3 = 3x$
The area of the figure is x^2+3x.

89. The area of the figure is $x(1 + 2x)$.
The areas of the smaller rectangles are
$x \cdot 1 = x$
$x \cdot 2x = 2x^2$
The area of the figure is $x+2x^2$.

Copyright © 2012 Pearson Education, Inc. Publishing as Prentice Hall.

91. $5a + 6a = (5 + 6)a = 11a$

93. $(5x)^2 + (2y)^2 = 5^2x^2 + 2^2y^2 = 25x^2 + 4y^2$

95. a. $(a+b)(a-b) = a(a-b) + b(a-b)$
$= a(a) + a(-b) + b(a) + b(-b)$
$= a^2 - ab + ab - b^2$
$= a^2 - b^2$

b. $(2x+3y)(2x-3y)$
$= 2x(2x-3y) + 3y(2x-3y)$
$= 2x(2x) + 2x(-3y) + 3y(2x) + 3y(-3y)$
$= 4x^2 - 6xy + 6xy - 9y^2$
$= 4x^2 - 9y^2$

c. $(4x+7)(4x-7)$
$= 4x(4x-7) + 7(4x-7)$
$= 4x(4x) + 4x(-7) + 7(4x) + 7(-7)$
$= 16x^2 - 28x + 28x - 49$
$= 16x^2 - 49$

d. answers may vary

Section 3.6 Practice

1. $(x+7)(x-5)$
$= (x)(x) + (x)(-5) + (7)(x) + (7)(-5)$
$= x^2 - 5x + 7x - 35$
$= x^2 + 2x - 35$

2. $(6x-1)(x-4)$
$= 6x(x) + 6x(-4) + (-1)(x) + (-1)(-4)$
$= 6x^2 - 24x - x + 4$
$= 6x^2 - 25x + 4$

3. $(2y^2+3)(y-4) = 2y^3 - 8y^2 + 3y - 12$

4. $(2x+9)^2 = (2x+9)(2x+9)$
$= (2x)(2x) + (2x)(9) + 9(2x) + 9(9)$
$= 4x^2 + 18x + 18x + 81$
$= 4x^2 + 36x + 81$

5. $(y+3)^2 = y^2 + 2(y)(3) + 3^2 = y^2 + 6y + 9$

6. $(r-s)^2 = r^2 - 2(r)(s) + s^2 = r^2 - 2rs + s^2$

7. $(6x+5)^2 = (6x)^2 + 2(6x)(5) + 5^2$
$= 36x^2 + 60x + 25$

8. $(x^2-3y)^2 = (x^2)^2 - 2(x^2)(3y) + (3y)^2$
$= x^4 - 6x^2y + 9y^2$

9. $(x+9)(x-9) = x^2 - 9^2 = x^2 - 81$

10. $(5+4y)(5-4y) = 5^2 - (4y)^2 = 25 - 16y^2$

11. $\left(x - \frac{1}{3}\right)\left(x + \frac{1}{3}\right) = x^2 - \left(\frac{1}{3}\right)^2 = x^2 - \frac{1}{9}$

12. $(3a-b)(3a+b) = (3a)^2 - b^2 = 9a^2 - b^2$

13. $(2x^2-6y)(2x^2+6y) = (2x^2)^2 - (6y)^2$
$= 4x^4 - 36y^2$

14. $(7x-1)^2 = (7x)^2 - 2(7x)(1) + 1^2$
$= 49x^2 - 14x + 1$

15. $(5y+3)(2y-5)$
$= (5y)(2y) + (5y)(-5) + (3)(2y) + (3)(-5)$
$= 10y^2 - 25y + 6y - 15$
$= 10y^2 - 19y - 15$

16. $(2a-1)(2a+1) = (2a)^2 - (1)^2 = 4a^2 - 1$

17. $\left(5y - \frac{1}{9}\right)^2 = (5y)^2 - 2(5y)\left(\frac{1}{9}\right) + \left(\frac{1}{9}\right)^2$
$= 25y^2 - \frac{10}{9}y + \frac{1}{81}$

Vocabulary and Readiness Check

1. $(x+4)^2 = x^2 + 2(x)(4) + 4^2$
$= x^2 + 8x + 16$
$\neq x^2 + 16$
The statement is false.

2. The statement is true.

3. $(x+4)(x-4) = x^2 - 4^2 = x^2 - 16 \neq x^2 + 16$
The statement is false.

Copyright © 2012 Pearson Education, Inc. Publishing as Prentice Hall.

4. $(x-1)(x^3+3x-1) = x(x^3+3x-1)-1(x^3+3x-1)$
$= x^4+3x^2-x-x^3-3x+1$
$= x^4-x^3+3x^2-4x+1$
The product is a polynomial of degree 4, not 5. The statement is false.

Exercise Set 3.6

1. $(x+3)(x+4) = x^2+4x+3x+12 = x^2+7x+12$

3. $(x-5)(x+10) = x^2+10x-5x-50$
$= x^2+5x-50$

5. $(5x-6)(x+2) = 5x^2+10x-6x-12$
$= 5x^2+4x-12$

7. $(y-6)(4y-1) = 4y^2-y-24y+6$
$= 4y^2-25y+6$

9. $(2x+5)(3x-1) = 6x^2-2x+15x-5$
$= 6x^2+13x-5$

11. $(y^2+7)(6y+4) = 6y^3+4y^2+42y+28$

13. $\left(x-\frac{1}{3}\right)\left(x+\frac{2}{3}\right) = x^2+\frac{2}{3}x-\frac{1}{3}x-\frac{2}{9}$
$= x^2+\frac{1}{3}x-\frac{2}{9}$

15. $(0.4-3a)(0.2-5a) = 0.08-2.0a-0.6a+15a^2$
$= 0.08-2.6a+15a^2$

17. $(x+5y)(2x-y) = 2x^2-xy+10xy-5y^2$
$= 2x^2+9xy-5y^2$

19. $(x+2)^2 = x^2+2(2)(x)+2^2 = x^2+4x+4$

21. $(2a-3)^2 = (2a)^2-2(2a)(3)+(3)^2$
$= 4a^2-12a+9$

23. $(3a-5)^2 = (3a)^2-2(3a)(5)+5^2$
$= 9a^2-30a+25$

25. $(x^2+0.5)^2 = (x^2)^2+2(x^2)(0.5)+(0.5)^2$
$= x^4+x^2+0.25$

27. $\left(y-\frac{2}{7}\right)^2 = y^2-2(y)\left(\frac{2}{7}\right)+\left(\frac{2}{7}\right)^2$
$= y^2-\frac{4}{7}y+\frac{4}{49}$

29. $(2x-1)^2 = (2x)^2-2(2x)(1)+(1)^2$
$= 4x^2-4x+1$

31. $(5x+9)^2 = (5x)^2+2(5x)(9)+9^2$
$= 25x^2+90x+81$

33. $(3x-7y)^2 = (3x)^2-2(3x)(7y)+(7y)^2$
$= 9x^2-42xy+49y^2$

35. $(4m+5n)^2 = (4m)^2+2(4m)(5n)+(5n)^2$
$= 16m^2+40mn+25n^2$

37. $(5x^4-3)^2 = (5x^4)^2-2(5x^4)(3)+(3)^2$
$= 25x^8-30x^4+9$

39. $(a-7)(a+7) = a^2-7^2 = a^2-49$

41. $(x+6)(x-6) = (x)^2-(6)^2 = x^2-36$

43. $(3x-1)(3x+1) = (3x)^2-1^2 = 9x^2-1$

45. $(x^2+5)(x^2-5) = (x^2)^2-(5)^2 = x^4-25$

47. $(2y^2-1)(2y^2+1) = (2y^2)^2-1^2 = 4y^4-1$

49. $(4-7x)(4+7x) = (4)^2-(7x)^2 = 16-49x^2$

51. $\left(3x-\frac{1}{2}\right)\left(3x+\frac{1}{2}\right) = (3x)^2-\left(\frac{1}{2}\right)^2 = 9x^2-\frac{1}{4}$

53. $(9x+y)(9x-y) = (9x)^2-(y)^2 = 81x^2-y^2$

55. $(2m+5n)(2m-5n) = (2m)^2-(5n)^2$
$= 4m^2-25n^2$

57. $(a+5)(a+4) = a^2+4a+5a+20 = a^2+9a+20$

59. $(a-7)^2 = a^2-2(a)(7)+7^2 = a^2-14a+49$

Copyright © 2012 Pearson Education, Inc. Publishing as Prentice Hall.

61. $(4a+1)(3a-1)=12a^2-4a+3a-1$
$=12a^2-a-1$

63. $(x+2)(x-2)=x^2-2^2=x^2-4$

65. $(3a+1)^2=(3a)^2+2(3a)(1)+(1)^2=9a^2+6a+1$

67. $(x+y)(4x-y)=4x^2-xy+4xy-y^2$
$=4x^2+3xy-y^2$

69. $\left(\frac{1}{3}a^2-7\right)\left(\frac{1}{3}a^2+7\right)=\left(\frac{1}{3}a^2\right)^2-7^2$
$=\frac{1}{9}a^4-49$

71. $(3b+7)(2b-5)=6b^2-15b+14b-35$
$=6b^2-b-35$

73. $(x^2+10)(x^2-10)=(x^2)^2-(10)^2=x^4-100$

75. $(4x+5)(4x-5)=(4x)^2-5^2=16x^2-25$

77. $(5x-6y)^2=(5x)^2-2(5x)(6y)+(6y)^2$
$=25x^2-60xy+36y^2$

79. $(2r-3s)(2r+3s)=(2r)^2-(3s)^2=4r^2-9s^2$

81. $(2x+1)^2=(2x)^2+2(2x)(1)+(1)^2=4x^2+4x+1$
The area of the rug is $(4x^2+4x+1)$ square feet.

83. $\frac{50b^{10}}{70b^5}=\frac{50}{70}\cdot\frac{b^{10}}{b^5}=\frac{5}{7}b^{10-5}=\frac{5}{7}b^5=\frac{5b^5}{7}$

85. $\frac{8a^{17}b^5}{-4a^7b^{10}}=\frac{8}{-4}\cdot\frac{a^{17}}{a^7}\cdot\frac{b^5}{b^{10}}$
$=-2a^{17-7}b^{5-10}$
$=-2a^{10}b^{-5}$
$=-\frac{2a^{10}}{b^5}$

87. $\frac{2x^4y^{12}}{3x^4y^4}=\frac{2}{3}\cdot\frac{x^4}{x^4}\cdot\frac{y^{12}}{y^4}$
$=\frac{2}{3}x^{4-4}y^{12-4}$
$=\frac{2}{3}x^0y^8$
$=\frac{2y^8}{3}$

89. $(a-b)^2=(a)^2-2(a)(b)+(b)^2=a^2-2ab+b^2$
which is choice c.

91. $(a+b)^2=(a)^2+2(a)(b)+(b)^2=a^2+2ab+b^2$
which is choice d.

93. Since $x^2\cdot x^2=x^4$ and $3x^2+7x^2=10x^2$, $(x^2+7)(x^2+3)=x^4+10x^2+21$ is a true statement.

95. $(x^2-1)^2-x^2=((x^2)^2-2(x^2)(1)+1^2)-x^2$
$=(x^4-2x^2+1)-x^2$
$=x^4-2x^2+1-x^2$
$=x^4-3x^2+1$
The area is (x^4-3x^2+1) square meters.

97. $(5x-3)^2-(x+1)^2$
$=((5x)^2-2(5x)(3)+3^2)-(x^2+2x+1^2)$
$=(25x^2-30x+9)+(-x^2-2x-1)$
$=25x^2-30x+9-x^2-2x-1$
$=24x^2-32x+8$
The area is $(24x^2-32x+8)$ square meters.

99. answers may vary

101. answers may vary

Integrated Review

1. $(5x^2)(7x^3)=(5\cdot 7)(x^2\cdot x^3)=35x^5$

2. $(4y^2)(-8y^7)=(4\cdot -8)(y^2\cdot y^7)=-32y^9$

3. $-4^2=-(4^2)=-16$

4. $(-4)^2=(-4)(-4)=16$

Copyright © 2012 Pearson Education, Inc. Publishing as Prentice Hall.

5. $(x-5)(2x+1) = 2x^2 + 1x - 10x - 5 = 2x^2 - 9x - 5$

6. $(3x-2)(x+5) = 3x^2 + 15x - 2x - 10$
$= 3x^2 + 13x - 10$

7. $(x-5) + (2x+1) = x - 5 + 2x + 1 = 3x - 4$

8. $(3x-2) + (x+5) = 3x - 2 + x + 5 = 4x + 3$

9. $\dfrac{7x^9y^{12}}{x^3y^{10}} = 7 \cdot x^{9-3} \cdot y^{12-10} = 7x^6y^2$

10. $\dfrac{20a^2b^8}{14a^2b^2} = \dfrac{20}{14}a^{2-2}b^{8-2} = \dfrac{10a^0b^6}{7} = \dfrac{10b^6}{7}$

11. $(12m^7n^6)^2 = 12^2 \cdot m^{7\cdot 2}n^{6\cdot 2} = 144m^{14}n^{12}$

12. $(4y^9z^{10})^3 = 4^3 \cdot y^{9\cdot 3}z^{10\cdot 3} = 64y^{27}z^{30}$

13. $(4y-3)(4y+3) = (4y)^2 - 3^2 = 16y^2 - 9$

14. $(7x-1)(7x+1) = (7x)^2 - 1^2 = 49x^2 - 1$

15. $(x^{-7}y^5)^9 = x^{-7\cdot 9}y^{5\cdot 9} = x^{-63}y^{45} = \dfrac{y^{45}}{x^{63}}$

16. $8^{-2} = \dfrac{1}{8^2} = \dfrac{1}{64}$

17. $(3^{-1}x^9)^3 = 3^{-1\cdot 3}x^{9\cdot 3} = 3^{-3}x^{27} = \dfrac{x^{27}}{3^3} = \dfrac{x^{27}}{27}$

18. $\dfrac{(r^7s^{-5})^6}{(2r^{-4}s^{-4})^4} = \dfrac{r^{7\cdot 6}s^{-5\cdot 6}}{2^4r^{-4\cdot 4}s^{-4\cdot 4}}$
$= \dfrac{r^{42}s^{-30}}{16r^{-16}s^{-16}}$
$= \dfrac{r^{42-(-16)}s^{-30-(-16)}}{16}$
$= \dfrac{r^{58}s^{-14}}{16}$
$= \dfrac{r^{58}}{16s^{14}}$

19. $(7x^2 - 2x + 3) - (5x^2 + 9)$
$= (7x^2 - 2x + 3) + (-5x^2 - 9)$
$= 7x^2 - 2x + 3 - 5x^2 - 9$
$= 2x^2 - 2x - 6$

20. $(10x^2 + 7x - 9) - (4x^2 - 6x + 2)$
$= (10x^2 + 7x - 9) + (-4x^2 + 6x - 2)$
$= 10x^2 + 7x - 9 - 4x^2 + 6x - 2$
$= 6x^2 + 13x - 11$

21. $0.7y^2 - 1.2 + 1.8y^2 - 6y + 1 = 2.5y^2 - 6y - 0.2$

22. $7.8x^2 - 6.8x - 3.3 + 0.6x^2 - 0.9$
$= 8.4x^2 - 6.8x - 4.2$

23. $(3y^2 - 6y + 1) - (y^2 + 2)$
$= (3y^2 - 6y + 1) + (-y^2 - 2)$
$= 3y^2 - 6y + 1 - y^2 - 2$
$= 2y^2 - 6y - 1$

24. $(z^2 + 5) - (3z^2 - 1) + \left(8z^2 + 2z - \dfrac{1}{2}\right)$
$= (z^2 + 5) + (-3z^2 + 1) + \left(8z^2 + 2z - \dfrac{1}{2}\right)$
$= z^2 + 5 - 3z^2 + 1 + 8z^2 + 2z - \dfrac{1}{2}$
$= 6z^2 + 2z + \dfrac{11}{2}$

25. $(x+4)^2 = x^2 + 2(x)(4) + 4^2 = x^2 + 8x + 16$

26. $(y-9)^2 = y^2 - 2(y)(9) + 9^2 = y^2 - 18x + 81$

27. $(x+4) + (x+4) = x + 4 + x + 4 = 2x + 8$

28. $(y-9) + (y-9) = y - 9 + y - 9 = 2y - 18$

29. $7x^2 - 6xy + 4(y^2 - xy) = 7x^2 - 6xy + 4y^2 - 4xy$
$= 7x^2 - 10xy + 4y^2$

30. $5a^2 - 3ab + 6(b^2 - a^2) = 5a^2 - 3ab + 6b^2 - 6a^2$
$= -a^2 - 3ab + 6b^2$

Copyright © 2012 Pearson Education, Inc. Publishing as Prentice Hall.

31. $(x-3)(x^2+5x-1) = x(x^2+5x-1)+(-3)(x^2+5x-1)$
$= x(x^2)+x(5x)+x(-1)+(-3)(x^2)+(-3)(5x)+(-3)(-1)$
$= x^3+5x^2-x-3x^2-15x+3$
$= x^3+2x^2-16x+3$

32. $(x+1)(x^2-3x-2) = x(x^2-3x-2)+1(x^2-3x-2)$
$= x(x^2)+x(-3x)+x(-2)+1(x^2)+1(-3x)+1(-2)$
$= x^3-3x^2-2x+x^2-3x-2$
$= x^3-2x^2-5x-2$

33. $(2x-7)(3x+10) = 6x^2+20x-21x-70$
$= 6x^2-x-70$

34. $(5x-1)(4x+5) = 20x^2+25x-4x-5$
$= 20x^2+21x-5$

35. $(2x-7)(x^2-6x+1) = 2x(x^2-6x+1)+(-7)(x^2-6x+1)$
$= 2x(x^2)+2x(-6x)+2x(1)+(-7)(x^2)+(-7)(-6x)+(-7)(1)$
$= 2x^3-12x^2+2x-7x^2+42x-7$
$= 2x^3-19x^2+44x-7$

36. $(5x-1)(x^2+2x-3) = 5x(x^2+2x-3)+(-1)(x^2+2x-3)$
$= 5x(x^2)+5x(2x)+5x(-3)+(-1)(x^2)+(-1)(2x)+(-1)(-3)$
$= 5x^3+10x^2-15x-x^2-2x+3$
$= 5x^3+9x^2-17x+3$

37. $\left(2x+\frac{5}{9}\right)\left(2x-\frac{5}{9}\right) = (2x)^2-\left(\frac{5}{9}\right)^2 = 4x^2-\frac{25}{81}$

38. $\left(12y+\frac{3}{7}\right)\left(12y-\frac{3}{7}\right) = (12y)^2-\left(\frac{3}{7}\right)^2$
$= 144y^2-\frac{9}{49}$

Section 3.7 Practice

1. $(25x^3+5x^2)\div 5x^2 = \frac{25x^3+5x^2}{5x^2} = \frac{25x^3}{5x^2}+\frac{5x^2}{5x^2} = 5x+1$

2. $\frac{24x^7+12x^2-4x}{4x^2} = \frac{24x^7}{4x^2}+\frac{12x^2}{4x^2}-\frac{4x}{4x^2}$
$= 6x^5+3-\frac{1}{x}$

Copyright © 2012 Pearson Education, Inc. Publishing as Prentice Hall.

3. $\dfrac{12x^3y^3-18xy+6y}{3xy}=\dfrac{12x^3y^3}{3xy}-\dfrac{18xy}{3xy}+\dfrac{6y}{3xy}$

$=4x^2y^2-6+\dfrac{2}{x}$

4.
$$\begin{array}{r} x+7 \\ x+5\overline{\big)\,x^2+12x+35} \\ \underline{x^2+\ 5x} \\ 7x+35 \\ \underline{7x+35} \\ 0 \end{array}$$

Thus, $\dfrac{x^2+12x+35}{x+5}=x+7.$

5.
$$\begin{array}{r} 4x+3 \\ 2x-1\overline{\big)\,8x^2+2x-7} \\ \underline{8x^2-4x} \\ 6x-7 \\ \underline{6x-3} \\ -4 \end{array}$$

Thus, $\dfrac{8x^2+2x-7}{2x-1}=4x+3+\dfrac{-4}{2x-1}$ or

$4x+3-\dfrac{4}{2x-1}.$

6. $15-2x^2=-2x^2+0x+15$

$$\begin{array}{r} -2x-6 \\ x-3\overline{\big)\,-2x^2+0x+15} \\ \underline{-2x^2+6x} \\ -6x+15 \\ \underline{-6x+18} \\ -3 \end{array}$$

Thus, $\dfrac{15-2x^2}{x-3}=-2x-6+\dfrac{-3}{x-3}$ or

$-2x-6-\dfrac{3}{x-3}.$

7. $\dfrac{5-x+9x^3}{3x+2}=\dfrac{9x^3+0x^2-x+5}{3x+2}$

$$\begin{array}{r} 3x^2-2x+1 \\ 3x+2\overline{\big)\,9x^3+0x^2-\ x+5} \\ \underline{9x^3+6x^2} \\ -6x^2-\ x \\ \underline{-6x^2-4x} \\ 3x+5 \\ \underline{3x+2} \\ 3 \end{array}$$

Thus, $\dfrac{5-x+9x^3}{3x+2}=3x^2-2x+1+\dfrac{3}{3x+2}.$

8.
$$\begin{array}{r} x^2+x+1 \\ x-1\overline{\big)\,x^3+0x^2+0x-1} \\ \underline{x^3-\ x^2} \\ x^2+0x \\ \underline{x^2-\ x} \\ x-1 \\ \underline{x-1} \\ 0 \end{array}$$

Thus, $\dfrac{x^3-1}{x-1}=x^2+x+1.$

Vocabulary and Readiness Check

1. In $\begin{array}{r} 3 \\ 6\overline{\big)\,18} \end{array}$, the 18 is the <u>dividend</u>, the 3 is the <u>quotient</u>, and the 6 is the <u>divisor</u>.

2. In $\begin{array}{r} x+2 \\ x+1\overline{\big)\,x^2+3x+2} \end{array}$, the $x+1$ is the <u>divisor</u>, the x^2+3x+2 is the <u>dividend</u>, and the $x+2$ is the <u>quotient</u>.

3. $\dfrac{a^6}{a^4}=a^{6-4}=a^2$

4. $\dfrac{p^8}{p^3}=p^{8-3}=p^5$

5. $\dfrac{y^2}{y}=\dfrac{y^2}{y^1}=y^{2-1}=y^1=y$

Copyright © 2012 Pearson Education, Inc. Publishing as Prentice Hall.

6. $\dfrac{a^3}{a}=\dfrac{a^3}{a^1}=a^{3-1}=a^2$

Exercise Set 3.7

1. $\dfrac{12x^4+3x^2}{x}=\dfrac{12x^4}{x}+\dfrac{3x^2}{x}=12x^3+3x$

3. $\dfrac{20x^3-30x^2+5x+5}{5}=\dfrac{20x^3}{5}-\dfrac{30x^2}{5}+\dfrac{5x}{5}+\dfrac{5}{5}$
$=4x^3-6x^2+x+1$

5. $\dfrac{15p^3+18p^2}{3p}=\dfrac{15p^3}{3p}+\dfrac{18p^2}{3p}=5p^2+6p$

7. $\dfrac{-9x^4+18x^5}{6x^5}=\dfrac{-9x^4}{6x^5}+\dfrac{18x^5}{6x^5}=\dfrac{-3}{2x}+3$

9. $\dfrac{-9x^5+3x^4-12}{3x^3}=\dfrac{-9x^5}{3x^3}+\dfrac{3x^4}{3x^3}-\dfrac{12}{3x^3}$
$=-3x^2+x-\dfrac{4}{x^3}$

11. $\dfrac{4x^4-6x^3+7}{-4x^4}=\dfrac{4x^4}{-4x^4}-\dfrac{6x^3}{-4x^4}+\dfrac{7}{-4x^4}$
$=-1+\dfrac{3}{2x}-\dfrac{7}{4x^4}$

13.
$$\begin{array}{r|l} & \phantom{x^2+{}}x+1 \\ \hline x+3 & x^2+4x+3 \\ & \underline{x^2+3x} \\ & \phantom{x^2+{}}x+3 \\ & \phantom{x^2+{}}\underline{x+3} \\ & \phantom{x^2+{}x+{}}0 \end{array}$$

$\dfrac{x^2+4x+3}{x+3}=x+1$

15.
$$\begin{array}{r|l} & \phantom{x^2+{}}2x+3 \\ \hline x+5 & 2x^2+13x+15 \\ & \underline{2x^2+10x} \\ & \phantom{2x^2+{}}3x+15 \\ & \phantom{2x^2+{}}\underline{3x+15} \\ & \phantom{2x^2+{}x+{}}0 \end{array}$$

$\dfrac{2x^2+13x+15}{x+5}=2x+3$

17.
$$\begin{array}{r|l} & \phantom{x^2-{}}2x+1 \\ \hline x-4 & 2x^2-7x+3 \\ & \underline{2x^2-8x} \\ & \phantom{2x^2-{}}x+3 \\ & \phantom{2x^2-{}}\underline{x-4} \\ & \phantom{2x^2-{}x-{}}7 \end{array}$$

$\dfrac{2x^2-7x+3}{x-4}=2x+1+\dfrac{7}{x-4}$

19.
$$\begin{array}{r|l} & \phantom{9a^3-{}}3a^2-3a+1 \\ \hline 3a+2 & 9a^3-3a^2-3a+4 \\ & \underline{9a^3+6a^2} \\ & \phantom{9a^3+{}}-9a^2-3a \\ & \phantom{9a^3+{}}\underline{-9a^2-6a} \\ & \phantom{9a^3+{}-9a^2-{}}3a+4 \\ & \phantom{9a^3+{}-9a^2-{}}\underline{3a+2} \\ & \phantom{9a^3+{}-9a^2-{}3a+{}}2 \end{array}$$

$\dfrac{9a^3-3a^2-3a+4}{3a+2}=3a^2-3a+1+\dfrac{2}{3a+2}$

21.
$$\begin{array}{r|l} & \phantom{8x^2+{}}4x+3 \\ \hline 2x+1 & 8x^2+10x+1 \\ & \underline{8x^2+4x} \\ & \phantom{8x^2+{}}6x+1 \\ & \phantom{8x^2+{}}\underline{6x+3} \\ & \phantom{8x^2+{}x+{}}-2 \end{array}$$

$\dfrac{8x^2+10x+1}{2x+1}=4x+3-\dfrac{2}{2x+1}$

23.
$$\begin{array}{r|l} & \phantom{2x^3+{}}2x^2+6x-5 \\ \hline x-2 & 2x^3+2x^2-17x+8 \\ & \underline{2x^3-4x^2} \\ & \phantom{2x^3+{}}6x^2-17x \\ & \phantom{2x^3+{}}\underline{6x^2-12x} \\ & \phantom{2x^3+{}6x^2-{}}-5x+8 \\ & \phantom{2x^3+{}6x^2-{}}\underline{-5x+10} \\ & \phantom{2x^3+{}6x^2-{}x+{}}-2 \end{array}$$

$\dfrac{2x^3+2x^2-17x+8}{x-2}=2x^2+6x-5-\dfrac{2}{x-2}$

Copyright © 2012 Pearson Education, Inc. Publishing as Prentice Hall.

25.
$$\begin{array}{r} x+6 \\ x-6\overline{\big)\,x^2+0x-36} \\ \underline{x^2-6x} \\ 6x-36 \\ \underline{6x-36} \\ 0 \end{array}$$

$$\frac{x^2-36}{x-6}=x+6$$

27.
$$\begin{array}{r} x^2+3x+9 \\ x-3\overline{\big)\,x^3+0x^2+0x-27} \\ \underline{x^3-3x^2} \\ 3x^2+0x \\ \underline{3x^2-9x} \\ 9x-27 \\ \underline{9x-27} \\ 0 \end{array}$$

$$\frac{x^3-27}{x-3}=x^2+3x+9$$

29. $1-3x^2=-3x^2+0x+1$

$$\begin{array}{r} -3x+6 \\ x+2\overline{\big)\,-3x^2+0x\ \ +1} \\ \underline{-3x^2-6x} \\ 6x+\ 1 \\ \underline{6x+12} \\ -11 \end{array}$$

$$\frac{1-3x^2}{x+2}=-3x+6-\frac{11}{x+2}$$

31.
$$\begin{array}{r} 2b-1 \\ 2b-1\overline{\big)\,-4b^2-4b-5} \\ \underline{4b^2-2b} \\ -2b-5 \\ \underline{-2b+1} \\ -6 \end{array}$$

$$\frac{-4b+4b^2-5}{2b-1}=2b-1-\frac{6}{2b-1}$$

33. $\dfrac{a^2b^2-ab^3}{ab}=\dfrac{a^2b^2}{ab}-\dfrac{ab^3}{ab}=ab-b^2$

35.
$$\begin{array}{r} 4x+9 \\ 2x-3\overline{\big)\,8x^2+\ 6x-27} \\ \underline{8x^2-12x} \\ 18x-27 \\ \underline{18x-27} \\ 0 \end{array}$$

$$\frac{8x^2+6x-27}{2x-3}=4x+9$$

37. $\dfrac{2x^2y+8x^2y^2-xy^2}{2xy}=\dfrac{2x^2y}{2xy}+\dfrac{8x^2y^2}{2xy}-\dfrac{xy^2}{2xy}$

$$=x+4xy-\frac{y}{2}$$

39.
$$\begin{array}{r} 2b^2+b+2 \\ b+4\overline{\big)\,2b^3+9b^2+6b-4} \\ \underline{2b^3+8b^2} \\ b^2+6b \\ \underline{b^2+4b} \\ 2b-4 \\ \underline{2b+8} \\ -12 \end{array}$$

$$\frac{2b^3+9b^2+6b-4}{b+4}=2b^2+b+2-\frac{12}{b+4}$$

41.
$$\begin{array}{r} y^2+5y+10 \\ y-2\overline{\big)\,y^3+3y^2+\ 0y+\ 4} \\ \underline{y^3-2y^2} \\ 5y^2+\ 0y \\ \underline{5y^2-10y} \\ 10y+\ 4 \\ \underline{10y-20} \\ 24 \end{array}$$

$$\frac{y^3+3y^2+4}{y-2}=y^2+5y+10+\frac{24}{y-2}$$

43.
$$\begin{array}{r} -6x-12 \\ x-2\overline{\big)\,-6x^2+\ 0x+\ 5} \\ \underline{-6x^2+12x} \\ -12x+\ 5 \\ \underline{-12x+24} \\ -19 \end{array}$$

$$\frac{5-6x^2}{x-2}=-6x-12-\frac{19}{x-2}$$

Copyright © 2012 Pearson Education, Inc. Publishing as Prentice Hall.

45.
$$\begin{array}{r} x^3 - x^2 + x \\ x^2 + x \overline{\big) x^5 + 0x^4 + 0x^3 + x^2} \\ \underline{x^5 + x^4} \\ -x^4 + 0x^3 \\ \underline{-x^4 - x^3} \\ x^3 + x^2 \\ \underline{x^3 + x^2} \\ 0 \end{array}$$

$$\frac{x^5 + x^2}{x^2 + x} = x^3 - x^2 + x$$

47. $\frac{12}{4} = 3$, so $12 = 4 \cdot 3$.

49. $\frac{20}{-5} = -4$, so $20 = -5 \cdot -4$.

51. $\frac{9x^2}{3x} = 3x$, so $9x^2 = 3x \cdot 3x$.

53. $\frac{36x^2}{4x} = 9x$, so $36x^2 = 4x \cdot 9x$.

55. $\frac{12x^3 + 4x - 16}{4} = \frac{12x^3}{4} + \frac{4x}{4} - \frac{16}{4}$
$= 3x^3 + x - 4$

The length of each side of the square is $(3x^3 + x - 4)$ feet.

57.
$$\begin{array}{r} 2x + 5 \\ 5x + 3 \overline{\big) 10x^2 + 31x + 15} \\ \underline{10x^2 + 6x} \\ 25x + 15 \\ \underline{25x + 15} \\ 0 \end{array}$$

The height of the parallelogram is $(2x + 5)$ meters.

59. answers may vary

61. $\frac{a+7}{7} = \frac{a}{7} + \frac{7}{7} = \frac{a}{7} + 1$ which is choice c.

Chapter 3 Vocabulary Check

1. A <u>term</u> is a number or the product of numbers and variables raised to powers.
2. The <u>FOIL</u> method may be used when multiplying two binomials.
3. A polynomial with exactly 3 terms is called a <u>trinomial</u>.
4. The <u>degree of polynomial</u> is the greatest degree of any term of the polynomial.
5. A polynomial with exactly 2 terms is called a <u>binomial</u>.
6. The <u>coefficient</u> of a term is its numerical factor.
7. The <u>degree of a term</u> is the sum of the exponents on the variables in the term.
8. A polynomial with exactly 1 term is called a <u>monomial</u>.
9. Monomials, binomials, and trinomials are all examples of <u>polynomials</u>.
10. The <u>distributive</u> property is used to multiply $2x(x - 4)$.

Chapter 3 Review

1. In 3^2, the base is 3 and the exponent is 2.
2. In $(-5)^4$, the base is -5 and the exponent is 4.
3. In -5^4, the base is 5 and the exponent is 4.
4. In x^6, the base is x and the exponent is 6.
5. $8^3 = 8 \cdot 8 \cdot 8 = 512$
6. $(-6)^2 = (-6)(-6) = 36$
7. $-6^2 = -(6 \cdot 6) = -36$
8. $-4^3 - 4^0 = -64 - 1 = -65$
9. $(3b)^0 = 1$
10. $\frac{8b}{8b} = 1$
11. $y^2 \cdot y^7 = y^{2+7} = y^9$

Copyright © 2012 Pearson Education, Inc. Publishing as Prentice Hall.

12. $x^9 \cdot x^5 = x^{9+5} = x^{14}$

13. $(2x^5)(-3x^6) = -6x^{5+6} = -6x^{11}$

14. $(-5y^3)(4y^4) = -20y^{3+4} = -20y^7$

15. $(x^4)^2 = x^{4\cdot 2} = x^8$

16. $(y^3)^5 = y^{3\cdot 5} = y^{15}$

17. $(3y^6)^4 = 3^4 \cdot y^{6\cdot 4} = 81y^{24}$

18. $(2x^3)^3 = 2^3 \cdot x^{3\cdot 3} = 8x^9$

19. $\dfrac{x^9}{x^4} = x^{9-4} = x^5$

20. $\dfrac{z^{12}}{z^5} = z^{12-5} = z^7$

21. $\dfrac{3x^4y^{10}}{12xy^6} = \dfrac{3}{12}x^{4-1}y^{10-6} = \dfrac{1}{4}x^3y^4 = \dfrac{x^3y^4}{4}$

22. $\dfrac{2x^7y^8}{8xy^2} = \dfrac{2}{8}x^{7-1}y^{8-2} = \dfrac{1}{4}x^6y^6 = \dfrac{x^6y^6}{4}$

23. $5a^7(2a^4)^3 = 5a^7 \cdot 2^3 \cdot (a^4)^3$
$= 5a^7 \cdot 8 \cdot a^{12}$
$= 40a^{7+12}$
$= 40a^{19}$

24. $(2x)^2(9x) = 4x^2(9x)$
$= 36x^{2+1}$
$= 36x^3$

25. $\dfrac{(4a^5b)^2}{-16ab^2} = \dfrac{4^2(a^5)^2b^2}{-16ab^2}$
$= \dfrac{16a^{10}b^2}{-16ab^2}$
$= \dfrac{16}{-16} \cdot a^{10-1} \cdot b^{2-2}$
$= -1 \cdot a^9 \cdot b^0$
$= -1 \cdot a^9 \cdot 1$
$= -a^9$

26. $\dfrac{(2x^3y)^4}{-16x^5y^4} = \dfrac{2^4(x^3)^4y^4}{-16x^5y^4}$
$= \dfrac{16x^{12}y^4}{-16x^5y^4}$
$= \dfrac{16}{-16} \cdot x^{12-5} \cdot y^{4-4}$
$= -1 \cdot x^7 \cdot y^0$
$= -1 \cdot x^7 \cdot 1$
$= -x^7$

27. $(-5a)^0 + 7^0 + 8^0 = 1+1+1 = 3$

28. $8x^0 + 9^0 = 8 \cdot 1 + 1 = 8 + 1 = 9$

29. $\left(\dfrac{3x^4}{4y}\right)^3 = \dfrac{3^3x^{4\cdot 3}}{4^3y^{1\cdot 3}} = \dfrac{27x^{12}}{64y^3}$; b

30. $\left(\dfrac{5a^6}{b^3}\right)^2 = \dfrac{5^2a^{6\cdot 2}}{b^{3\cdot 2}} = \dfrac{25a^{12}}{b^6}$; c

31. $7^{-2} = \dfrac{1}{7^2} = \dfrac{1}{49}$

32. $-7^{-2} = -\dfrac{1}{7^2} = -\dfrac{1}{49}$

33. $2x^{-4} = \dfrac{2}{x^4}$

34. $(2x)^{-4} = \dfrac{1}{(2x)^4} = \dfrac{1}{2^4x^4} = \dfrac{1}{16x^4}$

35. $\left(\dfrac{1}{5}\right)^{-3} = (5)^3 = 125$

36. $\left(\dfrac{-2}{3}\right)^{-2} = \left(\dfrac{3}{-2}\right)^2 = \dfrac{9}{4}$

37. $2^0 + 2^{-4} = 1 + \dfrac{1}{2^4} = 1 + \dfrac{1}{16} = \dfrac{17}{16}$

38. $6^{-1} - 7^{-1} = \dfrac{1}{6} - \dfrac{1}{7} = \dfrac{7}{42} - \dfrac{6}{42} = \dfrac{1}{42}$

Copyright © 2012 Pearson Education, Inc. Publishing as Prentice Hall.

39. $\dfrac{r^{-3}}{r^{-4}} = r^{-3-(-4)} = r^1 = r$

40. $\dfrac{y^{-2}}{y^{-5}} = y^{-2-(-5)} = y^3$

41. $\left(\dfrac{bc^{-2}}{bc^{-3}}\right)^4 = (b^{1-1}c^{-2-(-3)})^4 = (b^0c^1)^4 = c^4$

42. $\left(\dfrac{x^{-3}y^{-4}}{x^{-2}y^{-5}}\right)^{-3} = (x^{-3-(-2)}y^{-4-(-5)})^{-3}$

$= (x^{-1}y^1)^{-3}$

$= x^3y^{-3}$

$= x^3 \cdot \dfrac{1}{y^3}$

$= \dfrac{x^3}{y^3}$

43. $\dfrac{10a^3b^4c^0}{50ab^{11}c^3} = \dfrac{10}{50} \cdot a^{3-1} \cdot b^{4-11} \cdot c^{0-3}$

$= \dfrac{1}{5} \cdot a^2 \cdot b^{-7} \cdot c^{-3}$

$= \dfrac{1}{5} \cdot a^2 \cdot \dfrac{1}{b^7} \cdot \dfrac{1}{c^3}$

$= \dfrac{a^2}{5b^7c^3}$

44. $\dfrac{8a^0b^4c^5}{40a^6bc^{12}} = \dfrac{8}{40} \cdot a^{0-6} \cdot b^{4-1} \cdot c^{5-12}$

$= \dfrac{1}{5} \cdot a^{-6} \cdot b^3 \cdot c^{-7}$

$= \dfrac{1}{5} \cdot \dfrac{1}{a^6} \cdot b^3 \cdot \dfrac{1}{c^7}$

$= \dfrac{b^3}{5a^6c^7}$

45. $\dfrac{9x^{-4}y^{-6}}{x^2y^7} = 9x^{-4-2}y^{-6-7}$

$= 9x^{-6}y^{-13}$

$= 9 \cdot \dfrac{1}{x^6} \cdot \dfrac{1}{y^{13}}$

$= \dfrac{9}{x^6y^{13}}$

46. $\dfrac{3a^5b^{-5}}{a^{-5}b^5} = 3a^{5-(-5)}b^{(-5)-5}$

$= 3a^{10}b^{-10}$

$= 3 \cdot a^{10} \cdot \dfrac{1}{b^{10}}$

$= \dfrac{3a^{10}}{b^{10}}$

47. $0.00027 = 2.7 \times 10^{-4}$

48. $0.8868 = 8.868 \times 10^{-1}$

49. $80{,}800{,}000 = 8.08 \times 10^7$

50. $868{,}000 = 8.68 \times 10^5$

51. $127{,}000{,}000 = 1.27 \times 10^8$

52. $150{,}000 = 1.5 \times 10^5$

53. $8.67 \times 10^5 = 867{,}000$

54. $3.86 \times 10^{-3} = 0.00386$

55. $8.6 \times 10^{-4} = 0.00086$

56. $8.936 \times 10^5 = 893{,}600$

57. $1.43128 \times 10^{15} = 1{,}431{,}280{,}000{,}000{,}000$

58. $1 \times 10^{-10} = 0.0000000001$

59. $(8 \times 10^4)(2 \times 10^{-7}) = 16 \times 10^{-3} = 0.016$

60. $\dfrac{8 \times 10^4}{2 \times 10^{-7}} = 4 \times 10^{11} = 400{,}000{,}000{,}000$

61. The degree of $(y^5 + 7x - 8x^4)$ is 5.

62. The degree of $(9y^2 + 30y + 25)$ is 2.

63. The degree of $(-14x^2y - 28x^2y^3 - 42x^2y^2)$ is 2 + 3 or 5.

64. The degree of $(6x^2y^2z^2 + 5x^2y^3 - 12xyz)$ is 2 + 2 + 2 or 6.

Copyright © 2012 Pearson Education, Inc. Publishing as Prentice Hall.

65. $-16t^2+4000=-16(0)^2+4000$
$=0+4000$
$=4000$

$-16t^2+4000=-16(1)^2+4000$
$=-16+4000$
$=3984$

$-16t^2+4000=-16(3)^2+4000$
$=-144+4000$
$=3856$

$-16t^2+4000=-16(5)^2+4000$
$=-400+4000$
$=3600$

t	0 seconds	1 second	3 seconds	5 seconds
$-16t^2+4000$	4000 feet	3984 feet	3856 feet	3600 feet

66. $2x^2+20x=2(1)^2+20(1)=2+20=22$
$2x^2+20x=2(3)^2+20(3)=18+60=78$
$2x^2+20x=2(5.1)^2+20(5.1)=52.02+102$
$=154.02$
$2x^2+20x=2(10)^2+20(10)=200+200=400$

x	1	3	5.1	10
$2x^2+20x$	22	78	154.02	400

67. $7a^2-4a^2-a^2=(7-4-1)a^2=2a^2$

68. $9y+y-14y=(9+1-14)y=-4y$

69. $6a^2+4a+9a^2=6a^2+9a^2+4a=15a^2+4a$

70. $21x^2+3x+x^2+6=21x^2+x^2+3x+6$
$=22x^2+3x+6$

71. $4a^2b-3b^2-8q^2-10a^2b+7q^2=4a^2b-10a^2b-3b^2-8q^2+7q^2$
$=-6a^2b-3b^2-q^2$

72. $2s^{14}+3s^{13}+12s^{12}-s^{10}$ cannot be combined.

73. $(3x^2+2x+6)+(5x^2+x)=3x^2+2x+6+5x^2+x$
$=3x^2+5x^2+2x+x+6$
$=8x^2+3x+6$

Copyright © 2012 Pearson Education, Inc. Publishing as Prentice Hall.

74. $(2x^5+3x^4+4x^3+5x^2)+(4x^2+7x+6)=2x^5+3x^4+4x^3+5x^2+4x^2+7x+6$
$=2x^5+3x^4+4x^3+9x^2+7x+6$

75. $(-5y^2+3)-(2y^2+4)=(-5y^2+3)+(-2y^2-4)$
$=-5y^2+3-2y^2-4$
$=-7y^2-1$

76. $(2m^7+3x^4+7m^6)-(8m^7+4m^2+6x^4)=(2m^7+3x^4+7m^6)+(-8m^7-4m^2-6x^4)$
$=2m^7+3x^4+7m^6-8m^7-4m^2-6x^4$
$=-6m^7-3x^4+7m^6-4m^2$

77. $(3x^2-7xy+7y^2)-(4x^2-xy+9y^2)=(3x^2-7xy+7y^2)+(-4x^2+xy-9y^2)$
$=3x^2-7xy+7y^2-4x^2+xy-9y^2$
$=-x^2-6xy-2y^2$

78. $(8x^6-5xy-10y^2)-(7x^6-9xy-12y^2)=(8x^6-5xy-10y^2)+(-7x^6+9xy+12y^2)$
$=8x^6-5xy-10y^2-7x^6+9xy+12y^2$
$=x^6+4xy+2y^2$

79. $(-9x^2+6x+2)+(4x^2-x-1)=-9x^2+6x+2+4x^2-x-1$
$=-5x^2+5x+1$

80. $[(x^2+7x+9)+(x^2+4)]-(4x^2+8x-7)=(x^2+7x+9)+(x^2+4)+(-4x^2-8x+7)$
$=x^2+7x+9+x^2+4-4x^2-8x+7$
$=-2x^2-x+20$

81. $6(x+5)=6(x)+6(5)=6x+30$

82. $9(x-7)=9(x)+9(-7)=9x-63$

83. $4(2a+7)=4(2a)+4(7)=8a+28$

84. $9(6a-3)=9(6a)+9(-3)=54a-27$

85. $-7x(x^2+5)=(-7x)(x^2)+(-7x)(5)$
$=-7x^3-35x$

86. $-8y(4y^2-6)=(-8y)(4y^2)+(-8y)(-6)$
$=-32y^3+48y$

87. $-2(x^3-9x^2+x)=(-2)(x^3)+(-2)(-9x^2)+(-2)(x)$
$=-2x^3+18x^2-2x$

88. $-3a(a^2b+ab+b^2)=(-3a)(a^2b)+(-3a)(ab)+(-3a)(b^2)$
$=-3a^3b-3a^2b-3ab^2$

Copyright © 2012 Pearson Education, Inc. Publishing as Prentice Hall.

89. $(-2a)(3a^3-4a+1)=(-2a)(3a^3)+(-2a)(-4a)+(-2a)(1)$
$=-6a^4+8a^2-2a$

90. $(7b)(6b^3-4b+2)=(7b)(6b^3)+(7b)(-4b)+(7b)(2)$
$=42b^4-28b^2+14b$

91. $(2x+2)(x-7)=2x^2-14x+2x-14$
$=2x^2-12x-14$

92. $(2x-5)(3x+2)=6x^2+4x-15x-10$
$=6x^2-11x-10$

93. $(4a-1)(a+7)=4a^2+28a-a-7$
$=4a^2+27a-7$

94. $(6a-1)(7a+3)=42a^2+18a-7a-3$
$=42a^2+11a-3$

95. $(x+7)(x^3+4x-5)=x(x^3+4x-5)+7(x^3+4x-5)$
$=x(x^3)+x(4x)+x(-5)+7(x^3)+7(4x)+7(-5)$
$=x^4+4x^2-5x+7x^3+28x-35$
$=x^4+7x^3+4x^2+23x-35$

96. $(x+2)(x^5+x+1)=x(x^5+x+1)+2(x^5+x+1)$
$=x(x^5)+x(x)+x(1)+2(x^5)+2(x)+2(1)$
$=x^6+x^2+x+2x^5+2x+2$
$=x^6+2x^5+x^2+3x+2$

97. $(x^2+2x+4)(x^2+2x-4)=x^2(x^2+2x-4)+2x(x^2+2x-4)+4(x^2+2x-4)$
$=x^4+2x^3-4x^2+2x^3+4x^2-8x+4x^2+8x-16$
$=x^4+4x^3+4x^2-16$

98. $(x^3+4x+4)(x^3+4x-4)=x^3(x^3+4x-4)+4x(x^3+4x-4)+4(x^3+4x-4)$
$=x^6+4x^4-4x^3+4x^4+16x^2-16x+4x^3+16x-16$
$=x^6+8x^4+16x^2-16$

99. $(x+7)^3=(x+7)(x+7)^2$
$=(x+7)(x^2+2(x)(7)+7^2)$
$=(x+7)(x^2+14x+49)$
$=x(x^2+14x+49)+7(x^2+14x+49)$
$=x^3+14x^2+49x+7x^2+98x+343$
$=x^3+21x^2+147x+343$

Copyright © 2012 Pearson Education, Inc. Publishing as Prentice Hall.

100. $(2x-5)^3$
$= (2x-5)(2x-5)^2$
$= (2x-5)((2x)^2-(2)(2x)(5)+5^2)$
$= (2x-5)(4x^2-20x+25)$
$= 2x(4x^2-20x+25)+(-5)(4x^2-20x+25)$
$= 8x^3-40x^2+50x-20x^2+100x-125$
$= 8x^3-60x^2+150x-125$

101. $(x+7)^2 = x^2+2(x)(7)+7^2 = x^2+14x+49$

102. $(x-5)^2 = x^2-2(x)(5)+5^2 = x^2-10x+25$

103. $(3x-7)^2 = (3x)^2-2(3x)(7)+7^2$
$= 9x^2-42x+49$

104. $(4x+2)^2 = (4x)^2+2(4x)(2)+2^2$
$= 16x^2+16x+4$

105. $(5x-9)^2 = (5x)^2-2(5x)(9)+9^2$
$= 25x^2-90x+81$

106. $(5x+1)(5x-1) = (5x)^2-1^2 = 25x^2-1$

107. $(7x+4)(7x-4) = (7x)^2-4^2 = 49x^2-16$

108. $(a+2b)(a-2b) = a^2-(2b)^2 = a^2-4b^2$

109. $(2x-6)(2x+6) = (2x)^2-6^2 = 4x^2-36$

110. $(4a^2-2b)(4a^2+2b) = (4a^2)^2-(2b)^2$
$= 16a^4-4b^2$

111. $(3x-1)^2 = (3x)^2-2(3x)(1)+1^2$
$= 9x^2-6x+1$
The area is $(9x^2-6x+1)$ square meters.

112. $(5x+2)(x-1) = 5x^2-5x+2x-2$
$= 5x^2-3x-2$
The area is $(5x^2-3x-2)$ square miles.

113. $\dfrac{x^2+21x+49}{7x^2} = \dfrac{x^2}{7x^2}+\dfrac{21x}{7x^2}+\dfrac{49}{7x^2}$
$= \dfrac{1}{7}+\dfrac{3}{x}+\dfrac{7}{x^2}$

114. $\dfrac{5a^3b-15ab^2+20ab}{-5ab} = \dfrac{5a^3b}{-5ab}+\dfrac{-15ab^2}{-5ab}+\dfrac{20ab}{-5ab}$
$= -a^2+3b-4$

115.
$$\begin{array}{r|l} & a+1 \\ \hline a-2 & a^2-a+4 \\ & \underline{a^2-2a} \\ & \quad a+4 \\ & \quad \underline{a-2} \\ & \qquad 6 \end{array}$$

$\dfrac{a^2-a+4}{a-2} = a+1+\dfrac{6}{a-2}$

116.
$$\begin{array}{r|l} & 4x \\ \hline x+5 & 4x^2+20x+7 \\ & \underline{4x^2+20x} \\ & \qquad 7 \end{array}$$

$\dfrac{4x^2+20x+7}{x+5} = 4x+\dfrac{7}{x+5}$

117.
$$\begin{array}{r|l} & a^2+3a+8 \\ \hline a-2 & a^3+a^2+2a+6 \\ & \underline{a^3-2a^2} \\ & \quad 3a^2+2a \\ & \quad \underline{3a^2-6a} \\ & \qquad 8a+6 \\ & \qquad \underline{8a-16} \\ & \qquad\quad 22 \end{array}$$

$\dfrac{a^3+a^2+2a+6}{a-2} = a^2+3a+8+\dfrac{22}{a-2}$

118.
$$\begin{array}{r|l} & 3b^2-4b \\ \hline 3b-2 & 9b^3-18b^2+8b-1 \\ & \underline{9b^3-6b^2} \\ & \quad -12b^2+8b \\ & \quad \underline{-12b^2+8b} \\ & \qquad\quad -1 \end{array}$$

$\dfrac{9b^3-18b^2+8b-1}{3b-2} = 3b^2-4b-\dfrac{1}{3b-2}$

Copyright © 2012 Pearson Education, Inc. Publishing as Prentice Hall.

119.
$$\begin{array}{r} 2x^3 - x^2 + 0x + 2 \\ 2x-1\overline{\big)\,4x^4 - 4x^3 + x^2 + 4x - 3} \\ \underline{4x^4 - 2x^3} \\ -2x^3 + x^2 \\ \underline{-2x^3 + x^2} \\ 4x - 3 \\ \underline{4x - 2} \\ -1 \end{array}$$

$$\frac{4x^4 - 4x^3 + x^2 + 4x - 3}{2x-1} = 2x^3 - x^2 + 2 - \frac{1}{2x-1}$$

120.
$$\begin{array}{r} x^2 - 16x - 117 \\ x-6\overline{\big)\,-x^3 - 10x^2 - 21x + 18} \\ \underline{-x^3 + 6x^2} \\ -16x^2 - 21x \\ \underline{-16x^2 + 96x} \\ -117x + 18 \\ \underline{-117x + 702} \\ -684 \end{array}$$

$$\frac{-x^3 - 10x^2 - 21x + 18}{x-6} = -x^2 - 16x - 117 - \frac{684}{x-6}$$

121. $\dfrac{15x^3 - 3x^2 + 60}{3x^2} = \dfrac{15x^3}{3x^2} - \dfrac{3x^2}{3x^2} + \dfrac{60}{3x^2}$
$= 5x - 1 + \dfrac{20}{x^2}$

The width is $\left(5x - 1 + \dfrac{20}{x^2}\right)$ feet.

122. $\dfrac{21a^3b^6 + 3a - 3}{3} = \dfrac{21a^3b^6}{3} + \dfrac{3a}{3} - \dfrac{3}{3}$
$= 7a^3b^6 + a - 1$

The length of a side is $(7a^3b^6 + a - 1)$ units.

123. $3^3 = (3)(3)(3) = 27$

124. $\left(-\dfrac{1}{2}\right)^3 = \left(-\dfrac{1}{2}\right)\left(-\dfrac{1}{2}\right)\left(-\dfrac{1}{2}\right) = -\dfrac{1}{8}$

125. $(4xy^2)(x^3y^5) = 4x^{1+3}y^{2+5} = 4x^4y^7$

126. $\dfrac{18x^9}{27x^3} = \dfrac{2}{3}x^{9-3} = \dfrac{2x^6}{3}$

127. $\left(\dfrac{3a^4}{b^2}\right)^3 = \dfrac{3^3a^{4\cdot 3}}{b^{2\cdot 3}} = \dfrac{27a^{12}}{b^6}$

128. $(2x^{-4}y^3)^{-4} = 2^{-4}x^{-4\cdot -4}y^{3\cdot -4} = \dfrac{x^{16}}{16y^{12}}$

129. $\dfrac{a^{-3}b^6}{9^{-1}a^{-5}b^{-2}} = 9a^{-3-(-5)}b^{6-(-2)} = 9a^2b^8$

130. $(-y^2 - 4) + (3y^2 - 6) = -y^2 - 4 + 3y^2 - 6$
$= 2x^2 - 10$

131. $(6x + 2) + (5x - 7) = 6x + 2 + 5x - 7 = 11x - 5$

132. $(5x^2 + 2x - 6) - (-x - 4)$
$= (5x^2 + 2x - 6) + (x + 4)$
$= 5x^2 + 2x - 6 + x + 4$
$= 5x^2 + 3x - 2$

133. $(8y^2 - 3y + 1) - (3y^2 + 2)$
$= (8y^2 - 3y + 1) + (-3y^2 - 2)$
$= 8y^2 - 3y + 1 - 3y^2 - 2$
$= 5y^2 - 3y - 1$

134. $(2x + 5)(3x - 2) = 6x^2 - 4x + 15x - 10$
$= 6x^2 + 11x - 10$

135. $4x(7x^2 + 3) = 4x(7x^2) + 4x(3) = 28x^3 + 12x$

136. $(7x - 2)(4x - 9) = 28x^2 - 63x - 8x + 18$
$= 28x^2 - 71x + 18$

137. $(x - 3)(x^2 + 4x - 6)$
$= x(x^2 + 4x - 6) + (-3)(x^2 + 4x - 6)$
$= x^3 + 4x^2 - 6x - 3x^2 - 12x + 18$
$= x^3 + x^2 - 18x + 18$

138. $(5x + 4)^2 = (5x)^2 + 2(5x)(4) + 4^2$
$= 25x^2 + 40x + 16$

139. $(6x + 3)(6x - 3) = (6x)^2 - 3^2 = 36x^2 - 9$

Copyright © 2012 Pearson Education, Inc. Publishing as Prentice Hall.

140. $\dfrac{8a^4-2a^3+4a-5}{2a^3}=\dfrac{8a^4}{2a^3}-\dfrac{2a^3}{2a^3}+\dfrac{4a}{2a^3}-\dfrac{5}{2a^3}$
$=4a-1+\dfrac{2}{a^2}-\dfrac{5}{2a^3}$

141.
$$\begin{array}{r} x-3 \\ x+5\overline{\big)\,x^2+2x+10} \\ \underline{x^2+5x} \\ -3x+10 \\ \underline{-3x-15} \\ 25 \end{array}$$

$\dfrac{x^2+2x+10}{x+5}=x-3+\dfrac{25}{x+5}$

142.
$$\begin{array}{r} 2x^2+7x+5 \\ 2x-3\overline{\big)\,4x^3+8x^2-11x+\ 4} \\ \underline{4x^3-6x^2} \\ 14x^2-11x \\ \underline{14x^2-21x} \\ 10x+\ 4 \\ \underline{10x-15} \\ 19 \end{array}$$

$\dfrac{4x^3+8x^2-11x+4}{2x-3}=2x^2+7x+5+\dfrac{19}{2x-3}$

Chapter 3 Test

1. $2^5=2\cdot2\cdot2\cdot2\cdot2=32$

2. $(-3)^4=(-3)(-3)(-3)(-3)=81$

3. $-3^4=-(3\cdot3\cdot3\cdot3)=-81$

4. $4^{-3}=\dfrac{1}{4^3}=\dfrac{1}{64}$

5. $(3x^2)(-5x^9)=3(-5)(x^2\cdot x^9)$
$=-15x^{2+9}$
$=-15x^{11}$

6. $\dfrac{y^7}{y^2}=y^{7-2}=y^5$

7. $\dfrac{r^{-8}}{r^{-3}}=r^{-8-(-3)}=r^{-5}=\dfrac{1}{r^5}$

8. $\left(\dfrac{4x^2y^3}{x^3y^{-4}}\right)^2=\dfrac{4^2x^{2\cdot2}y^{3\cdot2}}{x^{3\cdot2}y^{-4\cdot2}}$
$=\dfrac{16x^4y^6}{x^6y^{-8}}$
$=16x^{4-6}y^{6-(-8)}$
$=16x^{-2}y^{14}$
$=\dfrac{16y^{14}}{x^2}$

9. $\dfrac{6^2x^{-4}y^{-1}}{6^3x^{-3}y^7}=\dfrac{6^2}{6^3}\cdot\dfrac{x^{-4}}{x^{-3}}\cdot\dfrac{y^{-1}}{y^7}$
$=6^{2-3}x^{-4-(-3)}y^{-1-7}$
$=6^{-1}x^{-1}y^{-8}$
$=\dfrac{1}{6}\cdot\dfrac{1}{x}\cdot\dfrac{1}{y^8}$
$=\dfrac{1}{6xy^8}$

10. $563{,}000=5.63\times10^5$

11. $0.0000863=8.63\times10^{-5}$

12. $1.5\times10^{-3}=0.0015$

13. $6.23\times10^4=62{,}300$

14. $(1.2\times10^5)(3\times10^{-7})=3.6\times10^{-2}=0.036$

15. a. $4xy^2+7xyz+x^3y-2$

Term	Numerical Coefficient	Degree of Terms
$4xy^2$	4	3
$7xyz$	7	3
x^3y	1	4
-2	-2	0

b. The degree of $4xy^2+7xyz+x^3y$ is 3 + 1 or 4.

Copyright © 2012 Pearson Education, Inc. Publishing as Prentice Hall.

16. $5x^2+4x-7x^2+11+8x$
$=(5-7)x^2+(4+8)x+11$
$=-2x^2+12x+11$

17. $(8x^3+7x^2+4x-7)+(8x^3-7x-6)$
$=8x^3+7x^2+4x-7+8x^3-7x-6$
$=8x^3+8x^3+7x^2+4x-7x-7-6$
$=16x^3+7x^2-3x-13$

18.
$$\begin{array}{r} 5x^3+\ x^2+5x-2 \\ -(8x^3-4x^2+\ x-7) \\ \hline \end{array} \qquad \begin{array}{r} 5x^3+\ x^2+5x-2 \\ -8x^3+4x^2-\ x+7 \\ \hline -3x^3+5x^2+4x+5 \end{array}$$

19. $[(8x^2+7x+5)+(x^3-8)]-(4x+2)$
$=(8x^2+7x+5)+(x^3-8)+(-4x-2)$
$=8x^2+7x+5+x^3-8-4x-2$
$=x^3+8x^2+3x-5$

20. $(3x+7)(x^2+5x+2)$
$=3x(x^2+5x+2)+7(x^2+5x+2)$
$=3x(x^2)+3x(5x)+3x(2)+7(x^2)+7(5x)+7(2)$
$=3x^3+15x^2+6x+7x^2+35x+14$
$=3x^3+22x^2+41x+14$

21. $3x^2(2x^2-3x+7)$
$=3x^2(2x^2)+3x^2(-3x)+3x^2(7)$
$=6x^4-9x^3+21x^2$

22. $(x+7)(3x-5)=3x^2-5x+21x-35$
$=3x^2+16x-35$

23. $\left(3x-\frac{1}{5}\right)\left(3x+\frac{1}{5}\right)=(3x)^2-\left(\frac{1}{5}\right)^2=9x^2-\frac{1}{25}$

24. $(4x-2)^2=(4x)^2-2(4x)(2)+2^2$
$=16x^2-16x+4$

25. $(8x+3)^2=(8x)^2+2(8x)(3)+(3)^2$
$=64x^2+48x+9$

26. $(x^2-9b)(x^2+9b)=(x^2)^2-(9b)^2=x^4-81b^2$

27. $-16t^2+1001=-16(0)^2+1001=1001$
At 0 seconds, the height of the object is 1001 feet.
$=-16(1)^2+1001=-16+1001$
$=985$
At 1 second, the height of the object is 985 feet.
$=-16(3)^2+1001=-144+1001$
$=857$
At 3 seconds, the height of the object is 857 feet.
$=-16(5)^2+1001=-400+1001$
$=601$
At 5 seconds, the height of the object is 601 feet.

28. $(2x-3)(2x+3)=(2x)^2-(3)^2$
$=4x^2-9$
The area is $(2x-3)(2x+3)$ or $(4x^2-9)$ square inches.

29. $\frac{4x^2+2xy-7x}{8xy}=\frac{4x^2}{8xy}+\frac{2xy}{8xy}-\frac{7x}{8xy}$
$=\frac{x}{2y}+\frac{1}{4}-\frac{7}{8y}$

30.
$$\begin{array}{r} x+2 \\ x+5\overline{)x^2+7x+10} \\ \underline{x^2+5x} \\ 2x+10 \\ \underline{2x+10} \\ 0 \end{array}$$

$\frac{x^2+7x+10}{x+5}=x+2$

31.
$$\begin{array}{r} 9x^2-6x+4 \\ 3x+2\overline{)27x^3+\ 0x^2+\ 0x-8} \\ \underline{27x^3+18x^2} \\ -18x^2+\ 0x \\ \underline{-18x^2-12x} \\ 12x-8 \\ \underline{12x+8} \\ -16 \end{array}$$

$\frac{27x^3-8}{3x+2}=9x^2-6x+4-\frac{16}{3x+2}$

Cumulative Review Chapters 1–3

1. a. The natural numbers in the set are 11 and 112.

Copyright © 2012 Pearson Education, Inc. Publishing as Prentice Hall.

b. The whole numbers in the set are 0, 11, and 112.

c. The integers in the set are –3, –2, 0, 11, and 112.

d. The rational numbers in the set are –3, –2, 0, $\frac{1}{4}$, 11, and 112.

e. The only irrational number in the set is $\sqrt{2}$.

f. The real numbers in the set are –2, 0, $\frac{1}{4}$, 112, –3, 11, and $\sqrt{2}$.

2. a. $|-7.2| = 7.2$

b. $|0| = 0$

c. $\left|-\frac{1}{2}\right| = \frac{1}{2}$

3. a. $3^2 = 3 \cdot 3 = 9$

b. $5^3 = 5 \cdot 5 \cdot 5 = 125$

c. $2^4 = 2 \cdot 2 \cdot 2 \cdot 2 = 16$

d. $7^1 = 7$

e. $\left(\frac{3}{7}\right)^2 = \frac{3}{7} \cdot \frac{3}{7} = \frac{9}{49}$

f. $(0.6)^2 = 0.6 \cdot 0.6 = 0.36$

4. a. $\frac{3}{4} \cdot \frac{7}{21} = \frac{1}{4} \cdot \frac{21}{21} = \frac{1}{4}$

b. $\frac{1}{2} \cdot 4\frac{5}{6} = \frac{1}{2} \cdot \frac{29}{6} = \frac{29}{12} = 2\frac{5}{12}$

5. $\frac{3}{2} \cdot \frac{1}{2} - \frac{1}{2} = \frac{3}{4} - \frac{1}{2} = \frac{3}{4} - \frac{2}{4} = \frac{1}{4}$

6. $\frac{2x - 7y}{x^2} = \frac{2(5) - 7(1)}{5^2} = \frac{10 - 7}{25} = \frac{3}{25}$

7. a. "The sum of a number and 3" is $x + 3$.

b. "The product of 3 and a number" is $3x$.

c. "The quotient of 7.3 and a number" is $7.3 \div x$ or $\frac{7.3}{x}$.

d. "10 decreased by a number" is $10 - x$.

e. "5 times a number, increased by 7" is $5x + 7$.

8.
$$\begin{aligned} 8 + 3(2 \cdot 6 - 1) &= 8 + 3(12 - 1) \\ &= 8 + 3(11) \\ &= 8 + 33 \\ &= 41 \end{aligned}$$

9. $11.4 + (-4.7) = 6.7$

10.
$$\begin{aligned} 5(1)^2 + 2 &\stackrel{?}{=} 1 - 8 \\ 5 + 2 &\stackrel{?}{=} 1 - 8 \\ 7 &= -7 \quad \text{False} \end{aligned}$$
No, $x = 1$ is not a solution.

11. a. $\frac{x - y}{12 + x} = \frac{2 - (-5)}{12 + 2} = \frac{2 + 5}{12 + 2} = \frac{7}{14} = \frac{1}{2}$

b. $x^2 - y = 2^2 - (-5) = 4 + 5 = 9$

12. a. $7 - 40 = 7 + (-40) = -33$

b. $-5 - (-10) = -5 + 10 = 5$

13. $\frac{-30}{-10} = 3$

14. $\frac{-48}{6} = -8$

15. $\frac{42}{-0.6} = -70$

16. $\frac{-30}{-0.2} = 150$

17. $5(3x + 2) = 5(3x) + 5(2) = 15x + 10$

18. $-3(2x - 3) = (-3)(2x) + (-3)(-3) = -6x + 9$

19.
$$\begin{aligned} &-2(y + 0.3z - 1) \\ &= (-2)(y) + (-2)(0.3z) + (-2)(-1) \\ &= -2y - 0.6z + 2 \end{aligned}$$

Copyright © 2012 Pearson Education, Inc. Publishing as Prentice Hall.

20. $4x(-x^2+6x-1)=4x(-x^2)+4x(6x)+4x(-1)$
$=-4x^3+24x^2-4x$

21. $-(9x+y-2z+6)=-9x-y+2z-6$

22. $-(-4xy+6y-2)=4xy-6y+2$

23. $6(2a-1)-(11a+6)=7$
$12a-6-11a-6=7$
$a-12=7$
$a=19$

24. $2x+\frac{1}{8}=x-\frac{3}{8}$
$8\left(2x+\frac{1}{8}\right)=8\left(x-\frac{3}{8}\right)$
$16x+1=8x-3$
$8x+1=-3$
$8x=-4$
$x=\frac{-4}{8}$
$x=-\frac{1}{2}$

25. $\frac{y}{7}=20$
$7\cdot\frac{y}{7}=20\cdot 7$
$y=140$

26. $10=5j-2$
$12=5j$
$\frac{12}{5}=j$

27. $0.25x+0.10(x-3)=1.1$
$0.25x+0.10x-0.30=1.1$
$0.35x-0.30=1.1$
$0.35x=1.4$
$x=4$

28. $\frac{7x+5}{3}=x+3$
$7x+5=3(x+3)$
$7x+5=3x+9$
$4x+5=9$
$4x=4$
$x=1$

29. $2(x+4)=4x-12$
$2x+8=4x-12$
$8=2x-12$
$20=2x$
$10=x$
The number is 10.

30. $(x+7)-2x=-x+7$

31. $30\cdot 2+2x=140$
$60+2x=140$
$2x=80$
$x=40$
The length is 40 feet.

32. $\frac{4(-3)+(-8)}{5+(-5)}=\frac{-12-8}{5-5}=\frac{-20}{0}=$ undefined

33. $\frac{120}{x}=\frac{15}{100}$
$15x=12{,}000$
$x=800$

34.

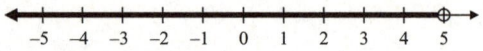

35. $-4x+7\ge -9$
$-4x\ge -16$
$x\le 4$
$\{x|x\le 4\}$

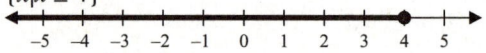

36. a. $(-5)^2=(-5)(-5)=25$

b. $-5^2=-(5\cdot 5)=-25$

c. $2\cdot 5^2=2\cdot 25=50$

37. a. $x^7\cdot x^4=x^{7+4}=x^{11}$

b. $\left(\frac{t}{2}\right)^4=\frac{t^4}{2^4}=\frac{t^4}{16}$

c. $(9y^5)^2=9^2y^{5\cdot 2}=81y^{10}$

38. $\frac{(z^2)^3\cdot z^7}{z^9}=\frac{z^6\cdot z^7}{z^9}=\frac{z^{6+7}}{z^9}=\frac{z^{13}}{z^9}=z^{13-9}=z^4$

Copyright © 2012 Pearson Education, Inc. Publishing as Prentice Hall.

39. $\left(\frac{3a^2}{b}\right)^{-3} = \frac{3^{-3}a^{2(-3)}}{b^{-3}} = \frac{3^{-3}a^{-6}}{b^{-3}} = \frac{b^3}{27a^6}$

40. $(5x^7)(-3x^9) = -15x^{7+9} = -15x^{16}$

41. $(5y^3)^{-2} = 5^{-2}y^{3\cdot -2} = 5^{-2}y^{-6} = \frac{1}{25y^6}$

42. $(-3)^{-2} = \frac{1}{(-3)^2} = \frac{1}{9}$

43. $9x^3 + x = (9+1)x^3 = 10x^3$

44. $(5y^2-6)-(y^2+2) = (5y^2-6)+(-y^2-2)$
$= 5y^2-6-y^2-2$
$= 4y^2-8$

45. $5x^2+6x-9x-3 = 5x^2-3x-3$

46. $(10x^2-3)(10x^2+3) = (10x^2)^2-(3)^2$
$= 100x^4-9$

47. $7x(x^2+2x+5) = 7x(x^2)+7x(2x)+7x(5)$
$= 7x^3+14x^2+35x$

48. $(10x^2+3)^2 = (10x^2)^2+2(10x^2)(3)+3^2$
$= 100x^4+60x^2+9$

49. $\frac{9x^5-12x^2+3x}{3x^2} = \frac{9x^5}{3x^2}-\frac{12x^2}{3x^2}+\frac{3x}{3x^2}$
$= 3x^3-4+\frac{1}{x}$

Copyright © 2012 Pearson Education, Inc. Publishing as Prentice Hall.

Chapter 4

Section 4.1 Practice

1. a. $45 = 3 \cdot 3 \cdot 5$
$75 = 3 \cdot 5 \cdot 5$
$\text{GCF} = 3 \cdot 5 = 15$

b. $32 = 2 \cdot 2 \cdot 2 \cdot 2 \cdot 2$
$33 = 3 \cdot 11$
There are no common prime factors; thus, the GCF is 1.

c. $14 = 2 \cdot 7$
$24 = 2 \cdot 2 \cdot 2 \cdot 3$
$60 = 2 \cdot 2 \cdot 3 \cdot 5$
$\text{GCF} = 2$

2. a. The GCF is y^4, since 4 is the smallest exponent to which y is raised.

b. The GCF is x^1 or x, since 1 is the smallest exponent on x.

3. a. $6x^2 = 2 \cdot 3 \cdot x^2$
$9x^4 = 3 \cdot 3 \cdot x^4$
$-12x^5 = -1 \cdot 2 \cdot 2 \cdot 3 \cdot x^5$
$\text{GCF} = 3 \cdot x^2 = 3x^2$

b. $-16y = -1 \cdot 2 \cdot 2 \cdot 2 \cdot 2 \cdot y$
$-20y^6 = -1 \cdot 2 \cdot 2 \cdot 5 \cdot y^6$
$40y^4 = 2 \cdot 2 \cdot 2 \cdot 5 \cdot y^4$
$\text{GCF} = 2 \cdot 2 \cdot y = 4y$

c. The GCF of a^5, a, and a^3 is a.
The GCF of b^4, b^3, and b^2 is b^2.
Thus, the GCF of a^5b^4, ab^3, and a^3b^2 is ab^2.

4. a. The GCF of terms $10y$ and 25 is 5.
$10y + 25 = 5 \cdot 2y + 5 \cdot 5 = 5(2y + 5)$

b. The GCF of x^4 and x^9 is x^4.
$x^4 - x^9 = x^4(1) - x^4(x^5) = x^4(1 - x^5)$

5. $-10x^3 + 8x^2 - 2x = 2x(-5x^2) + 2x(4x) + 2x(-1)$
$= 2x(-5x^2 + 4x - 1)$

6. $4x^3 + 12x = 4x(x^2 + 3)$

7. $\frac{2}{5}a^5 - \frac{4}{5}a^3 + \frac{1}{5}a^2 = \frac{1}{5}a^2(2a^3 - 4a + 1)$

8. $6a^3b + 3a^3b^2 + 9a^2b^4 = 3a^2b(2a + ab + 3b^3)$

9. $7(p + 2) + q(p + 2) = (p + 2)(7 + q)$

10. $ab + 7a + 2b + 14 = (ab + 7a) + (2b + 14)$
$= a(b + 7) + 2(b + 7)$
$= (b + 7)(a + 2)$

11. $28x^3 - 7x^2 + 12x - 3 = (28x^3 - 7x^2) + (12x - 3)$
$= 7x^2(4x - 1) + 3(4x - 1)$
$= (4x - 1)(7x^2 + 3)$

12. $2xy + 5y^2 - 4x - 10y$
$= (2xy + 5y^2) + (-4x - 10y)$
$= y(2x + 5y) - 2(2x + 5y)$
$= (2x + 5y)(y - 2)$

13. $3x^2 + 4xy + 3x + 4y = (3x^2 + 4xy) + (3x + 4y)$
$= x(3x + 4y) + 1(3x + 4y)$
$= (3x + 4y)(x + 1)$

14. $4x^3 + x - 20x^2 - 5 = x(4x^2 + 1) - 5(4x^2 + 1)$
$= (4x^2 + 1)(x - 5)$

15. $3xy - 4 + x - 12y = (3xy + x) + (-12y - 4)$
$= x(3y + 1) - 4(3y + 1)$
$= (3y + 1)(x - 4)$

16. $2x - 2 + x^3 - 3x^2 = 2(x - 1) + x^2(x - 3)$
There is no common binomial factor that can now be factored out. This polynomial is not factorable by grouping.

Vocabulary and Readiness Check

1. Since $5 \cdot 4 = 20$, the numbers 5 and 4 are called <u>factors</u> of 20.

2. The <u>greatest common factor</u> of a list of integers is the largest integer that is a factor of all the integers in the list.

Copyright © 2012 Pearson Education, Inc. Publishing as Prentice Hall.

3. The greatest common factor of a list of common variables raised to powers is the variable raised to the least exponent in the list.

4. The process of writing a polynomial as a product is called factoring.

5. A factored form of $7x + 21 + xy + 3y$ is $7(x + 3) + y(x + 3)$. false

6. A factored form of $3x^3 + 6x + x^2 + 2$ is $3x(x^2 + 2)$. false

7. $14 = 2 \cdot 7$

8. $15 = 3 \cdot 5$

9. The GCF of 18 and 3 is 3.

10. The GCF of 7 and 35 is 7.

11. The GCF of 20 and 15 is 5.

12. The GCF of 6 and 15 is 3.

Exercise Set 4.1

1. $32 = 2 \cdot 2 \cdot 2 \cdot 2 \cdot 2$
$36 = 2 \cdot 2 \cdot 3 \cdot 3$
$\text{GCF} = 2 \cdot 2 = 4$

3. $18 = 2 \cdot 3 \cdot 3$
$42 = 2 \cdot 3 \cdot 7$
$84 = 2 \cdot 2 \cdot 3 \cdot 7$
$\text{GCF} = 2 \cdot 3 = 6$

5. $24 = 2 \cdot 2 \cdot 2 \cdot 3$
$14 = 2 \cdot 7$
$21 = 3 \cdot 7$
GCF = 1 since there are no common prime factors.

7. The GCF of y^2, y^4, and y^7 is y^2.

9. The GCF of z^7, z^9, and z^{11} is z^7.

11. The GCF of x^{10}, x, and x^3 is x.
The GCF of y^2, y^2, and y^3 is y^2.
Thus, the GCF of $x^{10}y^2$, xy^2, and x^3y^3 is xy^2.

13. $14x = 2 \cdot 7 \cdot x$
$21 = 3 \cdot 7$
$\text{GCF} = 7$

15. $12y^4 = 2 \cdot 2 \cdot 3 \cdot y^4$
$20y^3 = 2 \cdot 2 \cdot 5 \cdot y^3$
$\text{GCF} = 2 \cdot 2 \cdot y^3 = 4y^3$

17. $-10x^2 = -1 \cdot 2 \cdot 5 \cdot x^2$
$15x^3 = 3 \cdot 5 \cdot x^3$
$\text{GCF} = 5 \cdot x^2 = 5x^2$

19. $12x^3 = 2 \cdot 2 \cdot 3 \cdot x^3$
$-6x^4 = -1 \cdot 2 \cdot 3 \cdot x^4$
$3x^5 = 3 \cdot x^5$
$\text{GCF} = 3 \cdot x^3 = 3x^3$

21. $-18x^2y = -1 \cdot 2 \cdot 3 \cdot 3 \cdot x^2 \cdot y$
$9x^3y^3 = 3 \cdot 3 \cdot x^3 \cdot y^3$
$36x^3y = 2 \cdot 2 \cdot 3 \cdot 3 \cdot x^3 \cdot y$
$\text{GCF} = 3 \cdot 3 \cdot x^2 \cdot y = 9x^2y$

23. $20a^6b^2c^8 = 2 \cdot 2 \cdot 5 \cdot a^6 \cdot b^2 \cdot c^8$
$50a^7b = 2 \cdot 5 \cdot 5 \cdot a^7 \cdot b$
$\text{GCF} = 2 \cdot 5 \cdot a^6 \cdot b = 10a^6b$

25. $3a + 6 = 3(a + 2)$

27. $30x - 15 = 15(2x - 1)$

29. $x^3 + 5x^2 = x^2(x+5)$

31. $6y^4 + 2y^3 = 2y^3(3y+1)$

33. $32xy - 18x^2 = 2x(16y - 9x)$

35. $4x - 8y + 4 = 4(x - 2y + 1)$

37. $6x^3 - 9x^2 + 12x = 3x(2x^2 - 3x + 4)$

39. $a^7b^6 - a^3b^2 + a^2b^5 - a^2b^2$
$= a^2b^2(a^5b^4 - a + b^3 - 1)$

41. $5x^3y - 15x^2y + 10xy = 5xy(x^2 - 3x + 2)$

43. $8x^5 + 16x^4 - 20x^3 + 12 = 4(2x^5 + 4x^4 - 5x^3 + 3)$

45. $\frac{1}{3}x^4 + \frac{2}{3}x^3 - \frac{4}{3}x^5 + \frac{1}{3}x = \frac{1}{3}x(x^3 + 2x^2 - 4x^4 + 1)$

Copyright © 2012 Pearson Education, Inc. Publishing as Prentice Hall.

47. $y(x^2+2)+3(x^2+2)=(x^2+2)(y+3)$

49. $z(y+4)+3(y+4)=(y+4)(z+3)$

51. $r(z^2-6)+(z-6)=r(z^2-6)+1(z^2-6)$
$=(z^2-6)(r+1)$

53. $-x-7=(-1)x+(-1)(7)=-1(x+7)$

55. $-2+z=-1(2)+(-1)(-z)=-1(2-z)$

57. $3a-b+2=(-1)(-3a)+(-1)(b)-(-1)(2)$
$=-1(-3a+b-2)$

59. $x^3+2x^2+5x+10=x^2(x+2)+5(x+2)$
$=(x+2)(x^2+5)$

61. $5x+15+xy+3y=5(x+3)+y(x+3)$
$=(x+3)(5+y)$

63. $6x^3-4x^2+15x-10=2x^2(3x-2)+5(3x-2)$
$=(3x-2)(2x^2+5)$

65. $5m^3+6mn+5m^2+6n$
$=m(5m^2+6n)+1(5m^2+6n)$
$=(5m^2+6n)(m+1)$

67. $2y-8+xy-4x=2(y-4)+x(y-4)$
$=(y-4)(2+x)$

69. $2x^3+x^2+8x+4=x^2(2x+1)+4(2x+1)$
$=(2x+1)(x^2+4)$

71. $3x-3+x^3-4x^2=3(x-1)+x^2(x-4)$
The polynomial is not factorable by grouping.

73. $4x^2-8xy-3x+6y=4x(x-2y)-3(x-2y)$
$=(x-2y)(4x-3)$

75. $5q^2-4pq-5q+4p=q(5q-4p)-1(5q-4p)$
$=(5q-4p)(q-1)$

77. $12x^2y-42x^2-4y+14$
$=2(6x^2y-21x^2-2y+7)$
$=2[3x^2(2y-7)-1(2y-7)]$
$=2(2y-7)(3x^2-1)$

79. $6a^2+9ab^2+6ab+9b^3$
$=3(2a^2+3ab^2+2ab+3b^3)$
$=3[a(2a+3b^2)+b(2a+3b^2)]$
$=3(2a+3b^2)(a+b)$

81. $(x+2)(x+5)=x^2+5x+2x+10=x^2+7x+10$

83. $(b+1)(b-4)=b^2-4b+b-4=b^2-3b-4$

85. $2\cdot 6=12$
$2+6=8$
2 and 6 have a product of 12 and a sum of 8.

87. $-1\cdot(-8)=8$
$-1+(-8)=-9$
-1 and -8 have a product of 8 and a sum of -9.

89. $-2\cdot 5=-10$
$-2+5=3$
-2 and 5 have a product of -10 and a sum of 3.

91. $-8\cdot 3=-24$
$-8+3=-5$
-8 and 3 have a product of -24 and a sum of -5.

93. $-2x+14=-2(x-7)$, which is choice d.

95. $(a+6)(a+2)$ is factored.

97. Since $5(2y+z)-b(2y+z)=(2y+z)(5-b)$, the given expression is not factored.

99. a. $-1264x^2+5056x+18{,}960$
$=-1264(1)^2+5056(1)+18{,}960$
$=-1264(1)+5056(1)+18{,}960$
$=-1264+5056+18{,}960$
$=22{,}752$
The cotton harvest in 2004 was approximately 22,752 thousand bales.

b. Let $x=4$ for 2007.
$-1264x^2+5056x+18{,}960$
$=-1264(4)^2+5056(4)+18{,}960$
$=-1264(16)+5056(4)+18{,}960$
$=-20{,}224+20{,}224+18{,}960$
$=18{,}960$
The cotton harvest in 2007 was approximately 18,960 thousand bales.

Copyright © 2012 Pearson Education, Inc. Publishing as Prentice Hall.

c. $-1264x^2 + 5056x + 18{,}960$
$= -1264(x^2 - 4x - 15)$
or $1264(-x^2 + 4x + 15)$

101. The area of the circle is πx^2. Since the sides of the square have length $2x$, the area of the square is $(2x)^2 = 4x^2$. The shaded region is the region inside the square but outside the circle. The area of the shaded region is $4x^2 - \pi x^2 = x^2(4 - \pi)$.

103. Area = width · length
$5x^5 - 5x^2 = 5x^2(x^3 - 1)$
The length is $(x^3 - 1)$ units.

105. answers may vary

107. answers may vary

Section 4.2 Practice

1.

Factors of 20	Sum of Factors
1, 20	21
2, 10	12
4, 5	9

Thus, $x^2 + 12x + 20 = (x + 10)(x + 2)$.

2. a.

Factors of 22	Sum of Factors
−1, −22	−23
−2, −11	−13

$x^2 - 23x + 22 = (x - 1)(x - 22)$

b.

Factors of 50	Sum of Factors
−1, −50	−51
−2, −25	−27
−5, −10	−15

$x^2 - 27x + 50 = (x - 2)(x - 25)$

3.

Factors of −36	Sum of Factors
−1, 36	35
1, −36	−35
−2, 18	16
2, −18	−16
−3, 12	9
3, −12	−9
−4, 9	5
4, −9	−5
−6, 6	0

$x^2 + 5x - 36 = (x + 9)(x - 4)$

4. a. Two factors of −40 whose sum is −3 are −8 and 5.
$q^2 - 3q - 40 = (q - 8)(q + 5)$

b. Two factors of −48 whose sum is 2 are 8 and −6.
$y^2 + 2y - 48 = (y + 8)(y - 6)$

5. Since there are no two numbers whose product is 15 and whose sum is 6, the polynomial is prime.

6. a. Two factors of $14y^2$ whose sum is $9y$ are $2y$ and $7y$.
$x^2 + 9xy + 14y^2 = (x + 2y)(x + 7y)$

b. Two factors of $30b^2$ whose sum is $-13b$ are $-3b$ and $-10b$.
$a^2 - 13ab + 30b^2 = (a - 3b)(a - 10b)$

7. Two factors of 12 whose sum is 8 are 6 and 2.
$x^4 + 8x^2 + 12 = (x^2 + 6)(x^2 + 2)$

8. $48 - 14x + x^2 = x^2 - 14x + 48$
Two factors of 48 whose sum is −14 are −6 and −8.
$x^2 - 14x + 48 = (x - 6)(x - 8)$

Copyright © 2012 Pearson Education, Inc. Publishing as Prentice Hall.

9. a. $4x^2-24x+36=4(x^2-6x+9)$
Two factors of 9 whose sum is –6 are –3 and –3.
$4(x^2-6x+9)=4(x-3)(x-3)$ or $4(x-3)^2$

b. $x^3+3x^2-4x=x(x^2+3x-4)$
Two factors of –4 whose sum is 3 are 4 and –1.
$x(x^2+3x-4)=x(x+4)(x-1)$

10. $5x^5-25x^4-30x^3=5x^3(x^2-5x-6)$
$=5x^3(x+1)(x-6)$

Vocabulary and Readiness Check

1. To factor x^2+7x+6, we look for two numbers whose product is 6 and whose sum is 7. true
2. We can write the factorization $(y+2)(y+4)$ also as $(y+4)(y+2)$. true
3. The factorization $(4x-12)(x-5)$ is completely factored. false
4. The factorization $(x+2y)(x+y)$ may also be written as $(x+2y)^2$. false
5. $x^2+9x+20=(x+4)(x+5)$
6. $x^2+12x+35=(x+5)(x+7)$
7. $x^2-7x+12=(x-4)(x-3)$
8. $x^2-13x+22=(x-2)(x-11)$
9. $x^2+4x+4=(x+2)(x+2)$
10. $x^2+10x+24=(x+6)(x+4)$

Exercise Set 4.2

1. Two factors of 6 whose sum is 7 are 6 and 1.
$x^2+7x+6=(x+6)(x+1)$

3. Two factors of 9 whose sum is –10 are –9 and –1.
$y^2-10y+9=(y-9)(y-1)$

5. Two factors of 9 whose sum is –6 are –3 and –3.
$x^2-6x+9=(x-3)(x-3)$ or $(x-3)^2$

7. Two factors of –18 whose sum is –3 are –6 and 3.
$x^2-3x-18=(x-6)(x+3)$

9. Two factors of –70 whose sum is 3 are 10 and –7.
$x^2+3x-70=(x+10)(x-7)$

11. Since there are no two numbers whose product is 2 and whose sum is 5, the polynomial is prime.

13. Two factors of $15y^2$ whose sum is $8y$ are $5y$ and $3y$.
$x^2+8xy+15y^2=(x+5y)(x+3y)$

15. Two factors of –15 whose sum is –2 are –5 and 3.
$a^4-2a^2-15=(a^2-5)(a^2+3)$

17. $13+14m+m^2=m^2+14m+13$
Two factors of 13 whose sum is 14 are 13 and 1.
$13+14m+m^2=m^2+14m+13=(m+13)(m+1)$

19. $10t-24+t^2=t^2+10t-24$
Two factors of –24 whose sum is 10 are –2 and 12.
$10t-24+t^2=t^2+10t-24=(t-2)(t+12)$

21. Two factors of $16b^2$ whose sum is $-10b$ are $-2b$ and $-8b$.
$a^2-10ab+16b^2=(a-2b)(a-8b)$

23. $2z^2+20z+32=2(z^2+10z+16)$
$=2(z+8)(z+2)$

25. $2x^3-18x^2+40x=2x(x^2-9x+20)$
$=2x(x-5)(x-4)$

27. $x^2-3xy-4y^2=(x-4y)(x+y)$

29. $x^2+15x+36=(x+12)(x+3)$

31. $x^2-x-2=(x-2)(x+1)$

Copyright © 2012 Pearson Education, Inc. Publishing as Prentice Hall.

33. $r^2 - 16r + 48 = (r-12)(r-4)$

35. $x^2 + xy - 2y^2 = (x+2y)(x-y)$

37. $3x^2 + 9x - 30 = 3(x^2 + 3x - 10) = 3(x+5)(x-2)$

39. $3x^2 - 60x + 108 = 3(x^2 - 20x + 36)$
$= 3(x-18)(x-2)$

41. $x^2 - 18x - 144 = (x-24)(x+6)$

43. $r^2 - 3r + 6$ is prime.

45. $x^2 - 8x + 15 = (x-5)(x-3)$

47. $6x^3 + 54x^2 + 120x = 6x(x^2 + 9x + 20)$
$= 6x(x+4)(x+5)$

49. $4x^2y + 4xy - 12y = 4y(x^2 + x - 3)$

51. $x^2 - 4x - 21 = (x-7)(x+3)$

53. $x^2 + 7xy + 10y^2 = (x+5y)(x+2y)$

55. $64 + 24t + 2t^2 = 2t^2 + 24t + 64$
$= 2(t^2 + 12t + 32)$
$= 2(t+8)(t+4)$

57. $x^3 - 2x^2 - 24x = x(x^2 - 2x - 24)$
$= x(x-6)(x+4)$

59. $2t^5 - 14t^4 + 24t^3 = 2t^3(t^2 - 7t + 12)$
$= 2t^3(t-4)(t-3)$

61. $5x^3y - 25x^2y^2 - 120xy^3$
$= 5xy(x^2 - 5xy - 24y^2)$
$= 5xy(x-8y)(x+3y)$

63. $162 - 45m + 3m^2 = 3m^2 - 45m + 162$
$= 3(m^2 - 15m + 54)$
$= 3(m-9)(m-6)$

65. $-x^2 + 12x - 11 = -1(x^2 - 12x + 11)$
$= -1(x-11)(x-1)$

67. $\frac{1}{2}y^2 - \frac{9}{2}y - 11 = \frac{1}{2}(y^2 - 9y - 22)$
$= \frac{1}{2}(y-11)(y+2)$

69. $x^3y^2 + x^2y - 20x = x(x^2y^2 + xy - 20)$
$= x(xy-4)(xy+5)$

71. $(2x+1)(x+5) = 2x^2 + 10x + x + 5$
$= 2x^2 + 11x + 5$

73. $(5y-4)(3y-1) = 15y^2 - 5y - 12y + 4$
$= 15y^2 - 17y + 4$

75. $(a+3b)(9a-4b) = 9a^2 - 4ab + 27ab - 12b^2$
$= 9a^2 + 23ab - 12b^2$

77. $(x-3)(x+8) = x^2 + 8x - 3x - 24 = x^2 + 5x - 24$

79. answers may vary

81. $P = 2l + 2w$
$= 2(x^2 + 10x) + 2(4x + 33)$
$= 2x^2 + 20x + 8x + 66$
$= 2x^2 + 28x + 66$
$= 2(x^2 + 14x + 33)$
$= 2(x+3)(x+11)$

83. $-16t^2 + 64t + 80 = -16(t^2 - 4t - 5)$
$= -16(t-5)(t+1)$

85. $x^2 + \frac{1}{2}x + \frac{1}{16} = \left(x + \frac{1}{4}\right)\left(x + \frac{1}{4}\right)$ or $\left(x + \frac{1}{4}\right)^2$

87. $z^2(x+1) - 3z(x+1) - 70(x+1)$
$= (x+1)(z^2 - 3z - 70)$
$= (x+1)(z-10)(z+7)$

89. The factors of c must sum to -16. Since c is positive, both factors must have the same sign.
$-1 + (-15) = -16$; $(-1)(-15) = 15$
$-2 + (-14) = -16$; $(-2)(-14) = 28$
$-3 + (-13) = -16$; $(-3)(-13) = 39$
$-4 + (-12) = -16$; $(-4)(-12) = 48$
$-5 + (-11) = -16$; $(-5)(-11) = 55$
$-6 + (-10) = -16$; $(-6)(-10) = 60$
$-7 + (-9) = -16$; $(-7)(-9) = 63$
$-8 + (-8) = -16$; $(-8)(-8) = 64$
The possible values of c are 15, 28, 39, 48, 55, 60, 63, and 64.

Copyright © 2012 Pearson Education, Inc. Publishing as Prentice Hall.

91. The factors of 20 must sum to b. Since 20 is positive and b is positive, both factors of 20 must be positive.
$1 + 20 = 21$; $(1)(20) = 20$
$2 + 10 = 12$; $(2)(10) = 20$
$4 + 5 = 9$; $(4)(5) = 20$
The possible values of b are 9, 12, and 21.

93. $x^{2n} + 8x^n - 20 = (x^n + 10)(x^n - 2)$

Section 4.3 Practice

1. a. Factors of $5x^2$: $5x^2 = 5x \cdot x$
Factors of 10: $10 = 1 \cdot 10$, $10 = 2 \cdot 5$
$5x^2 + 27x + 10 = (5x + 2)(x + 5)$

b. Factors of $4x^2$: $4x^2 = 4x \cdot x$, $4x^2 = 2x \cdot 2x$
Factors of 5: $5 = 1 \cdot 5$
$4x^2 + 12x + 5 = (2x + 5)(2x + 1)$

2. a. Factors of $2x^2$: $2x^2 = 2x \cdot x$
Factors of 12: $12 = -1 \cdot -12$, $12 = -2 \cdot -6$, $12 = -3 \cdot -4$
$2x^2 - 11x + 12 = (2x - 3)(x - 4)$

b. Factors of $6x^2$: $6x^2 = 6x \cdot x$, $6x^2 = 3x \cdot 2x$
Factors of 1: $1 = -1 \cdot -1$
$6x^2 - 5x + 1 = (3x - 1)(2x - 1)$

3. a. Factors of $3x^2$: $3x^2 = 3x \cdot x$
Factors of -5: $-5 = -5 \cdot 1$, $-5 = 1 \cdot -5$
$3x^2 + 14x - 5 = (3x - 1)(x + 5)$

b. Factors of $35x^2$: $35x^2 = 35x \cdot x$,
$35x^2 = 5x \cdot 7x$
Factors of -4: $-4 = -1 \cdot 4$, $-4 = 1 \cdot -4$, $-4 = -2 \cdot 2$
$35x^2 + 4x - 4 = (5x + 2)(7x - 2)$

4. a. Factors of $14x^2$: $14x^2 = 14x \cdot x$,
$14x^2 = 7x \cdot 2x$
Factors of $-2y^2$: $-2y^2 = -2y \cdot y$,
$-2y^2 = 2y \cdot -y$
$14x^2 - 3xy - 2y^2 = (7x + 2y)(2x - y)$

b. Factors of $12a^2$: $12a^2 = 12a \cdot a$,
$12a^2 = 6a \cdot 2a$, $12a^2 = 3a \cdot 4a$
Factors of $-3b^2$: $-3b^2 = -3b \cdot b$,
$-3b^2 = 3b \cdot -b$
$12a^2 - 16ab - 3b^2 = (6a + b)(2a - 3b)$

5. Factors of $2x^4$: $2x^4 = 2x^2 \cdot x^2$
Factors of -7: $-7 = -7 \cdot 1$, $-7 = 7 \cdot -1$
$2x^4 - 5x^2 - 7 = (2x^2 - 7)(x^2 + 1)$

6. a. $3x^3 + 17x^2 + 10x = x(3x^2 + 17x + 10)$
Factors of $3x^2$: $3x^2 = 3x \cdot x$
Factors of 10: $10 = 1 \cdot 10$, $10 = 2 \cdot 5$
$x(3x^2 + 17x + 10) = x(3x + 2)(x + 5)$

b. $6xy^2 + 33xy - 18x = 3x(2y^2 + 11y - 6)$
Factors of $2y^2$: $2y^2 = 2y \cdot y$
Factors of -6: $-6 = 1 \cdot -6$, $-6 = -1 \cdot 6$, $-6 = 2 \cdot -3$, $-6 = -2 \cdot 3$
$3x(2y^2 + 11y - 6) = 3x(2y - 1)(y + 6)$

7. $-5x^2 - 19x + 4 = -1(5x^2 + 19x - 4)$
Factors of $5x^2$: $5x^2 = 5x \cdot x$
Factors of -4: $-4 = -4 \cdot 1$, $-4 = 4 \cdot -1$, $-4 = 2 \cdot -2$
$-1(5x^2 + 19x - 4) = -1(x + 4)(5x - 1)$

Vocabulary and Readiness Check

1. $2x^2 + 5x + 3$ factors as $(2x + 3)(x + 1)$, which is choice d.

2. $7x^2 + 9x + 2$ factors as $(7x + 2)(x + 1)$, which is choice b.

3. $3x^2 + 31x + 10$ factors as $(3x + 1)(x + 10)$, which is choice c.

4. $5x^2 + 61x + 12$ factors as $(5x + 1)(x + 12)$, which is choice a.

Copyright © 2012 Pearson Education, Inc. Publishing as Prentice Hall.

Exercise Set 4.3

1. $5x^2 = 5x \cdot x$
$8 = 2 \cdot 4$
$5x^2 + 22x + 8 = (5x+2)(x+4)$

3. $50x^2 = 5x \cdot 10x$
$-2 = 2 \cdot -1$
$50x^2 + 15x - 2 = (5x+2)(10x-1)$

5. $20x^2 = 5x \cdot 4x$
$-6 = 2 \cdot -3$
$20x^2 - 7x - 6 = (5x+2)(4x-3)$

7. Factors of $2x^2$: $2x^2 = 2x \cdot x$
Factors of 15: $15 = 1 \cdot 15$, $15 = 3 \cdot 5$
$2x^2 + 13x + 15 = (2x+3)(x+5)$

9. Factors of $8y^2$: $8y^2 = 8y \cdot y$, $8y^2 = 4y \cdot 2y$.
Factors of 9: $9 = -1 \cdot -9$, $-3 \cdot -3$.
$8y^2 - 17y + 9 = (y-1)(8y-9)$

11. Factors of $2x^2$: $2x^2 = 2x \cdot x$
Factors of -5: $-5 = 1 \cdot -5$, $-5 = -1 \cdot 5$
$2x^2 - 9x - 5 = (2x+1)(x-5)$

13. Factors of $20r^2$: $20r^2 = 20r \cdot r$, $20r^2 = 10r \cdot 2r$,
$20r^2 = 5r \cdot 4r$.
Factors of -8: $-8 = -1 \cdot 8$, $-8 = -2 \cdot 4$,
$-8 = -4 \cdot 2$, $-8 = -8 \cdot 1$
$20r^2 + 27r - 8 = (4r-1)(5r+8)$

15. Factors of $10x^2$: $10x^2 = 10x \cdot x$, $10x^2 = 5x \cdot 2x$
Factors of 3: $3 = 1 \cdot 3$
$10x^2 + 31x + 3 = (10x+1)(x+3)$

17. $x + 3x^2 - 2 = 3x^2 + x - 2$
Factors of $3x^2$: $3x^2 = 3x \cdot x$
Factors of -2: $-2 = -1 \cdot 2$, $-2 = 2 \cdot -1$
$3x^2 + x - 2 = (3x-2)(x+1)$

19. Factors of $6x^2$: $6x^2 = 6x \cdot x$, $6x^2 = 3x \cdot 2x$
Factors of $5y^2$: $5y^2 = -5y \cdot -y$
$6x^2 - 13xy + 5y^2 = (3x-5y)(2x-y)$

21. Factors of $15m^2$: $15m^2 = 15m \cdot m$,
$15m^2 = 5m \cdot 3m$.
Factors of -15: $-15 = -1 \cdot 15$, $-15 = -3 \cdot 5$,
$-15 = -5 \cdot 3$, $-15 = -15 \cdot 1$
$15m^2 - 16m - 15 = (3m-5)(5m+3)$

23. $-9x + 20 + x^2 = x^2 - 9x + 20$
Factors of x^2: $x^2 = x \cdot x$
Factors of 20: $20 = -1 \cdot -20$, $20 = -2 \cdot -10$,
$20 = -4 \cdot -5$
$x^2 - 9x + 20 = (x-4)(x-5)$

25. Factors of $2x^2$: $2x^2 = 2x \cdot x$
Factors of -99: $-99 = -1 \cdot 99$, $-99 = -3 \cdot 33$,
$-99 = -9 \cdot 11$, $-99 = -11 \cdot 9$, $-99 = -33 \cdot 3$,
$-99 = -99 \cdot 1$
$2x^2 - 7x - 99 = (2x+11)(x-9)$

27. $-27t + 7t^2 - 4 = 7t^2 - 27t - 4$
Factors of $7t^2$: $7t^2 = 7t \cdot t$
Factors of -4: $-4 = -1 \cdot 4$, $-4 = 1 \cdot -4$,
$-4 = 2 \cdot -2$
$7t^2 - 27t - 4 = (7t+1)(t-4)$

29. Factors of $3a^2$: $3a^2 = 3a \cdot a$
Factors of $3b^2$: $3b^2 = b \cdot 3b$
$3a^2 + 10ab + 3b^2 = (3a+b)(a+3b)$

31. Factors of $49p^2$: $49p^2 = 49p \cdot p$,
$49p^2 = 7p \cdot 7p$
Factors of -2: $-2 = -1 \cdot 2$, $-2 = 1 \cdot -2$
$49p^2 - 7p - 2 = (7p+1)(7p-2)$

33. Factors of $18x^2$: $18x^2 = 18x \cdot x$,
$18x^2 = 9x \cdot 2x$, $18x^2 = 6x \cdot 3x$
Factors of -14: $-14 = -1 \cdot 14$, $-14 = -2 \cdot 7$,
$-14 = -7 \cdot 2$, $-14 = -14 \cdot 1$
$18x^2 - 9x - 14 = (6x-7)(3x+2)$

35. Factors of $2m^2$: $2m^2 = 2m \cdot m$
Factors of 10: $10 = 1 \cdot 10$, $10 = 2 \cdot 5$
$2m^2 + 17m + 10$ is prime.

Copyright © 2012 Pearson Education, Inc. Publishing as Prentice Hall.

37. Factors of $24x^2$: $24x^2 = 24x \cdot x$,
$24x^2 = 12x \cdot 2x$, $24x^2 = 8x \cdot 3x$, $24x^2 = 6x \cdot 4x$
Factors of 12: $12 = 1 \cdot 12$, $12 = 2 \cdot 6$, $12 = 3 \cdot 4$
$24x^2 + 41x + 12 = (3x+4)(8x+3)$

39. $12x^3 + 11x^2 + 2x = x(12x^2 + 11x + 2)$
Factor of $12x^2$: $12x^2 = 12x \cdot x$, $12x^2 = 6x \cdot 2x$,
$12x^2 = 4x \cdot 3x$
Factors of 2: $2 = 1 \cdot 2$
$12x^3 + 11x^2 + 2x = x(3x+2)(4x+1)$

41. $21b^2 - 48b - 45 = 3(7b^2 - 16b - 15)$
Factors of $7b^2$: $7b^2 = 7b \cdot b$
Factors of -15: $-15 = -1 \cdot 15$, $-15 = -3 \cdot 5$,
$-15 = -5 \cdot 3$, $-15 = -15 \cdot 1$
$21b^2 - 48b - 45 = 3(7b+5)(b-3)$

43. $7z + 12z^2 - 12 = 12z^2 + 7z - 12$
Factors of $12z^2$: $12z^2 = 12z \cdot z$, $12z^2 = 6z \cdot 2z$,
$12z^2 = 4z \cdot 3z$
Factors of -12: $-12 = -12 \cdot 1$, $-12 = 12 \cdot -1$,
$-12 = -6 \cdot 2$, $-12 = 6 \cdot -2$, $-12 = -4 \cdot 3$,
$-12 = 4 \cdot -3$
$12z^2 + 7z - 12 = (3z+4)(4z-3)$

45. $6x^2y^2 - 2xy^2 - 60y^2 = 2y^2(3x^2 - x - 30)$
Factors of $3x^2$: $3x^2 = 3x \cdot x$
Factors of -30: $-30 = -1 \cdot 30$, $-30 = -2 \cdot 15$,
$-30 = -3 \cdot 10$, $-30 = -5 \cdot 6$, $-30 = -6 \cdot 5$,
$-30 = -10 \cdot 3$, $-30 = -15 \cdot 2$, $-30 = -30 \cdot 1$
$6x^2y^2 - 2xy^2 - 60y^2 = 2y^2(3x-10)(x+3)$

47. Factors of $4x^2$: $4x^2 = 4x \cdot x$, $4x^2 = 2x \cdot 2x$
Factors of -21: $-21 = -1 \cdot 21$, $-21 = 1 \cdot -21$,
$-21 = -7 \cdot 3$, $-21 = 7 \cdot -3$
$4x^2 - 8x - 21 = (2x-7)(2x+3)$

49. $3x^2 - 42x + 63 = 3(x^2 - 14x + 21)$
$x^2 - 14x + 21$ is prime, so $3(x^2 - 14x + 21)$ is a factored form of $3x^2 - 42x + 63$.

51. Factors of $8x^2$: $8x^2 = 8x \cdot x$, $8x^2 = 4x \cdot 2x$
Factors of $-27y^2$: $-27y^2 = -27y \cdot y$,
$-27y^2 = 27y \cdot -y$, $-27y^2 = -9y \cdot 3y$,
$-27y^2 = 9y \cdot -3y$
$8x^2 + 6xy - 27y^2 = (4x+9y)(2x-3y)$

53. $-x^2 + 2x + 24 = -1(x^2 - 2x - 24)$
$= -1(x-6)(x+4)$

55. $4x^3 - 9x^2 - 9x = x(4x^2 - 9x - 9)$
Factors of $4x^2$: $4x^2 = 4x \cdot x$, $4x^2 = 2x \cdot 2x$
Factors of -9: $-9 = -1 \cdot 9$, $-9 = 1 \cdot -9$,
$-9 = 3 \cdot -3$
$4x^3 - 9x^2 - 9x = x(4x+3)(x-3)$

57. Factors of $24x^2$: $24x^2 = 24x \cdot x$,
$24x^2 = 12x \cdot 2x$, $24x^2 = 8x \cdot 3x$, $24x^2 = 6x \cdot 4x$
Factors of 9: $9 = -1 \cdot -9$, $9 = -3 \cdot -3$
$24x^2 - 58x + 9 = (4x-9)(6x-1)$

59. $40a^2b + 9ab - 9b = b(40a^2 + 9a - 9)$
Factors of $40a^2$: $40a^2 = 40a \cdot a$,
$40a^2 = 20a \cdot 2a$, $40a^2 = 10a \cdot 4a$,
$40a^2 = 8a \cdot 5a$
Factors of -9: $-9 = -1 \cdot 9$, $-9 = 1 \cdot -9$,
$-9 = 3 \cdot -3$
$40a^2b + 9ab - 9b = b(8a-3)(5a+3)$

61. $30x^3 + 38x^2 + 12x = 2x(15x^2 + 19x + 6)$
Factors of $15x^2$: $15x^2 = 15x \cdot x$, $15x^2 = 5x \cdot 3x$
Factors of 6: $6 = 1 \cdot 6$, $6 = 2 \cdot 3$
$30x^3 + 38x^2 + 12x = 2x(3x+2)(5x+3)$

63. $6y^3 - 8y^2 - 30y = 2y(3y^2 - 4y - 15)$
Factors of $3y^2$: $3y^2 = 3y \cdot y$
Factors of -15: $-15 = -1 \cdot 15$, $-15 = 1 \cdot -15$,
$-15 = -3 \cdot 5$, $-15 = 3 \cdot -5$
$6y^3 - 8y^2 - 30y = 2y(3y+5)(y-3)$

65. $10x^4 + 25x^3y - 15x^2y^2 = 5x^2(2x^2 + 5xy - 3y^2)$
Factors of $2x^2$: $2x^2 = 2x \cdot x$
Factors of $-3y^2$: $-3y^2 = -y \cdot 3y$,
$-3y^2 = -3y \cdot y$
$10x^4 + 25x^3y - 15x^2y^2 = 5x^2(2x-y)(x+3y)$

Copyright © 2012 Pearson Education, Inc. Publishing as Prentice Hall.

67. $-14x^2+39x-10=-1(14x^2-39x+10)$

Factors of $14x^2$: $14x^2=14x\cdot x$, $14x^2=7x\cdot 2x$
Factors of 10: $10=-1\cdot -10$, $10=-2\cdot -5$
$-14x^2+39x-10=-1(2x-5)(7x-2)$

69. $16p^4-40p^3+25p^2=p^2(16p^2-40p+25)$

Factors of $16p^2$: $16p^2=16p\cdot p$,
$16p^2=8p\cdot 2p$, $16p^2=4p\cdot 4p$
Factors of 25: $25=-1\cdot -25$, $25=-5\cdot -5$
$16p^4-40p^3+25p^2=p^2(4p-5)(4p-5)$ or
$p^2(4p-5)^2$

71. $-2x^2+9x+5=-1(2x^2-9x-5)$

Factors of $2x^2$: $2x^2=2x\cdot x$
Factors of -5: $-5=-1\cdot 5$, $-5=1\cdot -5$
$-2x^2+9x+5=-1(2x+1)(x-5)$

73. $-4+52x-48x^2=-48x^2+52x-4$
$=-4(12x^2-13x+1)$

Factors of $12x^2$: $12x^2=12x\cdot x$, $12x^2=6x\cdot 2x$,
$12x^2=4x\cdot 3x$
Factors of 1: $1=-1\cdot -1$
$-4+52x-48x^2=-4(12x-1)(x-1)$

75. Factors of $2t^4$: $2t^4=2t^2\cdot t^2$
Factors of -27: $-27=-1\cdot 27$, $-27=1\cdot -27$,
$-27=-3\cdot 9$, $-27=3\cdot -9$
$2t^4+3t^2-27=(2t^2+9)(t^2-3)$

77. Factors of $5x^2y^2$: $5x^2y^2=5xy\cdot xy$
Factors of 1: $1=1\cdot 1$
There is no combination that gives the correct middle term, so $5x^2y^2+20xy+1$ is prime.

79. $6a^5+37a^3b^2+6ab^4=a(6a^4+37a^2b^2+6b^4)$

Factors of $6a^4$: $6a^4=6a^2\cdot a^2$, $6a^4=3a^2\cdot 2a^2$
Factors of $6b^4$: $6b^4=6b^2\cdot b^2$, $6b^4=3b^2\cdot 2b^2$
$6a^5+37a^3b^2+6ab^4=a(6a^2+b^2)(a^2+6b^2)$

81. $(x-4)(x+4)=(x)^2-(4)^2=x^2-16$

83. $(x+2)^2=x^2+2x(2)+2^2=x^2+4x+4$

85. $(2x-1)^2=(2x)^2-2(2x)(1)+(1)^2=4x^2-4x+1$

87. No.
$4x^2=2\cdot 2\cdot x\cdot x$
$19x=19\cdot x$
$12=2\cdot 2\cdot 3$
There is no common factor (other than 1).

89. $(3x^2+1)+(6x+4)+(x^2+15x)=4x^2+21x+5$
$=(4x+1)(x+5)$

91. $4x^2+2x+\frac{1}{4}=\left(2x+\frac{1}{2}\right)\left(2x+\frac{1}{2}\right)$ or $\left(2x+\frac{1}{2}\right)^2$

93. $4x^2(y-1)^2+25x(y-1)^2+25(y-1)^2$
$=(y-1)^2(4x^2+25x+25)$
$=(y-1)^2(4x+5)(x+5)$

95. Factors of $3x^2$: $3x^2=3x\cdot x$
Factors of -5: $-5=-1\cdot 5$, $-5=1\cdot -5$
$(3x-1)(x+5)=3x^2+14x-5$
$(3x+1)(x-5)=3x^2-14x-5$
$(3x-5)(x+1)=3x^2-2x-5$
$(3x+5)(x-1)=3x^2+2x-5$
Since b is positive, the possible values are 2 and 14.

97. Note that $5+2=7$.
$(5x+2)(x+1)=5x(x+1)+2(x+1)$
$=5x^2+5x+2x+2$
$=5x^2+7x+2$
If $c=2$, then $5x^2+7x+c$ is factorable.

99. answers may vary

Section 4.4 Practice

1. a. $3\cdot 8=24$
$12\cdot 2=24$
$12+2=14$
$3x^2+14x+8=3x^2+12x+2x+8$
$=3x(x+4)+2(x+4)$
$=(x+4)(3x+2)$

Copyright © 2012 Pearson Education, Inc. Publishing as Prentice Hall.

b. $12 \cdot 5 = 60$
$15 \cdot 4 = 60$
$15 + 4 = 19$
$$\begin{aligned} 12x^2 + 19x + 5 &= 12x^2 + 15x + 4x + 5 \\ &= 3x(4x+5) + 1(4x+5) \\ &= (4x+5)(3x+1) \end{aligned}$$

2. a. $30x^2 - 26x + 4 = 2(15x^2 - 13x + 2)$
$15 \cdot 2 = 30$
$-3 \cdot -10 = 30$
$-3 + (-10) = -13$
$$\begin{aligned} 30x^2 - 26x + 4 &= 2(15x^2 - 13x + 2) \\ &= 2(15x^2 - 3x - 10x + 2) \\ &= 2[3x(5x-1) - 2(5x-1)] \\ &= 2(5x-1)(3x-2) \end{aligned}$$

b. $6x^2y - 7xy - 5y = y(6x^2 - 7x - 5)$
$6 \cdot -5 = -30$
$3 \cdot -10 = -30$
$3 + (-10) = -7$
$$\begin{aligned} 6x^2y - 7xy - 5y &= y(6x^2 - 7x - 5) \\ &= y(6x^2 + 3x - 10x - 5) \\ &= y[3x(2x+1) - 5(2x+1)] \\ &= y(2x+1)(3x-5) \end{aligned}$$

3. $12y^5 + 10y^4 - 42y^3 = 2y^3(6y^2 + 5y - 21)$
$6 \cdot -21 = -126$
$14 \cdot -9 = -126$
$14 + (-9) = 5$
$$\begin{aligned} &12y^5 + 10y^4 - 42y^3 \\ &= 2y^3(6y^2 + 5y - 21) \\ &= 2y^3(6y^2 + 14y - 9y - 21) \\ &= 2y^3[2y(3y+7) - 3(3y+7)] \\ &= 2y^3(3y+7)(2y-3) \end{aligned}$$

Vocabulary and Readiness Check

1. $a = 1, b = 6, c = 8$
$a \cdot c = 1 \cdot 8 = 8$
$4 \cdot 2 = 8$ and $4 + 2 = 6$; choice a.

2. $a = 1, b = 11, c = 24$
$a \cdot c = 1 \cdot 24 = 24$
$8 \cdot 3 = 24$ and $8 + 3 = 11$; choice c.

3. $a = 2, b = 13, c = 6$
$a \cdot c = 2 \cdot 6 = 12$
$12 \cdot 1 = 12$ and $12 + 1 = 13$; choice b.

4. $a = 4, b = 8, c = 3$
$a \cdot c = 4 \cdot 3 = 12$
$2 \cdot 6 = 12$ and $2 + 6 = 8$; choice d.

Exercise Set 4.4

1. $$\begin{aligned} x^2 + 3x + 2x + 6 &= x(x+3) + 2(x+3) \\ &= (x+3)(x+2) \end{aligned}$$

3. $$\begin{aligned} y^2 + 8y - 2y - 16 &= y(y+8) - 2(y+8) \\ &= (y+8)(y-2) \end{aligned}$$

5. $$\begin{aligned} 8x^2 - 5x - 24x + 15 &= x(8x-5) - 3(8x-5) \\ &= (8x-5)(x-3) \end{aligned}$$

7. $$\begin{aligned} 5x^4 - 3x^2 + 25x^2 - 15 &= x^2(5x^2 - 3) + 5(5x^2 - 3) \\ &= (5x^2 - 3)(x^2 + 5) \end{aligned}$$

9. a. $9 \cdot 2 = 18$
$9 + 2 = 11$
9 and 2 are numbers whose product is 18 and whose sum is 11.

b. $11x = 9x + 2x$

c. $$\begin{aligned} 6x^2 + 11x + 3 &= 6x^2 + 9x + 2x + 3 \\ &= 3x(2x+3) + 1(2x+3) \\ &= (2x+3)(3x+1) \end{aligned}$$

11. a. $-3 \cdot -20 = 60$
$-3 + (-20) = -23$
-3 and -20 are numbers whose product is 60 and whose sum is -23.

b. $-23x = -3x - 20x$

c. $$\begin{aligned} 15x^2 - 23x + 4 &= 15x^2 - 3x - 20x + 4 \\ &= 3x(5x-1) - 4(5x-1) \\ &= (5x-1)(3x-4) \end{aligned}$$

13. $21 \cdot 2 = 42$
$14 \cdot 3 = 42$
$14 + 3 = 17$
$$\begin{aligned} 21y^2 + 17y + 2 &= 21y^2 + 14y + 3y + 2 \\ &= 7y(3y+2) + 1(3y+2) \\ &= (3y+2)(7y+1) \end{aligned}$$

Copyright © 2012 Pearson Education, Inc. Publishing as Prentice Hall.

15. $7 \cdot -11 = -77$
$-11 \cdot 7 = -77$
$-11 + 7 = -4$
$$\begin{aligned} 7x^2 - 4x - 11 &= 7x^2 - 11x + 7x - 11 \\ &= x(7x-11) + 1(7x-11) \\ &= (7x-11)(x+1) \end{aligned}$$

17. $10 \cdot 2 = 20$
$-4 \cdot -5 = 20$
$-4 + (-5) = -9$
$$\begin{aligned} 10x^2 - 9x + 2 &= 10x^2 - 4x - 5x + 2 \\ &= 2x(5x-2) - 1(5x-2) \\ &= (5x-2)(2x-1) \end{aligned}$$

19. $2 \cdot 5 = 10$
$-5 \cdot -2 = 10$
$-5 + (-2) = -7$
$$\begin{aligned} 2x^2 - 7x + 5 &= 2x^2 - 5x - 2x + 5 \\ &= x(2x-5) - 1(2x-5) \\ &= (2x-5)(x-1) \end{aligned}$$

21. $12x + 4x^2 + 9 = 4x^2 + 12x + 9$
$4 \cdot 9 = 36$
$6 \cdot 6 = 36$
$6 + 6 = 12$
$$\begin{aligned} 4x^2 + 12x + 9 &= 4x^2 + 6x + 6x + 9 \\ &= 2x(2x+3) + 3(2x+3) \\ &= (2x+3)(2x+3) \text{ or } (2x+3)^2 \end{aligned}$$

23. $4 \cdot -21 = -84$
$6 \cdot -14 = -84$
$6 + (-14) = -8$
$$\begin{aligned} 4x^2 - 8x - 21 &= 4x^2 + 6x - 14x - 21 \\ &= 2x(2x+3) - 7(2x+3) \\ &= (2x+3)(2x-7) \end{aligned}$$

25. $10 \cdot 12 = 120$
$-8 \cdot -15 = 120$
$-8 + (-15) = -23$
$$\begin{aligned} 10x^2 - 23x + 12 &= 10x^2 - 8x - 15x + 12 \\ &= 2x(5x-4) - 3(5x-4) \\ &= (5x-4)(2x-3) \end{aligned}$$

27. $2x^3 + 13x^2 + 15x = x(2x^2 + 13x + 15)$
$2 \cdot 15 = 30$
$3 \cdot 10 = 30$
$3 + 10 = 13$
$$\begin{aligned} 2x^3 + 13x^2 + 15x &= x(2x^2 + 13x + 15) \\ &= x(2x^2 + 3x + 10x + 15) \\ &= x[x(2x+3) + 5(2x+3)] \\ &= x(2x+3)(x+5) \end{aligned}$$

29. $16y^2 - 34y + 18 = 2(8y^2 - 17y + 9)$
$8 \cdot 9 = 72$
$-9 \cdot -8 = 72$
$-9 + (-8) = -17$
$$\begin{aligned} 16y^2 - 34y + 18 &= 2(8y^2 - 17y + 9) \\ &= 2(8y^2 - 9y - 8y + 9) \\ &= 2[y(8y-9) - 1(8y-9)] \\ &= 2(8y-9)(y-1) \end{aligned}$$

31. $-13x + 6 + 6x^2 = 6x^2 - 13x + 6$
$6 \cdot 6 = 36$
$-9 \cdot -4 = 36$
$-9 + (-4) = -13$
$$\begin{aligned} 6x^2 - 13x + 6 &= 6x^2 - 9x - 4x + 6 \\ &= 3x(2x-3) - 2(2x-3) \\ &= (2x-3)(3x-2) \end{aligned}$$

33. $54a^2 - 9a - 30 = 3(18a^2 - 3a - 10)$
$18 \cdot -10 = -180$
$12 \cdot -15 = -180$
$12 + (-15) = -3$
$$\begin{aligned} 54a^2 - 9a - 30 &= 3(18a^2 - 3a - 10) \\ &= 3(18a^2 + 12a - 15a - 10) \\ &= 3[6a(3a+2) - 5(3a+2)] \\ &= 3(3a+2)(6a-5) \end{aligned}$$

35. $20a^3 + 37a^2 + 8a = a(20a^2 + 37a + 8)$
$20 \cdot 8 = 160$
$5 \cdot 32 = 160$
$5 + 32 = 37$
$$\begin{aligned} 20a^3 + 37a^2 + 8a &= a(20a^2 + 37a + 8) \\ &= a(20a^2 + 5a + 32a + 8) \\ &= a[5a(4a+1) + 8(4a+1)] \\ &= a(4a+1)(5a+8) \end{aligned}$$

Copyright © 2012 Pearson Education, Inc. Publishing as Prentice Hall.

37. $12x^3 - 27x^2 - 27x = 3x(4x^2 - 9x - 9)$

$4 \cdot -9 = -36$

$3 \cdot -12 = -36$

$3 + (-12) = -9$

$$\begin{aligned} 12x^3 - 27x^2 - 27x &= 3x(4x^2 - 9x - 9) \\ &= 3x(4x^2 + 3x - 12x - 9) \\ &= 3x[x(4x+3) - 3(4x+3)] \\ &= 3x(4x+3)(x-3) \end{aligned}$$

39. $3x^2y + 4xy^2 + y^3 = y(3x^2 + 4xy + y^2)$

$3 \cdot y^2 = 3y^2$

$y \cdot 3y = 3y^2$

$y + 3y = 4y$

$$\begin{aligned} 3x^2y + 4xy^2 + y^3 &= y(3x^2 + 4xy + y^2) \\ &= y(3x^2 + xy + 3xy + y^2) \\ &= y[x(3x+y) + y(3x+y)] \\ &= y(3x+y)(x+y) \end{aligned}$$

41. $20 \cdot 1 = 20$

There are no factors of 20 which sum to 7, so $20z^2 + 7z + 1$ is prime.

43. $24a^2 - 6ab - 30b^2 = 6(4a^2 - ab - 5b^2)$

$4 \cdot -5b^2 = -20b^2$

$4b \cdot -5b = -20b^2$

$4b + (-5b) = -b$

$$\begin{aligned} 24a^2 - 6ab - 30b^2 &= 6(4a^2 - ab - 5b^2) \\ &= 6(4a^2 + 4ab - 5ab - 5b^2) \\ &= 6[4a(a+b) - 5b(a+b)] \\ &= 6(a+b)(4a-5b) \end{aligned}$$

45. $15p^4 + 31p^3q + 2p^2q^2$

$= p^2(15p^2 + 31pq + 2q^2)$

$15 \cdot 2q^2 = 30q^2$

$q \cdot 30q = 30q^2$

$q + 30q = 31q$

$$\begin{aligned} &15p^4 + 31p^3q + 2p^2q^2 \\ &= p^2(15p^2 + 31pq + 2q^2) \\ &= p^2(15p^2 + pq + 30pq + 2q^2) \\ &= p^2[p(15p+q) + 2q(15p+q)] \\ &= p^2(15p+q)(p+2q) \end{aligned}$$

47. $35 + 12x + x^2 = x^2 + 12x + 35$

$1 \cdot 35 = 35$

$7 \cdot 5 = 35$

$7 + 5 = 12$

$$\begin{aligned} x^2 + 12x + 35 &= x^2 + 7x + 5x + 35 \\ &= x(x+7) + 5(x+7) \\ &= (x+7)(x+5) \text{ or } (7+x)(5+x) \end{aligned}$$

49. $6 - 11x + 5x^2 = 5x^2 - 11x + 6$

$5 \cdot 6 = 30$

$-6 \cdot -5 = 30$

$-6 + (-5) = -11$

$$\begin{aligned} 5x^2 - 11x + 6 &= 5x^2 - 6x - 5x + 6 \\ &= x(5x-6) - 1(5x-6) \\ &= (5x-6)(x-1) \text{ or } (6-5x)(1-x) \end{aligned}$$

51. $(x-2)(x+2) = x^2 - 2^2 = x^2 - 4$

53. $$\begin{aligned} (y+4)(y+4) &= (y+4)^2 \\ &= (y)^2 + 2(y)(4) + (4)^2 \\ &= y^2 + 8y + 16 \end{aligned}$$

55. $(9z+5)(9z-5) = (9z)^2 - 5^2 = 81z^2 - 25$

57. $$\begin{aligned} (4x-3)^2 &= (4x)^2 - 2(4x)(3) + (3)^2 \\ &= 16x^2 - 24x + 9 \end{aligned}$$

59. $5(2x^2 + 9x + 9) = 10x^2 + 45x + 45$

$2 \cdot 9 = 18$

$3 \cdot 6 = 18$

$3 + 6 = 9$

$$\begin{aligned} 5(2x^2 + 9x + 9) &= 5(2x^2 + 3x + 6x + 9) \\ &= 5[x(2x+3) + 3(2x+3)] \\ &= 5(2x+3)(x+3) \end{aligned}$$

The perimeter is $10x^2 + 45x + 45$ or $5(2x + 3)(x + 3)$.

61. $$\begin{aligned} x^{2n} + 2x^n + 3x^n + 6 &= x^n(x^n + 2) + 3(x^n + 2) \\ &= (x^n + 2)(x^n + 3) \end{aligned}$$

Copyright © 2012 Pearson Education, Inc. Publishing as Prentice Hall.

63. $3 \cdot -35 = -105$
$-5 \cdot 21 = -105$
$-5 + 21 = 16$

$$\begin{aligned} 3x^{2n} + 16x^n - 35 &= 3x^{2n} - 5x^n + 21x^n - 35 \\ &= x^n(3x^n - 5) + 7(3x^n - 5) \\ &= (3x^n - 5)(x^n + 7) \end{aligned}$$

65. answers may vary

Section 4.5 Practice

1. a. Since $36 = 6^2$ and $12x = 2 \cdot 6 \cdot x$, $x^2 + 12x + 36$ is a perfect square trinomial.

b. Since $100 = 10^2$ and $20x = 2 \cdot 10 \cdot x$, $x^2 + 20x + 100$ is a perfect square trinomial.

2. a. Since $9x^2 = (3x)^2$ and $25 = 5^2$, but $20x \neq 2 \cdot 3x \cdot 5$, the polynomial is not a perfect square trinomial.

b. Since $4x^2 = (2x)^2$, but 11 is not a perfect square, the polynomial is not a perfect square trinomial.

3. a. Since $25x^2 = (5x)^2$ and $1 = 1^2$, and $2 \cdot 5x \cdot 1 = 10x$ which is the opposite of $-10x$, the trinomial is a perfect square trinomial.

b. Since $9x^2 = (3x)^2$ and $49 = 7^2$, and $2 \cdot 3x \cdot 7 = 42x$ which is the opposite of $-42x$, the trinomial is a perfect square trinomial.

4. $$\begin{aligned} x^2 + 16x + 64 &= (x)^2 + 2 \cdot x \cdot 8 + 8^2 \\ &= (x+8)^2 \end{aligned}$$

5. $$\begin{aligned} 9r^2 + 24rs + 16s^2 &= (3r)^2 + 2 \cdot 3r \cdot 4s + (4s)^2 \\ &= (3r + 4s)^2 \end{aligned}$$

6. $$\begin{aligned} 9n^4 - 6n^2 + 1 &= (3n^2)^2 - 2 \cdot 3n^2 \cdot 1 + 1^2 \\ &= (3n^2 - 1)^2 \end{aligned}$$

7. Notice that this trinomial is not a perfect square trinomial.
$9x^2 = (3x)^2$ and $4 = 2^2$, but $2 \cdot 3x \cdot 2 = 12x$ and $12x$ is not the middle term $15x$.
Factor by grouping.
$9 \cdot 4 = 36$
$3 \cdot 12 = 36$
$3 + 12 = 15$

$$\begin{aligned} 9x^2 + 15x + 4 &= 9x^2 + 3x + 12x + 4 \\ &= 3x(3x+1) + 4(3x+1) \\ &= (3x+1)(3x+4) \end{aligned}$$

8. a. $$\begin{aligned} 8n^2 + 40n + 50 &= 2(4n^2 + 20n + 25) \\ &= 2[(2n)^2 + 2 \cdot 2n \cdot 5 + 5^2] \\ &= 2(2n+5)^2 \end{aligned}$$

b. $$\begin{aligned} &12x^3 - 84x^2 + 147x \\ &= 3x(4x^2 - 28x + 49) \\ &= 3x[(2x)^2 - 2 \cdot 2x \cdot 7 + 7^2] \\ &= 3x(2x-7)^2 \end{aligned}$$

9. $x^2 - 9 = x^2 - 3^2 = (x-3)(x+3)$

10. $a^2 - 16 = a^2 - 4^2 = (a-4)(a+4)$

11. $c^2 - \frac{9}{25} = c^2 - \left(\frac{3}{5}\right)^2 = \left(c - \frac{3}{5}\right)\left(c + \frac{3}{5}\right)$

12. $s^2 + 9$ is prime since it is the sum of two squares.

13. $9s^2 - 1 = (3s)^2 - 1^2 = (3s-1)(3s+1)$

14. $$\begin{aligned} 16x^2 - 49y^2 &= (4x)^2 - (7y)^2 \\ &= (4x - 7y)(4x + 7y) \end{aligned}$$

15. $$\begin{aligned} p^4 - 81 &= (p^2)^2 - 9^2 \\ &= (p^2 + 9)(p^2 - 9) \\ &= (p^2 + 9)(p+3)(p-3) \end{aligned}$$

16. $$\begin{aligned} 9x^3 - 25x &= x(9x^2 - 25) \\ &= x[(3x)^2 - 5^2] \\ &= x(3x-5)(3x+5) \end{aligned}$$

Copyright © 2012 Pearson Education, Inc. Publishing as Prentice Hall.

17. $48x^4 - 3 = 3(16x^4 - 1)$
$= 3[(4x^2)^2 - 1^2]$
$= 3(4x^2 + 1)(4x^2 - 1)$
$= 3(4x^2 + 1)(2x + 1)(2x - 1)$

18. $-9x^2 + 100 = -1(9x^2 - 100)$
$= -1[(3x)^2 - 10^2]$
$= -1(3x - 10)(3x + 10)$

19. $121 - m^2 = 11^2 - m^2 = (11 + m)(11 - m)$
or
$121 - m^2 = -m^2 + 121$
$= -1(m^2 - 121)$
$= -1(m^2 - 11^2)$
$= -1(m + 11)(m - 11)$

Calculator Explorations

	$x^2 - 2x + 1$	$x^2 - 2x - 1$	$(x-1)^2$
$x = 5$	16	14	16
$x = -3$	16	14	16
$x = 2.7$	2.89	0.89	2.89
$x = -12.1$	171.61	169.61	171.61
$x = 0$	1	1	1

Vocabulary and Readiness Check

1. A perfect square trinomial is a trinomial that is the square of a binomial.

2. The term $25y^2$ written as a square is $(5y)^2$.

3. The expression $x^2 + 10x + 25y^2$ is called a perfect square trinomial.

4. The expression $x^2 - 49$ is called a difference of two squares.

5. The factorization $(x + 5y)(x + 5y)$ may also be written as $(x + 5y)^2$.

6. The factorization $(x - 5y)(x + 5y)$ may also be written as $(x - 5y)^2$. false

7. The trinomial $x^2 - 6x - 9$ is a perfect square trinomial. false

8. The binomial $y^2 + 9$ factors as $(y + 3)^2$. false

9. $64 = 8^2$

10. $9 = 3^2$

11. $121a^2 = (11a)^2$

12. $81b^2 = (9b)^2$

13. $36p^4 = (6p^2)^2$

14. $4q^4 = (2q^2)^2$

Exercise Set 4.5

1. Since $64 = 8^2$ and $16x = 2 \cdot 8 \cdot x$, $x^2 + 16x + 64$ is a perfect square trinomial.

3. Since $25 = 5^2$ but $5y \neq 2 \cdot 5 \cdot y$, $y^2 + 5y + 25$ is not a perfect square trinomial.

5. Since $1 = 1^2$ and $-2m = -2 \cdot 1 \cdot m$, $m^2 - 2m + 1$ is a perfect square trinomial.

7. Since $49 = 7^2$ but $16a \neq 2 \cdot 7 \cdot a$, $a^2 - 16a + 49$ is not a perfect square trinomial.

9. $4x^2 = (2x)^2$ but $8y^2$ is not a perfect square, so $4x^2 + 12xy + 8y^2$ is not a perfect square trinomial.

11. $25a^2 = (5a)^2$, $16b^2 = (4b)^2$, and $40ab = 2 \cdot 5a \cdot 4b$, so $25a^2 - 40ab + 16b^2$ is a perfect square trinomial.

13. $x^2 + 22x + 121 = x^2 + 2 \cdot x \cdot 11 + 11^2$
$= (x + 11)^2$

15. $x^2 - 16x + 64 = x^2 - 2 \cdot x \cdot 8 + 8^2 = (x - 8)^2$

17. $16a^2 - 24a + 9 = (4a)^2 - 2 \cdot 4a \cdot 3 + 3^2$
$= (4a - 3)^2$

Copyright © 2012 Pearson Education, Inc. Publishing as Prentice Hall.

19. $x^4+4x^2+4=(x^2)^2+2\cdot x^2\cdot 2+2^2=(x^2+2)^2$

21. $2n^2-28n+98=2(n^2-14n+49)$
$=2[n^2-2\cdot n\cdot 7+7^2]$
$=2(n-7)^2$

23. $16y^2+40y+25=(4y)^2+2\cdot 4y\cdot 5+5^2$
$=(4y+5)^2$

25. $x^2y^2-10xy+25=(xy)^2-2\cdot xy\cdot 5+5^2$
$=(xy-5)^2$

27. $m^3+18m^2+81m=m(m^2+18m+81)$
$=m(m^2+2\cdot m\cdot 9+9^2)$
$=m(m+9)^2$

29. Since $1=1^2$ and $x^4=(x^2)^2$, but $6x^2\neq 2\cdot 1\cdot x^2$, the polynomial is not a perfect square trinomial. Since there are no factors of $1\cdot 1=1$ that sum to 6, the trinomial is prime.

31. $9x^2-24xy+16y^2=(3x)^2-2\cdot 3x\cdot 4y+(4y)^2$
$=(3x-4y)^2$

33. $x^2-4=x^2-(2)^2=(x+2)(x-2)$

35. $81-p^2=9^2-p^2=(9+p)(9-p)$
or
$81-p^2=-p^2+81$
$=-1(p^2-81)$
$=-1(p-9^2)$
$=-1(p+9)(p-9)$

37. $-4r^2+1=-1(4r^2-1)$
$=-1[(2r)^2-1^2]$
$=-1(2r+1)(2r-1)$

39. $9x^2-16=(3x)^2-4^2=(3x+4)(3x-4)$

41. $16r^2+1$ is the sum of two squares, which is prime.

43. $-36+x^2=-1(36-x^2)$
$=-1(6^2-x^2)$
$=-1(6+x)(6-x)$
or
$-36+x^2=x^2-36$
$=x^2-6^2$
$=(x-6)(x+6)$

45. $m^4-1=(m^2)^2-1^2$
$=(m^2+1)(m^2-1)$
$=(m^2+1)(m^2-1^2)$
$=(m^2+1)(m+1)(m-1)$

47. $x^2-169y^2=x^2-(13y)^2=(x^2+13y)(x-13y)$

49. $18r^2-8=2(9r^2-4)$
$=2[(3r)^2-2^2]$
$=2(3r+2)(3r-2)$

51. $9xy^2-4x=x(9y^2-4)$
$=x[(3y)^2-2^2]$
$=x(3y+2)(3y-2)$

53. $16x^4-64x^2=16x^2(x^2-4)$
$=16x^2(x^2-2^2)$
$=16x^2(x+2)(x-2)$

55. $xy^3-9xyz^2=xy(y^2-9z^2)$
$=xy[y^2-(3z)^2]$
$=xy(y-3z)(y+3z)$

57. $36x^2-64y^2=4(9x^2-16y^2)$
$=4[(3x)^2-(4y)^2]$
$=4(3x+4y)(3x-4y)$

59. $144-81x^2=9(16-9x^2)$
$=9[4^2-(3x)^2]$
$=9(4-3x)(4+3x)$

61. $25y^2-9=(5y)^2-3^2=(5y+3)(5y-3)$

63. $121m^2-100n^2=(11m)^2-(10n)^2$
$=(11m+10n)(11m-10n)$

Copyright © 2012 Pearson Education, Inc. Publishing as Prentice Hall.

65. $x^2y^2-1=(xy)^2-1^2=(xy+1)(xy-1)$

67. $x^2-\frac{1}{4}=x^2-\left(\frac{1}{2}\right)^2=\left(x-\frac{1}{2}\right)\left(x+\frac{1}{2}\right)$

69. $49-\frac{9}{25}m^2=7^2-\left(\frac{3}{5}m\right)^2$
$=\left(7+\frac{3}{5}m\right)\left(7-\frac{3}{5}m\right)$

71. $81a^2-25b^2=(9a)^2-(5b)^2=(9a+5b)(9a-5b)$

73. $x^2+14xy+49y^2=x^2+2\cdot x\cdot 7y+(7y)^2$
$=(x+7y)^2$

75. $32n^4-112n^2+98=2(16n^4-56n^2+49)$
$=2[(4n^2)^2-2\cdot 4n^2\cdot 7+7^2]$
$=2(4n^2-7)^2$

77. $x^6-81x^2=x^2(x^4-81)$
$=x^2[(x^2)^2-9^2]$
$=x^2(x^2+9)(x^2-9)$
$=x^2(x^2+9)[(x)^2-3^2]$
$=x^2(x^2+9)(x+3)(x-3)$

79. $64p^3q-81pq^3=pq(64p^2-81q^2)$
$=pq[(8p)^2-(9q)^2]$
$=pq(8p-9q)(8p+9q)$

81. $x-6=0$
$x-6+6=0+6$
$x=6$

83. $2m+4=0$
$2m+4-4=0-4$
$2m=-4$
$\frac{2m}{2}=\frac{-4}{2}$
$m=-2$

85. $5z-1=0$
$5z-1+1=0+1$
$5z=1$
$\frac{5z}{5}=\frac{1}{5}$
$z=\frac{1}{5}$

87. $x^2-\frac{2}{3}x+\frac{1}{9}=x^2-2\cdot x\cdot\frac{1}{3}+\left(\frac{1}{3}\right)^2=\left(x-\frac{1}{3}\right)^2$

89. $(x+2)^2-y^2=[(x+2)+y][(x+2)-y]$
$=(x+2+y)(x+2-y)$

91. $a^2(b-4)-16(b-4)=(b-4)(a^2-16)$
$=(b-4)(a^2-4^2)$
$=(b-4)(a-4)(a+4)$

93. $(x^2+6x+9)-4y^2$
$=(x^2+2\cdot x\cdot 3+3^2)-4y^2$
$=(x+3)^2-4y^2$
$=(x+3)^2-(2y)^2$
$=[(x+3)+2y][(x+3)-2y]$
$=(x+3+2y)(x+3-2y)$

95. $x^{2n}-100=(x^n)^2-10^2=(x^n+10)(x^n-10)$

97. $x^2=x^2$ and $16=4^2$.
$(x+4)^2=x^2+2\cdot x\cdot 4+4^2=x^2+8x+16$, so the number 8 makes the given expression a perfect square trinomial.

99. answers may vary

101. The difference of two squares is the result of multiplying binomials of the form $(x+a)$ and $(x-a)$, so multiplying $(x-6)$ by $(x+6)$ results in the difference of two squares.

103. $(a+b)^2=a^2+2\cdot a\cdot b+b^2=a^2+2ab+b^2$

105. a. $2704-16t^2=2704-16(3)^2$
$=2704-16\cdot 9$
$=2704-144$
$=2560$
After 3 seconds, the filter is 2560 feet above the river.

Copyright © 2012 Pearson Education, Inc. Publishing as Prentice Hall.

b. $2704-16t^2 = 2704-16(7)^2$
$= 2704-16\cdot 49$
$= 2704-784$
$= 1920$

After 7 seconds, the filter is 1920 feet above the river.

c. The filter lands in the river when its height is 0 feet.

$2704-16t^2 = 0$
$(52+4t)(52-4t) = 0$
$52+4t = 0$ or $52-4t = 0$
$4t = -52$ $\quad -4t = -52$
$t = -13$ $\quad t = 13$

Discard $t = -13$ since time cannot be negative. The filter lands in the river after 13 seconds.

d. $2704-16t^2 = 16(169-t^2)$
$= 16[(13)^2-t^2]$
$= 16(13-t)(13+t)$

107. a. $1600-16t^2 = 1600-16(3)^2$
$= 1600-16\cdot 9$
$= 1600-144$
$= 1456$

After 3 seconds, the height of the bolt is 1456 feet.

b. $1600-16t^2 = 1600-16(7)^2$
$= 1600-16(49)$
$= 1600-784$
$= 816$

After 7 seconds, the height of the bolt is 816 feet.

c. The bolt hits the ground when its height is 0 feet.

$1600-16t^2 = 0$
$(40-4t)(40+4t) = 0$
$40-4t = 0$ or $40+4t = 0$
$-4t = -40$ $\quad 4t = -40$
$t = 10$ $\quad t = -10$

Discard $t = -10$ since time cannot be negative. The bolt hits the ground after 10 seconds.

d. $1600-16t^2 = 16(100-t^2)$
$= 16(10^2-t^2)$
$= 16(10+t)(10-t)$

Integrated Review

1–74. Factoring methods may vary.

1. $x^2+x-12 = (x-3)(x+4)$

2. $x^2-10x+16 = (x-8)(x-2)$

3. $x^2+2x+1 = x^2+2\cdot x\cdot 1+1^2$
$= (x+1)^2$

4. $x^2-6x+9 = x^2-2\cdot x\cdot 3+3^2$
$= (x-3)^2$

5. $x^2-x-6 = (x+2)(x-3)$

6. $x^2+x-2 = (x+2)(x-1)$

7. $x^2+x-6 = (x+3)(x-2)$

8. $x^2+7x+12 = (x+3)(x+4)$

9. $x^2-7x+10 = (x-5)(x-2)$

10. $x^2-x-30 = (x-6)(x+5)$

11. $2x^2-98 = 2(x^2-49)$
$= 2(x^2-7^2)$
$= 2(x-7)(x+7)$

12. $3x^2-75 = 3(x^2-25)$
$= 3(x^2-5^2)$
$= 3(x-5)(x+5)$

13. $x^2+3x+5x+15 = x(x+3)+5(x+3)$
$= (x+3)(x+5)$

14. $3y-21+xy-7x = 3(y-7)+x(y-7)$
$= (y-7)(3+x)$

15. $x^2+6x-16 = (x+8)(x-2)$

16. $x^2-3x-28 = (x-7)(x+4)$

17. $4x^3+20x^2-56x = 4x(x^2+5x-14)$
$= 4x(x+7)(x-2)$

Copyright © 2012 Pearson Education, Inc. Publishing as Prentice Hall.

18. $6x^3 - 6x^2 - 120x = 6x(x^2 - x - 20)$
$= 6x(x-5)(x+4)$

19. $12x^2 + 34x + 24 = 2(6x^2 + 17x + 12)$
$= 2(3x+4)(2x+3)$

20. $24a^2 + 18ab - 15b^2 = 3(8a^2 + 6ab - 5b^2)$
$= 3(2a-b)(4a+5b)$

21. $4a^2 - b^2 = (2a)^2 - b^2 = (2a+b)(2a-b)$

22. $x^2 - 25y^2 = x^2 - (5y)^2 = (x+5y)(x-5y)$

23. $28 - 13x - 6x^2 = (4-3x)(7+2x)$

24. $20 - 3x - 2x^2 = (5-2x)(4+x)$

25. $4 - 2x + x^2$ is prime.

26. $a + a^2 - 3$ is prime.

27. $6y^2 + y - 15 = (3y+5)(2y-3)$

28. $4x^2 - x - 5 = (4x-5)(x+1)$

29. $18x^3 - 63x^2 + 9x = 9x(2x^2 - 7x + 1)$

30. $12a^3 - 24a^2 + 4a = 4a(3a^2 - 6a + 1)$

31. $16a^2 - 56a + 49 = (4a)^2 - 2 \cdot 4a \cdot 7 + 7^2$
$= (4a-7)^2$

32. $25p^2 - 70p + 49 = (5p)^2 - 2 \cdot 5p \cdot 7 + 7^2$
$= (5p-7)^2$

33. $14 + 5x - x^2 = (7-x)(2+x)$

34. $3 - 2x - x^2 = (3+x)(1-x)$

35. $3x^4y + 6x^3y - 72x^2y = 3x^2y(x^2 + 2x - 24)$
$= 3x^2y(x+6)(x-4)$

36. $2x^3y + 8x^2y^2 - 10xy^3 = 2xy(x^2 + 4xy - 5y^2)$
$= 2xy(x+5y)(x-y)$

37. $12x^3y + 243xy = 3xy(4x^2 + 81)$

38. $6x^3y^2 + 8xy^2 = 2xy^2(3x^2 + 4)$

39. $2xy - 72x^3y = 2xy(1 - 36x^2)$
$= 2xy[1 - (6x)^2]$
$= 2xy(1-6x)(1+6x)$

40. $2x^3 - 18x = 2x(x^2 - 9)$
$= 2x(x^2 - 3^2)$
$= 2x(x-3)(x+3)$

41. $x^3 + 6x^2 - 4x - 24 = x^2(x+6) - 4(x+6)$
$= (x+6)(x^2 - 4)$
$= (x+6)(x+2)(x-2)$

42. $x^3 - 2x^2 - 36x + 72 = x^2(x-2) - 36(x-2)$
$= (x-2)(x^2 - 36)$
$= (x-2)(x-6)(x+6)$

43. $6a^3 + 10a^2 = 2a^2(3a+5)$

44. $4n^2 - 6n = 2n(2n-3)$

45. $3x^3 - x^2 + 12x - 4 = x^2(3x-1) + 4(3x-1)$
$= (3x-1)(x^2+4)$

46. $x^3 - 2x^2 + 3x - 6 = x^2(x-2) + 3(x-2)$
$= (x-2)(x^2+3)$

47. $6x^2 + 18xy + 12y^2 = 6(x^2 + 3xy + 2y^2)$
$= 6(x+2y)(x+y)$

48. $12x^2 + 46xy - 8y^2 = 2(6x^2 + 23xy - 4y^2)$
$= 2(x+4y)(6x-y)$

49. $5(x+y) + x(x+y) = (x+y)(5+x)$

50. $7(x-y) + y(x-y) = (x-y)(7+y)$

51. $14t^2 - 9t + 1 = (7t-1)(2t-1)$

52. $3t^2 - 5t + 1$ is prime.

53. $-3x^2 - 2x + 5 = -1(3x^2 + 2x - 5)$
$= -1(3x+5)(x-1)$

Copyright © 2012 Pearson Education, Inc. Publishing as Prentice Hall.

54. $-7x^2-19x+6=-1(7x^2+19x-6)$
$=-1(7x-2)(x+3)$

55. $1-8a-20a^2=(1-10a)(1+2a)$

56. $1-7a-60a^2=(1+5a)(1-12a)$

57. $x^4-10x^2+9=(x^2-9)(x^2-1)$
$=(x-3)(x+3)(x-1)(x+1)$

58. $x^4-13x^2+36=(x^2-9)(x^2-4)$
$=(x-3)(x+3)(x-2)(x+2)$

59. $x^2-23x+120=(x-15)(x-8)$

60. $y^2+22y+96=(y+16)(y+6)$

61. $25p^2-70pq+49q^2=(5p)^2-2\cdot 5p\cdot 7q+(7q)^2$
$=(5p-7q)^2$

62. $16a^2-56ab+49b^2=(4a)^2-2\cdot 4a\cdot 7b+(7b)^2$
$=(4a-7b)^2$

63. $x^2-14x-48$ is prime.

64. $7x^2+24xy+9y^2=(7x+3y)(x+3y)$

65. $-x^2-x+30=-1(x^2+x-30)=-1(x-5)(x+6)$

66. $-x^2+6x-8=-1(x^2-6x+8)$
$=-1(x-2)(x-4)$

67. $3rs-s+12r-4=s(3r-1)+4(3r-1)$
$=(3r-1)(s+4)$

68. $x^3-2x^2+x-2=x^2(x-2)+1(x-2)$
$=(x-2)(x^2+1)$

69. $4x^2-8xy-3x+6y=4x(x-2y)-3(x-2y)$
$=(x-2y)(4x-3)$

70. $4x^2-2xy-7yz+14xz=2x(2x-y)-7z(y-2x)$
$=2x(2x-y)+7z(2x-y)$
$=(2x-y)(2x+7z)$

71. $x^2+9xy-36y^2=(x+12y)(x-3y)$

72. $3x^2+10xy-8y^2=(3x-2y)(x+4y)$

73. $x^4-14x^2-32=(x^2+2)(x^2-16)$
$=(x^2+2)(x+4)(x-4)$

74. $x^4-22x^2-75=(x^2+3)(x^2-25)$
$=(x^2+3)(x+5)(x-5)$

75. answers may vary

76. Yes; $9x^2+81y^2=9(x^2+9y^2)$

Section 4.6 Practice

1. $(x-7)(x+2)=0$
$x-7=0$ or $x+2=0$
$x=7$ $\quad$ $x=-2$
The solutions are 7 and –2.

2. $(x-10)(3x+1)=0$
$x-10=0$ or $3x+1=0$
$x=10$ $\quad$ $3x=-1$
$x=-\frac{1}{3}$
The solutions are 10 and $-\frac{1}{3}$.

3. a. $y(y+3)=0$
$y=0$ or $y+3=0$
$y=-3$
The solutions are 0 and –3.

b. $x(4x-3)=0$
$x=0$ or $4x-3=0$
$4x=3$
$x=\frac{3}{4}$
The solutions are 0 and $\frac{3}{4}$.

4. $x^2-3x-18=0$
$(x-6)(x+3)=0$
$x-6=0$ or $x+3=0$
$x=6$ $\quad$ $x=-3$
The solutions are 6 and –3.

Copyright © 2012 Pearson Education, Inc. Publishing as Prentice Hall.

5. $9x^2 - 24x = -16$

$9x^2 - 24x + 16 = 0$

$(3x-4)(3x-4) = 0$

$3x - 4 = 0$ or $3x - 4 = 0$

$3x = 4$ $\quad$ $3x = 4$

$x = \frac{4}{3}$ $\quad$ $x = \frac{4}{3}$

The solution is $\frac{4}{3}$.

6. a. $x(x-4) = 5$

$x^2 - 4x = 5$

$x^2 - 4x - 5 = 0$

$(x-5)(x+1) = 0$

$x - 5 = 0$ or $x + 1 = 0$

$x = 5$ $\quad$ $x = -1$

The solutions are 5 and –1.

b. $x(3x+7) = 6$

$3x^2 + 7x = 6$

$3x^2 + 7x - 6 = 0$

$(3x-2)(x+3) = 0$

$3x - 2 = 0$ or $x + 3 = 0$

$3x = 2$ $\quad$ $x = -3$

$x = \frac{2}{3}$

The solutions are $\frac{2}{3}$ and –3.

7. $2x^3 - 18x = 0$

$2x(x^2 - 9) = 0$

$2x(x-3)(x+3) = 0$

$2x = 0$ or $x - 3 = 0$ or $x + 3 = 0$

$x = 0$ $\quad$ $x = 3$ $\quad$ $x = -3$

The solutions are 0, 3, and –3.

8. $(x+3)(3x^2 - 20x - 7) = 0$

$x + 3 = 0$ or $3x^2 - 20x - 7 = 0$

$x = -3$ $\quad$ $(3x+1)(x-7) = 0$

$3x + 1 = 0$ or $x - 7 = 0$

$3x = -1$ $\quad$ $x = 7$

$x = -\frac{1}{3}$

The solutions are –3, $-\frac{1}{3}$, and 7.

Vocabulary and Readiness Check

1. An equation that can be written in the form $ax^2 + bx + c = 0$, (with $a \neq 0$), is called a <u>quadratic</u> equation.
2. If the product of two numbers is 0, then at least one of the numbers must be <u>0</u>.
3. The solutions to $(x - 3)(x + 5) = 0$ are <u>3, –5</u>.
4. If $a \cdot b = 0$, then <u>$a = 0$ or $b = 0$</u>.

Exercise Set 4.6

1. $(x - 2)(x + 1) = 0$

$x - 2 = 0$ or $x + 1 = 0$

$x = 2$ $\quad$ $x = -1$

The solutions are 2 and –1.

3. $(x - 6)(x - 7) = 0$

$x - 6 = 0$ or $x - 7 = 0$

$x = 6$ $\quad$ $x = 7$

The solutions are 6 and 7.

5. $(x + 9)(x + 17) = 0$

$x + 9 = 0$ or $x + 17 = 0$

$x = -9$ $\quad$ $x = -17$

The solutions are –9 and –17.

7. $x(x + 6) = 0$

$x = 0$ or $x + 6 = 0$

$x = -6$

The solutions are 0 and –6.

9. $3x(x - 8) = 0$

$3x = 0$ or $x - 8 = 0$

$x = 0$ $\quad$ $x = 8$

The solutions are 0 and 8.

11. $(2x + 3)(4x - 5) = 0$

$2x + 3 = 0$ or $4x - 5 = 0$

$2x = -3$ $\quad$ $4x = 5$

$x = -\frac{3}{2}$ $\quad$ $x = \frac{5}{4}$

The solutions are $-\frac{3}{2}$ and $\frac{5}{4}$.

Copyright © 2012 Pearson Education, Inc. Publishing as Prentice Hall.

13. $(2x-7)(7x+2)=0$

$2x-7=0$ or $7x+2=0$

$2x=7$ $\quad$ $7x=-2$

$x=\frac{7}{2}$ $\quad$ $x=-\frac{2}{7}$

The solutions are $\frac{7}{2}$ and $-\frac{2}{7}$.

15. $\left(x-\frac{1}{2}\right)\left(x+\frac{1}{3}\right)=0$

$x-\frac{1}{2}=0$ or $x+\frac{1}{3}=0$

$x=\frac{1}{2}$ $\quad$ $x=-\frac{1}{3}$

The solutions are $\frac{1}{2}$ and $-\frac{1}{3}$.

17. $(x+0.2)(x+1.5)=0$

$x+0.2=0$ or $x+1.5=0$

$x=-0.2$ $\quad$ $x=-1.5$

The solutions are -0.2 and -1.5.

19. $x^2-13x+36=0$

$(x-9)(x-4)=0$

$x-9=0$ or $x-4=0$

$x=9$ $\quad$ $x=4$

The solutions are 9 and 4.

21. $x^2+2x-8=0$

$(x+4)(x-2)=0$

$x+4=0$ or $x-2=0$

$x=-4$ $\quad$ $x=2$

The solutions are -4 and 2.

23. $x^2-7x=0$

$x(x-7)=0$

$x=0$ or $x-7=0$

$x=7$

The solutions are 0 and 7.

25. $x^2+20x=0$

$x(x+20)=0$

$x=0$ or $x+20=0$

$x=-20$

The solutions are 0 and -20.

27. $x^2=16$

$x^2-16=0$

$(x-4)(x+4)=0$

$x-4=0$ or $x+4=0$

$x=4$ $\quad$ $x=-4$

The solutions are 4 and -4.

29. $x^2-4x=32$

$x^2-4x-32=0$

$(x-8)(x+4)=0$

$x-8=0$ or $x+4=0$

$x=8$ $\quad$ $x=-4$

The solutions are 8 and -4.

31. $(x+4)(x-9)=4x$

$x^2-5x-36=4x$

$x^2-9x-36=0$

$(x+3)(x-12)=0$

$x+3=0$ or $x-12=0$

$x=-3$ $\quad$ $x=12$

The solutions are -3 and 12.

33. $x(3x-1)=14$

$3x^2-x=14$

$3x^2-x-14=0$

$3x^2-7x+6x-14=0$

$x(3x-7)+2(3x-7)=0$

$(3x-7)(x+2)=0$

$3x-7=0$ or $x+2=0$

$3x=7$ $\quad$ $x=-2$

$x=\frac{7}{3}$

The solutions are $\frac{7}{3}$ and -2.

35. $3x^2+19x-72=0$

$(3x-8)(x+9)=0$

$3x-8=0$ or $x+9=0$

$3x=8$ $\quad$ $x=-9$

$x=\frac{8}{3}$

The solutions are $\frac{8}{3}$ and -9.

Copyright © 2012 Pearson Education, Inc. Publishing as Prentice Hall.

37.
$$4x^3 - x = 0$$
$$x(4x^2 - 1) = 0$$
$$x[(2x)^2 - (1)^2] = 0$$
$$x(2x+1)(2x-1) = 0$$
$$x = 0 \quad \text{or} \quad 2x+1 = 0 \quad \text{or} \quad 2x-1 = 0$$
$$2x = -1 \qquad 2x = 1$$
$$x = -\frac{1}{2} \qquad x = \frac{1}{2}$$

The solutions are 0, $-\frac{1}{2}$, and $\frac{1}{2}$.

39.
$$4(x-7) = 6$$
$$4x - 28 = 6$$
$$4x = 34$$
$$x = \frac{34}{4}$$
$$x = \frac{17}{2}$$

The solution is $\frac{17}{2}$.

41.
$$(4x-3)(16x^2 - 24x + 9) = 0$$
$$(4x-3)[(4x)^2 - 2\cdot 4x\cdot 3 + (3)^2] = 0$$
$$(4x-3)(4x-3)^2 = 0$$
$$(4x-3)(4x-3)(4x-3) = 0$$
$$4x - 3 = 0$$
$$4x = 3$$
$$x = \frac{3}{4}$$

The solution is $\frac{3}{4}$.

43.
$$4y^2 - 1 = 0$$
$$(2y-1)(2y+1) = 0$$
$$2y - 1 = 0 \quad \text{or} \quad 2y + 1 = 0$$
$$2y = 1 \qquad 2y = -1$$
$$y = \frac{1}{2} \qquad y = -\frac{1}{2}$$

The solutions are $\frac{1}{2}$ and $-\frac{1}{2}$.

45.
$$(2x+3)(2x^2 - 5x - 3) = 0$$
$$(2x+3)(2x^2 + x - 6x - 3) = 0$$
$$(2x+3)[x(2x+1) - 3(2x+1)] = 0$$
$$(2x+3)(2x+1)(x-3) = 0$$
$$2x + 3 = 0 \quad \text{or} \quad 2x + 1 = 0 \quad \text{or} \quad x - 3 = 0$$
$$2x = -3 \qquad 2x = -1 \qquad x = 3$$
$$x = -\frac{3}{2} \qquad x = -\frac{1}{2}$$

The solutions are $-\frac{3}{2}$, $-\frac{1}{2}$, and 3.

47.
$$x^2 - 15 = -2x$$
$$x^2 + 2x - 15 = 0$$
$$(x+5)(x-3) = 0$$
$$x + 5 = 0 \quad \text{or} \quad x - 3 = 0$$
$$x = -5 \qquad x = 3$$

The solutions are –5 and 3.

49.
$$30x^2 - 11x = 30$$
$$30x^2 - 11x - 30 = 0$$
$$30x^2 + 25x - 36x - 30 = 0$$
$$5x(6x+5) - 6(6x+5) = 0$$
$$(6x+5)(5x-6) = 0$$
$$6x + 5 = 0 \quad \text{or} \quad 5x - 6 = 0$$
$$6x = -5 \qquad 5x = 6$$
$$x = -\frac{5}{6} \qquad x = \frac{6}{5}$$

The solutions are $-\frac{5}{6}$ and $\frac{6}{5}$.

51.
$$5x^2 - 6x - 8 = 0$$
$$(5x+4)(x-2) = 0$$
$$5x + 4 = 0 \quad \text{or} \quad x - 2 = 0$$
$$5x = -4 \qquad x = 2$$
$$x = -\frac{4}{5}$$

The solutions are $-\frac{4}{5}$ and 2.

53.
$$6y^2 - 22y - 40 = 0$$
$$2(3y^2 - 11y - 20) = 0$$
$$2(3y^2 + 4y - 15y - 20) = 0$$
$$2[y(3y+4) - 5(3y+4)] = 0$$
$$2(3y+4)(y-5) = 0$$
$$3y + 4 = 0 \quad \text{or} \quad y - 5 = 0$$
$$3y = -4 \qquad y = 5$$
$$y = -\frac{4}{3}$$

The solutions are $-\frac{4}{3}$ and 5.

Copyright © 2012 Pearson Education, Inc. Publishing as Prentice Hall.

55. $(y-2)(y+3)=6$
$y^2+y-6=6$
$y^2+y-12=0$
$(y+4)(y-3)=0$
$y+4=0$ or $y-3=0$
$y=-4$ $\quad$ $y=3$
The solutions are –4 and 3.

57. $x^3-12x^2+32x=0$
$x(x^2-12x+32)=0$
$x(x-8)(x-4)=0$
$x=0$ or $x-8=0$ or $x-4=0$
$x=8$ $\quad$ $x=4$
The solutions are 0, 8, and 4.

59. $x^2+14x+49=0$
$(x+7)(x+7)=0$
$x+7=0$
$x=-7$
The solution is –7.

61. $12y=8y^2$
$0=8y^2-12y$
$0=4y(2y-3)$
$4y=0$ or $2y-3=0$
$y=0$ $\quad$ $2y=3$
$y=\frac{3}{2}$
The solutions are 0 and $\frac{3}{2}$.

63. $7x^3-7x=0$
$7x(x^2-1)=0$
$7x(x-1)(x+1)=0$
$7x=0$ or $x-1=0$ or $x+1=0$
$x=0$ $\quad$ $x=1$ $\quad$ $x=-1$
The solutions are 0, 1, and –1.

65. $3x^2+8x-11=13-6x$
$3x^2+14x-11=13$
$3x^2+14x-24=0$
$3x^2+18x-4x-24=0$
$3x(x+6)-4(x+6)=0$
$(x+6)(3x-4)=0$
$x+6=0$ or $3x-4=0$
$x=-6$ $\quad$ $3x=4$
$x=\frac{4}{3}$
The solutions are –6 and $\frac{4}{3}$.

67. $3x^2-20x=-4x^2-7x-6$
$7x^2-13x+6=0$
$(7x-6)(x-1)=0$
$7x-6=0$ or $x-1=0$
$7x=6$ $\quad$ $x=1$
$x=\frac{6}{7}$
The solutions are $\frac{6}{7}$ and 1.

69. $\frac{3}{5}+\frac{4}{9}=\frac{3\cdot 9}{5\cdot 9}+\frac{4\cdot 5}{9\cdot 5}=\frac{27}{45}+\frac{20}{45}=\frac{47}{45}$

71. $\frac{7}{10}-\frac{5}{12}=\frac{7\cdot 6}{10\cdot 6}-\frac{5\cdot 5}{12\cdot 5}=\frac{42}{60}-\frac{25}{60}=\frac{17}{60}$

73. $\frac{4}{5}\cdot\frac{7}{8}=\frac{4\cdot 7}{5\cdot 8}=\frac{4\cdot 7}{5\cdot 4\cdot 2}=\frac{7}{5\cdot 2}=\frac{7}{10}$

75. The equation must first be written in standard form.
$x(x-2)=8$
$x^2-2x=8$
$x^2-2x-8=0$
$(x-4)(x+2)=0$
$x-4=0$ or $x+2=0$
$x=4$ $\quad$ $x=-2$
The solutions are 4 and –2.

77. answers may vary, for example $(x-6)(x+1)=0$

79. answers may vary, for example
$(x-5)(x-7)=0$
$x^2-7x-5x+35=0$
$x^2-12x+35=0$

Copyright © 2012 Pearson Education, Inc. Publishing as Prentice Hall.

81. a.

Time, x (seconds)	Height, y (feet)
0	$-16(0)^2+20(0)+300=300$
1	$-16(1)^2+20(1)+300=304$
2	$-16(2)^2+20(2)+300=276$
3	$-16(3)^2+20(3)+300=216$
4	$-16(4)^2+20(4)+300=124$
5	$-16(5)^2+20(5)+300=0$
6	$-16(6)^2+20(6)+300=-156$

b. When the compass strikes the ground, its height is 0 feet, so it strikes the ground after 5 seconds.

c. The maximum height listed in the table is 304 feet, after 1 second.

83.
$$(x-3)(3x+4)=(x+2)(x-6)$$
$$3x^2+4x-9x-12=x^2-6x+2x-12$$
$$3x^2-5x-12=x^2-4x-12$$
$$2x^2-x=0$$
$$x(2x-1)=0$$
$$x=0 \quad \text{or} \quad 2x-1=0$$
$$x=\frac{1}{2}$$

The solutions are 0 and $\frac{1}{2}$.

85.
$$(2x-3)(x+8)=(x-6)(x+4)$$
$$2x^2+16x-3x-24=x^2+4x-6x-24$$
$$2x^2+13x-24=x^2-2x-24$$
$$x^2+15x=0$$
$$x(x+15)=0$$
$$x=0 \quad \text{or} \quad x+15=0$$
$$x=-15$$

The solutions are 0 and –15.

Section 4.7 Practice

1. The diver will be at a height of 0 when he or she reaches the pool.
$$h=-16t^2+64$$
$$0=-16t^2+64$$
$$0=-16(t^2-4)$$
$$0=-16(t-2)(t+2)$$
$$t-2=0 \quad \text{or} \quad t+2=0$$
$$t=2 \qquad t=-2$$

It takes the diver 2 seconds to reach the pool.

2. Let x be the number.
$$x^2-2x=63$$
$$x^2-2x-63=0$$
$$(x-9)(x+7)=0$$
$$x-9=0 \quad \text{or} \quad x+7=0$$
$$x=9 \qquad x=-7$$

The numbers are 9 and –7.

3. Let x be the width. Then $x + 5$ is the length.
$$\text{Area} = \text{length} \cdot \text{width}$$
$$176=(x+5)\cdot x$$
$$176=x^2+5x$$
$$0=x^2+5x-176$$
$$0=(x+16)(x-11)$$
$$x+16=0 \quad \text{or} \quad x-11=0$$
$$x=-16 \qquad x=11$$

The width is 11 feet and the length is $x + 5 = 11 + 5 = 16$ feet.

4. Let x be the first odd integer. Then $x + 2$ is the next consecutive odd integer.
$$x(x+2)=x+(x+2)+23$$
$$x^2+2x=2x+25$$
$$x^2=25$$
$$x^2-25=0$$
$$(x+5)(x-5)=0$$
$$x+5=0 \quad \text{or} \quad x-5=0$$
$$x=-5 \qquad x=5$$
$$x+2=-3 \qquad x+2=7$$

The two consecutive odd integers are –5 and –3 or 5 and 7.

Copyright © 2012 Pearson Education, Inc. Publishing as Prentice Hall.

5. Let x be the length of one leg. Then $x-7$ is the length of the other leg.

$$x^2+(x-7)^2=13^2$$
$$x^2+x^2-14x+49=169$$
$$2x^2-14x-120=0$$
$$2(x^2-7x-60)=0$$
$$2(x+5)(x-12)=0$$
$$x+5=0 \quad \text{or} \quad x-12=0$$
$$x=-5 \qquad\qquad x=12$$

Discard $x=-5$ since length cannot be negative. If $x=12$, then $x-7=5$. The lengths of the legs are 5 meters and 12 meters.

Exercise Set 4.7

1. Let x be the width of the rectangle, then the length is $x+4$.

3. Let x be the first odd integer, then the next consecutive odd integer is $x+2$.

5. Let x be the base of the triangle, then the height is $4x+1$.

7. area $=(\text{side})^2$

$$121=x^2$$
$$0=x^2-121$$
$$0=(x-11)(x+11)$$
$$x-11=0 \quad \text{or} \quad x+11=0$$
$$x=11 \qquad\qquad x=-11$$

Discard $x=-11$ since length cannot be negative. The length of its sides are 11 units.

9. The perimeter is the sum of the lengths of the sides.

$$(x+5)+(x^2-3x)+(3x-8)+(x+3)=120$$
$$x^2+2x=120$$
$$x^2+2x-120=0$$
$$(x+12)(x-10)=0$$
$$x+12=0 \quad \text{or} \quad x-10=0$$
$$x=-12 \qquad\qquad x=10$$

For $x=-12$, $x+5$, $x+3$, and $3x-8$ are negative. Since lengths cannot be negative, $x=10$ is the only solution that works in this context.

$x+3=10+3=13$

$x+5=10+5=15$

$x^2-3x=(10)^2-3(10)=100-30=70$

$3x-8=3(10)-8=30-8=22$

The sides have lengths 13 cm, 15 cm, 70 cm, and 22 cm.

11. Area = base · height

$$96=(x+5)(x-5)$$
$$96=x^2-25$$
$$0=x^2-121$$
$$0=(x+11))(x-11)$$
$$x+11=0 \quad \text{or} \quad x-11=0$$
$$x=-11 \qquad\qquad x=11$$

For $x=-11$, $x+5$ and $x-5$ are negative. Since lengths cannot be negative, $x=11$ is the only solution that works in this context.

$x+5=11+5=16$

$x-5=11-5=6$

The base is 16 miles and the height is 6 miles.

13. The object will hit the ground when its height is 0.

$$0=-16t^2+64t+80$$
$$0=-16(t^2-4t-5)$$
$$0=-16(t-5)(t+1)$$
$$0=t-5 \quad \text{or} \quad 0=t+1$$
$$5=t \qquad\qquad -1=t$$

Since the time t cannot be negative, we discard $t=-1$. The object hits the ground after 5 seconds.

15. Let x be the width. Then $2x-7$ is the length.

Area = length · width

$$30=(2x-7)\cdot(x)$$
$$30=2x^2-7x$$
$$0=2x^2-7x-30$$
$$0=(2x+5)(x-6)$$
$$2x+5=0 \quad \text{or} \quad x-6=0$$
$$x=-\frac{5}{2} \qquad\qquad x=6$$

Discard $x=-\frac{5}{2}$ since length cannot be negative.

The width is 6 cm and the length is $2x-7=2(6)-7=12-7=5$ cm.

17. $D=\frac{1}{2}n(n-3)$

$$=\frac{1}{2}(12)(12-3)$$
$$=\frac{1}{2}(12)(9)$$
$$=54$$

A polygon with 12 sides has 54 diagonals.

Copyright © 2012 Pearson Education, Inc. Publishing as Prentice Hall.

19.
$$D=\frac{1}{2}n(n-3)$$
$$35=\frac{1}{2}n(n-3)$$
$$2\cdot 35=2\cdot\frac{1}{2}n(n-3)$$
$$70=n(n-3)$$
$$70=n^2-3n$$
$$0=n^2-3n-70$$
$$0=(n-10)(n+7)$$
$$n-10=0 \quad \text{or} \quad n+7=0$$
$$n=10 \qquad n=-7$$
Discard $n=-7$. The polygon has 10 sides.

21. Let x be the number.
$$x+x^2=132$$
$$x^2+x-132=0$$
$$(x+12)(x-11)=0$$
$$x+12=0 \quad \text{or} \quad x-11=0$$
$$x=-12 \qquad x=11$$
The number is -12 or 11.

23. Let x be the first number. Then $x+1$ is the next consecutive number.
$$x(x+1)=210$$
$$x^2+x=210$$
$$x^2+x-210=0$$
$$(x-14)(x+15)=0$$
$$x-14=0 \quad \text{or} \quad x+15=0$$
$$x=14 \qquad x=-15$$
Discard $x=-15$. The room numbers are 14 and $x+1=14+1=15$.

25. Use the Pythagorean theorem where x = hypotenuse, $(x-1)$ = one leg, and 5 = other leg.
$$x^2=(x-1)^2+5^2$$
$$x^2=x^2-2x+1+25$$
$$x^2=x^2-2x+26$$
$$0=-2x+26$$
$$2x=26$$
$$x=13$$
The length of the ladder is 13 feet.

27.
$$(x+3)^2=64$$
$$x^2+6x+9=64$$
$$x^2+6x-55=0$$
$$(x-5)(x+11)=0$$
$$x-5=0 \quad \text{or} \quad x+11=0$$
$$x=5 \qquad x=-11$$
Discard $x=-11$. The original square had sides of 5 inches.

29. Let x be the length of the shorter leg. Then the length of the other leg is $x+4$ and the length of the hypotenuse is $x+8$.
$$(x+8)^2=x^2+(x+4)^2$$
$$x^2+16x+64=x^2+x^2+8x+16$$
$$0=x^2-8x-48$$
$$0=(x-12)(x+4)$$
$$0=x-12 \quad \text{or} \quad 0=x+4$$
$$12=x \qquad -4=x$$
Discard $x=-4$ since length cannot be negative.
$x+4=12+4=16$
$x+8=12+8=20$
The lengths of the sides are 12 mm, 16 mm, and 20 mm.

31.
$$\text{area}=\frac{1}{2}\cdot\text{base}\cdot\text{height}$$
$$100=\frac{1}{2}\cdot 2x\cdot x$$
$$100=x^2$$
$$0=x^2-100$$
$$0=(x-10)(x+10)$$
$$x-10=0 \quad \text{or} \quad x+10=0$$
$$x=10 \qquad x=-10$$
Discard $x=-10$. The height is 10 kilometers.

33. Let x be the length of the shorter leg. Then $x+12$ is the length of the longer leg and $2x-12$ is the length of the hypotenuse.
$$(2x-12)^2=x^2+(x+12)^2$$
$$4x^2-48x+144=x^2+x^2+24x+144$$
$$2x^2-72x=0$$
$$2x(x-36)=0$$
$$2x=0 \quad \text{or} \quad x-36=0$$
$$x=0 \qquad x=36$$
Discard $x=0$ since the length must be positive. The shorter leg of the triangle has length 36 feet.

Copyright © 2012 Pearson Education, Inc. Publishing as Prentice Hall.

35. When the object reaches the ground, the height is 0.

$h = -16t^2 + 1444$

$0 = -16t^2 + 1444$

$0 = -4(4t^2 - 361)$

$0 = -4(2t - 19)(2t + 19)$

$2t - 19 = 0$ or $2t + 19 = 0$

$t = \frac{19}{2}$ $\quad$ $t = -\frac{19}{2}$

Discard $t = -\frac{19}{2}$ since time must be positive.

The object reaches the ground after $\frac{19}{2}$ or 9.5 seconds.

37. Use $A = P(1+r)^2$ when $P = 100$ and $A = 144$.

$144 = 100(1+r)^2$

$144 = 100(1 + 2r + r^2)$

$144 = 100 + 200r + 100r^2$

$0 = 100r^2 + 200r - 44$

$0 = 4(25r^2 + 50r - 11)$

$0 = 4(5r - 1)(5r + 11)$

$0 = 5r - 1$ or $0 = 5r + 11$

$1 = 5r$ $\quad$ $-11 = 5r$

$\frac{1}{5} = r$ $\quad$ $-\frac{11}{5} = r$

Discard $r = -\frac{11}{5}$ since the interest rate must be positive. The interest rate is $\frac{1}{5} = 0.20 = 20\%$.

39. Let x be the length. Then $x - 7$ is the width.

area = length · width

$120 = x(x - 7)$

$0 = x^2 - 7x - 120$

$0 = (x - 15)(x + 8)$

$x - 15 = 0$ or $x + 8 = 0$

$x = 15$ $\quad$ $x = -8$

Discard $x = -8$ since length is positive. The length is 15 miles and the width is $x - 7 = 15 - 7 = 8$ miles.

41. Let $C = 9500$ in $C = x^2 - 15x + 50$.

$9500 = x^2 - 15x + 50$

$0 = x^2 - 15x - 9450$

$0 = (x + 90)(x - 105)$

$0 = x + 90$ or $0 = x - 105$

$-90 = x$ $\quad$ $105 = x$

Discard $x = -90$ since the number of units manufactured cannot be negative.
105 units are manufactured at a cost of \$9500.

43. From the graph, there were approximately 1.9 million, or 1,900,000, visitors to Glacier National Park in 2002.

45. From the graph, there were approximately 1.9 million, or 1,900,000 visitors to Glacier National Park in 2005.

47. From the graph, the lines intersect at approximately 2003.

49. answers may vary

51. Since x is the width of the rectangle, the length is $x + 6$.

$x^2 + x(x + 6) = 176$

$x^2 + x^2 + 6x = 176$

$2x^2 + 6x - 176 = 0$

$2(x^2 + 3x - 88) = 0$

$2(x - 8)(x + 11) = 0$

$x - 8 = 0$ or $x + 11 = 0$

$x = 8$ $\quad$ $x = -11$

Discard $x = -11$ since length cannot be negative.
The side of the square is 8 meters.

53. Let x be one of the numbers. Since the numbers sum to 25, the other number is $25 - x$.

$x^2 + (25 - x)^2 = 325$

$x^2 + 625 - 50x + x^2 = 325$

$2x^2 - 50x + 300 = 0$

$2(x^2 - 25x + 150) = 0$

$2(x - 10)(x - 15) = 0$

$x - 10 = 0$ or $x - 15 = 0$

$x = 10$ $\quad$ $x = 15$

If $x = 10$, then $25 - x = 25 - 10 = 15$.
If $x = 15$, then $25 - x = 25 - 15 = 10$.
The numbers are 10 and 15.

Copyright © 2012 Pearson Education, Inc. Publishing as Prentice Hall.

55. Dimensions of total area:
length $= x + 6 + 4 + 4 = x + 14$
width $= x + 4 + 4 = x + 8$

$$\begin{aligned} \text{Total area} &= \text{pool area} + 576 \\ (x+14)(x+8) &= x(x+6)+576 \\ x^2+22x+112 &= x^2+6x+576 \\ 22x+112 &= 6x+576 \\ 16x &= 464 \\ x &= 29 \end{aligned}$$

$x + 6 = 29 + 6 = 35$
The pool is 29 meters by 35 meters.

Chapter 4 Vocabulary Check

1. An equation that can be written in the form $ax^2+bx+c=0$ (with a not 0) is called a <u>quadratic equation</u>.
2. <u>Factoring</u> is the process of writing an expression as a product.
3. The <u>greatest common factor</u> of a list of terms is the product of all common factors.
4. A trinomial that is the square of some binomial is called a <u>perfect square trinomial</u>.
5. In a right triangle, the side opposite the right angle is called the <u>hypotenuse</u>.
6. In a right triangle, each side adjacent to the right angle is called a <u>leg</u>.
7. The Pythagorean theorem states that $(\text{leg})^2+(\text{leg})^2=(\underline{\text{hypotenuse}})^2$.

Chapter 4 Review

1. $6x^2-15x=3x\cdot 2x-3x\cdot 5=3x(2x-5)$
2. $4x^5+2x-10x^4=2x\cdot 2x^4+2x\cdot 1-2x\cdot 5x^3$
$=2x(2x^4+1-5x^3)$
3. $5m + 30 = 5 \cdot m + 5 \cdot 6 = 5(m + 6)$
4. $20x^3+12x^2+24x=4x\cdot 5x^2+4x\cdot 3x+4x\cdot 6$
$=4x(5x^2+3x+6)$
5. $3x(2x + 3) - 5(2x + 3) = (2x + 3)(3x - 5)$
6. $5x(x+1)-(x+1)=5x(x+1)-1\cdot(x+1)$
$=(x+1)(5x-1)$
7. $3x^2-3x+2x-2=3x(x-1)+2(x-1)$
$=(x-1)(3x+2)$
8. $3a^2+9ab+3b^2+ab=3a(a+3b)+b(3b+a)$
$=(a+3b)(3a+b)$
9. $10a^2+5ab+7b^2+14ab$
$=5a(2a+b)+7b(b+2a)$
$=(2a+b)(5a+7b)$
10. $6x^2+10x-3x-5=2x(3x+5)-1(3x+5)$
$=(3x+5)(2x-1)$
11. $x^2+6x+8=(x+4)(x+2)$
12. $x^2-11x+24=(x-8)(x-3)$
13. x^2+x+2 is prime.
14. $x^2-5x-6=(x-6)(x+1)$
15. $x^2+2x-8=(x+4)(x-2)$
16. $x^2+4xy-12y^2=(x+6y)(x-2y)$
17. $x^2+8xy+15y^2=(x+5y)(x+3y)$
18. $72-18x-2x^2=2(36-9x-x^2)$
$=2(3-x)(12+x)$
or
$72-18x-2x^2=-2x^2-18x+72$
$=-2(x^2+9x-36)$
$=-2(x-3)(x+12)$
19. $32+12x-4x^2=4(8+3x-x^2)$
or
$32+12x-4x^2=-4x^2+12x+32$
$=-4(x^2-3x-8)$
20. $5y^3-50y^2+120y=5y(y^2-10y+24)$
$=5y(y-6)(y-4)$
21. To factor $x^2+2x-48$, think of two numbers whose product is –48 and whose sum is 2.

Copyright © 2012 Pearson Education, Inc. Publishing as Prentice Hall.

22. The first step to factor $3x^2+15x+30$ is to factor out the GCF, 3.

23. Factors of $2x^2$: $2x \cdot x$
Factors of 6: $6 = 1 \cdot 6$, $6 = 2 \cdot 3$
$2x^2+13x+6=(2x+1)(x+6)$

24. Factors of $4x^2$: $4x^2 = 4x \cdot x$, $4x^2 = 2x \cdot 2x$
Factors of -3: $-3 = -1 \cdot 3$, $-3 = 1 \cdot -3$
$4x^2+4x-3=(2x+3)(2x-1)$

25. Factors of $6x^2$: $6x^2 = 6x \cdot x$, $6x^2 = 3x \cdot 2x$
Factors of $-4y^2$: $-4y^2 = -4y \cdot y$,
$-4y^2 = 4y \cdot -y$, $-4y^2 = -2y \cdot 2y$
$6x^2+5xy-4y^2=(3x+4y)(2x-y)$

26. x^2-x+2 is prime.

27. $2 \cdot -39 = -78$
$3 \cdot -26 = -78$
$3 + (-26) = -23$
$$\begin{aligned} 2x^2-23x-39 &= 2x^2+3x-26x-39 \\ &= x(2x+3)-13(2x+3) \\ &= (2x+3)(x-13) \end{aligned}$$

28. $18 \cdot -20y^2 = -360y^2$
$15y \cdot -24y = -360y^2$
$15y + (-24y) = -9y$
$$\begin{aligned} 18x^2-9xy-20y^2 &= 18x^2+15xy-24xy-20y^2 \\ &= 3x(6x+5y)-4y(6x+5y) \\ &= (6x+5y)(3x-4y) \end{aligned}$$

29. $10y^3+25y^2-60y=5y(2y^2+5y-12)$
$2 \cdot -12 = -24$
$-3 \cdot 8 = -24$
$-3 + 8 = 5$
$$\begin{aligned} 10y^3+25y^2-60y &= 5y(2y^2+5y-12) \\ &= 5y(2y^2-3y+8y-12) \\ &= 5y[y(2y-3)+4(2y-3)] \\ &= 5y(2y-3)(y+4) \end{aligned}$$

30. $60y^3-39y^2+6y=3y(20y^2-13y+2)$
$20 \cdot 2 = 40$
$-5 \cdot -8 = 40$
$-5 + (-8) = -13$
$$\begin{aligned} 60y^3-39y^2+6y &= 3y(20y^2-13y+2) \\ &= 3y(20y^2-5y-8y+2) \\ &= 3y[5y(4y-1)-2(4y-1)] \\ &= 3y(4y-1)(5y-2) \end{aligned}$$

31. $(x^2-2)+(x^2-4x)+(3x^2-5x)$
$= x^2-2+x^2-4x+3x^2-5x$
$= 5x^2-9x-2$
$5 \cdot -2 = -10$
$1 \cdot -10 = -10$
$1 + (-10) = -9$
$$\begin{aligned} 5x^2-9x-2 &= 5x^2+x-10x-2 \\ &= x(5x+1)-2(5x+1) \\ &= (5x+1)(x-2) \end{aligned}$$
The perimeter is $5x^2-9x-2$ or $(5x+1)(x-2)$.

32. $$\begin{aligned} 2(2x^2+3)+2(6x^2-14x) &= 4x^2+6+12x^2-28x \\ &= 16x^2-28x+6 \\ &= 2(8x^2-14x+3) \end{aligned}$$
$8 \cdot 3 = 24$
$-2 \cdot -12 = 24$
$-2 + (-12) = -14$
$$\begin{aligned} 2(8x^2-14x+3) &= 2(8x^2-2x-12x+3) \\ &= 2[2x(4x-1)-3(4x-1)] \\ &= 2(4x-1)(2x-3) \end{aligned}$$
The perimeter is $16x^2-28x+6$ or $2(4x-1)(2x-3)$.

33. Since $9=3^2$ and $6x = 2 \cdot 3 \cdot x$, x^2+6x+9 is a perfect square trinomial.

34. Since $64=8^2$, but $8x \neq 2 \cdot 8 \cdot x$, $x^2+8x+64$ is not a perfect square trinomial.

35. $9m^2=(3m)^2$ and $16=4^2$, but $2 \cdot 3m \cdot 4 \neq 12m$, so $9m^2-12m+16$ is not a perfect square trinomial.

Copyright © 2012 Pearson Education, Inc. Publishing as Prentice Hall.

36. Since $4y^2 = (2y)^2$ and $49 = 7^2$ and $2 \cdot 2y \cdot 7 = 28y$, $4y^2 - 28y + 49$ is a perfect square trinomial.

37. Yes; $x^2 - 9$ or $x^2 - 3^2$ is the difference of two squares.

38. No; $x^2 + 16$ or $x^2 + 4^2$ is the sum of two squares.

39. Yes; $4x^2 - 25y^2$ or $(2x)^2 - (5y)^2$ is the difference of two squares.

40. No; $9a^3 - 1$ is not the difference of two squares [note: $9a^2 - 1$ or $(3a)^2 - 1^2$ is the difference of two squares.]

41. $x^2 - 81 = x^2 - 9^2 = (x-9)(x+9)$

42. $x^2 + 12x + 36 = x^2 + 2 \cdot x \cdot 6 + 6^2 = (x+6)^2$

43. $4x^2 - 9 = (2x)^2 - 3^2 = (2x-3)(2x+3)$

44. $9t^2 - 25s^2 = (3t)^2 - (5s)^2 = (3t-5s)(3t+5s)$

45. $16x^2 = (4x)^2$ and y^2 are both perfect squares, but they are added instead of subtracted, so the polynomial is prime.

46. $n^2 - 18n + 81 = n^2 - 2 \cdot n \cdot 9 + 9^2 = (n-9)^2$

47. $3r^2 + 36r + 108 = 3(r^2 + 12 + 36)$
$= 3(r^2 + 2 \cdot r \cdot 6 + 6^2)$
$= 3(r+6)^2$

48. $9y^2 - 42y + 49 = (3y)^2 - 2 \cdot 3y \cdot 7 + 7^2$
$= (3y-7)^2$

49. $5m^8 - 5m^6 = 5m^6(m^2 - 1)$
$= 5m^6(m^2 - 1^2)$
$= 5m^6(m+1)(m-1)$

50. $4x^2 - 28xy + 49y^2 = (2x)^2 - 2 \cdot 2x \cdot 7y + (7y)^2$
$= (2x - 7y)^2$

51. $3x^2y + 6xy^2 + 3y^3 = 3y(x^2 + 2xy + y^2)$
$= 3y(x+y)^2$

52. $16x^4 - 1 = (4x^2)^2 - 1^2$
$= (4x^2 - 1)(4x^2 + 1)$
$= [(2x)^2 - 1^2](4x^2 + 1)$
$= (2x-1)(2x+1)(4x^2+1)$

53. $(x + 6)(x - 2) = 0$
$x + 6 = 0$ or $x - 2 = 0$
$x = -6$ $\quad$ $x = 2$
The solutions are -6 and 2.

54. $(x - 7)(x + 11) = 0$
$x - 7 = 0$ or $x + 11 = 0$
$x = 7$ $\quad$ $x = -11$
The solutions are 7 and -11.

55. $3x(x + 1)(7x - 2) = 0$
$3x = 0$ or $x + 1 = 0$ or $7x - 2 = 0$
$x = 0$ $\quad$ $x = -1$ $\quad$ $x = \frac{2}{7}$
The solutions are 0, -1, and $\frac{2}{7}$.

56. $4(5x + 1)(x + 3) = 0$
$5x + 1 = 0$ or $x + 3 = 0$
$x = -\frac{1}{5}$ $\quad$ $x = -3$
The solutions are $-\frac{1}{5}$ and -3.

57. $x^2 + 8x + 7 = 0$
$(x+7)(x+1) = 0$
$x + 7 = 0$ or $x + 1 = 0$
$x = -7$ $\quad$ $x = -1$
The solutions are -7 and -1.

58. $x^2 - 2x - 24 = 0$
$(x+4)(x-6) = 0$
$x + 4 = 0$ or $x - 6 = 0$
$x = -4$ $\quad$ $x = 6$
The solutions are -4 and 6.

Copyright © 2012 Pearson Education, Inc. Publishing as Prentice Hall.

59. $x^2+10x=-25$
$x^2+10x+25=0$
$(x+5)^2=0$
$x+5=0$
$x=-5$
The solution is –5.

60. $x(x-10)=-16$
$x^2-10x=-16$
$x^2-10x+16=0$
$(x-2)(x-8)=0$
$x-2=0$ or $x-8=0$
$x=2$ $x=8$
The solutions are 2 and 8.

61. $(3x-1)(9x^2+3x+1)=0$
$3x-1=0$ or $9x^2+3x+1=0$
$x=\frac{1}{3}$ $9x^2+3x+1>0$
The solution is $\frac{1}{3}$.

62. $56x^2-5x-6=0$
$(7x+2)(8x-3)=0$
$7x+2=0$ or $8x-3=0$
$x=-\frac{2}{7}$ $x=\frac{3}{8}$
The solutions are $-\frac{2}{7}$ and $\frac{3}{8}$.

63. $m^2=6m$
$m^2-6m=0$
$m(m-6)=0$
$m=0$ or $m-6=0$
$m=6$
The solutions are 0 and 6.

64. $r^2=25$
$r^2-25=0$
$(r-5)(r+5)=0$
$r-5=0$ or $r+5=0$
$r=5$ $r=-5$
The solutions are 5 and –5.

65. $(x-4)(x-5)=0$
$x^2-9x+20=0$

66. $[x-(-1)][x-(-1)]=0$
$(x+1)(x+1)=0$
$x^2+2x+1=0$

67. Let x be the width. Then $2x$ is the length.
perimeter $=2\cdot$ length $+2\cdot$ width
$24=2(2x)+2x$
$24=4x+2x$
$24=6x$
$4=x$
$8=2x$
The dimensions are 4 inches by 8 inches. The choice is c.

68. Let x be the width. Then $3x+1$ is the length.
area = length · width
$80=(3x+1)\cdot x$
$80=3x^2+x$
$0=3x^2+x-80$
$0=(3x+16)(x-5)$
$3x+16=0$ or $x-5=0$
$x=-\frac{16}{3}$ $x=5$
Discard $x=-\frac{16}{3}$ since length cannot be negative. The dimensions are 5 meters by $3x+1=3(5)+1=15+1=16$ meters.
The choice is d.

69. area $=$ side^2
$81=x^2$
$0=x^2-81$
$0=(x-9)(x+9)$
$x-9=0$ or $x+9=0$
$x=9$ $x=-9$
Discard $x=-9$ since length cannot be negative.
The length of each side is 9 units.

70. The perimeter is the sum of the sides.
$(2x+3)+(3x+1)+(x^2-3x)+(x+3)=47$
$x^2+3x+7=47$
$x^2+3x-40=0$
$(x-5)(x+8)=0$
$x-5=0$ or $x+8=0$
$x=5$ $x=-8$
Discard $x=-8$ since, for example, $x+3$ would be negative.
$2x+3=2(5)+3=10+3=13$
$3x+1=3(5)+1=15+1=16$

Copyright © 2012 Pearson Education, Inc. Publishing as Prentice Hall.

$x^2 - 3x = 5^2 - 3(5) = 25 - 15 = 10$

$x + 3 = 5 + 3 = 8$

The lengths of the sides are 13 units, 16 units, 10 units, and 8 units.

71. Let x be the width. Then $2x - 15$ is the length.

area = length · width

$500 = (2x - 15) \cdot x$

$500 = 2x^2 - 15x$

$0 = 2x^2 - 15x - 500$

$0 = (2x + 25)(x - 20)$

$2x + 25 = 0$ or $x - 20 = 0$

$x = -\frac{25}{2}$ $\quad x = 20$

Discard $x = -\frac{25}{2}$ since length cannot be negative. The width is 20 inches and the length is $2 \cdot 20 - 15 = 40 - 15 = 25$ inches.

72. Let x be the height. Then $4x$ is the base.

$\text{area} = \frac{1}{2} \cdot \text{base} \cdot \text{height}$

$162 = \frac{1}{2} \cdot 4x \cdot x$

$162 = 2x^2$

$0 = 2x^2 - 162$

$0 = 2(x^2 - 81)$

$0 = 2(x - 9)(x + 9)$

$x - 9 = 0$ or $x + 9 = 0$

$x = 9$ $\quad x = -9$

Discard $x = -9$ since length cannot be negative. The height is 9 yards and the base is $4x = 4(9) = 36$ yards.

73. Let x be the first positive integer. Then $x + 1$ is the next consecutive integer.

$x(x + 1) = 380$

$x^2 + x = 380$

$x^2 + x - 380 = 0$

$(x + 20)(x - 19) = 0$

$x + 20 = 0$ or $x - 19 = 0$

$x = -20$ $\quad x = 19$

Discard $x = -20$ since it is not positive. The integers are 19 and $19 + 1 = 20$.

74. Let x be the first positive even integer. Then $x + 2$ is the next consecutive even integer.

$x(x + 2) = 440$

$x^2 + 2x = 440$

$x^2 + 2x - 440 = 0$

$(x + 22)(x - 20) = 0$

$x + 22 = 0$ or $x - 20 = 0$

$x = -22$ $\quad x = 20$

Discard $x = -22$ since it is not positive. The integers are 20 and $20 + 2 = 22$.

75. a.

$h = -16t^2 + 440t$

$2800 = -16t^2 + 440t$

$16t^2 - 440t + 2800 = 0$

$8(2t^2 - 55t + 350) = 0$

$8(2t - 35)(t - 10) = 0$

$2t - 35 = 0$ or $t - 10 = 0$

$t = \frac{35}{2}$ or 17.5 $\quad t = 10$

The rocket reaches a height of 2800 feet at 10 seconds on the way up and at 17.5 seconds on the way down.

b. The height is 0 when the rocket reaches the ground.

$h = -16t^2 + 440t$

$0 = -16t^2 + 440t$

$0 = -8t(2t - 55)$

$-8t = 0$ or $2t - 55 = 0$

$t = 0$ $\quad 2t = 55$

$t = 27.5$

The rocket reaches the ground again after 27.5 seconds.

76. Let x be the length of the longer leg. Then $x + 8$ is the length of the hypotenuse and $x - 8$ is the length of the shorter leg.

$(x + 8)^2 = (x - 8)^2 + x^2$

$x^2 + 16x + 64 = x^2 - 16x + 64 + x^2$

$0 = x^2 - 32x$

$0 = x(x - 32)$

$x = 0$ or $x - 32 = 0$

$x = 32$

The longer leg is 32 centimeters.

77. $6x + 24 = 6 \cdot x + 6 \cdot 4 = 6(x + 4)$

78. $7x - 63 = 7 \cdot x - 7 \cdot 9 = 7(x - 9)$

Copyright © 2012 Pearson Education, Inc. Publishing as Prentice Hall.

79. $11x(4x-3)-6(4x-3)=(4x-3)(11x-6)$

80. $2x(x-5)-(x-5)=2x(x-5)-1(x-5)$
$=(x-5)(2x-1)$

81. $3x^3-4x^2+6x-8=x^2(3x-4)+2(3x-4)$
$=(3x-4)(x^2+2)$

82. $xy+2x-y-2=x(y+2)-1(y+2)$
$=(y+2)(x-1)$

83. $2x^2+2x-24=2(x^2+x-12)=2(x+4)(x-3)$

84. $3x^3-30x^2+27x=3x(x^2-10x+9)$
$=3x(x-9)(x-1)$

85. $4x^2-81=(2x)^2-9^2=(2x+9)(2x-9)$

86. $2x^2-18=2(x^2-9)$
$=2(x^2-3^2)$
$=2(x-3)(x+3)$

87. $16x^2-24x+9=(4x)^2-2\cdot 4x\cdot 3+3^2=(4x-3)^2$

88. $5x^2+20x+20=5(x^2+4x+4)$
$=5(x^2+2\cdot x\cdot 2+2^2)$
$=5(x+2)^2$

89. $2x^2-x-28=0$
$(2x+7)(x-4)=0$
$2x+7=0$ or $x-4=0$
$x=-\frac{7}{2}$ $x=4$
The solutions are $-\frac{7}{2}$ and 4.

90. $x^2-2x=15$
$x^2-2x-15=0$
$(x+3)(x-5)=0$
$x+3=0$ or $x-5=0$
$x=-3$ $x=5$
The solutions are –3 and 5.

91. $2x(x+7)(x+4)=0$
$2x=0$ or $x+7=0$ or $x+4=0$
$x=0$ $x=-7$ $x=-4$

92. $x(x-5)=-6$
$x^2-5x=-6$
$x^2-5x+6=0$
$(x-3)(x-2)=0$
$x-3=0$ or $x-2=0$
$x=3$ $x=2$
The solutions are 3 and 2.

93. $x^2=16x$
$x^2-16x=0$
$x(x-16)=0$
$x=0$ or $x-16=0$
$x=16$
The solutions are 0 and 16.

94. The perimeter is the sum of the sides.
$(x^2+3)+2x+(4x+5)=48$
$x^2+6x+8=48$
$x^2+6x-40=0$
$(x-4)(x+10)=0$
$x-4=0$ or $x+10=0$
$x=4$ $x=-10$
Discard $x=-10$ since length, such as $2x=2(-10)=-20$ cannot be negative.
$x^2+3=4^2+3=16+3=19$
$2x=2\cdot 4=8$
$4x+5=4\cdot 4+5=16+5=21$
The lengths are 19 inches, 8 inches, and 21 inches.

95. Let x be the length. Then $x-4$ is the width.
area = length · width
$12=x(x-4)$
$12=x^2-4x$
$0=x^2-4x-12$
$0=(x-6)(x+2)$
$x-6=0$ or $x+2=0$
$x=6$ $x=-2$
Discard $x=-2$ since length cannot be negative.
The length is 6 inches and the width is $x-4=6-4=2$ inches.

Chapter 4 Test

1. $9x^2-3x=3x(3x-1)$

2. $x^2+11x+28=(x+7)(x+4)$

3. $49-m^2=7^2-m^2=(7-m)(7+m)$

Copyright © 2012 Pearson Education, Inc. Publishing as Prentice Hall.

4. $y^2+22y+121=y^2+2\cdot y\cdot 11+11^2$
$=(y+11)^2$

5. $x^4-16=(x^2)^2-(4)^2$
$=(x^2+4)(x^2-4)$
$=(x^2+4)[(x)^2-(2)^2]$
$=(x^2+4)(x+2)(x-2)$

6. $4(a+3)-y(a+3)=(a+3)(4-y)$

7. x^2+4 is prime.

8. $y^2-8y-48=(y-12)(y+4)$

9. $3a^2+3ab-7a-7b=3a(a+b)-7(a+b)$
$=(a+b)(3a-7)$

10. $3x^2-5x+2=(3x-2)(x-1)$

11. $180-5x^2=5(36-x^2)$
$=5(6^2-x^2)$
$=5(6-x)(6+x)$

12. $3x^3-21x^2+30x=3x(x^2-7x+10)$
$=3x(x-5)(x-2)$

13. $6t^2-t-5=6t^2+5t-6t-5$
$=t(6t+5)-1(6t+5)$
$=(6t+5)(t-1)$

14. $xy^2-7y^2-4x+28=y^2(x-7)-4(x-7)$
$=(x-7)(y^2-4)$
$=(x-7)(y-2)(y+2)$

15. $x-x^5=x(1-x^4)$
$=x[1^2-(x^2)^2]$
$=x(1+x^2)(1-x^2)$
$=x(1+x^2)(1+x)(1-x)$

16. $x^2+14xy+24y^2=x^2+12xy+2xy+24y^2$
$=x(x+12y)+2y(x+12y)$
$=(x+12y)(x+2y)$

17. $(x-3)(x+9)=0$
$x-3=0$ or $x+9=0$
$x=3$ $\quad$ $x=-9$
The solutions are 3 and –9.

18. $x^2+5x=14$
$x^2+5x-14=0$
$(x+7)(x-2)=0$
$x+7=0$ or $x-2=0$
$x=-7$ $\quad$ $x=2$
The solutions are –7 and 2.

19. $x(x+6)=7$
$x^2+6x=7$
$x^2+6x-7=0$
$(x+7)(x-1)=0$
$x+7=0$ or $x-1=0$
$x=-7$ $\quad$ $x=1$
The solutions are –7 and 1.

20. $3x(2x-3)(3x+4)=0$
$3x=0$ or $2x-3=0$ or $3x+4=0$
$x=0$ $\quad$ $x=\frac{3}{2}$ $\quad$ $x=-\frac{4}{3}$
The solutions are 0, $\frac{3}{2}$, and $-\frac{4}{3}$.

21. $5t^3-45t=0$
$5t(t^2-9)=0$
$5t[(t)^2-(3)^2]=0$
$5t(t+3)(t-3)=0$
$5t=0$ or $t+3=0$ or $t-3=0$
$t=0$ $\quad$ $t=-3$ $\quad$ $t=3$
The solutions are 0, –3, and 3.

22. $t^2-2t-15=0$
$(t+3)(t-5)=0$
$t+3=0$ or $t-5=0$
$t=-3$ $\quad$ $t=5$
The solutions are –3 and 5.

23. $6x^2=15x$
$6x^2-15x=0$
$3x(2x-5)=0$

Copyright © 2012 Pearson Education, Inc. Publishing as Prentice Hall.

$3x = 0$ or $2x - 5 = 0$
$x = 0$ $\quad x = \frac{5}{2}$

The solutions are 0 and $\frac{5}{2}$.

24. Let x be the height. Then the base is $x + 9$.

$$\text{area} = \frac{1}{2} \cdot \text{base} \cdot \text{height}$$
$$68 = \frac{1}{2} \cdot (x+9) \cdot x$$
$$2 \cdot 68 = 2 \cdot \frac{1}{2} \cdot (x+9) \cdot x$$
$$136 = x(x+9)$$
$$136 = x^2 + 9x$$
$$0 = x^2 + 9x - 136$$
$$0 = (x+17)(x-8)$$
$x + 17 = 0$ or $x - 8 = 0$
$x = -17$ $\quad x = 8$

Discard $x = -17$ since length cannot be negative. The height is 8 feet and the base is $x + 9 = 8 + 9 = 17$ feet.

25.
$$(x-1)(x+2) = 54$$
$$x^2 + 2x - x - 2 = 54$$
$$x^2 + x - 56 = 0$$
$$(x+8)(x-7) = 0$$
$x + 8 = 0$ or $x - 7 = 0$
$x = -8$ $\quad x = 7$

Discard $x = -8$ since length cannot be negative.
$x - 1 = 7 - 1 = 6$
$x + 2 = 7 + 2 = 9$
The width of the rectangle is 6 units and the length is 9 units.

26. The object is at a height of 0 when it reaches the ground.

$$h = -16t^2 + 784$$
$$0 = -16t^2 + 784$$
$$0 = -16(t^2 - 49)$$
$$0 = -16(t-7)(t+7)$$
$t - 7 = 0$ or $t + 7 = 0$
$t = 7$ $\quad t = -7$

Discard $t = -7$ since time cannot be negative. The object reaches the ground after 7 seconds.

27. Let x be the length of the shorter leg. Then $x + 10$ is the length of the hypotenuse and $x + 10 - 5 = x + 5$ is the length of the longer leg.

$$x^2 + (x+5)^2 = (x+10)^2$$
$$x^2 + x^2 + 10x + 25 = x^2 + 20x + 100$$
$$x^2 - 10x - 75 = 0$$
$$(x-15)(x+5) = 0$$
$x - 15 = 0$ or $x + 5 = 0$
$x = 15$ $\quad x = -5$

Discard $x = -5$ since length cannot be negative. The shorter leg is 15 cm, the longer leg is $15 + 5 = 20$ cm, and the hypotenuse is $15 + 10 = 25$ cm.

28. The height of the object is 0 when it reaches the ground.

$$h = -16t^2 + 1089$$
$$0 = -16t^2 + 1089$$
$$0 = -1(16t^2 - 1089)$$
$$0 = -1[(4t)^2 - 33^2]$$
$$0 = -1(4t - 33)(4t + 33)$$
$4t - 33 = 0$ or $4t + 33 = 0$
$t = \frac{33}{4}$ or 8.25 $\quad t = -\frac{33}{4}$

Discard $t = -\frac{33}{4}$ since time cannot be negative.
The object reaches the ground after 8.25 seconds.

Cumulative Review Chapters 1–4

1. a. Nine is less than or equal to eleven is translated as $9 \le 11$.

b. Eight is greater than one is translated as $8 > 1$.

c. Three is not equal to four is translated as $3 \ne 4$.

2. a. $|-5| = 5$
$|-3| = 3$
$|-5| > |-3|$ since $5 > 3$

b. $|0| = 0$
$|-2| = 2$
$|0| < |-2|$ since $0 < 2$

3. Replace x with 2.
$$3x + 10 = 8x$$
$$3(2) + 10 \stackrel{?}{=} 8(2)$$
$$6 + 10 \stackrel{?}{=} 16$$
$$16 = 16 \quad \text{True}$$
Since a true statement results, 2 is a solution of the equation.

Copyright © 2012 Pearson Education, Inc. Publishing as Prentice Hall.

4. Replace x with 20 and y with 10.

$$\frac{x}{y}+5x=\frac{20}{10}+5(20)=2+100=102$$

5. $-4-8=-4+(-8)=-12$

6. Replace x with -20 and y with 10.

$$\frac{x}{y}+5x=\frac{-20}{10}+5(-20)=-2+(-100)=-102$$

7. Replace x with -2 and y with -4.

a. $\frac{3x}{2y}=\frac{3(-2)}{2(-4)}=\frac{-6}{-8}=\frac{-2\cdot 3}{-2\cdot 4}=\frac{3}{4}$

b.
$$\begin{aligned} x^3-y^2 &= (-2)^3-(-4)^2 \\ &= -8-16 \\ &= -8+(-16) \\ &= -24 \end{aligned}$$

c. $\frac{x-y}{-x}=\frac{-2-(-4)}{-(-2)}=\frac{-2+4}{2}=\frac{2}{2}=1$

8. Replace x with -20 and y with -10.

$$\frac{x}{y}+5x=\frac{-20}{-10}+5(-20)=2+(-100)=-98$$

9. $2x+3x+5+2=(2x+3x)+(5+2)=5x+7$

10.
$$\begin{aligned} 5-2(3x-7) &= 5-2\cdot 3x-2(-7) \\ &= 5-6x+14 \\ &= 19-6x \end{aligned}$$

11. $-5a-3+a+2=-5a+a-3+2=-4a-1$

12.
$$\begin{aligned} 5(x-6)+9(-2x+1) &= 5\cdot x-5\cdot 6+9\cdot(-2x)+9\cdot 1 \\ &= 5x-30-18x+9 \\ &= -13x-21 \end{aligned}$$

13. $2.3x+5x-6=7.3x-6$

14.
$$\begin{aligned} 0.8y+0.2(y-1) &= 1.8 \\ 10[0.8y+0.2(y-1)] &= 10(1.8) \\ 8y+2(y-1) &= 18 \\ 8y+2y-2 &= 18 \\ 10y-2 &= 18 \\ 10y-2+2 &= 18+2 \\ 10y &= 20 \\ \frac{10y}{10} &= \frac{20}{10} \\ y &= 2 \end{aligned}$$

The solution is 2.

15.
$$\begin{aligned} -3x &= 33 \\ \frac{-3x}{-3} &= \frac{33}{-3} \\ x &= -11 \end{aligned}$$

The solution is -11.

16.
$$\begin{aligned} \frac{x}{-7} &= -4 \\ -7\cdot\frac{x}{-7} &= -7\cdot -4 \\ x &= 28 \end{aligned}$$

The solution is 28.

17.
$$\begin{aligned} 3(x-4) &= 3x-12 \\ 3x-12 &= 3x-12 \\ -3x+3x-12 &= -3x+3x-12 \\ -12 &= -12 \end{aligned}$$

Since this results in a true statement, the solution is every real number.

18.
$$\begin{aligned} -\frac{2}{3}x &= -22 \\ -\frac{3}{2}\cdot -\frac{2}{3}x &= -\frac{3}{2}\cdot -22 \\ x &= 33 \end{aligned}$$

The solution is 33.

19.
$$\begin{aligned} V &= lwh \\ \frac{V}{wh} &= \frac{lwh}{wh} \\ \frac{V}{wh} &= l \text{ or } l=\frac{V}{wh} \end{aligned}$$

Copyright © 2012 Pearson Education, Inc. Publishing as Prentice Hall.

20.
$$\begin{aligned} 3x+2y&=-7\\ -3x+3x+2y&=-3x-7\\ 2y&=-3x-7\\ \frac{2y}{2}&=\frac{-3x-7}{2}\\ y&=\frac{-3x-7}{2}\end{aligned}$$

or

$$y=-\frac{3}{2}x-\frac{7}{2}$$

21. $(5^3)^6=5^{3\cdot 6}=5^{18}$

22. $5^2+5^1=25+5=30$

23. $(y^8)^2=y^{8\cdot 2}=y^{16}$

24. $y^8\cdot y^2=y^{8+2}=y^{10}$

25.
$$\begin{aligned}\frac{(2x^3)^4x}{x^7}&=\frac{2^4x^{3\cdot 4}\cdot x^1}{x^7}\\ &=\frac{16x^{12}\cdot x^1}{x^7}\\ &=\frac{16x^{12+1}}{x^7}\\ &=\frac{16x^{13}}{x^7}\\ &=16x^{13-7}\\ &=16x^6\end{aligned}$$

26. $3^{-2}=\frac{1}{3^2}=\frac{1}{9}$

27. $(y^{-3}z^6)^{-6}=y^{-3(-6)}z^{6(-6)}=y^{18}z^{-36}=\frac{y^{18}}{z^{36}}$

28. $\frac{x^{-3}}{x^{-7}}=x^{-3-(-7)}=x^{-3+7}=x^4$

29. $\frac{x^{-7}}{(x^4)^3}=\frac{x^{-7}}{x^{4\cdot 3}}=\frac{x^{-7}}{x^{12}}=\frac{1}{x^{12-(-7)}}=\frac{1}{x^{12+7}}=\frac{1}{x^{19}}$

30. $\frac{(5a^7)^2}{a^5}=\frac{5^2a^{7\cdot 2}}{a^5}=\frac{25a^{14}}{a^5}=25a^{14-5}=25a^9$

31. $-3x+7x=(-3+7)x=4x$

32.
$$\begin{aligned}\frac{2}{3}x+23+\frac{1}{6}x-100&=\left(\frac{2}{3}x+\frac{1}{6}x\right)+(23-100)\\ &=\left(\frac{4}{6}x+\frac{1}{6}x\right)+(23-100)\\ &=\frac{5}{6}x-77\end{aligned}$$

33.
$$\begin{aligned}11x^2+5+2x^2-7&=11x^2+2x^2+5-7\\ &=13x^2-2\end{aligned}$$

34.
$$\begin{aligned}0.2x-1.1+2.3-0.7x&=0.2x-0.7x-1.1+2.3\\ &=-0.5x+1.2\end{aligned}$$

35.
$$\begin{aligned}&(2x-y)^2\\ &=(2x-y)(2x-y)\\ &=2x(2x)+2x(-y)+(-y)(2x)+(-y)(-y)\\ &=4x^2-2xy-2xy+y^2\\ &=4x^2-4xy+y^2\end{aligned}$$

36.
$$\begin{aligned}&(3x-7y)^2\\ &=(3x-7y)(3x-7y)\\ &=3x(3x)+3x(-7y)+(-7y)(3x)+(-7y)(-7y)\\ &=9x^2-21xy-21xy+49y^2\\ &=9x^2-42xy+49y^2\end{aligned}$$

37. $(t+2)^2=t^2+2(t)(2)+2^2=t^2+4t+4$

38.
$$\begin{aligned}(x-13)^2&=x^2-2(x)(13)+13^2\\ &=x^2-26x+169\end{aligned}$$

39.
$$\begin{aligned}(x^2-7y)^2&=(x^2)^2-2(x^2)(7y)+(7y)^2\\ &=x^4-14x^2y+49y^2\end{aligned}$$

40.
$$\begin{aligned}(7x+y)^2&=(7x)^2+2(7x)(y)+(y)^2\\ &=49x^2+14xy+y^2\end{aligned}$$

41.
$$\begin{aligned}\frac{8x^2y^2-16xy+2x}{4xy}&=\frac{8x^2y^2}{4xy}-\frac{16xy}{4xy}+\frac{2x}{4xy}\\ &=2xy-4+\frac{1}{2y}\end{aligned}$$

42.
$$\begin{aligned}z^3+7z+z^2+7&=z(z^2+7)+1(z^2+7)\\ &=(z^2+7)(z+1)\end{aligned}$$

43. $5(x+3)+y(x+3)=(x+3)(5+y)$

Copyright © 2012 Pearson Education, Inc. Publishing as Prentice Hall.

44. $2x^3+2x^2-84x=2x(x^2+x-42)$
$=2x(x^2+7x-6x-42)$
$=2x[x(x+7)-6(x+7)]$
$=2x(x+7)(x-6)$

45. $x^4+5x^2+6=x^4+2x^2+3x^2+6$
$=x^2(x^2+2)+3(x^2+2)$
$=(x^2+2)(x^2+3)$

46. $-4x^2-23x+6=-4x^2+x-24x+6$
$=x(-4x+1)+6(-4x+1)$
$=(-4x+1)(x+6)$

or

$-4x^2-23x+6=-1(4x^2+23x-6)$
$=-1(4x^2-x+24x-6)$
$=-1[x(4x-1)+6(4x-1)]$
$=-1(4x-1)(x+6)$

47. $6x^2-2x-20=2(3x^2-x-10)$
$=2(3x^2-6x+5x-10)$
$=2[3x(x-2)+5(x-2)]$
$=2(x-2)(3x+5)$

48. $9xy^2-16x=x(9y^2-16)$
$=x[(3y)^2-4^2]$
$=x(3y+4)(3y-4)$

49. The height is 0 when the diver reaches the ocean.
$h=-16t^2+144$
$0=-16t^2+144$
$0=-16(t^2-9)$
$0=-16(t-3)(t+3)$
$t-3=0$ or $t+3=0$
$t=3$ $\quad$ $t=-3$
Discard $t=-3$ since time cannot be negative.
The diver reaches the ocean after 3 seconds.

50. $x^2-13x=-36$
$x^2-13x+36=0$
$x^2-9x-4x+36=0$
$x(x-9)-4(x-9)=0$
$(x-9)(x-4)=0$
$x-9=0$ or $x-4=0$
$x=9$ $\quad$ $x=4$
The solutions are 9 and 4.

Copyright © 2012 Pearson Education, Inc. Publishing as Prentice Hall.

Chapter 5

Section 5.1 Practice

1. a. $\frac{x-3}{5x+1}=\frac{4-3}{5(4)+1}=\frac{1}{20+1}=\frac{1}{21}$

   b. $\frac{x-3}{5x+1}=\frac{(-3)-3}{5(-3)+1}=\frac{-6}{-15+1}=\frac{-6}{-14}=\frac{3}{7}$

2. a. $x+8=0$
   $x=-8$

   When $x=-8$, the expression $\frac{x}{x+8}$ is undefined.

   b. $x^2+5x+4=0$
   $(x+4)(x+1)=0$
   $x+4=0$ or $x+1=0$
   $x=-4$ $\quad$ $x=-1$
   When $x=-4$ or $x=-1$, the expression $\frac{x-3}{x^2+5x+4}$ is undefined.

   c. The denominator of $\frac{x^2-3x+2}{5}$ is never 0, so there are no values of x for which this expression is undefined.

3. $\frac{x^4+x^3}{5x+5}=\frac{x^3(x+1)}{5(x+1)}=\frac{x^3}{5}$

4. $\frac{x^2+11x+18}{x^2+x-2}=\frac{(x+9)(x+2)}{(x-1)(x+2)}=\frac{x+9}{x-1}$

5. $\frac{x^2+10x+25}{x^2+5x}=\frac{(x+5)(x+5)}{x(x+5)}=\frac{x+5}{x}$

6. $\frac{x+5}{x^2-25}=\frac{x+5}{(x+5)(x-5)}=\frac{1}{x-5}$

7. a. $\frac{x+4}{4+x}=\frac{x+4}{x+4}=1$

   b. $\frac{x-4}{4-x}=\frac{x-4}{(-1)(x-4)}=\frac{1}{-1}=-1$

8. $-\frac{3x+7}{x-6}=\frac{-(3x+7)}{x-6}=\frac{-3x-7}{x-6}$ or
   $\frac{3x+7}{-(x-6)}=\frac{3x+7}{-x+6}$

Vocabulary and Readiness Check

1. A <u>rational expression</u> is an expression that can be written in the form $\frac{P}{Q}$ where P and Q are polynomials and $Q\neq 0$.

2. The expression $\frac{x+3}{3+x}$ simplifies to <u>1</u>.

3. The expression $\frac{x-3}{3-x}$ simplifies to <u>-1</u>.

4. A rational expression is undefined for values that make the denominator <u>0</u>.

5. The expression $\frac{7x}{x-2}$ is undefined for $x=$ <u>2</u>.

6. The process of writing a rational expression in lowest terms is called <u>simplifying</u>.

7. For a rational expression, $-\frac{a}{b}=\frac{-a}{\underline{b}}=\frac{a}{\underline{-b}}$.

8. $\frac{x}{x+7}$ cannot be simplified.

9. Since $3+x=x+3$, $\frac{3+x}{x+3}$ can be simplified.

10. Since $5-x=-1(x-5)$, $\frac{5-x}{x-5}$ can be simplified.

11. $\frac{x+2}{x+8}$ cannot be simplified.

Exercise Set 5.1

1. $\frac{x+5}{x+2}=\frac{2+5}{2+2}=\frac{7}{4}$

Copyright © 2012 Pearson Education, Inc. Publishing as Prentice Hall.

3. $\dfrac{y^3}{y^2-1}=\dfrac{(-2)^3}{(-2)^2-1}=\dfrac{-8}{4-1}=-\dfrac{8}{3}$

5. $\dfrac{x^2+8x+2}{x^2-x-6}=\dfrac{2^2+8(2)+2}{2^2-2-6}$
$=\dfrac{4+16+2}{4-2-6}$
$=\dfrac{22}{-4}$
$=-\dfrac{11}{2}$

7. a. $A=\dfrac{3x+400}{x}=\dfrac{3(1)+400}{1}=403$
The cost of producing 1 DVD is \$403.

b. $A=\dfrac{3(100)+400}{100}=\dfrac{300+400}{100}=\dfrac{700}{100}=7$
The average cost for producing 100 DVDs is \$7.

c. decrease; answers may vary

9. $2x=0$
$x=0$
$\dfrac{7}{2x}$ is undefined for $x=0$.

11. $x+2=0$
$x=-2$
$\dfrac{x+3}{x+2}$ is undefined for $x=-2$.

13. $2x-5=0$
$2x=5$
$x=\dfrac{5}{2}$
$\dfrac{x-4}{2x-5}$ is undefined for $x=\dfrac{5}{2}$.

15. $15x^2+30x=0$
$15x(x+2)=0$
$15x=0$ or $x+2=0$
$x=0$ $\quad$ $x=-2$
$\dfrac{9x^3+4}{15x^2+30x}$ is undefined for $x=0$ and $x=-2$.

17. Since $4\neq 0$, there are no values of x for which $\dfrac{x^2-5x-2}{4}$ is undefined.

19. $x^2-5x-6=0$
$(x-6)(x+1)=0$
$x-6=0$ or $x+1=0$
$x=6$ $\quad$ $x=-1$
$\dfrac{3x^2+9}{x^2-5x-6}$ is undefined for $x=6$ and $x=-1$.

21. $3x^2+13x+14=0$
$(x+2)(3x+7)=0$
$x+2=0$ or $3x+7=0$
$x=-2$ $\quad$ $3x=-7$
$x=-\dfrac{7}{3}$
$\dfrac{x}{3x^2+13x+14}$ is undefined for $x=-2$ and $x=-\dfrac{7}{3}$.

23. $\dfrac{x+7}{7+x}=\dfrac{x+7}{x+7}=1$

25. $\dfrac{x-7}{7-x}=\dfrac{-1(-x+7)}{7-x}=\dfrac{-1(7-x)}{7-x}=-1$

27. $\dfrac{2}{8x+16}=\dfrac{2}{8(x+2)}=\dfrac{1}{4(x+2)}$

29. $\dfrac{x-2}{x^2-4}=\dfrac{x-2}{(x+2)(x-2)}=\dfrac{1}{x+2}$

31. $\dfrac{2x-10}{3x-30}=\dfrac{2(x-5)}{3(x-10)}$ can't be simplified.

33. $\dfrac{-5a-5b}{a+b}=\dfrac{-5(a+b)}{a+b}=-5$

35. $\dfrac{7x+35}{x^2+5x}=\dfrac{7(x+5)}{x(x+5)}=\dfrac{7}{x}$

37. $\dfrac{x+5}{x^2-4x-45}=\dfrac{x+5}{(x-9)(x+5)}=\dfrac{1}{x-9}$

Copyright © 2012 Pearson Education, Inc. Publishing as Prentice Hall.

39. $\dfrac{5x^2+11x+2}{x+2}=\dfrac{(5x+1)(x+2)}{x+2}=5x+1$

41. $\dfrac{x^3+7x^2}{x^2+5x-14}=\dfrac{x^2(x+7)}{(x-2)(x+7)}=\dfrac{x^2}{x-2}$

43. $\dfrac{14x^2-21x}{2x-3}=\dfrac{7x(2x-3)}{2x-3}=7x$

45. $\dfrac{x^2+7x+10}{x^2-3x-10}=\dfrac{(x+5)(x+2)}{(x+2)(x-5)}=\dfrac{x+5}{x-5}$

47. $\dfrac{3x^2+7x+2}{3x^2+13x+4}=\dfrac{(3x+1)(x+2)}{(3x+1)(x+4)}=\dfrac{x+2}{x+4}$

49. $\dfrac{2x^2-8}{4x-8}=\dfrac{2(x^2-4)}{4(x-2)}=\dfrac{2(x+2)(x-2)}{2\cdot 2(x-2)}=\dfrac{x+2}{2}$

51. $\dfrac{4-x^2}{x-2}=\dfrac{-(x^2-4)}{x-2}=\dfrac{-(x+2)(x-2)}{x-2}=-(x+2)$

53. $\dfrac{x^2-1}{x^2-2x+1}=\dfrac{(x+1)(x-1)}{(x-1)^2}=\dfrac{x+1}{x-1}$

55. $\dfrac{x^2+xy+2x+2y}{x+2}=\dfrac{x(x+y)+2(x+y)}{x+2}$
$=\dfrac{(x+2)(x+y)}{x+2}$
$=x+y$

57. $\dfrac{5x+15-xy-3y}{2x+6}=\dfrac{5(x+3)-y(x+3)}{2(x+3)}$
$=\dfrac{(x+3)(5-y)}{2(x+3)}$
$=\dfrac{5-y}{2}$

59. $\dfrac{2xy+5x-2y-5}{3xy+4x-3y-4}=\dfrac{x(2y+5)-1(2y+5)}{x(3y+4)-1(3y+4)}$
$=\dfrac{(2y+5)(x-1)}{(3y+4)(x-1)}$
$=\dfrac{2y+5}{3y+4}$

61. $-\dfrac{x-10}{x+8}=\dfrac{-(x-10)}{x+8}$
$=\dfrac{-x+10}{x+8}$
$=\dfrac{x-10}{-(x+8)}$
$=\dfrac{x-10}{-x-8}$

63. $-\dfrac{5y-3}{y-12}=\dfrac{-(5y-3)}{y-12}$
$=\dfrac{-5y+3}{y-12}$
$=\dfrac{5y-3}{-(y-12)}$
$=\dfrac{5y-3}{-y+12}$

65. $\dfrac{9-x^2}{x-3}=\dfrac{9-x^2}{-(3-x)}$
$=\dfrac{(3+x)(3-x)}{-(3-x)}$
$=\dfrac{3+x}{-1}$
$=-3-x$

The given answer is correct.

67. $\dfrac{7-34x-5x^2}{25x^2-1}=\dfrac{(7+x)(1-5x)}{(5x+1)(5x-1)}$
$=\dfrac{(7+x)(1-5x)}{-(5x+1)(1-5x)}$
$=\dfrac{7+x}{-5x-1}$

The given answer is correct.

69. $\dfrac{1}{3}\cdot\dfrac{9}{11}=\dfrac{1\cdot 9}{3\cdot 11}=\dfrac{1\cdot 3\cdot 3}{3\cdot 11}=\dfrac{3}{11}$

71. $\dfrac{1}{3}\div\dfrac{1}{4}=\dfrac{1}{3}\cdot\dfrac{4}{1}=\dfrac{1\cdot 4}{3\cdot 1}=\dfrac{4}{3}$

73. $\dfrac{13}{20}\div\dfrac{2}{9}=\dfrac{13}{20}\cdot\dfrac{9}{2}=\dfrac{13\cdot 9}{20\cdot 2}=\dfrac{117}{40}$

75. $\dfrac{5a-15}{5}=\dfrac{5(a-3)}{5}=a-3$

The given answer is correct.

Copyright © 2012 Pearson Education, Inc. Publishing as Prentice Hall.

77. $\frac{1+2}{1+3}=\frac{3}{4}\neq\frac{2}{3}$
The given answer is incorrect.

79. answers may vary

81. answers may vary

83. Use $C=\frac{DA}{A+12}$ with $A = 8$ and $D = 1000$.
$C=\frac{DA}{A+12}=\frac{1000\cdot 8}{8+12}=\frac{8000}{20}=400$
The child should receive a dose of 400 milligrams.

85. Use $C=\frac{100W}{L}$ with $W = 5$ and $L = 6.4$.
$C=\frac{100W}{L}=\frac{100(5)}{6.4}=\frac{500}{6.4}=78.125$
The cephalic index is 78.125. The skull is medium.

87. Use $S=\frac{h+d+2t+3r}{b}$ with $h = 187$, $d = 44$, $t = 0$, $r = 37$, and $b = 524$.
$S=\frac{187+44+2(0)+3(37)}{524}=\frac{342}{524}\approx 0.6527$
Pujols' slugging percentage was about 65.3%.

Section 5.2 Practice

1. a. $\frac{16y}{3}\cdot\frac{1}{x^2}=\frac{16y\cdot 1}{3\cdot x^2}=\frac{16y}{3x^2}$

b. $\frac{-5a^3}{3b^3}\cdot\frac{2b^2}{15a}=\frac{-5a^3\cdot 2b^2}{3b^3\cdot 15a}$
$=\frac{-5\cdot a^3\cdot 2\cdot b^2}{3\cdot b^3\cdot 3\cdot 5\cdot a}$
$=-\frac{2a^2}{9b}$

2. $\frac{3x+6}{14}\cdot\frac{7x^2}{x^3+2x^2}=\frac{3(x+2)\cdot 7x^2}{2\cdot 7\cdot x^2(x+2)}=\frac{3}{2}$

3. $\frac{4x+8}{7x^2-14x}\cdot\frac{3x^2-5x-2}{9x^2-1}$
$=\frac{4(x+2)}{7x(x-2)}\cdot\frac{(3x+1)(x-2)}{(3x+1)(3x-1)}$
$=\frac{4(x+2)(3x+1)(x-2)}{7x(x-2)(3x+1)(3x-1)}$
$=\frac{4(x+2)}{7x(3x-1)}$

4. $\frac{7x^2}{6}\div\frac{x}{2y}=\frac{7x^2}{6}\cdot\frac{2y}{x}=\frac{7\cdot x\cdot x\cdot 2\cdot y}{2\cdot 3\cdot x}=\frac{7xy}{3}$

5. $\frac{(x-4)^2}{6}\div\frac{3x-12}{2}=\frac{(x-4)^2}{6}\cdot\frac{2}{3x-12}$
$=\frac{(x-4)(x-4)\cdot 2}{2\cdot 3\cdot 3\cdot(x-4)}$
$=\frac{x-4}{9}$

6. $\frac{10x+4}{x^2-4}\div\frac{5x^3+2x^2}{x+2}=\frac{10x+4}{x^2-4}\cdot\frac{x+2}{5x^3+2x^2}$
$=\frac{2(5x+2)(x+2)}{(x-2)(x+2)\cdot x^2(5x+2)}$
$=\frac{2}{x^2(x-2)}$

7. $\frac{3x^2-10x+8}{7x-14}\div\frac{9x-12}{21}$
$=\frac{3x^2-10x+8}{7x-14}\cdot\frac{21}{9x-12}$
$=\frac{(3x-4)(x-2)\cdot 3\cdot 7}{7(x-2)\cdot 3(3x-4)}$
$=1$

8. a. $\frac{x+3}{x}\cdot\frac{7}{x+3}=\frac{(x+3)\cdot 7}{x\cdot(x+3)}=\frac{7}{x}$

b. $\frac{x+3}{x}\div\frac{7}{x+3}=\frac{x+3}{x}\cdot\frac{x+3}{7}=\frac{(x+3)^2}{7x}$

Copyright © 2012 Pearson Education, Inc. Publishing as Prentice Hall.

c. $\dfrac{3-x}{x^2+6x+5}\cdot\dfrac{2x+10}{x^2-7x+12}$
$=\dfrac{(3-x)\cdot 2(x+5)}{(x+5)(x+1)\cdot(x-4)(x-3)}$
$=\dfrac{-1(x-3)\cdot 2}{(x+1)(x-4)(x-3)}$
$=-\dfrac{2}{(x+1)(x-4)}$

9. $288 \text{ sq in.} = \dfrac{288 \text{ sq in.}}{1}\cdot\dfrac{1 \text{ sq ft}}{144 \text{ sq in.}} = 2 \text{ sq ft}$

10. $3.5 \text{ sq ft} = \dfrac{3.5 \text{ sq ft}}{1}\cdot\dfrac{144 \text{ sq in.}}{1 \text{ sq ft}} = 504 \text{ sq in.}$

11. $61{,}000 \text{ sq yd} = 61{,}000 \text{ sq yd}\cdot\dfrac{9 \text{ sq ft}}{1 \text{ sq yd}}$
$= 549{,}000 \text{ sq ft}$

12. 102.7 feet/second
$=\dfrac{102.7 \text{ feet}}{1 \text{ second}}\cdot\dfrac{3600 \text{ seconds}}{1 \text{ hour}}\cdot\dfrac{1 \text{ mile}}{5280 \text{ feet}}$
$=\dfrac{102.7\cdot 3600}{5280}$ miles/hour
≈ 70.0 miles/hour

Vocabulary and Readiness Check

1. The expressions $\dfrac{x}{2y}$ and $\dfrac{2y}{x}$ are called <u>reciprocals</u>.

2. $\dfrac{a}{b}\cdot\dfrac{c}{d}=\dfrac{a\cdot c}{\underline{b\cdot d}}$

3. $\dfrac{a}{b}\div\dfrac{c}{d}=\dfrac{a\cdot d}{\underline{b\cdot c}}$

4. $\dfrac{x}{7}\cdot\dfrac{x}{6}=\dfrac{x^2}{\underline{42}}$

5. $\dfrac{x}{7}\div\dfrac{x}{6}=\dfrac{6}{\underline{7}}$

Exercise Set 5.2

1. $\dfrac{3x}{y^2}\cdot\dfrac{7y}{4x}=\dfrac{3\cdot 7\cdot x\cdot y}{4\cdot x\cdot y\cdot y}=\dfrac{21}{4y}$

3. $\dfrac{8x}{2}\cdot\dfrac{x^5}{4x^2}=\dfrac{2\cdot 4\cdot x\cdot x\cdot x\cdot x\cdot x\cdot x}{2\cdot 4\cdot x\cdot x}=x^4$

5. $-\dfrac{5a^2b}{30a^2b^2}\cdot b^3=-\dfrac{5a^2b}{30a^2b^2}\cdot\dfrac{b^3}{1}$
$=-\dfrac{5\cdot a^2b\cdot b\cdot b^2}{5\cdot 6\cdot a^2\cdot b^2}$
$=-\dfrac{b\cdot b}{6}$
$=-\dfrac{b^2}{6}$

7. $\dfrac{x}{2x-14}\cdot\dfrac{x^2-7x}{5}=\dfrac{x}{2(x-7)}\cdot\dfrac{x(x-7)}{5}$
$=\dfrac{x\cdot x}{2\cdot 5}$
$=\dfrac{x^2}{10}$

9. $\dfrac{6x+6}{5}\cdot\dfrac{10}{36x+36}=\dfrac{6(x+1)}{5}\cdot\dfrac{10}{36(x+1)}$
$=\dfrac{6\cdot 10}{5\cdot 36}$
$=\dfrac{6\cdot 2\cdot 5}{5\cdot 6\cdot 2\cdot 3}$
$=\dfrac{1}{3}$

11. $\dfrac{(m+n)^2}{m-n}\cdot\dfrac{m}{m^2+mn}=\dfrac{(m+n)(m+n)}{m-n}\cdot\dfrac{m}{m(m+n)}$
$=\dfrac{m+n}{m-n}$

13. $\dfrac{x^2-25}{x^2-3x-10}\cdot\dfrac{x+2}{x}=\dfrac{(x+5)(x-5)}{(x+2)(x-5)}\cdot\dfrac{x+2}{x}$
$=\dfrac{x+5}{x}$

15. $\dfrac{x^2+6x+8}{x^2+x-20}\cdot\dfrac{x^2+2x-15}{x^2+8x+16}$
$=\dfrac{(x+4)(x+2)}{(x+5)(x-4)}\cdot\dfrac{(x+5)(x-3)}{(x+4)(x+4)}$
$=\dfrac{(x+2)(x-3)}{(x-4)(x+4)}$

Copyright © 2012 Pearson Education, Inc. Publishing as Prentice Hall.

17. $\frac{5x^7}{2x^5} \div \frac{15x}{4x^3} = \frac{5x^7}{2x^5} \cdot \frac{4x^3}{15x}$
$= \frac{5 \cdot x^5 \cdot x \cdot x}{2 \cdot x^5} \cdot \frac{2 \cdot 2 \cdot x^3}{5 \cdot 3 \cdot x}$
$= \frac{2 \cdot x \cdot x^3}{3}$
$= \frac{2x^4}{3}$

19. $\frac{8x^2}{y^3} \div \frac{4x^2y^3}{6} = \frac{8x^2}{y^3} \cdot \frac{6}{4x^2y^3}$
$= \frac{2 \cdot 4 \cdot 6 \cdot x \cdot x}{4 \cdot x \cdot x \cdot y \cdot y \cdot y \cdot y \cdot y \cdot y}$
$= \frac{2 \cdot 6}{y \cdot y \cdot y \cdot y \cdot y \cdot y}$
$= \frac{12}{y^6}$

21. $\frac{(x-6)(x+4)}{4x} \div \frac{2x-12}{8x^2} = \frac{(x-6)(x+4)}{4x} \cdot \frac{8x^2}{2x-12}$
$= \frac{(x-6)(x+4)}{4x} \cdot \frac{4x \cdot 2 \cdot x}{2(x-6)}$
$= \frac{x(x+4)}{1}$
$= x(x+4)$

23. $\frac{3x^2}{x^2-1} \div \frac{x^5}{(x+1)^2} = \frac{3x^2}{x^2-1} \cdot \frac{(x+1)^2}{x^5}$
$= \frac{3x^2}{(x-1)(x+1)} \cdot \frac{(x+1)(x+1)}{x^2 \cdot x^3}$
$= \frac{3(x+1)}{x^3(x-1)}$

25. $\frac{m^2-n^2}{m+n} \div \frac{m}{m^2+nm} = \frac{m^2-n^2}{m+n} \cdot \frac{m^2+nm}{m}$
$= \frac{(m+n)(m-n)}{m+n} \cdot \frac{m(m+n)}{m}$
$= (m-n)(m+n)$
$= m^2-n^2$

27. $\frac{x+2}{7-x} \div \frac{x^2-5x+6}{x^2-9x+14} = \frac{x+2}{7-x} \cdot \frac{x^2-9x+14}{x^2-5x+6}$
$= \frac{x+2}{7-x} \cdot \frac{(x-7)(x-2)}{(x-3)(x-2)}$
$= \frac{x+2}{-(x-7)} \cdot \frac{(x-7)(x-2)}{(x-3)(x-2)}$
$= -\frac{x+2}{x-3}$

29. $\frac{x^2+7x+10}{x-1} \div \frac{x^2+2x-15}{x-1}$
$= \frac{x^2+7x+10}{x-1} \cdot \frac{x-1}{x^2+2x-15}$
$= \frac{(x+5)(x+2)}{x-1} \cdot \frac{x-1}{(x+5)(x-3)}$
$= \frac{x+2}{x-3}$

31. $\frac{5x-10}{12} \div \frac{4x-8}{8} = \frac{5x-10}{12} \cdot \frac{8}{4x-8}$
$= \frac{5(x-2)}{2 \cdot 6} \cdot \frac{4 \cdot 2}{4(x-2)}$
$= \frac{5}{6}$

33. $\frac{x^2+5x}{8} \cdot \frac{9}{3x+15} = \frac{x(x+5)}{8} \cdot \frac{3 \cdot 3}{3(x+5)} = \frac{3x}{8}$

35. $\frac{7}{6p^2+q} \div \frac{14}{18p^2+3q} = \frac{7}{6p^2+q} \cdot \frac{18p^2+3q}{14}$
$= \frac{7}{6p^2+q} \cdot \frac{3(6p^2+q)}{7 \cdot 2}$
$= \frac{3}{2}$

37. $\frac{3x+4y}{x^2+4xy+4y^2} \cdot \frac{x+2y}{2} = \frac{3x+4y}{(x+2y)^2} \cdot \frac{x+2y}{2}$
$= \frac{3x+4y}{2(x+2y)}$

39. $\frac{(x+2)^2}{x-2} \div \frac{x^2-4}{2x-4} = \frac{(x+2)^2}{x-2} \cdot \frac{2x-4}{x^2-4}$
$= \frac{(x+2)(x+2)}{x-2} \cdot \frac{2(x-2)}{(x-2)(x+2)}$
$= \frac{2(x+2)}{x-2}$

Copyright © 2012 Pearson Education, Inc. Publishing as Prentice Hall.

41. $\frac{x^2-4}{24x} \div \frac{2-x}{6xy} = \frac{x^2-4}{24x} \cdot \frac{6xy}{2-x}$
$= \frac{(x+2)(x-2)}{6 \cdot 4 \cdot x} \cdot \frac{6 \cdot x \cdot y}{-1(x-2)}$
$= \frac{y(x+2)}{-4}$
$= -\frac{y(x+2)}{4}$

43. $\frac{a^2+7a+12}{a^2+5a+6} \cdot \frac{a^2+8a+15}{a^2+5a+4}$
$= \frac{(a+4)(a+3)}{(a+3)(a+2)} \cdot \frac{(a+5)(a+3)}{(a+4)(a+1)}$
$= \frac{(a+5)(a+3)}{(a+2)(a+1)}$

45. $\frac{5x-20}{3x^2+x} \cdot \frac{3x^2+13x+4}{x^2-16}$
$= \frac{5(x-4)}{x(3x+1)} \cdot \frac{(x+4)(3x+1)}{(x+4)(x-4)}$
$= \frac{5}{x}$

47. $\frac{8n^2-18}{2n^2-5n+3} \div \frac{6n^2+7n-3}{n^2-9n+8}$
$= \frac{8n^2-18}{2n^2-5n+3} \cdot \frac{n^2-9n+8}{6n^2+7n-3}$
$= \frac{2(4n^2-9)}{(2n-3)(n-1)} \cdot \frac{(n-8)(n-1)}{(3n-1)(2n+3)}$
$= \frac{2(2n-3)(2n+3)}{(2n-3)(n-1)} \cdot \frac{(n-8)(n-1)}{(3n-1)(2n+3)}$
$= \frac{2(n-8)}{3n-1}$

49. 10 square feet
$= \frac{10 \text{ square feet}}{1} \cdot \frac{144 \text{ square inches}}{1 \text{ square foot}}$
$= 1440$ square inches

51. $45 \text{ square feet} = \frac{45 \text{ square feet}}{1} \cdot \frac{1 \text{ square yard}}{9 \text{ square feet}}$
$= 5$ square yards

53. $3 \text{ cubic yards} = \frac{3 \text{ cubic yards}}{1} \cdot \frac{27 \text{ cubic feet}}{1 \text{ cubic yard}}$
$= 81$ cubic feet

55. $\frac{50 \text{ miles}}{1 \text{ hour}} = \frac{50 \text{ miles}}{1 \text{ hour}} \cdot \frac{5280 \text{ feet}}{1 \text{ mile}} \cdot \frac{1 \text{ hour}}{3600 \text{ seconds}}$
≈ 73 feet per second

57. 6.3 square yards
$= \frac{6.3 \text{ square yards}}{1} \cdot \frac{9 \text{ square feet}}{1 \text{ square yard}}$
$= 56.7$ square feet

59. 133,500 square yards
$= \frac{133{,}500 \text{ square yards}}{1} \cdot \frac{9 \text{ square feet}}{1 \text{ square yard}}$
$= 1{,}201{,}500$ square feet

61. 359.2 feet per second
$= \frac{359.2 \text{ feet}}{1 \text{ second}} \cdot \frac{3600 \text{ seconds}}{1 \text{ hour}} \cdot \frac{1 \text{ mile}}{5280 \text{ feet}}$
≈ 244.9 miles per hour

63. $\frac{1}{5} + \frac{4}{5} = \frac{1+4}{5} = \frac{5}{5} = 1$

65. $\frac{9}{9} - \frac{19}{9} = \frac{9-19}{9} = \frac{-10}{9} = -\frac{10}{9}$

67. $\frac{6}{5} + \left(\frac{1}{5} - \frac{8}{5}\right) = \frac{6}{5} + \left(-\frac{7}{5}\right) = -\frac{1}{5}$

69. $\frac{4}{a} \cdot \frac{1}{b} = \frac{4 \cdot 1}{a \cdot b} = \frac{4}{ab}$
The statement is true.

71. $\frac{x}{5} \cdot \frac{x+3}{4} = \frac{x^2+3x}{20} \neq \frac{2x+3}{20}$
The statement is false.

73. $\frac{2x}{x^2-25} \cdot \frac{x+5}{9x} = \frac{2 \cdot x}{(x+5)(x-5)} \cdot \frac{x+5}{9 \cdot x}$
$= \frac{2}{9(x-5)}$

The area is $\frac{2}{9(x-5)}$ square feet.

Copyright © 2012 Pearson Education, Inc. Publishing as Prentice Hall.

75. $\left(\dfrac{x^2-y^2}{x^2+y^2} \div \dfrac{x^2-y^2}{3x}\right) \cdot \dfrac{x^2+y^2}{6}$

$= \dfrac{x^2-y^2}{x^2+y^2} \cdot \dfrac{3x}{x^2-y^2} \cdot \dfrac{x^2+y^2}{6}$

$= \dfrac{3x}{6}$

$= \dfrac{x}{2}$

77. $\left(\dfrac{2a+b}{b^2} \cdot \dfrac{3a^2-2ab}{ab+2b^2}\right) \div \dfrac{a^2-3ab+2b^2}{5ab-10b^2}$

$= \dfrac{2a+b}{b^2} \cdot \dfrac{3a^2-2ab}{ab+2b^2} \cdot \dfrac{5ab-10b^2}{a^2-3ab+2b^2}$

$= \dfrac{2a+b}{b^2} \cdot \dfrac{a(3a-2b)}{b(a+2b)} \cdot \dfrac{5 \cdot b(a-2b)}{(a-b)(a-2b)}$

$= \dfrac{5a(2a+b)(3a-2b)}{b^2(a-b)(a+2b)}$

79. answers may vary

81. $\$2000 \text{ US} = \dfrac{\$2000 \text{ US}}{1} \cdot \dfrac{1 \text{ euro}}{\$1.3245 \text{ US}}$

$= \dfrac{2000}{1.3245}$ euros

≈ 1510.004 euros

On that day, \$2000 US was worth 1510 euros.

Section 5.3 Practice

1. $\dfrac{8x}{3y} + \dfrac{x}{3y} = \dfrac{8x+x}{3y} = \dfrac{9x}{3y} = \dfrac{3x}{y}$

2. $\dfrac{3x}{3x-7} - \dfrac{7}{3x-7} = \dfrac{3x-7}{3x-7} = \dfrac{1}{1}$ or 1

3. $\dfrac{2x^2+5x}{x+2} - \dfrac{4x+6}{x+2} = \dfrac{2x^2+5x-(4x+6)}{x+2}$

$= \dfrac{2x^2+5x-4x-6}{x+2}$

$= \dfrac{2x^2+x-6}{x+2}$

$= \dfrac{(2x-3)(x+2)}{x+2}$

$= 2x-3$

4. a. $\dfrac{2}{9}, \dfrac{7}{15}$

$9 = 3^2$ and $15 = 3 \cdot 5$

$\text{LCD} = 3^2 \cdot 5 = 9 \cdot 5 = 45$

b. $\dfrac{5}{6x^3}, \dfrac{11}{8x^5}$

$6x^3 = 2 \cdot 3 \cdot x^3$ and $8x^5 = 2^3 \cdot x^5$

$\text{LCD} = 3 \cdot 2^3 \cdot x^5 = 3 \cdot 8 \cdot x^5 = 24x^5$

5. $\dfrac{3a}{a+5}, \dfrac{7a}{a-5}$

$\text{LCD} = (a+5)(a-5)$

6. $\dfrac{7x^2}{(x-4)^2}, \dfrac{5x}{3x-12}$

$(x-4)^2 = (x-4)(x-4)$ and $3x-12 = 3(x-4)$

$\text{LCD} = 3(x-4)(x-4) = 3(x-4)^2$

7. $\dfrac{y+5}{y^2+2y-3}, \dfrac{y+4}{y^2-3y+2}$

$y^2+2y-3 = (y+3)(y-1)$

$y^2-3y+2 = (y-2)(y-1)$

$\text{LCD} = (y+3)(y-1)(y-2)$

8. $\dfrac{6}{x-4}, \dfrac{9}{4-x}$

$(4-x) = -(x-4)$

$\text{LCD} = (x-4)$ or $(4-x)$

9. $\dfrac{2x}{5y} = \dfrac{2x}{5y} \cdot 1 = \dfrac{2x}{5y} \cdot \dfrac{4x^2y}{4x^2y} = \dfrac{8x^3y}{20x^2y^2}$

10. $\dfrac{3}{x^2-25} = \dfrac{3}{(x-5)(x+5)}$

$= \dfrac{3}{(x-5)(x+5)} \cdot \dfrac{x-3}{x-3}$

$= \dfrac{3(x-3)}{(x-5)(x+5)(x-3)}$

$= \dfrac{3x-9}{(x-5)(x+5)(x-3)}$

Copyright © 2012 Pearson Education, Inc. Publishing as Prentice Hall.

Vocabulary and Readiness Check

1. $\frac{7}{11}+\frac{2}{11}=\frac{9}{\underline{11}}$

2. $\frac{7}{11}-\frac{2}{11}=\frac{5}{\underline{11}}$

3. $\frac{a}{b}+\frac{c}{b}=\frac{a+c}{\underline{b}}$

4. $\frac{a}{b}-\frac{c}{b}=\frac{a-c}{\underline{b}}$

5. $\frac{5}{x}-\frac{6+x}{x}=\frac{5-(6+x)}{\underline{x}}$

Exercise Set 5.3

1. $\frac{a}{13}+\frac{9}{13}=\frac{a+9}{13}$

3. $\frac{4m}{3n}+\frac{5m}{3n}=\frac{4m+5m}{3n}=\frac{9m}{3n}=\frac{3m}{n}$

5. $\frac{4m}{m-6}-\frac{24}{m-6}=\frac{4m-24}{m-6}=\frac{4(m-6)}{m-6}=4$

7. $\frac{9}{3+y}+\frac{y+1}{3+y}=\frac{9+y+1}{3+y}=\frac{y+10}{3+y}$

9. $\frac{5x^2+4x}{x-1}-\frac{2x+3}{x-1}=\frac{5x^2+4x-(2x+3)}{x-1}$
$=\frac{5x^2+2x-3}{x-1}$
$=\frac{(5x-3)(x+1)}{x-1}$

11. $\frac{4a}{a^2+2a-15}-\frac{12}{a^2+2a-15}=\frac{4a-12}{a^2+2a-15}$
$=\frac{4(a-3)}{(a+5)(a-3)}$
$=\frac{4}{a+5}$

13. $\frac{2x+3}{x^2-x-30}-\frac{x-2}{x^2-x-30}=\frac{2x+3-(x-2)}{x^2-x-30}$
$=\frac{x+5}{(x-6)(x+5)}$
$=\frac{1}{x-6}$

15. $2x=2\cdot x$
$4x^3=2^2\cdot x^3$
$\text{LCD}=2^2\cdot x^3=4x^3$

17. $8x=2\cdot 2\cdot 2\cdot x$
$2x+4=2(x+2)$
$\text{LCD}=2\cdot 2\cdot 2\cdot x(x+2)=8x(x+2)$

19. $x+3=x+3$
$x-2=x-2$
$\text{LCD}=(x+3)(x-2)$

21. $x+6=x+6$
$3x+18=3(x+6)$
$\text{LCD}=3(x+6)$

23. $(x-6)^2=(x-6)(x-6)$
$5x-30=5(x-6)$
$\text{LCD}=5(x-6)(x-6)=5(x-6)^2$

25. $3x+3=3(x+1)$
$2x^2+4x+2=2(x^2+2x+1)=2(x+1)^2$
$\text{LCD}=2\cdot 3\cdot (x+1)^2=6(x+1)^2$

27. $8-x=-(x-8)$
$\text{LCD}=x-8$ or $8-x$

29. $x^2+3x-4=(x-1)(x+4)$
$x^2+2x-3=(x-1)(x+3)$
$\text{LCD}=(x-1)(x+4)(x+3)$

31. $3x^2+4x+1=(3x+1)(x+1)$
$2x^2-x-1=(2x+1)(x-1)$
$\text{LCD}=(3x+1)(x+1)(2x+1)(x-1)$

33. $x^2-16=(x+4)(x-4)$
$2x^3-8x^2=2x^2(x-4)$
$\text{LCD}=2x^2(x+4)(x-4)$

Copyright © 2012 Pearson Education, Inc. Publishing as Prentice Hall.

35. $\frac{3}{2x}=\frac{3\cdot 2x}{2x\cdot 2x}=\frac{6x}{4x^2}$

37. $\frac{6}{3a}=\frac{6\cdot 4b^2}{3a\cdot 4b^2}=\frac{24b^2}{12ab^2}$

39. $\frac{9}{2x+6}=\frac{9}{2(x+3)}=\frac{9y}{2(x+3)y}=\frac{9y}{2y(x+3)}$

41. $\frac{9a+2}{5a+10}=\frac{9a+2}{5(a+2)}=\frac{(9a+2)\cdot b}{5(a+2)\cdot b}=\frac{9ab+2b}{5b(a+2)}$

43. $\frac{x}{x^3+6x^2+8x}=\frac{x}{x(x^2+6x+8)}$
$=\frac{x}{x(x+4)(x+2)}$
$=\frac{x(x+1)}{x(x+4)(x+2)(x+1)}$
$=\frac{x^2+x}{x(x+4)(x+2)(x+1)}$

45. $\frac{9y-1}{15x^2-30}=\frac{(9y-1)\cdot 2}{(15x^2-30)\cdot 2}=\frac{18y-2}{30x^2-60}$

47. $\frac{5x}{7}+\frac{9x}{7}=\frac{5x+9x}{7}=\frac{14x}{7}=2x$

49. $\frac{x+3}{4}\div\frac{2x-1}{4}=\frac{x+3}{4}\cdot\frac{4}{2x-1}=\frac{x+3}{2x-1}$

51. $\frac{x^2}{x-6}-\frac{5x+6}{x-6}=\frac{x^2-(5x+6)}{x-6}$
$=\frac{x^2-5x-6}{x-6}$
$=\frac{(x-6)(x+1)}{x-6}$
$=x+1$

53. $\frac{x^2+5x}{x^2-25}\cdot\frac{3x-15}{x^2}=\frac{x(x+5)}{(x+5)(x-5)}\cdot\frac{3(x-5)}{x^2}=\frac{3}{x}$

55. $\frac{x^3+7x^2}{3x^3-x^2}\div\frac{5x^2+36x+7}{9x^2-1}$
$=\frac{x^3+7x^2}{3x^3-x^2}\cdot\frac{9x^2-1}{5x^2+36x+7}$
$=\frac{x^2(x+7)}{x^2(3x-1)}\cdot\frac{(3x-1)(3x+1)}{(5x+1)(x+7)}$
$=\frac{3x+1}{5x+1}$

57. $\frac{2}{3}+\frac{5}{7}=\frac{2\cdot 7}{3\cdot 7}+\frac{5\cdot 3}{7\cdot 3}=\frac{14}{21}+\frac{15}{21}=\frac{29}{21}$

59. $\frac{2}{6}-\frac{3}{4}=\frac{2\cdot 2}{6\cdot 2}-\frac{3\cdot 3}{4\cdot 3}=\frac{4}{12}-\frac{9}{12}=-\frac{5}{12}$

61. $\frac{1}{12}+\frac{3}{20}=\frac{1\cdot 5}{12\cdot 5}+\frac{3\cdot 3}{20\cdot 3}=\frac{5}{60}+\frac{9}{60}=\frac{14}{60}=\frac{7}{30}$

63. $4a-20=4(a-5)$
$(a-5)^2=(a-5)(a-5)$
$\text{LCD}=4(a-5)(a-5)=4(a-5)^2$; d

65. answers may vary

67. The perimeter of a square is 4 times the side length.
$4\cdot\frac{5}{x-2}=\frac{4}{1}\cdot\frac{5}{x-2}=\frac{20}{x-2}$
The perimeter is $\frac{20}{x-2}$ meters.

69. answers may vary

71. $88=2^3\cdot 11$
$4332=2^2\cdot 3\cdot 19^2$
$\text{LCD}=2^3\cdot 3\cdot 11\cdot 19^2=95{,}304$ Earth days

73. answers may vary

75. answers may vary

Section 5.4 Practice

1. a. $\frac{y}{5}-\frac{3y}{15}=\frac{y\cdot 3}{5\cdot 3}-\frac{3y}{15}=\frac{3y-3y}{15}=\frac{0}{15}=0$

Copyright © 2012 Pearson Education, Inc. Publishing as Prentice Hall.

b. $\frac{5}{8x}+\frac{11}{10x^2}=\frac{5\cdot 5x}{8x\cdot 5x}+\frac{11\cdot 4}{10x^2\cdot 4}$
$=\frac{25x}{40x^2}+\frac{44}{40x^2}$
$=\frac{25x+44}{40x^2}$

2. $\frac{10x}{x^2-9}-\frac{5}{x+3}=\frac{10x}{(x-3)(x+3)}-\frac{5(x-3)}{(x+3)(x-3)}$
$=\frac{10x-5(x-3)}{(x+3)(x-3)}$
$=\frac{10x-5x+15}{(x+3)(x-3)}$
$=\frac{5x+15}{(x+3)(x-3)}$
$=\frac{5(x+3)}{(x+3)(x-3)}$
$=\frac{5}{x-3}$

3. $\frac{5}{7x}+\frac{2}{x+1}=\frac{5(x+1)}{7x(x+1)}+\frac{2(7x)}{(x+1)(7x)}$
$=\frac{5(x+1)+2(7x)}{7x(x+1)}$
$=\frac{5x+5+14x}{7x(x+1)}$
$=\frac{19x+5}{7x(x+1)}$

4. $\frac{10}{x-6}-\frac{15}{6-x}=\frac{10}{x-6}-\frac{15}{-(x-6)}$
$=\frac{10}{x-6}-\frac{-15}{x-6}$
$=\frac{10-(-15)}{x-6}$
$=\frac{25}{x-6}$

5. $2+\frac{x}{x+5}=\frac{2}{1}+\frac{x}{x+5}$
$=\frac{2(x+5)}{1(x+5)}+\frac{x}{x+5}$
$=\frac{2x+10+x}{x+5}$
$=\frac{3x+10}{x+5}$

6. $\frac{4}{3x^2+2x}-\frac{3x}{12x+8}=\frac{4}{x(3x+2)}-\frac{3x}{4(3x+2)}$
$=\frac{4(4)}{x(3x+2)(4)}-\frac{3x(x)}{4(3x+2)(x)}$
$=\frac{16-3x^2}{4x(3x+2)}$

7. $\frac{6x}{x^2+4x+4}+\frac{x}{x^2-4}$
$=\frac{6x}{(x+2)(x+2)}+\frac{x}{(x+2)(x-2)}$
$=\frac{6x(x-2)}{(x+2)(x+2)(x-2)}+\frac{x(x+2)}{(x+2)(x-2)(x+2)}$
$=\frac{6x(x-2)+x(x+2)}{(x+2)^2(x-2)}$
$=\frac{6x^2-12x+x^2+2x}{(x+2)^2(x-2)}$
$=\frac{7x^2-10x}{(x+2)^2(x-2)}$
$=\frac{x(7x-10)}{(x+2)^2(x-2)}$

Vocabulary and Readiness Check

1. $\frac{3}{7x}+\frac{5}{7}=\frac{3}{7x}+\frac{5\cdot x}{7\cdot x}=\frac{3+5x}{7x}$; b

2. $\frac{1}{x}+\frac{2}{x^2}=\frac{1\cdot x}{x\cdot x}+\frac{2}{x^2}=\frac{x+2}{x^2}$; a

Exercise Set 5.4

1. $\frac{4}{2x}+\frac{9}{3x}=\frac{4\cdot 3}{2x\cdot 3}+\frac{9\cdot 2}{3x\cdot 2}=\frac{12}{6x}+\frac{18}{6x}=\frac{30}{6x}=\frac{5}{x}$

3. $\frac{15a}{b}+\frac{6b}{5}=\frac{15a\cdot 5}{b\cdot 5}+\frac{6b\cdot b}{5\cdot b}$
$=\frac{75a}{5b}+\frac{6b^2}{5b}$
$=\frac{75a+6b^2}{5b}$

5. $\frac{3}{x}+\frac{5}{2x^2}=\frac{3\cdot 2x}{x\cdot 2x}+\frac{5}{2x^2}=\frac{6x}{2x^2}+\frac{5}{2x^2}=\frac{6x+5}{2x^2}$

Copyright © 2012 Pearson Education, Inc. Publishing as Prentice Hall.

7. $\frac{6}{x+1}+\frac{10}{2x+2}=\frac{6}{x+1}+\frac{10}{2(x+1)}$
$=\frac{6}{x+1}+\frac{5}{x+1}$
$=\frac{11}{x+1}$

9. $\frac{3}{x+2}-\frac{2x}{x^2-4}=\frac{3(x-2)}{(x+2)(x-2)}-\frac{2x}{(x+2)(x-2)}$
$=\frac{3x-6-2x}{(x+2)(x-2)}$
$=\frac{x-6}{(x+2)(x-2)}$

11. $\frac{3}{4x}+\frac{8}{x-2}=\frac{3(x-2)}{4x(x-2)}+\frac{8(4x)}{(x-2)(4x)}$
$=\frac{3x-6+32x}{4x(x-2)}$
$=\frac{35x-6}{4x(x-2)}$

13. $\frac{6}{x-3}+\frac{8}{3-x}=\frac{6}{x-3}+\frac{8}{-(x-3)}$
$=\frac{6}{x-3}+\frac{-8}{x-3}$
$=\frac{6+(-8)}{x-3}$
$=\frac{-2}{x-3}$
$=-\frac{2}{x-3}$

15. $\frac{9}{x-3}+\frac{9}{3-x}=\frac{9}{x-3}+\frac{9}{-(x-3)}$
$=\frac{9}{x-3}+\frac{-9}{x-3}$
$=\frac{0}{x-3}$
$=0$

17. $\frac{-8}{x^2-1}-\frac{7}{1-x^2}=\frac{-8}{x^2-1}-\frac{7}{-(x^2-1)}$
$=\frac{-8}{x^2-1}-\frac{-7}{x^2-1}$
$=\frac{-8-(-7)}{x^2-1}$
$=\frac{-1}{x^2-1}$
$=-\frac{1}{x^2-1}$

19. $\frac{5}{x}+2=\frac{5}{x}+\frac{2}{1}=\frac{5}{x}+\frac{2\cdot x}{1\cdot x}=\frac{5+2x}{x}$

21. $\frac{5}{x-2}+6=\frac{5}{x-2}+\frac{6}{1}$
$=\frac{5}{x-2}+\frac{6(x-2)}{1(x-2)}$
$=\frac{5+6x-12}{x-2}$
$=\frac{6x-7}{x-2}$

23. $\frac{y+2}{y+3}-2=\frac{y+2}{y+3}-\frac{2}{1}$
$=\frac{y+2}{y+3}-\frac{2(y+3)}{1(y+3)}$
$=\frac{y+2}{y+3}-\frac{2y+6}{y+3}$
$=\frac{y+2-(2y+6)}{y+3}$
$=\frac{y+2-2y-6}{y+3}$
$=\frac{-y-4}{y+3}$
$=\frac{-(y+4)}{y+3}$
$=-\frac{y+4}{y+3}$

25. $\frac{-x+2}{x}-\frac{x-6}{4x}=\frac{4(-x+2)}{4x}-\frac{x-6}{4x}$
$=\frac{-4x+8-(x-6)}{4x}$
$=\frac{-4x+8-x+6}{4x}$
$=\frac{-5x+14}{4x}$ or $-\frac{5x-14}{4x}$

Copyright © 2012 Pearson Education, Inc. Publishing as Prentice Hall.

27. $\frac{5x}{x+2}-\frac{3x-4}{x+2}=\frac{5x-(3x-4)}{x+2}$
$=\frac{5x-3x+4}{x+2}$
$=\frac{2x+4}{x+2}$
$=\frac{2(x+2)}{x+2}$
$=2$

29. $\frac{3x^4}{7}-\frac{4x^2}{21}=\frac{3x^4\cdot 3}{7\cdot 3}-\frac{4x^2}{21}$
$=\frac{9x^4}{21}-\frac{4x^2}{21}$
$=\frac{9x^4-4x^2}{21}$

31. $\frac{1}{x+3}-\frac{1}{(x+3)^2}=\frac{1\cdot(x+3)}{(x+3)(x+3)}-\frac{1}{(x+3)(x+3)}$
$=\frac{x+3-1}{(x+3)^2}$
$=\frac{x+2}{(x+3)^2}$

33. $\frac{4}{5b}+\frac{1}{b-1}=\frac{4(b-1)}{5b(b-1)}+\frac{1\cdot 5b}{(b-1)(5b)}$
$=\frac{4b-4}{5b(b-1)}+\frac{5b}{5b(b-1)}$
$=\frac{4b-4+5b}{5b(b-1)}$
$=\frac{9b-4}{5b(b-1)}$

35. $\frac{2}{m}+1=\frac{2}{m}+\frac{m}{m}=\frac{2+m}{m}$

37. $\frac{2x}{x-7}-\frac{x}{x-2}=\frac{2x(x-2)}{(x-7)(x-2)}-\frac{x(x-7)}{(x-2)(x-7)}$
$=\frac{2x^2-4x-(x^2-7x)}{(x-7)(x-2)}$
$=\frac{2x^2-4x-x^2+7x}{(x-7)(x-2)}$
$=\frac{x^2+3x}{(x-7)(x-2)}$
$=\frac{x(x+3)}{(x-7)(x-2)}$

39. $\frac{6}{1-2x}-\frac{4}{2x-1}=\frac{6}{1-2x}-\frac{4}{-(1-2x)}$
$=\frac{6}{1-2x}-\frac{-4}{1-2x}$
$=\frac{6+4}{1-2x}$
$=\frac{10}{1-2x}$

41. $\frac{7}{(x+1)(x-1)}+\frac{8}{(x+1)^2}$
$=\frac{7(x+1)}{(x+1)(x-1)(x+1)}+\frac{8(x-1)}{(x+1)^2(x-1)}$
$=\frac{7x+7+8x-8}{(x+1)^2(x-1)}$
$=\frac{15x-1}{(x+1)^2(x-1)}$

43. $\frac{x}{x^2-1}-\frac{2}{x^2-2x+1}$
$=\frac{x(x-1)}{(x-1)(x+1)(x-1)}-\frac{2(x+1)}{(x-1)^2(x+1)}$
$=\frac{x^2-x-2x-2}{(x-1)^2(x+1)}$
$=\frac{x^2-3x-2}{(x-1)^2(x+1)}$

45. $\frac{3a}{2a+6}-\frac{a-1}{a+3}=\frac{3a}{2(a+3)}-\frac{(a-1)(2)}{(a+3)(2)}$
$=\frac{3a-(2a-2)}{2(a+3)}$
$=\frac{3a-2a+2}{2(a+3)}$
$=\frac{a+2}{2(a+3)}$

47. $\frac{y-1}{2y+3}+\frac{3}{(2y+3)^2}=\frac{(y-1)(2y+3)}{(2y+3)^2}+\frac{3}{(2y+3)^2}$
$=\frac{2y^2+3y-2y-3+3}{(2y+3)^2}$
$=\frac{2y^2+y}{(2y+3)^2}$
$=\frac{y(2y+1)}{(2y+3)^2}$

Copyright © 2012 Pearson Education, Inc. Publishing as Prentice Hall.

49. $\frac{5}{2-x}+\frac{x}{2x-4}=\frac{5}{-(x-2)}+\frac{x}{2(x-2)}$
$=\frac{-5(2)}{(x-2)(2)}+\frac{x}{2(x-2)}$
$=\frac{-10+x}{2(x-2)}$
$=\frac{x-10}{2(x-2)}$

51. $\frac{15}{x^2+6x-19}+\frac{2}{x+3}=\frac{15}{(x+3)^2}+\frac{2(x+3)}{(x+3)(x+3)}$
$=\frac{15+2x+6}{(x+3)^2}$
$=\frac{2x+21}{(x+3)^2}$

53. $\frac{13}{x^2-5x+6}-\frac{5}{x-3}$
$=\frac{13}{(x-3)(x-2)}-\frac{5(x-2)}{(x-3)(x-2)}$
$=\frac{13-(5x-10)}{(x-3)(x-2)}$
$=\frac{13-5x+10}{(x-3)(x-2)}$
$=\frac{-5x+23}{(x-3)(x-2)}$

55. $\frac{70}{m^2-100}+\frac{7}{2(m+10)}$
$=\frac{70\cdot 2}{(m-10)(m+10)2}+\frac{7(m-10)}{2(m+10)(m-10)}$
$=\frac{140+7m-70}{2(m-10)(m+10)}$
$=\frac{7m+70}{2(m-10)(m+10)}$
$=\frac{7(m+10)}{2(m-10)(m+10)}$
$=\frac{7}{2(m-10)}$

57. $\frac{x+8}{x^2-5x-6}+\frac{x+1}{x^2-4x-5}$
$=\frac{x+8}{(x+1)(x-6)}+\frac{x+1}{(x+1)(x-5)}$
$=\frac{(x+8)(x-5)+(x+1)(x-6)}{(x+1)(x-6)(x-5)}$
$=\frac{x^2+3x-40+x^2-5x-6}{(x+1)(x-6)(x-5)}$
$=\frac{2x^2-2x-46}{(x+1)(x-6)(x-5)}$
$=\frac{2(x^2-x-23)}{(x+1)(x-6)(x-5)}$

59. $\frac{5}{4n^2-12n+8}-\frac{3}{3n^2-6n}$
$=\frac{5\cdot 3n}{4(n-2)(n-1)3n}-\frac{3(n-1)\cdot 4}{3n(n-2)(n-1)4}$
$=\frac{15n-12n+12}{12n(n-2)(n-1)}$
$=\frac{3n+12}{12n(n-2)(n-1)}$
$=\frac{3(n+4)}{12n(n-2)(n-1)}$
$=\frac{n+4}{4n(n-2)(n-1)}$

61. $\frac{15x}{x+8}\cdot\frac{2x+16}{3x}=\frac{5\cdot 3x}{x+8}\cdot\frac{2(x+8)}{3x}=\frac{5\cdot 2}{1}=10$

63. $\frac{8x+7}{3x+5}-\frac{2x-3}{3x+5}=\frac{8x+7-(2x-3)}{3x+5}$
$=\frac{8x+7-2x+3}{3x+5}$
$=\frac{6x+10}{3x+5}$
$=\frac{2(3x+5)}{3x+5}$
$=2$

65. $\frac{5a+10}{18}\div\frac{a^2-4}{10a}=\frac{5a+10}{18}\cdot\frac{10a}{a^2-4}$
$=\frac{5(a+2)}{9\cdot 2}\cdot\frac{2\cdot 5a}{(a+2)(a-2)}$
$=\frac{5\cdot 5a}{9(a-2)}$
$=\frac{25a}{9(a-2)}$

Copyright © 2012 Pearson Education, Inc. Publishing as Prentice Hall.

67. $$\frac{5}{x^2-3x+2}+\frac{1}{x-2}=\frac{5}{(x-2)(x-1)}+\frac{1(x-1)}{(x-2)(x-1)}$$
$$=\frac{5+x-1}{(x-2)(x-1)}$$
$$=\frac{x+4}{(x-2)(x-1)}$$

69. $3x+5=7$
$3x=2$
$x=\frac{2}{3}$

71. $2x^2-x-1=0$
$(2x+1)(x-1)=0$
$2x+1=0$ or $x-1=0$
$2x=-1$ $\quad x=1$
$x=-\frac{1}{2}$

73. $4(x+6)+3=-3$
$4x+24+3=-3$
$4x+27=-3$
$4x=-30$
$x=\frac{-30}{4}$
$x=-\frac{15}{2}$

75. $$\frac{3}{x}-\frac{2x}{x^2-1}+\frac{5}{x+1}=\frac{3}{x}-\frac{2x}{(x-1)(x+1)}+\frac{5}{x+1}$$
$$=\frac{3(x-1)(x+1)}{x(x-1)(x+1)}-\frac{2x\cdot x}{(x-1)(x+1)x}+\frac{5\cdot x(x-1)}{(x+1)x(x-1)}$$
$$=\frac{3x^2-3-2x^2+5x^2-5x}{x(x-1)(x+1)}$$
$$=\frac{6x^2-5x-3}{x(x-1)(x+1)}$$

Copyright © 2012 Pearson Education, Inc. Publishing as Prentice Hall.

77. $$\frac{5}{x^2-4}+\frac{2}{x^2-4x+4}-\frac{3}{x^2-x-6}=\frac{5}{(x+2)(x-2)}+\frac{2}{(x-2)^2}-\frac{3}{(x+2)(x-3)}$$
$$=\frac{5(x-2)(x-3)+2(x+2)(x-3)-3(x-2)^2}{(x+2)(x-2)^2(x-3)}$$
$$=\frac{5(x^2-5x+6)+2(x^2-x-6)-3(x^2-4x+4)}{(x-2)^2(x+2)(x-3)}$$
$$=\frac{5x^2-25x+30+2x^2-2x-12-3x^2+12x-12}{(x-2)^2(x+2)(x-3)}$$
$$=\frac{4x^2-15x+6}{(x-2)^2(x+2)(x-3)}$$

79. $$\frac{9}{x^2+9x+14}-\frac{3x}{x^2+10x+21}+\frac{x+4}{x^2+5x+6}=\frac{9}{(x+2)(x+7)}-\frac{3x}{(x+3)(x+7)}+\frac{x+4}{(x+2)(x+3)}$$
$$=\frac{9(x+3)}{(x+2)(x+7)(x+3)}-\frac{3x(x+2)}{(x+3)(x+7)(x+2)}+\frac{(x+4)(x+7)}{(x+2)(x+3)(x+7)}$$
$$=\frac{9x+27-3x^2-6x+x^2+7x+4x+28}{(x+2)(x+7)(x+3)}$$
$$=\frac{-2x^2+14x+55}{(x+2)(x+7)(x+3)}$$

81. $$\frac{3}{x+4}-\frac{1}{x-4}=\frac{3(x-4)-1(x+4)}{(x+4)(x-4)}$$
$$=\frac{3x-12-x-4}{(x+4)(x-4)}$$
$$=\frac{2x-16}{(x+4)(x-4)}$$
$$=\frac{2(x-8)}{(x+4)(x-4)}$$

The other piece of the board measures $\frac{2(x-8)}{(x+4)(x-4)}$ inches.

83. $1-\frac{G}{P}=\frac{P}{P}-\frac{G}{P}=\frac{P-G}{P}$

85. answers may vary

87. $$90-\frac{40}{x}=\frac{90}{1}-\frac{40}{x}$$
$$=\frac{90x}{x}-\frac{40}{x}$$
$$=\frac{90x-40}{x}$$

The complement measures $\left(\frac{90x-40}{x}\right)^\circ$.

Copyright © 2012 Pearson Education, Inc. Publishing as Prentice Hall.

89. answers may vary

Section 5.5 Practice

1. The LCD is 20.

$$\frac{x}{4}+\frac{4}{5}=\frac{1}{20}$$
$$20\left(\frac{x}{4}+\frac{4}{5}\right)=20\left(\frac{1}{20}\right)$$
$$20\left(\frac{x}{4}\right)+20\left(\frac{4}{5}\right)=20\left(\frac{1}{20}\right)$$
$$5x+16=1$$
$$5x=-15$$
$$x=-3$$

The solution is -3.

2. The LCD is 15.

$$\frac{x+2}{3}-\frac{x-1}{5}=\frac{1}{15}$$
$$15\left(\frac{x+2}{3}-\frac{x-1}{5}\right)=15\left(\frac{1}{15}\right)$$
$$15\left(\frac{x+2}{3}\right)-15\left(\frac{x-1}{5}\right)=15\left(\frac{1}{15}\right)$$
$$5(x+2)-3(x-1)=1$$
$$5x+10-3x+3=1$$
$$2x+13=1$$
$$2x=-12$$
$$x=-6$$

The solution is -6.

3. The LCD is x.

$$2+\frac{6}{x}=x+7$$
$$x\left(2+\frac{6}{x}\right)=x(x+7)$$
$$x(2)+x\left(\frac{6}{x}\right)=x\cdot x+x\cdot 7$$
$$2x+6=x^2+7x$$
$$0=x^2+5x-6$$
$$0=(x+6)(x-1)$$
$$x+6=0 \quad \text{or} \quad x-1=0$$
$$x=-6 \qquad\qquad x=1$$

Neither -6 nor 1 makes the denominator 0 in the original equation. Both -6 and 1 are solutions.

Copyright © 2012 Pearson Education, Inc. Publishing as Prentice Hall.

4. The LCD is $(x + 3)(x - 3)$.

$$\frac{2}{x+3}+\frac{3}{x-3}=\frac{-2}{x^2-9}$$
$$(x+3)(x-3)\left(\frac{2}{x+3}+\frac{3}{x-3}\right)=(x+3)(x-3)\left(\frac{-2}{x^2-9}\right)$$
$$(x+3)(x-3)\cdot\frac{2}{x+3}+(x+3)(x-3)\cdot\frac{3}{x-3}=(x+3)(x-3)\left(\frac{-2}{x^2-9}\right)$$
$$2(x-3)+3(x+3)=-2$$
$$2x-6+3x+9=-2$$
$$5x+3=-2$$
$$5x=-5$$
$$x=-1$$

The solution is -1.

5. The LCD is $x - 1$.

$$\frac{5x}{x-1}=\frac{5}{x-1}+3$$
$$(x-1)\left(\frac{5x}{x-1}\right)=(x-1)\left(\frac{5}{x-1}+3\right)$$
$$(x-1)\cdot\frac{5x}{x-1}=(x-1)\cdot\frac{5}{x-1}+(x-1)\cdot 3$$
$$5x=5+3(x-1)$$
$$5x=5+3x-3$$
$$2x=2$$
$$x=1$$

1 makes the denominator 0 in the original equation. Therefore 1 is *not* a solution and this equation has no solution.

6. The LCD is $x + 3$.

$$x-\frac{6}{x+3}=\frac{2x}{x+3}+2$$
$$(x+3)\left(x-\frac{6}{x+3}\right)=(x+3)\left(\frac{2x}{x+3}+2\right)$$
$$(x+3)(x)-(x+3)\left(\frac{6}{x+3}\right)=(x+3)\left(\frac{2x}{x+3}\right)+(x+3)(2)$$
$$(x+3)(x)-6=2x+2(x+3)$$
$$x^2+3x-6=2x+2x+6$$
$$x^2-x-12=0$$
$$(x-4)(x+3)=0$$
$$x-4=0 \quad \text{or} \quad x+3=0$$
$$x=4 \qquad\qquad x=3$$

$x = 3$ can't be a solution of the original equation. The only solution is 4.

Copyright © 2012 Pearson Education, Inc. Publishing as Prentice Hall.

7. The LCD is abx.

$$\frac{1}{a}+\frac{1}{b}=\frac{1}{x}$$
$$abx\left(\frac{1}{a}+\frac{1}{b}\right)=abx\left(\frac{1}{x}\right)$$
$$abx\left(\frac{1}{a}\right)+abx\left(\frac{1}{b}\right)=abx\left(\frac{1}{x}\right)$$
$$bx+ax=ab$$
$$bx=ab-ax$$
$$bx=a(b-x)$$
$$\frac{bx}{b-x}=\frac{a(b-x)}{b-x}$$
$$\frac{bx}{b-x}=a$$

Vocabulary and Readiness Check

1. $4\left(\frac{3x}{2}+5\right)=4\left(\frac{1}{4}\right)$

$$4\left(\frac{3x}{2}\right)+4\cdot 5=4\left(\frac{1}{4}\right)$$
$$6x+20=1$$

The correct choice is c.

2. $5x\left(\frac{1}{x}-\frac{3}{5x}\right)=5x(2)$

$$5x\left(\frac{1}{x}\right)-5x\left(\frac{3}{5x}\right)=5x(2)$$
$$5-3=10x$$

The correct choice is b.

3. The LCD of $\frac{9}{x}$, $\frac{3}{4}$, and $\frac{1}{12}=\frac{1}{3\cdot 4}$ is $12x$; b.

4. The LCD of $\frac{8}{3x}$, $\frac{1}{x}$, and $\frac{7}{9}=\frac{7}{3\cdot 3}$ is $9x$; d.

5. The LCD of $\frac{9}{x-1}$ and $\frac{7}{(x-1)^2}$ is $(x-1)^2$; a.

6. The LCD of $\frac{1}{x-2}$, $\frac{3}{x^2-4}=\frac{3}{(x+2)(x-2)}$, and $8=\frac{8}{1}$ is x^2-4; c.

Exercise Set 5.5

1. The LCD is 5.

$$\frac{x}{5}+3=9$$
$$5\left(\frac{x}{5}+3\right)=5(9)$$
$$x+15=45$$
$$x=30$$

Check: $\frac{x}{5}+3=9$

$$\frac{30}{5}+3\stackrel{?}{=}9$$
$$6+3\stackrel{?}{=}9$$
$$9=9 \quad \text{True}$$

The solution is 30.

3. The LCD is 12.

$$\frac{x}{2}+\frac{5x}{4}=\frac{x}{12}$$
$$12\left(\frac{x}{2}+\frac{5x}{4}\right)=12\left(\frac{x}{12}\right)$$
$$6x+15x=x$$
$$21x=x$$
$$20x=0$$
$$x=0$$

Check: $\frac{x}{2}+\frac{5x}{4}=\frac{x}{12}$

$$\frac{0}{2}+\frac{5(0)}{4}\stackrel{?}{=}\frac{0}{12}$$
$$0+0\stackrel{?}{=}0$$
$$0=0 \quad \text{True}$$

The solution is 0.

5. The LCD is x.

$$2-\frac{8}{x}=6$$
$$x\left(2-\frac{8}{x}\right)=x(6)$$
$$2x-8=6x$$
$$-8=4x$$
$$-2=x$$

Check: $2-\frac{8}{x}=6$

$$2-\frac{8}{-2}\stackrel{?}{=}6$$
$$2-(-4)\stackrel{?}{=}6$$
$$2+4=6$$
$$6=6 \quad \text{True}$$

The solution is -2.

Copyright © 2012 Pearson Education, Inc. Publishing as Prentice Hall.

7. The LCD is x.

$$2+\frac{10}{x}=x+5$$
$$x\left(2+\frac{10}{x}\right)=x(x+5)$$
$$2x+10=x^2+5x$$
$$0=x^2+3x-10$$
$$0=(x+5)(x-2)$$
$$x+5=0 \quad \text{or} \quad x-2=0$$
$$x=-5 \qquad x=2$$

Check -5: $2+\frac{10}{x}=x+5$

$$2+\frac{10}{-5}\stackrel{?}{=}-5+5$$
$$2+(-2)\stackrel{?}{=}-5+5$$
$$0=0 \quad \text{True}$$

Check 2: $2+\frac{10}{x}=x+5$

$$2+\frac{10}{2}\stackrel{?}{=}2+5$$
$$2+5\stackrel{?}{=}2+5$$
$$7=7 \quad \text{True}$$

The solutions are -5 and 2.

9. The LCD is $5 \cdot 2 = 10$.

$$\frac{a}{5}=\frac{a-3}{2}$$
$$10\left(\frac{a}{5}\right)=10\left(\frac{a-3}{2}\right)$$
$$2a=5(a-3)$$
$$2a=5a-15$$
$$-3a=-15$$
$$a=5$$

Check: $\frac{a}{5}=\frac{a-3}{2}$

$$\frac{5}{5}\stackrel{?}{=}\frac{5-3}{2}$$
$$1\stackrel{?}{=}\frac{2}{2}$$
$$1=1 \quad \text{True}$$

The solution is 5.

11. The LCD is $5 \cdot 2 = 10$.

$$\frac{x-3}{5}+\frac{x-2}{2}=\frac{1}{2}$$
$$10\left(\frac{x-3}{5}+\frac{x-2}{2}\right)=10\left(\frac{1}{2}\right)$$
$$2(x-3)+5(x-2)=\frac{10}{2}$$
$$2x-6+5x-10=5$$
$$7x-16=5$$
$$7x=21$$
$$x=3$$

Check: $\frac{x-3}{5}+\frac{x-2}{2}=\frac{1}{2}$

$$\frac{3-3}{5}+\frac{3-2}{2}\stackrel{?}{=}\frac{1}{2}$$
$$\frac{0}{5}+\frac{1}{2}\stackrel{?}{=}\frac{1}{2}$$
$$\frac{1}{2}=\frac{1}{2} \quad \text{True}$$

The solution is 3.

13. The LCD is $2a - 5$.

$$\frac{3}{2a-5}=-1$$
$$(2a-5)\left(\frac{3}{2a-5}\right)=(2a-5)(-1)$$
$$3=-2a+5$$
$$-2=-2a$$
$$1=a$$

Check: $\frac{3}{2a-5}=-1$

$$\frac{3}{2(1)-5}\stackrel{?}{=}-1$$
$$\frac{3}{2-5}\stackrel{?}{=}-1$$
$$\frac{3}{-3}\stackrel{?}{=}-1$$
$$-1=-1 \quad \text{True}$$

The solution is 1.

Copyright © 2012 Pearson Education, Inc. Publishing as Prentice Hall.

15. The LCD is $y-4$.

$$\frac{4y}{y-4}+5=\frac{5y}{y-4}$$
$$(y-4)\left(\frac{4y}{y-4}+5\right)=(y-4)\left(\frac{5y}{y-4}\right)$$
$$4y+5(y-4)=5y$$
$$4y+5y-20=5y$$
$$9y-20=5y$$
$$-20=-4y$$
$$5=y$$

Check: $\frac{4y}{y-4}+5=\frac{5y}{y-4}$

$$\frac{4\cdot 5}{5-4}+5\stackrel{?}{=}\frac{5\cdot 5}{5-4}$$
$$\frac{20}{1}+5\stackrel{?}{=}\frac{25}{1}$$
$$25=25 \quad \text{True}$$

The solution is 5.

17. The LCD is $a-3$.

$$2+\frac{3}{a-3}=\frac{a}{a-3}$$
$$(a-3)\left(2+\frac{3}{a-3}\right)=(a-3)\frac{a}{a-3}$$
$$2(a-3)+3=a$$
$$2a-6+3=a$$
$$2a-3=a$$
$$-3=-a$$
$$3=a$$

Check: $2+\frac{3}{a-3}=\frac{a}{a-3}$

$$2+\frac{3}{3-3}\stackrel{?}{=}\frac{3}{3-3}$$
$$2+\frac{3}{0}\stackrel{?}{=}\frac{3}{0}$$

Since $\frac{3}{0}$ is undefined, $a=3$ does not check and the equation has no solution.

19. The LCD is $(x+3)(x-3)=x^2-9$.

$$\frac{1}{x+3}+\frac{6}{x^2-9}=1$$
$$(x^2-9)\left(\frac{1}{x+3}+\frac{6}{x^2-9}\right)=1(x^2-9)$$
$$(x-3)+6=x^2-9$$
$$x+3=x^2-9$$
$$0=x^2-x-12$$
$$0=(x-4)(x+3)$$

$x-4=0$ or $x+3=0$

$x=4$ $\quad$ $x=-3$

Check 4: $\frac{1}{x+3}+\frac{6}{x^2-9}=1$

$$\frac{1}{4+3}+\frac{6}{4^2-9}\stackrel{?}{=}1$$
$$\frac{1}{7}+\frac{6}{16-9}\stackrel{?}{=}1$$
$$\frac{1}{7}+\frac{6}{7}\stackrel{?}{=}1$$
$$\frac{7}{7}=1 \quad \text{True}$$

Check -3: $\frac{1}{x+3}+\frac{6}{x^2-9}=1$

$$\frac{1}{-3+3}+\frac{6}{(-3)^2-9}\stackrel{?}{=}1$$
$$\frac{1}{0}+\frac{6}{0}\stackrel{?}{=}1$$

Since $\frac{1}{0}$ and $\frac{6}{0}$ are undefined, $x=-3$ does not check. The solution is 4.

21. The LCD is $y+4$.

$$\frac{2y}{y+4}+\frac{4}{y+4}=3$$
$$(y+4)\left(\frac{2y}{y+4}+\frac{4}{y+4}\right)=(y+4)(3)$$
$$2y+4=3y+12$$
$$4=y+12$$
$$-8=y$$

Check: $\frac{2y}{y+4}+\frac{4}{y+4}=3$

$$\frac{2(-8)}{-8+4}+\frac{4}{-8+4}\stackrel{?}{=}3$$
$$\frac{-16}{-4}+\frac{4}{-4}\stackrel{?}{=}3$$
$$4-1\stackrel{?}{=}3$$
$$3=3 \quad \text{True}$$

The solution is -8.

Copyright © 2012 Pearson Education, Inc. Publishing as Prentice Hall.

23. The LCD is $(x+2)(x-2)=x^2-4$.

$$\frac{2x}{x+2}-2=\frac{x-8}{x-2}$$
$$(x^2-4)\left(\frac{2x}{x+2}-2\right)=(x^2-4)\left(\frac{x-8}{x-2}\right)$$
$$2x(x-2)-2(x^2-4)=(x+2)(x-8)$$
$$2x^2-4x-2x^2+8=x^2-8x+2x-16$$
$$-4x+8=x^2-6x-16$$
$$0=x^2-2x-24$$
$$0=(x-6)(x+4)$$
$$x-6=0 \quad \text{or} \quad x+4=0$$
$$x=6 \qquad\qquad x=-4$$

Check 6: $\frac{2x}{x+2}-2=\frac{x-8}{x-2}$

$$\frac{2(6)}{6+2}-2\stackrel{?}{=}\frac{6-8}{6-2}$$
$$\frac{12}{8}-2\stackrel{?}{=}\frac{-2}{4}$$
$$\frac{3}{2}-\frac{4}{2}\stackrel{?}{=}-\frac{1}{2}$$
$$-\frac{1}{2}=-\frac{1}{2} \quad \text{True}$$

Check –4: $\frac{2x}{x+2}-2=\frac{x-8}{x-2}$

$$\frac{2(-4)}{-4+2}-2\stackrel{?}{=}\frac{-4-8}{-4-2}$$
$$\frac{-8}{-2}-2\stackrel{?}{=}\frac{-12}{-6}$$
$$4-2\stackrel{?}{=}2$$
$$2=2 \quad \text{True}$$

The solutions are 6 and –4.

25. The LCD is $2y$.

$$\frac{2}{y}+\frac{1}{2}=\frac{5}{2y}$$
$$2y\left(\frac{2}{y}+\frac{1}{2}\right)=2y\left(\frac{5}{2y}\right)$$
$$2(2)+y(1)=5$$
$$4+y=5$$
$$y=1$$

The solution $y = 1$ checks.

27. The LCD is $(a-6)(a-1)$.

$$\frac{a}{a-6}=\frac{-2}{a-1}$$
$$(a-6)(a-1)\left(\frac{a}{a-6}\right)=(a-6)(a-1)\left(\frac{-2}{a-1}\right)$$
$$a(a-1)=-2(a-6)$$
$$a^2-a=-2a+12$$
$$a^2+a-12=0$$
$$(a+4)(a-3)=0$$
$$a+4=0 \quad \text{or} \quad a-3=0$$
$$a=-4 \qquad\qquad a=3$$

The solutions $a = -4$ and $a = 3$ check.

29. The LCD is $2x \cdot 3 = 6x$.

$$\frac{11}{2x}+\frac{2}{3}=\frac{7}{2x}$$
$$6x\left(\frac{11}{2x}+\frac{2}{3}\right)=6x\left(\frac{7}{2x}\right)$$
$$3(11)+2x(2)=3(7)$$
$$33+4x=21$$
$$4x=-12$$
$$x=-3$$

The solution $x = -3$ checks.

31. The LCD is $(x-2)(x+2)$.

$$\frac{2}{x-2}+1=\frac{x}{x+2}$$
$$(x-2)(x+2)\left(\frac{2}{x-2}+1\right)=(x-2)(x+2)\left(\frac{x}{x+2}\right)$$
$$2(x+2)+1(x-2)(x+2)=x(x-2)$$
$$2x+4+x^2-4=x^2-2x$$
$$4x=0$$
$$x=0$$

The solution $x = 0$ checks.

33. The LCD is 6.

$$\frac{x+1}{3}-\frac{x-1}{6}=\frac{1}{6}$$
$$6\left(\frac{x+1}{3}-\frac{x-1}{6}\right)=6\left(\frac{1}{6}\right)$$
$$2(x+1)-(x-1)=1$$
$$2x+2-x+1=1$$
$$x+3=1$$
$$x=-2$$

The solution $x = -2$ checks.

Copyright © 2012 Pearson Education, Inc. Publishing as Prentice Hall.

35. The LCD is $6(t-4)$.

$$\frac{t}{t-4}=\frac{t+4}{6}$$

$$6(t-4)\left(\frac{t}{t-4}\right)=6(t-4)\left(\frac{t+4}{6}\right)$$

$$6t=(t-4)(t+4)$$

$$6t=t^2-16$$

$$0=t^2-6t-16$$

$$0=(t-8)(t+2)$$

$t-8=0$ or $t+2=0$

$t=8$ $\quad$ $t=-2$

The solutions $t=8$ and $t=-2$ check.

37. $2y+2=2(y+1)$

$4y+4=4(y+1)=2\cdot 2(y+1)$

The LCD is $4(y+1)$.

$$\frac{y}{2y+2}+\frac{2y-16}{4y+4}=\frac{2y-3}{y+1}$$

$$4(y+1)\left(\frac{y}{2(y+1)}+\frac{2y-16}{4(y+1)}\right)=4(y+1)\left(\frac{2y-3}{y+1}\right)$$

$$2y+2y-16=4(2y-3)$$

$$4y-16=8y-12$$

$$-16=4y-12$$

$$-4=4y$$

$$-1=y$$

The solution $y=-1$ makes the denominators $2y+2$, $4y+4$, and $y+1$ zero, so the equation has no solution.

39. $r^2+5r-14=(r+7)(r-2)$

The LCD is $(r+7)(r-2)$.

$$\frac{4r-4}{r^2+5r-14}+\frac{2}{r+7}=\frac{1}{r-2}$$

$$(r+7)(r-2)\left(\frac{4r-4}{(r+7)(r-2)}+\frac{2}{r+7}\right)=(r+7)(r-2)\left(\frac{1}{r-2}\right)$$

$$4r-4+2(r-2)=1(r+7)$$

$$4r-4+2r-4=r+7$$

$$6r-8=r+7$$

$$5r=15$$

$$r=3$$

The solution $r=3$ checks.

Copyright © 2012 Pearson Education, Inc. Publishing as Prentice Hall.

41. $x^2+x-6=(x+3)(x-2)$

The LCD is $(x+3)(x-2)$.

$$\frac{x+1}{x+3}=\frac{x^2-11x}{x^2+x-6}-\frac{x-3}{x-2}$$

$$(x+3)(x-2)\left(\frac{x+1}{x+3}\right)=(x+3)(x-2)\left(\frac{x^2-11x}{x^2+x-6}-\frac{x-3}{x-2}\right)$$

$$(x-2)(x+1)=x^2-11x-(x+3)(x-3)$$

$$x^2-x-2=x^2-11x-(x^2-9)$$

$$x^2-x-2=x^2-11x-x^2+9$$

$$x^2-x-2=-11x+9$$

$$x^2+10x-11=0$$

$$(x+11)(x-1)=0$$

$x+11=0$ or $x-1=0$

$x=-11$ $\quad$ $x=1$

The solutions $x=-11$ and $x=1$ check.

43. The LCD is I.

$$R=\frac{E}{I}$$

$$I(R)=I\left(\frac{E}{I}\right)$$

$$IR=E$$

$$I=\frac{E}{R}$$

45. The LCD is $B+E$.

$$T=\frac{2U}{B+E}$$

$$(B+E)T=(B+E)\left(\frac{2U}{B+E}\right)$$

$$BT+ET=2U$$

$$BT=2U-ET$$

$$B=\frac{2U-ET}{T}$$

47. The LCD is h^2.

$$B=\frac{705w}{h^2}$$

$$h^2(B)=h^2\left(\frac{705w}{h^2}\right)$$

$$Bh^2=705w$$

$$\frac{Bh^2}{705}=w$$

Copyright © 2012 Pearson Education, Inc. Publishing as Prentice Hall.

49. The LCD is G.

$$N = R + \frac{V}{G}$$
$$G(N) = G\left(R + \frac{V}{G}\right)$$
$$GN = GR + V$$
$$GN - GR = V$$
$$G(N - R) = V$$
$$G = \frac{V}{N - R}$$

51. The LCD is πr.

$$\frac{C}{\pi r} = 2$$
$$\pi r\left(\frac{C}{\pi r}\right) = \pi r(2)$$
$$C = 2\pi r$$
$$\frac{C}{2\pi} = r$$

53. The LCD is $3xy$.

$$\frac{1}{y} + \frac{1}{3} = \frac{1}{x}$$
$$3xy\left(\frac{1}{y} + \frac{1}{3}\right) = 3xy\left(\frac{1}{x}\right)$$
$$3x + xy = 3y$$
$$x(3 + y) = 3y$$
$$x = \frac{3y}{3 + y}$$

55. The reciprocal of x is $\frac{1}{x}$.

57. The reciprocal of x, added to the reciprocal of 2 is $\frac{1}{x} + \frac{1}{2}$.

59. $\frac{1}{3}$ of the tank is filled in 1 hour.

61. $a^2 + 4a + 3 = (a + 3)(a + 1)$

$a^2 + a - 6 = (a + 3)(a - 2)$

$a^2 - a - 2 = (a + 1)(a - 2)$

The LCD is $(a + 3)(a + 1)(a - 2)$.

Copyright © 2012 Pearson Education, Inc. Publishing as Prentice Hall.

$$\frac{4}{a^2+4a+3}+\frac{2}{a^2+a-6}-\frac{3}{a^2-a-2}=0$$

$$(a+3)(a+1)(a-2)\left(\frac{4}{a^2+4a+3}+\frac{2}{a^2+a-6}-\frac{3}{a^2-a-2}\right)=(a+3)(a+1)(a-2)(0)$$

$$4(a-2)+2(a+1)-3(a+3)=0$$

$$4a-8+2a+2-3a-9=0$$

$$3a-15=0$$

$$3a=15$$

$$a=5$$

The solution $a = 5$ checks.

63. $\frac{20x}{3}+\frac{32x}{6}=180$

The LCD is 6.

$$6\left(\frac{20x}{3}+\frac{32x}{6}\right)=6(180)$$

$$2(20x)+32x=1080$$

$$40x+32x=1080$$

$$72x=1080$$

$$x=15$$

$$\frac{20x}{3}=\frac{20(15)}{3}=100°$$

$$\frac{32x}{6}=\frac{32(15)}{6}=80°$$

The angles measure 100° and 80°.

65. $\frac{450}{x}+\frac{150}{x}=90$

The LCD is x.

$$x\left(\frac{450}{x}+\frac{150}{x}\right)=x(90)$$

$$450+150=90x$$

$$600=90x$$

$$\frac{600}{90}=x$$

$$\frac{20}{3}=x$$

$$\frac{450}{x}=450\div\frac{20}{3}=\frac{450}{1}\cdot\frac{3}{20}=\frac{1350}{20}=67.5$$

$$\frac{150}{x}=150\div\frac{20}{3}=\frac{150}{1}\cdot\frac{3}{20}=\frac{450}{20}=22.5$$

The angles measure 22.5° and 67.5°.

67. No; multiplying both terms in the expressions by 4 changes the value of the original expressions.

Copyright © 2012 Pearson Education, Inc. Publishing as Prentice Hall.

Integrated Review

1. $\frac{1}{x}+\frac{2}{3}=\frac{1\cdot 3}{x\cdot 3}+\frac{2\cdot x}{3\cdot x}=\frac{3}{3x}+\frac{2x}{3x}=\frac{3+2x}{3x}$

This is an expression.

2. $\frac{3}{a}+\frac{5}{6}=\frac{3\cdot 6}{a\cdot 6}+\frac{5\cdot a}{6\cdot a}=\frac{18}{6a}+\frac{5a}{6a}=\frac{18+5a}{6a}$

This is an expression.

3.
$$\frac{1}{x}+\frac{2}{3}=\frac{3}{x}$$
$$3x\left(\frac{1}{x}+\frac{2}{3}\right)=3x\left(\frac{3}{x}\right)$$
$$3+2x=9$$
$$2x=6$$
$$x=3$$

This is an equation.

4.
$$\frac{3}{a}+\frac{5}{6}=1$$
$$6a\left(\frac{3}{a}+\frac{5}{6}\right)=6a(1)$$
$$18+5a=6a$$
$$18=a$$

This is an equation.

5.
$$\frac{2}{x+1}-\frac{1}{x}=\frac{2\cdot x}{(x+1)\cdot x}-\frac{1\cdot(x+1)}{x\cdot(x+1)}$$
$$=\frac{2x}{x(x+1)}-\frac{x+1}{x(x+1)}$$
$$=\frac{2x-x-1}{x(x+1)}$$
$$=\frac{x-1}{x(x+1)}$$

This is an expression.

6.
$$\frac{4}{x-3}-\frac{1}{x}=\frac{4\cdot x}{(x-3)\cdot x}-\frac{1\cdot(x-3)}{x\cdot(x-3)}$$
$$=\frac{4x}{x(x-3)}-\frac{x-3}{x(x-3)}$$
$$=\frac{4x-x+3}{x(x-3)}$$
$$=\frac{3x+3}{x(x-3)}$$
$$=\frac{3(x+1)}{x(x-3)}$$

This is an expression.

7.
$$\frac{2}{x+1}-\frac{1}{x}=1$$
$$x(x+1)\left(\frac{2}{x+1}-\frac{1}{x}\right)=x(x+1)(1)$$
$$2x-(x+1)=x(x+1)$$
$$2x-x-1=x^2+x$$
$$x-1=x^2+x$$
$$0=x^2+1$$

There are no solutions. This is an equation.

8.
$$\frac{4}{x-3}-\frac{1}{x}=\frac{6}{x(x-3)}$$
$$x(x-3)\left(\frac{4}{x-3}-\frac{1}{x}\right)=x(x-3)\left(\frac{6}{x(x-3)}\right)$$
$$4x-1(x-3)=6$$
$$4x-x+3=6$$
$$3x=3$$
$$x=1$$

This is an equation.

9. $\frac{15x}{x+8}\cdot\frac{2x+16}{3x}=\frac{3\cdot 5x}{x+8}\cdot\frac{2(x+8)}{3x}=5\cdot 2=10$

This is an expression.

10.
$$\frac{9z+5}{15}\cdot\frac{5z}{81z^2-25}=\frac{9z+5}{3\cdot 5}\cdot\frac{5z}{(9z+5)(9z-5)}$$
$$=\frac{z}{3(9z-5)}$$

This is an expression.

11. $\frac{2x+3}{x-3}+\frac{3x+6}{x-3}=\frac{2x+1+3x+6}{x-3}=\frac{5x+7}{x-3}$

This is an expression.

12. $\frac{4p-3}{2p+7}+\frac{3p+8}{2p+7}=\frac{4p-3+3p+8}{2p+7}=\frac{7p+5}{2p+7}$

This is an expression.

13.
$$\frac{x+5}{7}=\frac{8}{2}$$
$$14\left(\frac{x+5}{7}\right)=14\left(\frac{8}{2}\right)$$
$$2(x+5)=7(8)$$
$$2x+10=56$$
$$2x=46$$
$$x=23$$

This is an equation.

Copyright © 2012 Pearson Education, Inc. Publishing as Prentice Hall.

14.
$$\frac{1}{2}=\frac{x+1}{8}$$
$$8\left(\frac{1}{2}\right)=8\left(\frac{x+1}{8}\right)$$
$$4=x+1$$
$$3=x$$
This is an equation.

15.
$$\frac{5a+10}{18}\div\frac{a^2-4}{10a}=\frac{5a+10}{18}\cdot\frac{10a}{a^2-4}$$
$$=\frac{5(a+2)}{2\cdot 9}\cdot\frac{2\cdot 5a}{(a+2)(a-2)}$$
$$=\frac{5\cdot 5a}{9(a-2)}$$
$$=\frac{25a}{9(a-2)}$$
This is an expression.

16.
$$\frac{9}{x^2-1}\div\frac{12}{3x+3}=\frac{9}{x^2-1}\cdot\frac{3x+3}{12}$$
$$=\frac{9}{(x-1)(x+1)}\cdot\frac{3(x+1)}{3\cdot 4}$$
$$=\frac{9}{4(x-1)}$$

17.
$$\frac{x+2}{3x-1}+\frac{5}{(3x-1)^2}=\frac{(x+2)(3x-1)}{(3x-1)(3x-1)}+\frac{5}{(3x-1)^2}$$
$$=\frac{(x+2)(3x-1)+5}{(3x-1)^2}$$
$$=\frac{3x^2-x+6x-2+5}{(3x-1)^2}$$
$$=\frac{3x^2+5x+3}{(3x-1)^2}$$
This is an expression.

18.
$$\frac{4}{(2x-5)^2}+\frac{x+1}{2x-5}=\frac{4}{(2x-5)^2}+\frac{(x+1)(2x-5)}{(2x-5)(2x-5)}$$
$$=\frac{4+(x+1)(2x-5)}{(2x-5)^2}$$
$$=\frac{4+2x^2-5x+2x-5}{(2x-5)^2}$$
$$=\frac{2x^2-3x-1}{(2x-5)^2}$$
This is an expression.

19.
$$\frac{x-7}{x}-\frac{x+2}{5x}=\frac{(x-7)\cdot 5}{x\cdot 5}-\frac{x+2}{5x}$$
$$=\frac{5x-35-x-2}{5x}$$
$$=\frac{4x-37}{5x}$$
This is an expression.

20.
$$\frac{10x-9}{x}-\frac{x-4}{3x}=\frac{(10x-9)\cdot 3}{x\cdot 3}-\frac{x-4}{3x}$$
$$=\frac{3(10x-9)-x+4}{3x}$$
$$=\frac{30x-27-x+4}{3x}$$
$$=\frac{29x-23}{3x}$$
This is an expression.

21.
$$\frac{3}{x+3}=\frac{5}{x^2-9}-\frac{2}{x-3}$$
$$(x^2-9)\left(\frac{3}{x+3}\right)=(x^2-9)\left(\frac{5}{x^2-9}-\frac{2}{x-3}\right)$$
$$3(x-3)=5-2(x+3)$$
$$3x-9=5-2x-6$$
$$3x-9=-2x-1$$
$$5x=8$$
$$x=\frac{8}{5}$$
This is an equation.

22.
$$\frac{9}{x^2-4}+\frac{2}{x+2}=\frac{-1}{x-2}$$
$$(x^2-4)\left(\frac{9}{x^2-4}+\frac{2}{x+2}\right)=(x^2-4)\left(\frac{-1}{x-2}\right)$$
$$9+2(x-2)=-1(x+2)$$
$$9+2x-4=-x-2$$
$$2x+5=-x-2$$
$$3x=-7$$
$$x=-\frac{7}{3}$$
This is an equation.

23. answers may vary

24. answers may vary

Copyright © 2012 Pearson Education, Inc. Publishing as Prentice Hall.

Section 5.6 Practice

1.
$$\frac{3}{8}=\frac{63}{x}$$
$$3\cdot x=63\cdot 8$$
$$3x=504$$
$$\frac{3x}{3}=\frac{504}{3}$$
$$x=168$$

2.
$$\frac{2x+1}{7}=\frac{x-3}{5}$$
$$5(2x+1)=7(x-3)$$
$$10x+5=7x-21$$
$$3x+5=-21$$
$$3x=-26$$
$$\frac{3x}{3}=\frac{-26}{3}$$
$$x=-\frac{26}{3}$$

3.
$$\frac{26}{250}=\frac{x}{50{,}000}$$
$$26(50{,}000)=250x$$
$$1{,}300{,}000=250x$$
$$5200=x$$
5200 people are expected to have a flu shot.

4.
$$\frac{12}{x}=\frac{9}{15}$$
$$12\cdot 15=9\cdot x$$
$$180=9x$$
$$20=x$$
The missing length is 20 units.

5.
$$\frac{x}{2}-\frac{1}{3}=\frac{x}{6}$$
$$6\left(\frac{x}{2}-\frac{1}{3}\right)=6\left(\frac{x}{6}\right)$$
$$6\left(\frac{x}{2}\right)-6\left(\frac{1}{3}\right)=6\left(\frac{x}{6}\right)$$
$$3x-2=x$$
$$-2=-2x$$
$$\frac{-2}{-2}=\frac{-2x}{-2}$$
$$1=x$$
The number is 1.

Copyright © 2012 Pearson Education, Inc. Publishing as Prentice Hall.

6.

	Hours to Complete Total Job	Part of Job Completed in 1 Hour
Andrew	2	$\frac{1}{2}$
Tim	3	$\frac{1}{3}$
Together	x	$\frac{1}{x}$

$$\frac{1}{2}+\frac{1}{3}=\frac{1}{x}$$
$$6x\left(\frac{1}{2}\right)+6x\left(\frac{1}{3}\right)=6x\left(\frac{1}{x}\right)$$
$$3x+2x=6$$
$$5x=6$$
$$x=\frac{6}{5}=1\frac{1}{5}$$

Together they can sort one batch in $1\frac{1}{5}$ hours.

7. Let x be the speed of the motorcycle. Then the car's speed is $x + 15$.

	distance	= **rate**	· **time**
Car	600	$x + 15$	$\frac{600}{x+15}$
Motorcycle	450	x	$\frac{450}{x}$

$$\frac{600}{x+15}=\frac{450}{x}$$
$$600x=450(x+15)$$
$$600x=450x+6750$$
$$150x=6750$$
$$x=45$$
$$x + 15 = 45 + 15 = 60$$

The speed of the motorcycle is 45 mph. The speed of the car is 60 mph.

Vocabulary and Readiness Check

1. The time to complete the job working together will be less than both of the individual times. The answer is c.

2. The time to fill the pond with both pipes on at the same time will be less than both of the individual times. The answer is a.

3. A number: x

The reciprocal of the number: $\frac{1}{x}$

The reciprocal of the number, decreased by 3: $\frac{1}{x}-3$

Copyright © 2012 Pearson Education, Inc. Publishing as Prentice Hall.

4. A number: y

The reciprocal of the number: $\frac{1}{y}$

The reciprocal of the number, increased by 2:

$\frac{1}{y}+2$

5. A number: z
The sum of the number and 5: $z + 5$
The reciprocal of the sum of the number and 5:

$\frac{1}{z+5}$

6. A number: x
The difference of the number and 1: $x - 1$
The reciprocal of the difference of the number and 1: $\frac{1}{x-1}$

7. A number: y
Twice the number: $2y$

Eleven divided by twice the number: $\frac{11}{2y}$

8. A number: z
Triple the number: $3z$

Negative 10 divided by triple the number: $\frac{-10}{3y}$

Exercise Set 5.6

1. The LCD is 6.

$$\frac{2}{3}=\frac{x}{6}$$
$$6\left(\frac{2}{3}\right)=6\left(\frac{x}{6}\right)$$
$$4=x$$

3. $\frac{x}{10}=\frac{5}{9}$
$$9x=50$$
$$x=\frac{50}{9}$$

5. $\frac{x+1}{2x+3}=\frac{2}{3}$
$$3(x+1)=2(2x+3)$$
$$3x+3=4x+6$$
$$3=x+6$$
$$-3=x$$

7. $\frac{9}{5}=\frac{12}{3x+2}$
$$9(3x+2)=12\cdot 5$$
$$27x+18=60$$
$$27x=42$$
$$x=\frac{42}{27}=\frac{14}{9}$$

9. Let x be the elephant's weight on Pluto.
$$\frac{100}{3}=\frac{4100}{x}$$
$$100x=3(4100)$$
$$100x=12{,}300$$
$$x=123$$
The elephant weighs 123 pounds on Pluto.

11. Let y be the number of calories in 42.6 grams.
$$\frac{110}{28.4}=\frac{y}{42.6}$$
$$42.6(110)=28.4y$$
$$4686=28.4y$$
$$165=y$$
There are 165 calories in 42.6 grams of Crispy Rice.

13. $\frac{16}{10}=\frac{34}{y}$
$$16y=34(10)$$
$$16y=340$$
$$y=\frac{340}{16}$$
$$y=21.25$$

15. $\frac{y}{8}=\frac{20}{28}$
$$28y=20(8)$$
$$28y=160$$
$$y=\frac{160}{28}=\frac{40}{7}=5\frac{5}{7}\text{ ft}$$

17. Let x be the number.
$$3\left(\frac{1}{x}\right)=9\left(\frac{1}{6}\right)$$
$$\frac{3}{x}=\frac{9}{6}$$
$$3(6)=9x$$
$$18=9x$$
$$2=x$$
The number is 2.

Copyright © 2012 Pearson Education, Inc. Publishing as Prentice Hall.

19. Let x be the number.

$$\frac{2x+3}{x+1}=\frac{3}{2}$$
$$2(2x+3)=3(x+1)$$
$$4x+6=3x+3$$
$$x=-3$$

The number is -3.

21. Let x be the time in hours that it takes them to complete the job working together.

The experienced surveyor completes $\frac{1}{4}$ of the job in 1 hour. The apprentice surveyor completes $\frac{1}{5}$ of the job in 1 hour. Together, they complete $\frac{1}{x}$ of the job in 1 hour.

$$\frac{1}{4}+\frac{1}{5}=\frac{1}{x}$$

The LCD is $4 \cdot 5 \cdot x = 20x$.

$$20x\left(\frac{1}{4}+\frac{1}{5}\right)=20x\left(\frac{1}{x}\right)$$
$$5x+4x=20$$
$$9x=20$$
$$x=\frac{20}{9}$$

It takes them $\frac{20}{9}=2\frac{2}{9}$ hours to survey the roadbed together.

23. Let x be the time in minutes that it takes the two belts to complete the job working together. The first belt completes $\frac{1}{2}$ of the job in 1 minute. The smaller belt completes $\frac{1}{6}$ of the job in 1 minute. Together, they complete $\frac{1}{x}$ of the job in 1 minute.

$$\frac{1}{2}+\frac{1}{6}=\frac{1}{x}$$

The LCD is $6x$.

$$6x\left(\frac{1}{2}+\frac{1}{6}\right)=6x\left(\frac{1}{x}\right)$$
$$3x+x=6$$
$$4x=6$$
$$x=\frac{6}{4}=\frac{3}{2}=1\frac{1}{2}$$

It will take $1\frac{1}{2}$ minutes to move the cans to the storage when both belts are used.

Copyright © 2012 Pearson Education, Inc. Publishing as Prentice Hall.

25. Let r be her jogging rate.

	distance	=	rate	·	time
Trip to park	12		r		$\frac{12}{r}$
Return trip	18		r		$\frac{18}{r}$

Since the return trip took 1 hour longer, $\frac{18}{r}=1+\frac{12}{r}$. The LCD is r.

$$r\left(\frac{18}{r}\right)=r\left(1+\frac{12}{r}\right)$$
$$18=r+12$$
$$6=r$$

Her jogging rate is 6 miles per hour.

27. Let x be the speed for the first portion.

	distance	=	rate	·	time
1st portion	20		x		$\frac{20}{x}$
2nd portion	16		$x-2$		$\frac{16}{x-2}$

$$\frac{20}{x}=\frac{16}{x-2}$$

The LCD is $x(x-2)$.

$$x(x-2)\left(\frac{20}{x}\right)=x(x-2)\left(\frac{16}{x-2}\right)$$
$$20(x-2)=16x$$
$$20x-40=16x$$
$$4x=40$$
$$x=10$$

The cyclist's speed for first portion is 10 mph and the speed for the second portion is 8 mph.

29. $40 \text{ students}\cdot\frac{9 \text{ square feet}}{1 \text{ student}}=360 \text{ square feet}$

40 students need a minimum of 360 square feet.

31. Let n be the number.

$$\frac{1}{4}=\frac{n}{8}$$
$$8\left(\frac{1}{4}\right)=8\left(\frac{n}{8}\right)$$
$$2=n$$

The number is 2.

Copyright © 2012 Pearson Education, Inc. Publishing as Prentice Hall.

33. Let x be the time in hours that it takes Marcus and Tony to do the job working together.

Marcus lays $\frac{1}{6}$ of a slab in 1 hour. Tony lays $\frac{1}{4}$ of a slab in 1 hour. Together, they lay $\frac{1}{x}$ of the slab in 1 hour.

$$\frac{1}{6}+\frac{1}{4}=\frac{1}{x}$$

The LCD is $12x$.

$$12x\left(\frac{1}{6}+\frac{1}{4}\right)=12x\left(\frac{1}{x}\right)$$
$$2x+3x=12$$
$$5x=12$$
$$x=\frac{12}{5}$$

It will take Tony and Marcus $\frac{12}{5}$ hours to lay the slab, so the labor estimate should be $\frac{12}{5}(\$45)=\$108$.

35. Let w be the speed of the wind.

	distance	= **rate**	· **time**
With wind	400	$230+w$	$\frac{400}{230+w}$
Against wind	336	$230-w$	$\frac{336}{230-w}$

$$\frac{400}{230+w}=\frac{336}{230-w}$$
$$400(230-w)=336(230+w)$$
$$92{,}000-400w=77{,}280+336w$$
$$14{,}720=736w$$
$$20=w$$

The speed of the wind is 20 mph.

37. $\frac{2}{3}=\frac{25}{y}$

$$2y=3(25)$$
$$2y=75$$
$$y=\frac{75}{2}$$

The unknown length is $y=\frac{75}{2}=37.5$ feet.

39. Let x be the speed of the slower train. In 3.5 hours, the slower train travels $3.5x$ miles and the faster train travels $3.5(x+10)$ miles.

$$3.5x+3.5(x+10)=322$$
$$3.5x+3.5x+35=322$$
$$7x+35=322$$
$$7x=287$$
$$x=41$$

The slower train travels 41 mph and the faster train travels 41 + 10 = 51 mph.

Copyright © 2012 Pearson Education, Inc. Publishing as Prentice Hall.

41. Let x be the number.

$$\frac{2}{x-3}-\frac{4}{x+3}=8\cdot\frac{1}{x^2-9}$$

The LCD is $x^2-9=(x+3)(x-3)$.

$$(x+3)(x-3)\left(\frac{2}{x-3}-\frac{4}{x+3}\right)=(x^2-9)\left(\frac{8}{x^2-9}\right)$$

$$\begin{aligned}2(x+3)-4(x-3)&=8\\2x+6-4x+12&=8\\-2x+18&=8\\-2x&=-10\\x&=5\end{aligned}$$

The solution $x = 5$ checks, so the number is 5.

43. Let x be the rate in still air.

	distance	=	**rate**	·	**time**
With wind	630		$x+35$		$\frac{630}{x+35}$
Against wind	455		$x-35$		$\frac{455}{x-35}$

$$\begin{aligned}\frac{630}{x+35}&=\frac{455}{x-35}\\630(x-35)&=455(x+35)\\630x-22{,}050&=455x+15{,}925\\175x&=37{,}975\\x&=217\end{aligned}$$

The plane flies at a rate of 217 mph in still air.

45. Let x be the number of gallons of water needed.

$$\begin{aligned}\frac{8\text{ tsp}}{2\text{ gal}}&=\frac{36\text{ tsp}}{x\text{ gal}}\\\frac{8}{2}&=\frac{36}{x}\\8x&=36(2)\\8x&=72\\x&=9\end{aligned}$$

9 gallons of water are needed to mix with a box of weed killer.

47. Let x be the rate of the wind.

	distance	=	**rate**	·	**time**
with wind	48		$16+x$		$\frac{48}{16+x}$
into wind	16		$16-x$		$\frac{16}{16-x}$

Copyright © 2012 Pearson Education, Inc. Publishing as Prentice Hall.

$$\frac{48}{16+x}=\frac{16}{16-x}$$
$$42(16-x)=16(16+x)$$
$$768-48x=256+16x$$
$$512=64x$$
$$8=x$$

The rate of the wind is 8 mph.

49. Let x be the rate of the slower hiker. Then the rate of the faster hiker is $x + 1.1$. In 2 hours, the slower hiker walks $2x$ miles, while the faster hiker walks $2(x + 1.1)$ miles.

$$2x+2(x+1.1)=11$$
$$2x+2x+2.2=11$$
$$4x+2.2=11$$
$$4x=8.8$$
$$x=2.2$$

$x + 1.1 = 2.2 + 1.1 = 3.3$

The hikers walk 2.2 miles per hour and 3.3 miles per hour.

51. Let x be the time it takes for the second worker to do the same job alone.

$$\frac{1}{3}+\frac{1}{x}=\frac{1}{\frac{3}{2}}$$
$$\frac{1}{3}+\frac{1}{x}=\frac{2}{3}$$
$$3x\left(\frac{1}{3}+\frac{1}{x}\right)=3x\left(\frac{2}{3}\right)$$
$$x+3=2x$$
$$3=x$$

It will take the second worker 3 hours to get the job done.

53. $$\frac{20 \text{ feet}}{6 \text{ inches}}=\frac{x \text{ feet}}{8 \text{ inches}}$$
$$\frac{20}{6}=\frac{x}{8}$$
$$8(20)=6x$$
$$160=6x$$
$$\frac{160}{6}=x$$
$$\frac{80}{3}=x$$

The missing dimension is $\frac{80}{3}=26\frac{2}{3}$ feet.

55. Let x be the number of other nuts.

$$\frac{3}{2}=\frac{324}{x}$$
$$3x=2(324)$$
$$3x=648$$
$$x=216$$

There should be 216 other nuts in the can.

Copyright © 2012 Pearson Education, Inc. Publishing as Prentice Hall.

57. Let t be the time in hours that the jet plane travels.

	distance	=	rate	·	time
jet plane	$500t$		500		t
propeller plane	$200(t+2)$		200		$t+2$

$$500t = 200(t+2)$$
$$500t = 200t + 400$$
$$300t = 400$$
$$t = \frac{400}{300}$$
$$t = \frac{4}{3}$$
$$\text{distance} = 500t = 500\left(\frac{4}{3}\right) = 666\frac{2}{3}$$

The planes are $666\frac{2}{3}$ miles from the starting point.

59. Let x be the time that it takes the third pipe to fill the pool alone.

$$\frac{1}{20} + \frac{1}{15} + \frac{1}{x} = \frac{1}{6}$$
$$60x\left(\frac{1}{20} + \frac{1}{15} + \frac{1}{x}\right) = 60x\left(\frac{1}{6}\right)$$
$$3x + 4x + 60 = 10x$$
$$7x + 60 = 10x$$
$$60 = 3x$$
$$20 = x$$

It will take the third pump 20 hours to do the job alone.

61. Let r be the motorcycle's speed.

	distance	=	rate	·	time
car	280		$r+10$		$\frac{280}{r+10}$
motorcycle	240		r		$\frac{240}{r}$

$$\frac{280}{r+10} = \frac{240}{r}$$
$$280r = 240(r+10)$$
$$280r = 240r + 2400$$
$$40r = 2400$$
$$r = 60$$
$$r + 10 = 60 + 10 = 70$$

The motorcycle's speed was 60 miles per hour and the car's speed was 70 miles per hour.

Copyright © 2012 Pearson Education, Inc. Publishing as Prentice Hall.

63. Let x be the time for the third cook to prepare the same number of pies.

$$\frac{1}{6}+\frac{1}{7}+\frac{1}{x}=\frac{1}{2}$$
$$42x\left(\frac{1}{6}+\frac{1}{7}+\frac{1}{x}\right)=42x\left(\frac{1}{2}\right)$$
$$7x+6x+42=21x$$
$$13x+42=21x$$
$$42=8x$$
$$\frac{42}{8}=x$$
$$\frac{42}{8}=\frac{21}{4}=5\frac{1}{4}$$

It will take the third cook $5\frac{1}{4}$ hours to prepare the pies working alone.

65. Let x be the number.

$$\frac{x}{3}-1=\frac{5}{3}$$
$$3\left(\frac{x}{3}-1\right)=3\left(\frac{5}{3}\right)$$
$$x-3=5$$
$$x=8$$

The number is 8.

67. Let x be the speed of the second car.

	distance	= rate	· time
one car	224	$x+14$	$\frac{224}{x+14}$
second car	175	x	$\frac{175}{x}$

$$\frac{224}{x+14}=\frac{175}{x}$$
$$224x=175(x+14)$$
$$224x=175x+2450$$
$$49x=2450$$
$$x=50$$
$$x+14=50+14=64$$

The speed of one car is 64 miles per hour and the speed of the second car is 50 miles per hour.

69. Let x be the speed of the plane in still air.

	distance	= rate	· time
with wind	2160	$x+30$	$\frac{2160}{x+30}$
against wind	1920	$x-30$	$\frac{1920}{x-30}$

$$\frac{2160}{x+30}=\frac{1920}{x-30}$$
$$2160(x-30)=1920(x+30)$$
$$2160x-64{,}800=1920x+57{,}600$$
$$240x=122{,}400$$
$$x=510$$

The speed of the plane in still air is 510 miles per hour.

71. $\frac{x}{3.75}=\frac{12}{9}$

$$9x=12(3.75)$$
$$9x=45$$
$$x=5$$

The missing length is $x = 5$.

73. $\frac{x}{9}=\frac{24}{16}$

$$16x=9(24)$$
$$16x=216$$
$$x=\frac{216}{16}$$
$$x=\frac{27}{2}$$

The missing length is $x=\frac{27}{2}=13.5$.

75. $\dfrac{\frac{3}{4}+\frac{1}{4}}{\frac{3}{8}+\frac{13}{8}}=\dfrac{\frac{4}{4}}{\frac{16}{8}}=\dfrac{1}{2}$

77. $\dfrac{\frac{2}{5}+\frac{1}{5}}{\frac{7}{10}+\frac{7}{10}}=\dfrac{\frac{3}{5}}{\frac{14}{10}}=\dfrac{3}{5}\div\dfrac{14}{10}=\dfrac{3}{5}\cdot\dfrac{10}{14}=\dfrac{3\cdot2\cdot5}{5\cdot2\cdot7}=\dfrac{3}{7}$

79. Let x be the time in minutes that it takes for the faster pump to fill the tank, so it fills $\frac{1}{x}$ of the tank in 1 minute. It takes the slower pump $3x$ minutes to fill the tank, so the slower pump fills $\frac{1}{3x}$ of the tank in 1 minute. Together, the pumps fill $\frac{1}{21}$ of the tank in 1 minute.

$$\frac{1}{x}+\frac{1}{3x}=\frac{1}{21}$$
$$21x\left(\frac{1}{x}+\frac{1}{3x}\right)=21x\left(\frac{1}{21}\right)$$
$$21+7=x$$
$$28=x$$

Copyright © 2012 Pearson Education, Inc. Publishing as Prentice Hall.

$3x = 3(28) = 84$
The faster pump fills the tank in 28 minutes, while the slower pump takes 84 minutes.

81. answers may vary

83. $D = RT$

$$\frac{D}{T} = \frac{RT}{T}$$

$$\frac{D}{T} = R \text{ or } R = \frac{D}{T}$$

85. Let t be the time it takes for the hyena to overtake the giraffe.

$$0.5 + 32t = 40t$$
$$0.5 = 8t$$
$$0.0625 = t$$
$$0.0625 \text{ hr} \cdot \frac{60 \text{ min}}{1 \text{ hr}} = 3.75 \text{ min}$$

It will take the hyena 3.75 minutes to overtake the giraffe.

Section 5.7 Practice

1. $\dfrac{\frac{3}{7}}{\frac{5}{9}} = \dfrac{3}{7} \div \dfrac{5}{9} = \dfrac{3}{7} \cdot \dfrac{9}{5} = \dfrac{27}{35}$

2.
$$\frac{\frac{3}{4} - \frac{2}{3}}{\frac{1}{2} + \frac{3}{8}} = \frac{\frac{3(3)}{4(3)} - \frac{2(4)}{3(4)}}{\frac{1(4)}{2(4)} + \frac{3}{8}}$$
$$= \frac{\frac{9}{12} - \frac{8}{12}}{\frac{4}{8} + \frac{3}{8}}$$
$$= \frac{\frac{1}{12}}{\frac{7}{8}}$$
$$= \frac{1}{12} \cdot \frac{8}{7}$$
$$= \frac{1 \cdot 2 \cdot 4}{3 \cdot 4 \cdot 7}$$
$$= \frac{2}{21}$$

3.
$$\frac{\frac{2}{5} - \frac{1}{x}}{\frac{2x}{15} - \frac{1}{3}} = \frac{\frac{2(x)}{5(x)} - \frac{1(5)}{x(5)}}{\frac{2x}{15} - \frac{1(5)}{3(5)}}$$
$$= \frac{\frac{2x}{5x} - \frac{5}{5x}}{\frac{2x}{15} - \frac{5}{15}}$$
$$= \frac{\frac{2x-5}{5x}}{\frac{2x-5}{15}}$$
$$= \frac{2x-5}{5x} \cdot \frac{15}{2x-5}$$
$$= \frac{2x-15}{5x} \cdot \frac{3 \cdot 5}{2x-15}$$
$$= \frac{3}{x}$$

4. The LCD is 24.

$$\frac{\frac{3}{4} - \frac{2}{3}}{\frac{1}{2} + \frac{3}{8}} = \frac{24\left(\frac{3}{4} - \frac{2}{3}\right)}{24\left(\frac{1}{2} + \frac{3}{8}\right)}$$
$$= \frac{24\left(\frac{3}{4}\right) - 24\left(\frac{2}{3}\right)}{24\left(\frac{1}{2}\right) + 24\left(\frac{3}{8}\right)}$$
$$= \frac{18-16}{12+9}$$
$$= \frac{2}{21}$$

5. The LCD is y.

$$\frac{1 + \frac{x}{y}}{\frac{2x+1}{y}} = \frac{y\left(1 + \frac{x}{y}\right)}{y\left(\frac{2x+1}{y}\right)} = \frac{y(1) + y\left(\frac{x}{y}\right)}{y\left(\frac{2x+1}{y}\right)} = \frac{y+x}{2x+1}$$

6. The LCD is $6xy$.

$$\frac{\frac{5}{6y} + \frac{y}{x}}{\frac{y}{3} - x} = \frac{6xy\left(\frac{5}{6y} + \frac{y}{x}\right)}{6xy\left(\frac{y}{3} - x\right)}$$
$$= \frac{6xy\left(\frac{5}{6y}\right) + 6xy\left(\frac{y}{x}\right)}{6xy\left(\frac{y}{3}\right) - 6xy(x)}$$
$$= \frac{5x + 6y^2}{2xy^2 - 6x^2y}$$
$$= \frac{5x + 6y^2}{2xy(y - 3x)}$$

Copyright © 2012 Pearson Education, Inc. Publishing as Prentice Hall.

Vocabulary and Readiness Check

1. The LCD for $\frac{1}{4}=\frac{1}{2\cdot 2}$, $\frac{1}{2}$, $\frac{1}{3}$, and $\frac{1}{2}$ is $2\cdot 2\cdot 3=12$; c.

2. The LCD for $\frac{3}{5}$, $\frac{2}{3}$, $\frac{1}{10}=\frac{1}{2\cdot 5}$, and $\frac{1}{6}=\frac{1}{2\cdot 3}$ is $2\cdot 3\cdot 5=30$; b.

3. The LCD for $\frac{5}{2x^2}$, $\frac{3}{16x}=\frac{3}{2\cdot 2\cdot 2\cdot 2\cdot x}$, $\frac{x}{8}=\frac{x}{2\cdot 2\cdot 2}$, and $\frac{3}{4x}=\frac{3}{2\cdot 2\cdot x}$ is $2\cdot 2\cdot 2\cdot 2\cdot x^2=16x^2$; a.

4. The LCD for $\frac{11}{6}=\frac{11}{2\cdot 3}$, $\frac{10}{x^2}$, $\frac{7}{9}=\frac{7}{3\cdot 3}$, and $\frac{5}{x}$ is $2\cdot 3\cdot 3\cdot x^2=18x^2$; c.

Exercise Set 5.7

1. $\dfrac{\frac{1}{2}}{\frac{3}{4}}=\dfrac{1}{2}\div\dfrac{3}{4}=\dfrac{1}{2}\cdot\dfrac{4}{3}=\dfrac{1\cdot 4}{2\cdot 3}=\dfrac{2}{3}$

3. $\dfrac{-\frac{4x}{9}}{-\frac{2x}{3}}=-\dfrac{4x}{9}\div-\dfrac{2x}{3}=-\dfrac{4x}{9}\cdot\dfrac{3}{-2x}=\dfrac{4x\cdot 3}{9\cdot 2x}=\dfrac{2}{3}$

5. $\dfrac{\frac{1+x}{6}}{\frac{1+x}{3}}=\dfrac{1+x}{6}\div\dfrac{1+x}{3}=\dfrac{1+x}{6}\cdot\dfrac{3}{1+x}=\dfrac{3(1+x)}{6(1+x)}=\dfrac{1}{2}$

7. $\dfrac{\frac{1}{2}+\frac{2}{3}}{\frac{5}{9}-\frac{5}{6}}=\dfrac{18\left(\frac{1}{2}+\frac{2}{3}\right)}{18\left(\frac{5}{9}-\frac{5}{6}\right)}=\dfrac{9+12}{10-15}=\dfrac{21}{-5}=-\dfrac{21}{5}$

9. $\dfrac{2+\frac{7}{10}}{1+\frac{3}{5}}=\dfrac{10\left(2+\frac{7}{10}\right)}{10\left(1+\frac{3}{5}\right)}=\dfrac{20+7}{10+6}=\dfrac{27}{16}$

11. $\dfrac{\frac{1}{3}}{\frac{1}{2}-\frac{1}{4}}=\dfrac{12\left(\frac{1}{3}\right)}{12\left(\frac{1}{2}-\frac{1}{4}\right)}=\dfrac{4}{6-3}=\dfrac{4}{3}$

13. $\dfrac{-\frac{2}{9}}{-\frac{14}{3}}=-\dfrac{2}{9}\div\left(-\dfrac{14}{3}\right)=-\dfrac{2}{9}\cdot\left(-\dfrac{3}{14}\right)=\dfrac{2\cdot 3}{9\cdot 14}=\dfrac{1}{21}$

15. $\dfrac{-\frac{5}{12x^2}}{\frac{25}{16x^3}}=\dfrac{-5}{12x^2}\div\dfrac{25}{16x^3}$
$=-\dfrac{5}{12x^2}\cdot\dfrac{16x^3}{25}$
$=-\dfrac{5\cdot 16x^3}{12x^2\cdot 25}$
$=-\dfrac{4x}{15}$

17. $\dfrac{\frac{m}{n}-1}{\frac{m}{n}+1}=\dfrac{n\left(\frac{m}{n}-1\right)}{n\left(\frac{m}{n}+1\right)}=\dfrac{m-n}{m+n}$

19. $\dfrac{\frac{1}{5}-\frac{1}{x}}{\frac{7}{10}+\frac{1}{x^2}}=\dfrac{10x^2\left(\frac{1}{5}-\frac{1}{x}\right)}{10x^2\left(\frac{7}{10}+\frac{1}{x^2}\right)}$
$=\dfrac{2x^2-10x}{7x^2+10}$
$=\dfrac{2x(x-5)}{7x^2+10}$

21. $\dfrac{1+\frac{1}{y-2}}{y+\frac{1}{y-2}}=\dfrac{(y-2)\left(1+\frac{1}{y-2}\right)}{(y-2)\left(y+\frac{1}{y-2}\right)}$
$=\dfrac{y-2+1}{(y-2)y+1}$
$=\dfrac{y-1}{y^2-2y+1}$
$=\dfrac{y-1}{(y-1)^2}$
$=\dfrac{1}{y-1}$

23. $\dfrac{\frac{4y-8}{16}}{\frac{6y-12}{4}}=\dfrac{16\left(\frac{4y-8}{16}\right)}{16\left(\frac{6y-12}{4}\right)}=\dfrac{4y-8}{24y-48}=\dfrac{4(y-2)}{24(y-2)}=\dfrac{1}{6}$

25. $\dfrac{\frac{x}{y}+1}{\frac{x}{y}-1}=\dfrac{y\left(\frac{x}{y}+1\right)}{y\left(\frac{x}{y}-1\right)}=\dfrac{x+y}{x-y}$

27. $\dfrac{1}{2+\frac{1}{3}}=\dfrac{3(1)}{3\left(2+\frac{1}{3}\right)}=\dfrac{3}{6+1}=\dfrac{3}{7}$

Copyright © 2012 Pearson Education, Inc. Publishing as Prentice Hall.

29. $$\frac{\frac{ax+ab}{x^2-b^2}}{\frac{x+b}{x-b}} = \frac{ax+ab}{x^2-b^2} \div \frac{x+b}{x-b}$$
$$= \frac{ax+ab}{x^2-b^2} \cdot \frac{x-b}{x+b}$$
$$= \frac{a(x+b)}{(x+b)(x-b)} \cdot \frac{x-b}{x+b}$$
$$= \frac{a}{x+b}$$

31. $$\frac{-\frac{3+y}{4}}{\frac{8+y}{28}} = \frac{28\left(\frac{-3+y}{4}\right)}{28\left(\frac{8+y}{28}\right)}$$
$$= \frac{-21+7y}{8+y}$$
$$= \frac{7y-21}{8+y}$$
$$= \frac{7(y-3)}{8+y}$$

33. $$\frac{3+\frac{12}{x}}{1-\frac{16}{x^2}} = \frac{x^2\left(3+\frac{12}{x}\right)}{x^2\left(1-\frac{16}{x^2}\right)}$$
$$= \frac{3x^2+12x}{x^2-16}$$
$$= \frac{3x(x+4)}{(x+4)(x-4)}$$
$$= \frac{3x}{x-4}$$

35. $$\frac{\frac{8}{x+4}+2}{\frac{12}{x+4}-2} = \frac{(x+4)\left(\frac{8}{x+4}+2\right)}{(x+4)\left(\frac{12}{x+4}-2\right)}$$
$$= \frac{8+2x+8}{12-2x-8}$$
$$= \frac{2x+16}{-2x+4}$$
$$= \frac{2(x+8)}{2(-x+2)}$$
$$= -\frac{x+8}{x-2}$$

37. $$\frac{\frac{s}{r}+\frac{r}{s}}{\frac{s}{r}-\frac{r}{s}} = \frac{rs\left(\frac{s}{r}+\frac{r}{s}\right)}{rs\left(\frac{s}{r}-\frac{r}{s}\right)} = \frac{s^2+r^2}{s^2-r^2}$$

39. $$\frac{\frac{6}{x-5}+\frac{x}{x-2}}{\frac{3}{x-6}-\frac{2}{x-5}} = \frac{\frac{6(x-2)}{(x-5)(x-2)}+\frac{x(x-5)}{(x-2)(x-5)}}{\frac{3(x-5)}{(x-6)(x-5)}-\frac{2(x-6)}{(x-5)(x-6)}}$$
$$= \frac{\frac{6x-12+x^2-5x}{(x-5)(x-2)}}{\frac{3x-15-2x+12}{(x-6)(x-5)}}$$
$$= \frac{\frac{x^2+x-12}{(x-5)(x-2)}}{\frac{x-3}{(x-6)(x-5)}}$$
$$= \frac{(x+4)(x-3)}{(x-5)(x-2)} \cdot \frac{(x-6)(x-5)}{x-3}$$
$$= \frac{(x-6)(x+4)}{x-2}$$

41. The longest bar corresponds to Lindsay Davenport, so Lindsay Davenport has won the most prize money in her career.

43. $22 - 20.3 = 1.7$
The approximate spread in lifetime prize money between Lindsay Davenport and Venus Williams is \$1.7 million.

45. answers may vary

47. $$\frac{\frac{1}{3}+\frac{3}{4}}{2} = \frac{\frac{1\cdot 4}{3\cdot 4}+\frac{3\cdot 3}{4\cdot 3}}{2} = \frac{\frac{4}{12}+\frac{9}{12}}{2} = \frac{13}{12}\cdot\frac{1}{2} = \frac{13}{24}$$

49. $$\frac{1}{\frac{1}{R_1}+\frac{1}{R_2}} = \frac{R_1R_2(1)}{R_1R_2\left(\frac{1}{R_1}+\frac{1}{R_2}\right)} = \frac{R_1R_2}{R_2+R_1}$$

51. $$\frac{x^{-1}+2^{-1}}{x^{-2}-4^{-1}} = \frac{\frac{1}{x}+\frac{1}{2}}{\frac{1}{x^2}-\frac{1}{4}}$$
$$= \frac{\frac{1\cdot 2}{x\cdot 2}+\frac{1\cdot x}{2\cdot x}}{\frac{1\cdot 4}{x^2\cdot 4}-\frac{1\cdot x^2}{4\cdot x^2}}$$
$$= \frac{\frac{2+x}{2x}}{\frac{4-x^2}{4x^2}}$$
$$= \frac{2+x}{2x}\cdot\frac{4x^2}{4-x^2}$$
$$= \frac{2+x}{2x}\cdot\frac{4x^2}{(2-x)(2+x)}$$
$$= \frac{2x}{2-x}$$

Copyright © 2012 Pearson Education, Inc. Publishing as Prentice Hall.

53. $\dfrac{y^{-2}}{1-y^{-2}} = \dfrac{\frac{1}{y^2}}{1-\frac{1}{y^2}} = \dfrac{y^2\left(\frac{1}{y^2}\right)}{y^2\left(1-\frac{1}{y^2}\right)} = \dfrac{1}{y^2-1}$

55. $t = \dfrac{d}{r} = \dfrac{\frac{20x}{3}}{\frac{5x}{9}} = \dfrac{20x}{3}\cdot\dfrac{9}{5x} = 12$

The time is 12 hours.

Chapter 5 Vocabulary Check

1. A rational expression is an expression that can be written in the form $\dfrac{P}{Q}$, where P and Q are polynomials and Q is not 0.
2. In a complex fraction, the numerator or denominator or both may contain fractions.
3. For a rational expression, $-\dfrac{a}{b} = \dfrac{-a}{b} = \dfrac{a}{-b}$.
4. A rational expression is undefined when the denominator is 0.
5. The process of writing a rational expression in lowest terms is called simplifying.
6. The expressions $\dfrac{2x}{7}$ and $\dfrac{7}{2x}$ are called reciprocals.
7. The least common denominator of a list of rational expressions is a polynomial of least degree whose factors include all factors of the denominators in the list.
8. A unit fraction is a fraction that equals 1.
9. A ratio is the quotient of two numbers.
10. $\dfrac{x}{2} = \dfrac{7}{16}$ is an example of a proportion.
11. If $\dfrac{a}{b} = \dfrac{c}{d}$, then ad and bc are called cross products.
12. A rate is a special type of ratio where different measurements are used.

Chapter 5 Review

1.
$$x^2-4=0$$
$$(x-2)(x+2)=0$$
$$x-2=0 \quad \text{or} \quad x+2=0$$
$$x=2 \qquad\qquad x=-2$$

The expression $\dfrac{x+5}{x^2-4}$ is undefined for $x=2$ and $x=-2$.

2.
$$4x^2-4x-15=0$$
$$(2x-5)(2x+3)=0$$
$$2x-5=0 \quad \text{or} \quad 2x+3=0$$
$$2x=5 \qquad\qquad 2x=3$$
$$x=\frac{5}{2} \qquad\qquad x=-\frac{3}{2}$$

The expression $\dfrac{5x+9}{4x^2-4x-15}$ is undefined for $x=\dfrac{5}{2}$ and $x=-\dfrac{3}{2}$.

3. Replace z with -2.
$$\frac{2-z}{z+5} = \frac{2-(-2)}{-2+5} = \frac{4}{3}$$

4. Replace x with 5 and y with 7.
$$\frac{x^2+xy-y^2}{x+y} = \frac{(5)^2+(5)(7)-(7)^2}{5+7} = \frac{25+35-49}{12} = \frac{11}{12}$$

5. $\dfrac{2x+6}{x^2+3x} = \dfrac{2(x+3)}{x(x+3)} = \dfrac{2}{x}$

6. $\dfrac{3x-12}{x^2-4x} = \dfrac{3(x-4)}{x(x-4)} = \dfrac{3}{x}$

7. $\dfrac{x+2}{x^2-3x-10} = \dfrac{x+2}{(x-5)(x+2)} = \dfrac{1}{x-5}$

8. $\dfrac{x+4}{x^2+5x+4} = \dfrac{x+4}{(x+4)(x+1)} = \dfrac{1}{x+1}$

Copyright © 2012 Pearson Education, Inc. Publishing as Prentice Hall.

9. $$\frac{x^3-4x}{x^2+3x+2}=\frac{x(x^2-4)}{(x+2)(x+1)}=\frac{x(x-2)(x+2)}{(x+2)(x+1)}=\frac{x(x-2)}{x+1}$$

10. $$\frac{5x^2-125}{x^2+2x-15}=\frac{5(x^2-25)}{(x+5)(x-3)}=\frac{5(x-5)(x+5)}{(x+5)(x-3)}=\frac{5(x-5)}{x-3}$$

11. $$\frac{x^2-x-6}{x^2-3x-10}=\frac{(x-3)(x+2)}{(x-5)(x+2)}=\frac{x-3}{x-5}$$

12. $$\frac{x^2-2x}{x^2+2x-8}=\frac{x(x-2)}{(x+4)(x-2)}=\frac{x}{x+4}$$

13. $$\frac{x^2+xa+xb+ab}{x^2-xc+bx-bc}=\frac{(x^2+xa)+(xb+ab)}{(x^2-xc)+(bx-bc)}=\frac{x(x+a)+b(x+a)}{x(x-c)+b(x-c)}=\frac{(x+b)(x+a)}{(x+b)(x-c)}=\frac{x+a}{x-c}$$

14. $$\frac{x^2+5x-2x-10}{x^2-3x-2x+6}=\frac{(x^2+5x)+(-2x-10)}{(x^2-3x)+(-2x+6)}=\frac{x(x+5)-2(x+5)}{x(x-3)-2(x-3)}=\frac{(x-2)(x+5)}{(x-2)(x-3)}=\frac{x+5}{x-3}$$

15. $$\frac{15x^3y^2}{z}\cdot\frac{z}{5xy^3}=\frac{3\cdot5\cdot x\cdot x^2\cdot y^2\cdot z}{z\cdot5\cdot x\cdot y\cdot y^2}=\frac{3x^2}{y}$$

16. $$\frac{-y^3}{8}\cdot\frac{9x^2}{y^3}=-\frac{y^3\cdot9x^2}{8\cdot y^3}=-\frac{9x^2}{8}$$

17. $$\frac{x^2-9}{x^2-4}\cdot\frac{x-2}{x+3}=\frac{(x-3)(x+3)}{(x-2)(x+2)}\cdot\frac{x-2}{x+3}=\frac{x-3}{x+2}$$

18. $$\frac{2x+5}{x-6}\cdot\frac{2x}{-x+6}=\frac{2x+5}{x-6}\cdot\frac{2x}{-(x-6)}=-\frac{2x(2x+5)}{(x-6)^2}$$

19. $$\frac{x^2-5x-24}{x^2-x-12}\div\frac{x^2-10x+16}{x^2+x-6}=\frac{x^2-5x-24}{x^2-x-12}\cdot\frac{x^2+x-6}{x^2-10x+16}=\frac{(x-8)(x+3)}{(x-4)(x+3)}\cdot\frac{(x+3)(x-2)}{(x-8)(x-2)}=\frac{x+3}{x-4}$$

20. $$\frac{4x+4y}{xy^2}\div\frac{3x+3y}{x^2y}=\frac{4x+4y}{xy^2}\cdot\frac{x^2y}{3x+3y}=\frac{4(x+y)}{xy^2}\cdot\frac{x^2y}{3(x+y)}=\frac{4x}{3y}$$

21. $$\frac{x^2+x-42}{x-3}\cdot\frac{(x-3)^2}{x+7}=\frac{(x+7)(x-6)}{x-3}\cdot\frac{(x-3)^2}{x+7}=(x-6)(x-3)$$

22. $$\frac{2a+2b}{3}\cdot\frac{a-b}{a^2-b^2}=\frac{2(a+b)}{3}\cdot\frac{a-b}{(a-b)(a+b)}=\frac{2}{3}$$

23. $$\frac{2x^2-9x+9}{8x-12}\div\frac{x^2-3x}{2x}=\frac{2x^2-9x+9}{8x-12}\cdot\frac{2x}{x^2-3x}=\frac{(2x-3)(x-3)}{4(2x-3)}\cdot\frac{2x}{x(x-3)}=\frac{1}{2}$$

24. $$\frac{x^2-y^2}{x^2+xy}\div\frac{3x^2-2xy-y^2}{3x^2+6x}=\frac{x^2-y^2}{x^2+xy}\cdot\frac{3x^2+6x}{3x^2-2xy-y^2}=\frac{(x-y)(x+y)}{x(x+y)}\cdot\frac{3x(x+2)}{(3x+y)(x-y)}=\frac{3(x+2)}{3x+y}$$

Copyright © 2012 Pearson Education, Inc. Publishing as Prentice Hall.

25. $$\frac{x}{x^2+9x+14}+\frac{7}{x^2+9x+14}=\frac{x+7}{x^2+9x+14}=\frac{x+7}{(x+7)(x+2)}=\frac{1}{x+2}$$

26. $$\frac{x}{x^2+2x-15}+\frac{5}{x^2+2x-15}=\frac{x+5}{x^2+2x-15}=\frac{x+5}{(x+5)(x-3)}=\frac{1}{x-3}$$

27. $$\frac{4x-5}{3x^2}-\frac{2x+5}{3x^2}=\frac{4x-5-2x-5}{3x^2}=\frac{2x-10}{3x^2}=\frac{2(x-5)}{3x^2}$$

28. $$\frac{9x+7}{6x^2}-\frac{3x+4}{6x^2}=\frac{9x+7-3x-4}{6x^2}=\frac{6x+3}{6x^2}=\frac{3(2x+1)}{6x^2}=\frac{2x+1}{2x^2}$$

29. The LCD is $2 \cdot 7 \cdot x$ or $14x$.

30. $x^2-5x-24=(x-8)(x+3)$
$x^2+11x+24=(x+8)(x+3)$
The LCD is $(x+3)(x+8)(x-8)$.

31. $$\frac{5}{7x}=\frac{5\cdot 2x^2y}{7x\cdot 2x^2y}=\frac{10x^2y}{14x^3y}$$

32. $$\frac{9}{4y}=\frac{9\cdot 4y^2x}{4y\cdot 4y^2x}=\frac{36y^2x}{16y^3x}$$

33. $$\frac{x+2}{x^2+11x+18}=\frac{x+2}{(x+2)(x+9)}=\frac{(x+2)(x-5)}{(x+2)(x+9)(x-5)}=\frac{x^2-3x-10}{(x+2)(x-5)(x+9)}$$

34. $$\frac{3x-5}{x^2+4x+4}=\frac{3x-5}{(x+2)^2}=\frac{(3x-5)(x+3)}{(x+2)^2(x+3)}=\frac{3x^2+4x-15}{(x+2)^2(x+3)}$$

35. $$\frac{4}{5x^2}+\frac{6}{y}=\frac{4y}{5x^2y}+\frac{6\cdot 5x^2}{y\cdot 5x^2}=\frac{4y+30x^2}{5x^2y}$$

36. $$\frac{2}{x-3}-\frac{4}{x-1}=\frac{2(x-1)-4(x-3)}{(x-3)(x-1)}=\frac{2x-2-4x+12}{(x-3)(x-1)}=\frac{-2x+10}{(x-3)(x-1)}=-\frac{2(x-5)}{(x-3)(x-1)}$$

37. $$\frac{4}{x+3}-2=\frac{4-2(x+3)}{x+3}=\frac{4-2x-6}{x+3}=\frac{-2x-2}{x+3}=-\frac{2(x+1)}{x+3}$$

Copyright © 2012 Pearson Education, Inc. Publishing as Prentice Hall.

38. $$\begin{aligned}\frac{3}{x^2+2x-8}+\frac{2}{x^2-3x+2} &= \frac{3}{(x+4)(x-2)}+\frac{2}{(x-2)(x-1)}\\ &= \frac{3(x-1)+2(x+4)}{(x+4)(x-2)(x-1)}\\ &= \frac{3x-3+2x+8}{(x+4)(x-2)(x-1)}\\ &= \frac{5x+5}{(x+4)(x-2)(x-1)}\\ &= \frac{5(x+1)}{(x+4)(x-2)(x-1)}\end{aligned}$$

39. $$\begin{aligned}\frac{2x-5}{6x+9}-\frac{4}{2x^2+3x} &= \frac{2x-5}{3(2x+3)}-\frac{4}{x(2x+3)}\\ &= \frac{x(2x-5)-4(3)}{3x(2x+3)}\\ &= \frac{2x^2-5x-12}{3x(2x+3)}\\ &= \frac{(2x+3)(x-4)}{3x(2x+3)}\\ &= \frac{x-4}{3x}\end{aligned}$$

40. $$\begin{aligned}\frac{x-1}{x^2-2x+1}-\frac{x+1}{x-1} &= \frac{x-1}{(x-1)(x-1)}-\frac{x+1}{x-1}\\ &= \frac{1}{x-1}-\frac{x+1}{x-1}\\ &= \frac{1-x-1}{x-1}\\ &= -\frac{x}{x-1}\end{aligned}$$

41. $$\begin{aligned}\frac{n}{10} &= 9-\frac{n}{5}\\ 10\left(\frac{n}{10}\right) &= 10\left(9-\frac{n}{5}\right)\\ n &= 90-2n\\ 3n &= 90\\ n &= 30\end{aligned}$$

The solution is 30.

Copyright © 2012 Pearson Education, Inc. Publishing as Prentice Hall.

42.

$$\frac{2}{x+1}-\frac{1}{x-2}=-\frac{1}{2}$$

$$2(x+1)(x-2)\left(\frac{2}{x+1}-\frac{1}{x-2}\right)=2(x+1)(x-2)\left(-\frac{1}{2}\right)$$

$$2\cdot 2(x-2)-2(x+1)=-1(x+1)(x-2)$$

$$4x-8-2x-2=-(x^2-x-2)$$

$$2x-10=-x^2+x+2$$

$$x^2+x-12=0$$

$$(x+4)(x-3)=0$$

$x+4=0$ or $x-3=0$

$x=-4$ $\quad$ $x=3$

The solutions are –4 and 3.

43.

$$\frac{y}{2y+2}+\frac{2y-16}{4y+4}=\frac{y-3}{y+1}$$

$$\frac{y}{2(y+1)}+\frac{2y-16}{4(y+1)}=\frac{y-3}{y+1}$$

$$4(y+1)\left(\frac{y}{2(y+1)}+\frac{2y-16}{4(y+1)}\right)=4(y+1)\left(\frac{y-3}{y+1}\right)$$

$$2y+2y-16=4(y-3)$$

$$4y-16=4y-12$$

$$0y=4$$

$0=4$; no solution

44.

$$\frac{2}{x-3}-\frac{4}{x+3}=\frac{8}{x^2-9}$$

$$(x^2-9)\left(\frac{2}{x-3}-\frac{4}{x+3}\right)=(x^2-9)\left(\frac{8}{x^2-9}\right)$$

$$2(x+3)-4(x-3)=8$$

$$2x+6-4x+12=8$$

$$-2x+18=8$$

$$-2x=-10$$

$$x=5$$

The solution is 5.

45.

$$\frac{x-3}{x+1}-\frac{x-6}{x+5}=0$$

$$(x+1)(x+5)\left(\frac{x-3}{x+1}-\frac{x-6}{x+5}\right)=(x+1)(x+5)0$$

$$(x+5)(x-3)-(x+1)(x-6)=0$$

$$x^2+2x-15-x^2+5x+6=0$$

$$7x-9=0$$

$$7x=9$$

$$x=\frac{9}{7}$$

The solution is $\frac{9}{7}$.

Copyright © 2012 Pearson Education, Inc. Publishing as Prentice Hall.

46.
$$x+5=\frac{6}{x}$$
$$x(x+5)=x\left(\frac{6}{x}\right)$$
$$x^2+5x=6$$
$$x^2+5x-6=0$$
$$(x+6)(x-1)=0$$
$$x+6=0 \quad \text{or} \quad x-1=0$$
$$x=-6 \qquad x=1$$
The solutions are –6 and 1.

47.
$$\frac{2}{x-1}=\frac{3}{x+3}$$
$$2(x+3)=3(x-1)$$
$$2x+6=3x-3$$
$$9=x$$
The solution is 9.

48.
$$\frac{4}{y-3}=\frac{2}{y-3}$$
$$4(y-3)=2(y-3)$$
$$4y-12=2y-6$$
$$2y=6$$
$$y=3$$
$y = 3$ does not check. There is no solution.

49.
$$\frac{300}{20}=\frac{x}{45}$$
$$300(45)=20x$$
$$13{,}500=20x$$
$$675=x$$
The machine will process 675 parts in 45 minutes.

50.
$$\frac{90}{8}=\frac{x}{3}$$
$$90(3)=8x$$
$$270=8x$$
$$33.75=x$$
Mr. Visconti charges \$33.75 for 3 hours of consulting.

51. Let n be the number.
$$5\left(\frac{1}{n}\right)=\frac{3}{2}\left(\frac{1}{n}\right)+\frac{7}{6}$$
$$\frac{5}{n}=\frac{3}{2n}+\frac{7}{6}$$
$$6n\left(\frac{5}{n}\right)=6n\left(\frac{3}{2n}+\frac{7}{6}\right)$$
$$30=9+7n$$
$$21=7n$$
$$3=n$$
The number is 3.

52. Let n be the number.
$$\frac{1}{n}=\frac{1}{4-n}$$
$$n(4-n)\left(\frac{1}{n}\right)=n(4-n)\left(\frac{1}{4-n}\right)$$
$$4-n=n$$
$$4=2n$$
$$2=n$$
The number is 2.

53. Let x be the speed of the faster car.

	distance	=	**rate**	·	**time**
Faster car	90		x		$\frac{90}{x}$
Slower car	60		$x-10$		$\frac{60}{x-10}$

The times are equal.
$$\frac{90}{x}=\frac{60}{x-10}$$
$$90(x-10)=60x$$
$$90x-900=60x$$
$$30x=900$$
$$x=30$$
$x - 10 = 20$
The faster car is traveling at 30 mph and the slower car is traveling at 20 mph.

54. Let x be the speed of the boat in still water.

	distance	=	**rate**	·	**time**
Upstream	48		$x-4$		$\frac{48}{x-4}$
Downstream	72		$x+4$		$\frac{72}{x+4}$

The times are equal.

Copyright © 2012 Pearson Education, Inc. Publishing as Prentice Hall.

$$\frac{48}{x-4}=\frac{72}{x+4}$$
$$48(x+4)=72(x-4)$$
$$48x+192=72x-288$$
$$480=24x$$
$$20=x$$

The speed of the boat in still water is 20 mph.

55. Let x be the time for Maria alone. Together, Mark and Maria complete $\frac{1}{5}$ of the job in 1 hour. Individually, they complete $\frac{1}{7}$ and $\frac{1}{x}$ of the job in 1 hour.

$$\frac{1}{5}=\frac{1}{7}+\frac{1}{x}$$
$$35x\left(\frac{1}{5}\right)=35x\left(\frac{1}{7}+\frac{1}{x}\right)$$
$$7x=5x+35$$
$$2x=35$$
$$x=\frac{35}{2}=17\frac{1}{2}$$

It will take Maria $17\frac{1}{2}$ hours to manicure Mr. Sturgeon's lawn alone.

56. Let x be the time for the pipes to fill the pond working together. Pipe A fills $\frac{1}{20}$ of the pond in 1 day, pipe B fills $\frac{1}{15}$ of the pond in 1 day, and together they fill $\frac{1}{x}$ of the pond in one day.

$$\frac{1}{20}+\frac{1}{15}=\frac{1}{x}$$
$$60x\left(\frac{1}{20}+\frac{1}{15}\right)=60x\left(\frac{1}{x}\right)$$
$$3x+4x=60$$
$$7x=60$$
$$x=\frac{60}{7}=8\frac{4}{7}$$

It takes $8\frac{4}{7}$ days to fill the pond using both pipes.

57.
$$\frac{2}{3}=\frac{10}{x}$$
$$2\cdot x=3\cdot 10$$
$$2x=30$$
$$x=15$$

The missing length is $x = 15$.

58.
$$\frac{12}{4}=\frac{18}{x}$$
$$12\cdot x=4\cdot 18$$
$$12x=72$$
$$x=6$$

The missing length is $x = 6$.

59.
$$\frac{\frac{5x}{27}}{-\frac{10xy}{21}}=\frac{5x}{27}\div-\frac{10xy}{21}$$
$$=\frac{5x}{27}\cdot\frac{-21}{10xy}$$
$$=-\frac{5\cdot x\cdot 3\cdot 7}{3\cdot 9\cdot 2\cdot 5\cdot x\cdot y}$$
$$=-\frac{7}{18y}$$

60.
$$\frac{\frac{3}{5}+\frac{2}{7}}{\frac{1}{5}+\frac{5}{6}}=\frac{\frac{3\cdot 7}{5\cdot 7}+\frac{2\cdot 5}{7\cdot 5}}{\frac{1\cdot 6}{5\cdot 6}+\frac{5\cdot 5}{6\cdot 5}}$$
$$=\frac{\frac{21+10}{35}}{\frac{6+25}{30}}$$
$$=\frac{\frac{31}{35}}{\frac{31}{30}}$$
$$=\frac{31}{35}\div\frac{31}{30}$$
$$=\frac{31}{35}\cdot\frac{30}{31}$$
$$=\frac{31\cdot 5\cdot 6}{5\cdot 7\cdot 31}$$
$$=\frac{6}{7}$$

61.
$$\frac{3-\frac{1}{y}}{2-\frac{1}{y}}=\frac{y\left(3-\frac{1}{y}\right)}{y\left(2-\frac{1}{y}\right)}=\frac{3y-1}{2y-1}$$

Copyright © 2012 Pearson Education, Inc. Publishing as Prentice Hall.

62. $$\frac{\frac{6}{x+2}+4}{\frac{8}{x+2}-4}=\frac{(x+2)\left(\frac{6}{x+2}+4\right)}{(x+2)\left(\frac{8}{x+2}-4\right)}$$
$$=\frac{6+4(x+2)}{8-4(x+2)}$$
$$=\frac{6+4x+8}{8-4x-8}$$
$$=\frac{4x+14}{-4x}$$
$$=-\frac{2(2x+7)}{2\cdot 2x}$$
$$=-\frac{2x+7}{2x}$$

63. $$\frac{4x+12}{8x^2+24x}=\frac{4(x+3)}{8x(x+3)}=\frac{1}{2x}$$

64. $$\frac{x^3-6x^2+9x}{x^2+4x-21}=\frac{x(x^2-6x+9)}{(x+7)(x-3)}$$
$$=\frac{x(x-3)(x-3)}{(x+7)(x-3)}$$
$$=\frac{x(x-3)}{x+7}$$

65. $$\frac{x^2+9x+20}{x^2-25}\cdot\frac{x^2-9x+20}{x^2+8x+16}$$
$$=\frac{(x+4)(x+5)}{(x-5)(x+5)}\cdot\frac{(x-4)(x-5)}{(x+4)(x+4)}$$
$$=\frac{x-4}{x+4}$$

66. $$\frac{x^2-x-72}{x^2-x-30}\div\frac{x^2+6x-27}{x^2-9x+18}$$
$$=\frac{x^2-x-72}{x^2-x-30}\cdot\frac{x^2-9x+18}{x^2+6x-27}$$
$$=\frac{(x-9)(x+8)}{(x-6)(x+5)}\cdot\frac{(x-6)(x-3)}{(x+9)(x-3)}$$
$$=\frac{(x-9)(x+8)}{(x+5)(x+9)}$$

67. $$\frac{x}{x^2-36}+\frac{6}{x^2-36}=\frac{x+6}{x^2-36}$$
$$=\frac{x+6}{(x+6)(x-6)}$$
$$=\frac{1}{x-6}$$

68. $$\frac{5x-1}{4x}-\frac{3x-2}{4x}=\frac{5x-1-3x+2}{4x}=\frac{2x+1}{4x}$$

69. $$\frac{4}{3x^2+8x-3}+\frac{2}{3x^2-7x+2}$$
$$=\frac{4}{(3x-1)(x+3)}+\frac{2}{(3x-1)(x-2)}$$
$$=\frac{4(x-2)}{(3x-1)(x+3)(x-2)}+\frac{2(x+3)}{(3x-1)(x-2)(x+3)}$$
$$=\frac{4x-8+2x+6}{(3x-1)(x+3)(x-2)}$$
$$=\frac{6x-2}{(3x-1)(x+3)(x-2)}$$
$$=\frac{2(3x-1)}{(3x-1)(x+3)(x-2)}$$
$$=\frac{2}{(x+3)(x-2)}$$

70. $$\frac{3x}{x^2+9x+14}-\frac{6x}{x^2+4x-21}$$
$$=\frac{3x}{(x+7)(x+2)}-\frac{6x}{(x+7)(x-3)}$$
$$=\frac{3x(x-3)}{(x+7)(x+2)(x-3)}-\frac{6x(x+2)}{(x+7)(x-3)(x+2)}$$
$$=\frac{3x^2-9x-6x^2-12x}{(x+7)(x+2)(x-3)}$$
$$=\frac{-3x^2-21x}{(x+7)(x+2)(x-3)}$$
$$=\frac{-3x(x+7)}{(x+7)(x+2)(x-3)}$$
$$=-\frac{3x}{(x+2)(x-3)}$$

71. $$\frac{4}{a-1}+2=\frac{3}{a-1}$$
$$(a-1)\left(\frac{4}{a-1}+2\right)=(a-1)\left(\frac{3}{a-1}\right)$$
$$4+2(a-1)=3$$
$$4+2a-2=3$$
$$2+2a=3$$
$$2a=1$$
$$a=\frac{1}{2}$$

The solution is $\frac{1}{2}$.

Copyright © 2012 Pearson Education, Inc. Publishing as Prentice Hall.

72.
$$\frac{x}{x+3}+4=\frac{x}{x+3}$$
$$(x+3)\left(\frac{x}{x+3}+4\right)=(x+3)\left(\frac{x}{x+3}\right)$$
$$x+4(x+3)=x$$
$$x+4x+12=x$$
$$5x+12=x$$
$$4x+12=0$$
$$4x=-12$$
$$x=-3$$
-3 makes the denominator $x+3$ zero. Therefore there is no solution.

73. Let n be the number.
$$\frac{2n}{3}-\frac{1}{6}=\frac{n}{2}$$
$$6\left(\frac{2n}{3}-\frac{1}{6}\right)=6\left(\frac{n}{2}\right)$$
$$4n-1=3n$$
$$n-1=0$$
$$n=1$$
The number is 1.

74. Let x be the time to paint the shed together. Mr. Crocker can paint $\frac{1}{3}$ of the shed in one day, and his son can paint $\frac{1}{4}$. Together, they can paint $\frac{1}{x}$.
$$\frac{1}{3}+\frac{1}{4}=\frac{1}{x}$$
$$12x\left(\frac{1}{3}+\frac{1}{4}\right)=12x\left(\frac{1}{x}\right)$$
$$4x+3x=12$$
$$7x=12$$
$$x=\frac{12}{7} \text{ or } 1\frac{5}{7}$$
It will take $1\frac{5}{7}$ days to paint the shed when they work together.

75.
$$\frac{x}{10}=\frac{3}{5}$$
$$5x=3(10)$$
$$5x=30$$
$$x=6$$
The missing length is $x=6$.

76.
$$\frac{x}{4}=\frac{18}{6}$$
$$6x=4(18)$$
$$6x=72$$
$$x=12$$
The missing length is $x=12$.

77.
$$\frac{\frac{1}{4}}{\frac{1}{3}+\frac{1}{2}}=\frac{12\left(\frac{1}{4}\right)}{12\left(\frac{1}{3}+\frac{1}{2}\right)}=\frac{3}{4+6}=\frac{3}{10}$$

78.
$$\frac{4+\frac{2}{x}}{6+\frac{3}{x}}=\frac{x\left(4+\frac{2}{x}\right)}{x\left(6+\frac{3}{x}\right)}=\frac{4x+2}{6x+3}=\frac{2(2x+1)}{3(2x+1)}=\frac{2}{3}$$

Chapter 5 Test

1.
$$x^2+4x+3=0$$
$$(x+1)(x+3)=0$$
$$x+1=0 \quad \text{or} \quad x+3=0$$
$$x=-1 \qquad\qquad x=-3$$
The expression $\frac{x+5}{x^2+4x+3}$ is undefined for $x=-1$ and $x=-3$.

2. a.
$$C=\frac{100x+3000}{x}=\frac{100(200)+3000}{200}=\$115$$

b.
$$C=\frac{100x+3000}{x}$$
$$=\frac{100(1000)+3000}{1000}$$
$$=\$103$$

3.
$$\frac{3x-6}{5x-10}=\frac{3(x-2)}{5(x-2)}=\frac{3}{5}$$

4.
$$\frac{x+6}{x^2+12x+36}=\frac{x+6}{(x+6)(x+6)}=\frac{1}{x+6}$$

5.
$$\frac{7-x}{x-7}=\frac{-(x-7)}{x-7}=-1$$

6.
$$\frac{y-x}{x^2-y^2}=\frac{-(x-y)}{(x-y)(x+y)}=-\frac{1}{x+y}$$

Copyright © 2012 Pearson Education, Inc. Publishing as Prentice Hall.

7. $$\frac{2m^3-2m^2-12m}{m^2-5m+6}=\frac{2m(m^2-m-6)}{(m-3)(m-2)}=\frac{2m(m-3)(m+2)}{(m-3)(m-2)}=\frac{2m(m+2)}{m-2}$$

8. $$\frac{ay+3a+2y+6}{ay+3a+5y+15}=\frac{(ay+3a)+(2y+6)}{(ay+3a)+(5y+15)}=\frac{a(y+3)+2(y+3)}{a(y+3)+5(y+3)}=\frac{(a+2)(y+3)}{(a+5)(y+3)}=\frac{a+2}{a+5}$$

9. $$\frac{x^2-13x+42}{x^2+10x+21}\div\frac{x^2-4}{x^2+x-6}=\frac{x^2-13x+42}{x^2+10x+21}\cdot\frac{x^2+x-6}{x^2-4}=\frac{(x-6)(x-7)}{(x+3)(x+7)}\cdot\frac{(x+3)(x-2)}{(x+2)(x-2)}=\frac{(x-6)(x-7)}{(x+7)(x+2)}$$

10. $$\frac{3}{x-1}\cdot(5x-5)=\frac{3}{x-1}\cdot 5(x-1)=15$$

11. $$\frac{y^2-5y+6}{2y+4}\cdot\frac{y+2}{2y-6}=\frac{(y-3)(y-2)}{2(y+2)}\cdot\frac{y+2}{2(y-3)}=\frac{y-2}{4}$$

12. $$\frac{5}{2x+5}-\frac{6}{2x+5}=\frac{5-6}{2x+5}=-\frac{1}{2x+5}$$

13. $$\frac{5a}{a^2-a-6}-\frac{2}{a-3}=\frac{5a}{(a-3)(a+2)}-\frac{2}{a-3}=\frac{5a}{(a-3)(a+2)}-\frac{2(a+2)}{(a-3)(a+2)}=\frac{5a-2(a+2)}{(a-3)(a+2)}=\frac{5a-2a-4}{(a-3)(a+2)}=\frac{3a-4}{(a-3)(a+2)}$$

14. $$\frac{6}{x^2-1}+\frac{3}{x+1}=\frac{6}{(x-1)(x+1)}+\frac{3}{x+1}=\frac{6}{(x-1)(x+1)}+\frac{3(x-1)}{(x+1)(x-1)}=\frac{6+3x-3}{(x+1)(x-1)}=\frac{3x+3}{(x+1)(x-1)}=\frac{3(x+1)}{(x+1)(x-1)}=\frac{3}{x-1}$$

15. $$\frac{x^2-9}{x^2-3x}\div\frac{x^2+4x+1}{2x+10}=\frac{x^2-9}{x^2-3x}\cdot\frac{2x+10}{x^2+4x+1}=\frac{(x-3)(x+3)}{x(x-3)}\cdot\frac{2(x+5)}{x^2+4x+1}=\frac{2(x+3)(x+5)}{x(x^2+4x+1)}$$

16. $$\frac{x+2}{x^2+11x+18}+\frac{5}{x^2-3x-10}=\frac{x+2}{(x+9)(x+2)}+\frac{5}{(x-5)(x+2)}=\frac{(x+2)(x-5)}{(x+9)(x+2)(x-5)}+\frac{5(x+9)}{(x-5)(x+2)(x+9)}=\frac{x^2-3x-10+5x+45}{(x+9)(x+2)(x-5)}=\frac{x^2+2x+35}{(x+9)(x+2)(x-5)}$$

Copyright © 2012 Pearson Education, Inc. Publishing as Prentice Hall.

17. $\dfrac{4y}{y^2+6y+5}-\dfrac{3}{y^2+5y+4}$

$$=\frac{4y}{(y+5)(y+1)}-\frac{3}{(y+1)(y+4)}$$
$$=\frac{4y(y+4)-3(y+5)}{(y+1)(y+5)(y+4)}$$
$$=\frac{4y^2+16y-3y-15}{(y+1)(y+5)(y+4)}$$
$$=\frac{4y^2+13y-15}{(y+1)(y+5)(y+4)}$$

18. The LCD is $3\cdot 5\cdot y=15y$.

$$\frac{4}{y}-\frac{5}{3}=-\frac{1}{5}$$
$$15y\left(\frac{4}{y}-\frac{5}{3}\right)=15y\left(-\frac{1}{5}\right)$$
$$60-25y=-3y$$
$$60=22y$$
$$\frac{60}{22}=y$$
$$\frac{30}{11}=y$$

The solution is $\dfrac{30}{11}$.

19. $\dfrac{5}{y+1}=\dfrac{4}{y+2}$

$$5(y+2)=4(y+1)$$
$$5y+10=4y+4$$
$$y=-6$$

The solution is -6.

20. The LCD is $2(a-3)$.

$$\frac{a}{a-3}=\frac{3}{a-3}-\frac{3}{2}$$
$$2(a-3)\left(\frac{a}{a-3}\right)=2(a-3)\left(\frac{3}{a-3}-\frac{3}{2}\right)$$
$$2a=6-3(a-3)$$
$$2a=6-3a+9$$
$$5a=15$$
$$a=3$$

Since $a=3$ causes the denominator $a-3$ to be 0, the equation has no solution.

21. The LCD is $x^2-25=(x+5)(x-5)$.

$$\frac{10}{x^2-25}=\frac{3}{x+5}+\frac{1}{x-5}$$
$$(x^2-25)\left(\frac{10}{x^2-25}\right)=(x+5)(x-5)\left(\frac{3}{x+5}+\frac{1}{x-5}\right)$$
$$10=3(x-5)+1(x+5)$$
$$10=3x-15+x+5$$
$$10=4x-10$$
$$20=4x$$
$$5=x$$

Since $x=5$ causes the denominators x^2-25 and $x-5$ to be 0, the equation has no solution.

22.
$$x-\frac{14}{x-1}=4-\frac{2x}{x-1}$$
$$(x-1)\left(x-\frac{14}{x-1}\right)=(x-1)\left(4-\frac{2x}{x-1}\right)$$
$$x(x-1)-14=4(x-1)-2x$$
$$x^2-x-14=4x-4-2x$$
$$x^2-x-14=2x-4$$
$$x^2-3x-10=0$$
$$(x-5)(x+2)=0$$
$$x-5=0 \quad \text{or} \quad x+2=0$$
$$x=5 \qquad\qquad x=-2$$

The solutions are 5 and -2.

23.
$$\frac{\frac{5x^2}{yz^2}}{\frac{10x}{z^3}}=\frac{5x^2}{yz^2}\div\frac{10x}{z^3}$$
$$=\frac{5x^2}{yz^2}\cdot\frac{z^3}{10x}$$
$$=\frac{5\cdot x\cdot x\cdot z\cdot z^2}{y\cdot z^2\cdot 2\cdot 5\cdot x}$$
$$=\frac{xz}{2y}$$

24.
$$\frac{\frac{b}{a}-\frac{a}{b}}{\frac{1}{b}+\frac{1}{a}}=\frac{\left(\frac{b}{a}-\frac{a}{b}\right)ab}{\left(\frac{1}{b}+\frac{1}{a}\right)ab}$$
$$=\frac{b^2-a^2}{a+b}$$
$$=\frac{(b-a)(b+a)}{a+b}$$
$$=b-a$$

Copyright © 2012 Pearson Education, Inc. Publishing as Prentice Hall.

25. $\dfrac{5-\frac{1}{y^2}}{\frac{1}{y}+\frac{2}{y^2}}=\dfrac{y^2\left(5-\frac{1}{y^2}\right)}{y^2\left(\frac{1}{y}+\frac{2}{y^2}\right)}=\dfrac{5y^2-1}{y+2}$

26. Let n be the number.

$$n+5\left(\frac{1}{n}\right)=6$$
$$n+\frac{5}{n}=6$$
$$n\left(n+\frac{5}{n}\right)=n(6)$$
$$n^2+5=6n$$
$$n^2-6n+5=0$$
$$(n-5)(n-1)=0$$
$$n-5=0 \quad \text{or} \quad n-1=0$$
$$n=5 \qquad\qquad n=1$$

The number is 1 or 5.

27. Let x be the speed of the boat in still water. Let $x + 2$ be the speed of the boat going downstream. Let $x - 2$ be the speed of the boat going upstream.

	distance	=	rate	·	time
Upstream	14		$x-2$		$\frac{14}{x-2}$
Downstream	16		$x+2$		$\frac{16}{x+2}$

$$\frac{14}{x-2}=\frac{16}{x+2}$$
$$14(x+2)=16(x-2)$$
$$14x+28=16x-32$$
$$60=2x$$
$$30=x$$

The speed of the boat in still water is 30 mph.

28. Let x be the time in hours that it takes for both inlet pipes together to fill the tank.

The first pipe fills $\frac{1}{12}$ of the tank in 1 hour, the second pipe fills $\frac{1}{15}$ of the tank in 1 hour, and the two pipes fill $\frac{1}{x}$ of the tank in 1 hour.

$$\frac{1}{12}+\frac{1}{15}=\frac{1}{x}$$

The LCD is $60x$.

$$60x\left(\frac{1}{12}+\frac{1}{15}\right)=60x\left(\frac{1}{x}\right)$$
$$5x+4x=60$$
$$9x=60$$
$$x=\frac{60}{9}$$
$$x=\frac{20}{3}$$

It takes both pipes $\frac{20}{3}=6\frac{2}{3}$ hours to fill the tank.

29.
$$\frac{8}{x}=\frac{10}{15}$$
$$8(15)=10x$$
$$120=10x$$
$$12=x$$

30.
$$\frac{3}{85}=\frac{x}{510}$$
$$3(510)=85x$$
$$1530=85x$$
$$18=x$$

18 defective bulbs should be found in 510 bulbs.

Cumulative Review Chapters 1–5

1. a. The quotient of 15 and a number is 4 is written as $\frac{15}{x}=4$.

b. Three subtracted from 12 is a number is written as $12 - 3 = x$.

c. 17 added to four times a number is 21 is written as $4x + 17 = 21$.

2. a. The difference of 12 and a number is –45 is written as $12 - x = -45$.

b. The product of 12 and a number is –45 is written as $12x = -45$.

c. A number less 10 is twice the number is written as $x - 10 = 2x$.

3. a. $3 + (-7) + (-8) = (-4) + (-8) = -12$

b. $[7 + (-10)] + [-2 + (-4)] = (-3) + (-6) = -9$

4. a. $28 - 6 - 30 = 22 - 30 = -8$

b. $7 - 2 - 22 = 5 - 22 = -17$

Copyright © 2012 Pearson Education, Inc. Publishing as Prentice Hall.

5. $3(x + y) = 3 \cdot x + 3 \cdot y$ illustrates the distributive property.

6. $3 + y = y + 3$ illustrates the commutative property of addition.

7. $(x + 7) + 9 = x + (7 + 9)$ illustrates the associative property of addition.

8. $(x \cdot 7) \cdot 9 = x \cdot (7 \cdot 9)$ illustrates the associative property of multiplication.

9.
$$\begin{aligned} 3-x&=7\\ 2-x-3&=7-3\\ -x&=4\\ (-1)(-x)&=(-1)4\\ x&=-4 \end{aligned}$$
The solution is -4.

10.
$$\begin{aligned} 7x-6&=6x-6\\ 7x-6+6&=6x-6+6\\ 7x&=6x\\ 7x-6x&=6x-6x\\ x&=0 \end{aligned}$$
The solution is 0.

11. Let x be the length of the shorter piece. Then $4x$ is the length of the longer piece.
$$\begin{aligned} x+4x&=10\\ 5x&=10\\ x&=2 \end{aligned}$$
$4x = 4(2) = 8$
The shorter piece is 2 feet. The longer piece is 8 feet.

12. Let n be the first even integer. Then $n + 2$ is the second.
$$\begin{aligned} n+(n+2)&=382\\ 2n+2&=382\\ 2n&=380\\ n&=190 \end{aligned}$$
$n + 2 = 192$
The two consecutive even integers are 190 and 192.

13.
$$\begin{aligned} y&=mx+b\\ y-b&=mx\\ \frac{y-b}{m}&=x \end{aligned}$$

14.
$$\begin{aligned} 3x-2y&=6\\ 3x&=6+2y\\ x&=\frac{6+2y}{3} \end{aligned}$$

15.
$$\begin{aligned} x+4&\le -6\\ x+4-4&\le -6-4\\ x&\le -10 \end{aligned}$$
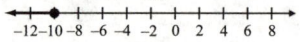

16.
$$\begin{aligned} -3x+7&>-x+9\\ 7&>2x+9\\ -2&>2x\\ -1&>x \end{aligned}$$
$\{x|x < -1\}$

17. $\dfrac{x^5}{x^2} = x^{5-2} = x^3$

18. $\dfrac{y^{14}}{y^{14}} = 1$

19. $\dfrac{4^7}{4^3} = 4^{7-3} = 4^4 = 256$

20. $(x^5y^2)^3 = x^{5\cdot 3}y^{2\cdot 3} = x^{15}y^6$

21. $\dfrac{(-3)^5}{(-3)^2} = (-3)^{5-2} = (-3)^3 = -27$

22. $\dfrac{x^{19}y^5}{xy} = \dfrac{x^{19}}{x}\cdot\dfrac{y^5}{y} = x^{19-1}\cdot y^{5-1} = x^{18}y^4$

23. $\dfrac{2x^5y^2}{xy} = \dfrac{2}{1}\cdot\dfrac{x^5}{x}\cdot\dfrac{y^2}{y} = 2x^{5-1}y^{2-1} = 2x^4y$

24.
$$\begin{aligned} (-3a^2b)(5a^3b)&=-3\cdot 5\cdot a^2\cdot a^3\cdot b\cdot b\\ &=-15a^{2+3}b^{1+1}\\ &=-15a^5b^2 \end{aligned}$$

25. $2x^{-3} = 2\cdot\dfrac{1}{x^3} = \dfrac{2}{x^3}$

26. $7^{-2} = \dfrac{1}{7^2} = \dfrac{1}{49}$

Copyright © 2012 Pearson Education, Inc. Publishing as Prentice Hall.

27. $(-2)^{-4} = \dfrac{1}{(-2)^4} = \dfrac{1}{16}$

28. $5z^{-7} = 5 \cdot \dfrac{1}{z^7} = \dfrac{5}{z^7}$

29. $5x(2x^3+6) = 5x(2x^3) + 5x(6) = 10x^4 + 30x$

30. $(x+9)^2 = (x)^2 + 2(x)(9) + (9)^2 = x^2 + 18x + 81$

31. $-3x^2(5x^2+6x-1)$
$= -3x^2(5x^2) + (-3x^2)(6x) - (-3x^2)(1)$
$= -15x^4 - 18x^3 + 3x^2$

32. $(2x+1)(2x-1) = (2x)^2 - (1)^2 = 4x^2 - 1$

33. $\dfrac{4x^2+7+8x^3}{2x+3} = 4x^2 - 4x + 6 - \dfrac{11}{2x+3}$

$$\begin{array}{r} 4x^2 - 4x + 6 \\ 2x+3\overline{\big)\,8x^3 + 4x^2 + 0x^1 + 7} \\ \underline{8x^3 + 12x^2} \\ -8x^2 + 0x^1 \\ \underline{-8x^2 - 12x} \\ 12x + 7 \\ \underline{12x + 18} \\ -11 \end{array}$$

34. $\dfrac{4x^3-9x+2}{x-4} = 4x^2 + 16x + 55 + \dfrac{222}{x-4}$

$$\begin{array}{r} 4x^2 + 16x + 55 \\ x-4\overline{\big)\,4x^3 - 0x^2 - 9x + 2} \\ \underline{4x^3 - 16x^2} \\ 16x^2 - 9x \\ \underline{16x^2 - 64x} \\ 55x + 2 \\ \underline{55x - 220} \\ 222 \end{array}$$

35. $x^2 + 7x + 12 = (x+4)(x+3)$

36. $-2a^2 + 10a + 12 = -2(a^2 - 5a - 6)$
$= -2(a-6)(a+1)$

37. $25x^2 + 20xy + 4y^2 = (5x+2y)(5x+2y)$
$= (5x+2y)^2$

38. $x^2 - 4 = (x-2)(x+2)$

39. $x^2 - 9x - 22 = 0$
$(x-11)(x+2) = 0$
$x - 11 = 0$ or $x + 2 = 0$
$x = 11$ $\quad$ $x = -2$
The solutions are 11 and –2.

40. $3x^2 + 5x = 2$
$3x^2 + 5x - 2 = 0$
$(3x-1)(x+2) = 0$
$3x - 1 = 0$ or $x + 2 = 0$
$3x = 1$ $\quad$ $x = -2$
$x = \dfrac{1}{3}$

The solutions are $\dfrac{1}{3}$ and –2.

41. $\dfrac{x^2+x}{3x} \cdot \dfrac{6}{5x+5} = \dfrac{x(x+1)}{3x} \cdot \dfrac{2 \cdot 3}{5(x+1)} = \dfrac{2}{5}$

42. $\dfrac{2x^2-50}{4x^4-20x^3} = \dfrac{2(x^2-25)}{4x^3(x-5)}$
$= \dfrac{2(x-5)(x+5)}{2 \cdot 2x^3(x-5)}$
$= \dfrac{x+5}{2x^3}$

43. $\dfrac{3x^2+2x}{x-1} - \dfrac{10x-5}{x-1} = \dfrac{3x^2+2x-10x+5}{x-1}$
$= \dfrac{3x^2-8x+5}{x-1}$
$= \dfrac{(3x-5)(x-1)}{x-1}$
$= 3x - 5$

44. $7x^6 - 7x^5 + 7x^4 = 7x^4(x^2 - x + 1)$

45. $\dfrac{6x}{x^2-4} - \dfrac{3}{x+2} = \dfrac{6x}{(x+2)(x-2)} - \dfrac{3(x-2)}{(x+2)(x-2)}$
$= \dfrac{6x-3x+6}{(x+2)(x-2)}$
$= \dfrac{3x+6}{(x+2)(x-2)}$
$= \dfrac{3(x+2)}{(x+2)(x-2)}$
$= \dfrac{3}{x-2}$

Copyright © 2012 Pearson Education, Inc. Publishing as Prentice Hall.

46. $4x^2+12x+9=(2x+3)(2x+3)=(2x+3)^2$

47.
$$\frac{t-4}{2}-\frac{t-3}{9}=\frac{5}{18}$$
$$18\left(\frac{t-4}{2}-\frac{t-3}{9}\right)=18\left(\frac{5}{18}\right)$$
$$9(t-4)-2(t-3)=5$$
$$9t-36-2t+6=5$$
$$7t-30=5$$
$$7t=35$$
$$t=5$$
The solution is 5.

48.
$$\frac{6x^2-18x}{3x^2-2x}\cdot\frac{15x-10}{x^2-9}=\frac{6x(x-3)}{x(3x-2)}\cdot\frac{5(3x-2)}{(x-3)(x+3)}$$
$$=\frac{30}{x+3}$$

49. Let x be the time in hours for Sam and Frank to complete the tour together. Sam completes $\frac{1}{3}$ of a tour in 1 hour, and Frank completes $\frac{1}{7}$ of a tour in 1 hour. Together, they complete $\frac{1}{x}$ of a tour in 1 hour.
$$\frac{1}{3}+\frac{1}{7}=\frac{1}{x}$$
$$21x\left(\frac{1}{3}+\frac{1}{7}\right)=21x\left(\frac{1}{x}\right)$$
$$7x+3x=21$$
$$10x=21$$
$$x=\frac{21}{10}=2\frac{1}{10}$$
Sam and Frank can complete a quality control tour together in $2\frac{1}{10}$ hours.

50.
$$\frac{\frac{m}{3}+\frac{n}{6}}{\frac{m+n}{12}}=\frac{12\left(\frac{m}{3}+\frac{n}{6}\right)}{12\left(\frac{m+n}{12}\right)}=\frac{4m+2n}{m+n}=\frac{2(2m+n)}{m+n}$$

Copyright © 2012 Pearson Education, Inc. Publishing as Prentice Hall.

Chapter 6

Section 6.1 Practice

1. **a.** Look for the shortest bar, which is the bar representing the Africa/Middle East region. To approximate the number, move from the right edge of this bar vertically downward. This region has approximately 145 million Internet users.

   b. The Asia/Oceania/Australia region has approximately 785 million Internet users. The Africa/Middle East region has approximately 145 million Internet users. Subtract, 785 – 145 = 640, to find that approximately 640 million more Internet users are in the Asia/Oceania/Australia region.

2. **a.** Locate 40 along the time axis and move vertically upward to the line. Then move horizontally to the left until the pulse rate axis is reached. The pulse rate is 70 beats per minute 40 minutes after lighting a cigarette.

   b. Read the pulse rate for time = 0. The pulse rate is 60 beats per minute when the cigarette in being lit.

   c. Read the time when the pulse rate is the highest, i.e., when the line is at the peak. The pulse rate is the highest 5 minutes after lighting the cigarette.

3. **a.** Point (4, 2) lies in quadrant I.

   b. Point (–1, –3) lies in quadrant III.

   c. Point (2, –2) lies in quadrant IV.

   d. Point (–5, 1) lies in quadrant II.

   e. Point (0, 3) lies on the y-axis.

   f. Point (3, 0) lies on the x-axis.

   g. Point (0, –4) lies on the y-axis.

   h. Point $\left(-2\frac{1}{2}, 0\right)$ lies on the x-axis.

   i. Point $\left(1, -3\frac{3}{4}\right)$ lies in quadrant IV.

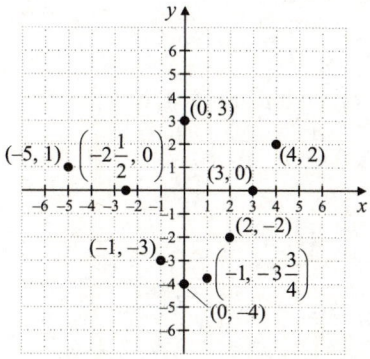

4. **a.** (2003, 1376), (2004, 1817), (2005, 1264), (2006, 1106), (2007, 1093), (2008, 1621)

   b.

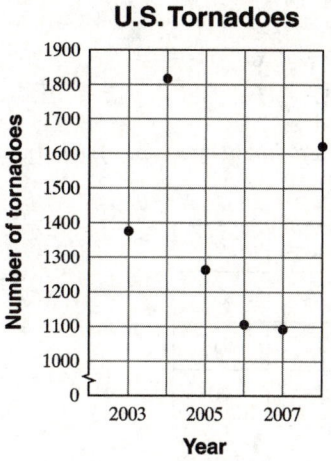

   c. The number of tornadoes varies greatly from year to year.

5. **a.** $x + 2y = 8$
   In (0,), the x-coordinate is 0.
   $0 + 2y = 8$
   $2y = 8$
   $y = 4$
   The ordered pair is (0, 4).

   b. $x + 2y = 8$
   In (, 3) the y-coordinate is 3.
   $x + 2(3) = 8$
   $x + 6 = 8$
   $x = 2$
   The ordered pair is (2, 3).

Copyright © 2012 Pearson Education, Inc. Publishing as Prentice Hall.

c. $x + 2y = 8$
In (–4,), the x-coordinate is –4.
$-4 + 2y = 8$
$2y = 12$
$y = 6$
The ordered pair is (–4, 6).

6. a. Let $x = -3$
$y = -2x$
$y = -2(-3)$
$y = 6$

b. Let $y = 0$.
$y = -2x$
$0 = -2x$
$0 = x$

c. Let $y = 10$.
$y = -2x$
$10 = -2x$
$-5 = x$

The ordered pairs are (–3, 6), (0, 0), and (–5, 10).

	x	y
a.	–3	6
b.	0	0
c.	–5	10

7. a. Let $x = -3$.
$y = \frac{1}{3}x - 1$
$y = \frac{1}{3}(-3) - 1$
$y = -1 - 1$
$y = -2$

b. Let $x = 0$.
$y = \frac{1}{3}x - 1$
$y = \frac{1}{3}(0) - 1$
$y = 0 - 1$
$y = -1$

c. Let $y = 0$.
$0 = \frac{1}{3}x - 1$
$1 = \frac{1}{3}x$
$3(1) = x$
$3 = x$

The ordered pairs are (–3, –2), (0, –1), and (3, 0).

	x	y
a.	–3	–2
b.	0	–1
c.	3	0

8. When $x = 1$,
$y = -50x + 400$
$y = -50(1) + 400$
$y = -50 + 400$
$y = 350$

When $x = 2$,
$y = -50x + 400$
$y = -50(2) + 400$
$y = -100 + 400$
$y = 300$

When $x = 3$,
$y = -50x + 400$
$y = -50(3) + 400$
$y = -150 + 400$
$y = 250$

When $x = 4$,
$y = -50x + 400$
$y = -50(4) + 400$
$y = -200 + 400$
$y = 200$

When $x = 5$,
$y = -50x + 400$
$y = -50(5) + 400$
$y = -250 + 400$
$y = 150$

When $x = 6$,
$y = -50x + 400$
$y = -50(6) + 400$
$y = -300 + 400$
$y = 100$

Copyright © 2012 Pearson Education, Inc. Publishing as Prentice Hall.

When $x = 7$,
$y = -50x + 400$
$y = -50(7) + 400$
$y = -350 + 400$
$y = 50$

The completed table is shown.

x	1	2	3	4	5	6	7
y	350	300	250	200	150	100	50

Vocabulary and Readiness Check

1. The horizontal axis is called the <u>x-axis</u>.
2. The vertical axis is called the <u>y-axis</u>.
3. The intersection of the horizontal axis and the vertical axis is a point called the <u>origin</u>.
4. The axes divide the plane into regions, called <u>quadrants</u>. There are <u>four</u> of these regions.
5. In the ordered pair of numbers (–2, 5), the number –2 is called the <u>x-coordinate</u> and the number 5 is called the <u>y-coordinate</u>.
6. Each ordered pair of numbers corresponds to <u>one</u> point in the plane.
7. An ordered pair is a <u>solution</u> of an equation in two variables if replacing the variables by the coordinates of the ordered pair results in a true statement.
8. The graph of paired data as points in a rectangular coordinate system is called a <u>scatter diagram</u>.

Exercise Set 6.1

1. The tallest bar corresponds to France, so France is the most popular tourist destination.
3. Find the bars that have heights greater than 40. France, U.S., Spain, Italy, and China have more than 40 million tourists per year.
5. The height of the bar is near 43, so approximately 43 million tourists go to Italy each year.
7. The line is at about 71,000 for year 2009. Thus, the attendance was 71,000 in 2009.
9. The highest point corresponds to 2005, and is at a height of about 78,100.
11. The line is at about 50 for year 1986. Thus, there were approximately 50 students per computer in 1986.
13. The greatest decrease was from 1984 to 1986.
15. The line falls below 15 in 1994.
17. 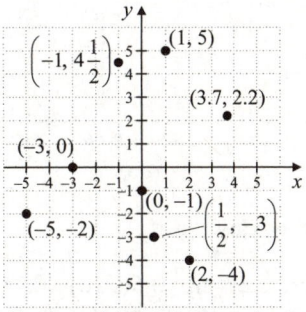

(1, 5) and (3.7, 2.2) are in quadrant I.
$\left(-1, 4\frac{1}{2}\right)$ is in quadrant II.
(–5, –2) is in quadrant III.
(2, –4) and $\left(\frac{1}{2}, -3\right)$ are in quadrant IV.
(–3, 0) lies on the x-axis.
(0, –1) lies on the y-axis.

19. Point A is at the origin, so its coordinates are (0, 0).
21. Point C is 3 units right and 2 units up from the origin, so its coordinates are (3, 2).
23. Point E is 2 units left and 2 units down from the origin, so its coordinates are (–2, –2).
25. Point G is 2 units right and 1 unit down from the origin, so its coordinates are (2, –1).
27. Point B is on the y-axis and 3 units down from the origin, so its coordinates are (0, –3).
29. Point D is 1 unit right and 3 units up from the origin, so its coordinates are (1, 3).
31. Point F is 3 units left and 1 unit down from the origin, so its coordinates are (–3, –1).
33. **a.** The ordered pairs are (2003, 9.17), (2004, 9.22), (2005, 8.83), (2006, 9.14), (2007, 9.63), (2008, 9.79).

Copyright © 2012 Pearson Education, Inc. Publishing as Prentice Hall.

b. The ordered pair (2006, 9.14) indicates that the domestic box office in 2006 was \$9.14 billion.

c.

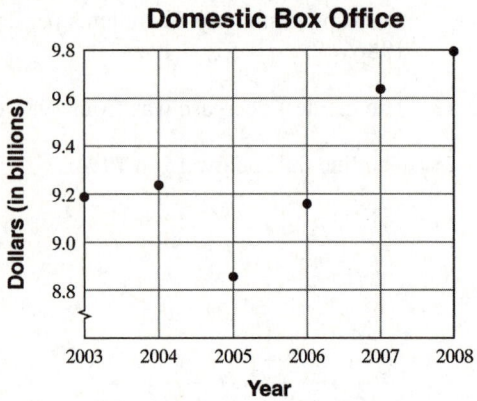

d. answers may vary

35. a. The ordered pairs are (0.50, 10), (0.75, 12), (1.00, 15), (1.25, 16), (1.50, 18), (1.50, 19), (1.75, 19), and (2.00, 20).

b. The ordered pair (1.25, 16) indicates that when Minh studied 1.25 hours, her quiz score was 16.

c.

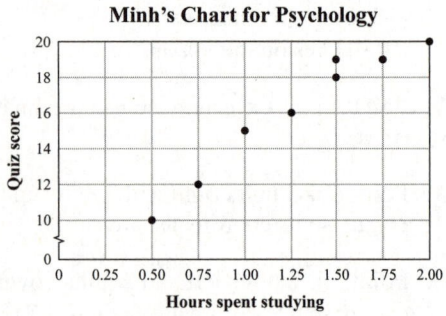

d. answers may vary

37. $x - 4y = 4$

In (, –2), the y-coordinate is –2.

$x - 4(-2) = 4$

$x + 8 = 4$

$x = -4$

In (4,), the x-coordinate is 4.

$4 - 4y = 4$

$-4y = 0$

$y = 0$

The completed coordinates are (–4, –2) and (4, 0).

39. $y = \frac{1}{4}x - 3$

In (–8,), the x-coordinate is –8.

$y = \frac{1}{4}(-8) - 3 = -2 - 3 = -5$

In (, 1), the y-coordinate is 1.

$1 = \frac{1}{4}x - 3$

$4 = \frac{1}{4}x$

$16 = x$

The completed coordinates are (–8, –5) and (16, 1).

41. $y = -7x$

$-\frac{1}{7}y = x$

$x = -\frac{1}{7}y$	$y = -7x$
0	$-7(0) = 0$
-1	$-7(-1) = 7$
$-\frac{1}{7}(2) = -\frac{2}{7}$	2

43. $x = -y + 2$

$x + y = 2$

$y = -x + 2$

$x = -y + 2$	$y = -x + 2$
0	$-0 + 2 = 2$
$-0 + 2 = 2$	0
-3	$-(-3) + 2 = 3 + 2 = 5$

45. $y = \frac{1}{2}x$

$2y = x$

$x = 2y$	$y = \frac{1}{2}x$
0	$\frac{1}{2}(0) = 0$
-6	$\frac{1}{2}(-6) = -3$
$2(1) = 2$	1

Copyright © 2012 Pearson Education, Inc. Publishing as Prentice Hall.

47. $x+3y=6$ $\quad$ $3y=-x+6$

$x=-3y+6$ $\quad$ $y=-\frac{1}{3}x+2$

$x=-3y+6$	$y=-\frac{1}{3}x+2$
0	$-\frac{1}{3}(0)+2=0+2=2$
$-3(0)+6=0+6=6$	0
$-3(1)+6=-3+6=3$	1

49. $y=2x-12$

$y+12=2x$

$\frac{1}{2}(y+12)=x$

$x=\frac{1}{2}(y+12)$	$y=2x-12$
0	$2(0)-12=0-12=-12$
$\frac{1}{2}(-2+12)=\frac{1}{2}(10)=5$	-2
3	$2(3)-12=6-12=-6$

51. $2x+7y=5$ $\quad$ $7y=-2x+5$

$2x=-7y+5$ $\quad$ $y=\frac{-2x+5}{7}$

$x=\frac{-7y+5}{2}$

$x=\frac{-7y+5}{2}$	$y=\frac{-2x+5}{7}$
0	$\frac{-2(0)+5}{7}=\frac{5}{7}$
$\frac{-7(0)+5}{2}=\frac{5}{2}$	0
$\frac{-7(1)+5}{2}=\frac{-2}{2}=-1$	1

53. $x=-5y$

$y=0$: $x=-5(0)=0$

$y=1$: $x=-5(1)=-5$

$x=10$: $10=-5y$

$-2=y$

The ordered pairs are (0, 0), (–5, 1), and (10, –2).

x	y
0	0
–5	1
10	–2

Copyright © 2012 Pearson Education, Inc. Publishing as Prentice Hall.

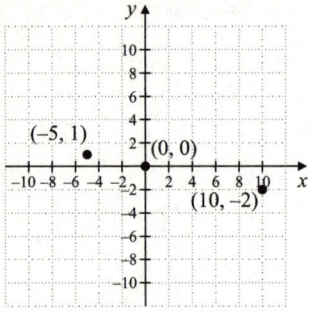

55. $y = \frac{1}{3}x + 2$

$x = 0$: $y = \frac{1}{3}(0) + 2 = 0 + 2 = 2$

$x = -3$: $y = \frac{1}{3}(-3) + 2 = -1 + 2 = 1$

$y = 0$: $0 = \frac{1}{3}x + 2$

$-2 = \frac{1}{3}x$

$-6 = x$

The ordered pairs are (0, 2), (−3, 1), and (−6, 0).

x	y
0	2
−3	1
−6	0

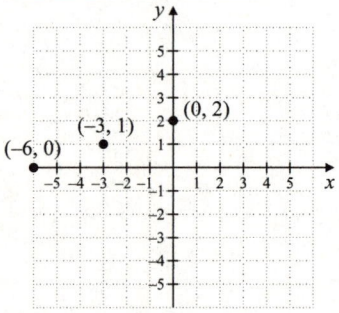

57. a.

x	100	200	300
$y = 80x + 5000$	80(100) + 5000 = 8000 + 5000 = 13,000	80(200) + 5000 = 16,000 + 5000 = 21,000	80(300) + 5000 = 24,000 + 5000 = 29,000

b. Find x when $y = 8600$.

$8600 = 80x + 5000$

$3600 = 80x$

$45 = x$

45 computer desks can be produced for $8600.

Copyright © 2012 Pearson Education, Inc. Publishing as Prentice Hall.

59. a.

x	1	3	5
$y = 0.2x + 5.39$	$0.2(1) + 5.39$ $= 0.2 + 5.39$ $= 5.59$	$0.2(3) + 5.39$ $= 0.6 + 5.39$ $= 5.99$	$0.2(5) + 5.39$ $= 1 + 5.39$ $= 6.39$

b. Find x when $y = 6.40$.

$6.40 = 0.2x + 5.39$
$1.01 = 0.2x$
$5.05 = x$
$5 \approx x$
$2000 + 5 = 2005$
The average cinema admission price was \$6.40 in 2005.

c. Find x when $y = 8.00$.

$8.00 = 0.2x + 5.39$
$2.61 = 0.2x$
$13.05 = x$
$13 \approx x$
$2000 + 13 = 2013$
The average cinema admission price is predicted to be \$8.00 in 2013.

61. $x + y = 5$
$y = 5 - x$

63. $2x + 4y = 5$
$4y = 5 - 2x$
$y = \dfrac{5 - 2x}{4}$

65. $10x = -5y$
$-2x = y$
$y = -2x$

67. False; point $(-1, 5)$ lies in quadrant II.

69. True

71. Points in quadrant III are to the left and down from the origin, so the x- and y-coordinates are negative. (negative, negative) corresponds to quadrant III.

73. Points in quadrant IV are to the right and down from the origin, so the x-coordinate is positive and the y-coordinate is negative. (positive, negative) corresponds to quadrant IV.

75. The origin corresponds to (0, 0).

77. If the x-coordinate of a point is 0, the point is neither to the left nor to the right of the origin, so it is on the y-axis.

79. no; answers may vary

81. answers may vary

83. answers may vary

Copyright © 2012 Pearson Education, Inc. Publishing as Prentice Hall.

85. A point four units to right of the y-axis and seven units below the x-axis has coordinates (4, –7).

87. The length of the rectangle is $3 - (-1) = 4$ and the width of the rectangle is $5 - (-4) = 9$.

$$\begin{aligned}\text{Perimeter} &= 2(\text{length}) + 2(\text{width})\\ &= 2(4) + 2(9)\\ &= 8 + 18\\ &= 26\end{aligned}$$

The perimeter is 26 units.

89. The revenues are approximately \$47 billion, \$53 billion, \$59 billion, and \$63 billion.

Section 6.2 Practice

1. $x + 3y = 6$

$x = 0$: $0 + 3y = 6$
$3y = 6$
$y = 2$

$x = 3$: $3 + 3y = 6$
$3y = 3$
$y = 1$

$y = 0$: $x + 3(0) = 6$
$x + 0 = 6$
$x = 6$

The ordered pairs are (0, 2), (3, 1), and (6, 0).

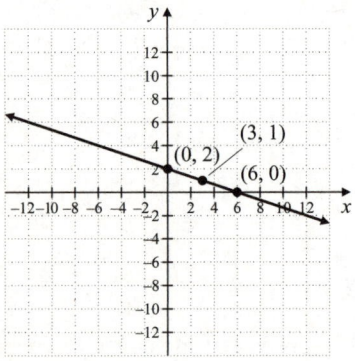

2. $-2x + 4y = 8$

$x = -2$: $-2(-2) + 4y = 8$
$4 + 4y = 8$
$4y = 4$
$y = 1$

$x = 0$: $-2(0) + 4y = 8$
$0 + 4y = 8$
$4y = 8$
$y = 2$

$x = 2$: $-2(2) + 4y = 8$
$-4 + 4y = 8$
$4y = 12$
$y = 3$

The ordered pairs are (–2, 1), (0, 2), and (2, 3).

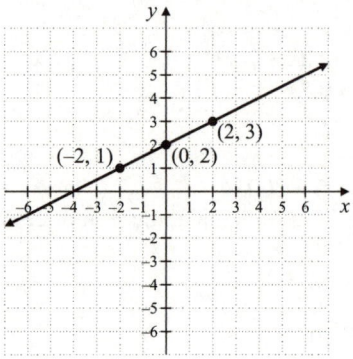

3. $y = 2x$

$x = -2$: $y = 2(-2) = -4$
$x = 0$: $y = 2(0) = 0$
$x = 3$: $y = 2(3) = 6$

The ordered pairs are (–2, –4), (0, 0), and (3, 6).

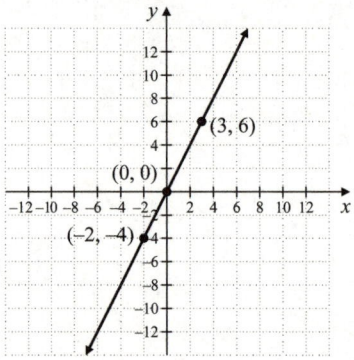

4. $y = -\frac{1}{2}x + 4$

$x = -6$: $y = -\frac{1}{2}(-6) + 4 = 3 + 4 = 7$

$x = 0$: $-\frac{1}{2}(0) + 4 = 0 + 4 = 4$

$x = 4$: $y = -\frac{1}{2}(4) + 4 = -2 + 4 = 2$

The ordered pairs are (–6, 7), (0, 4), and (4, 2).

Copyright © 2012 Pearson Education, Inc. Publishing as Prentice Hall.

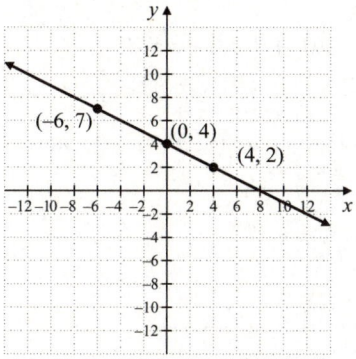

5. The equation $x = 3$ can be written in standard form as $x + 0y = 3$. No matter what value replaces y, x is always 3. It is a vertical line. Plot points (3, 2), (3, 0), and (3, –4), for example.

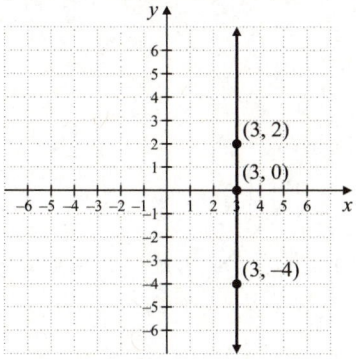

6. $x = 2015 - 2003 = 12$
Find 12 on the x-axis. Move vertically upward to the line and then horizontally to the left. In 2015, we predict that there will be 2840 thousand registered nurses.

Calculator Explorations

1. $y = -3x + 7$

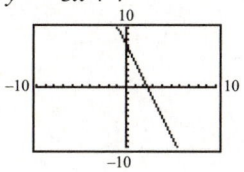

2. $y = -x + 5$

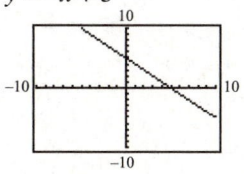

3. $y = 2.5x - 7.9$

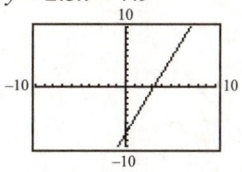

4. $y = -1.3x + 5.2$

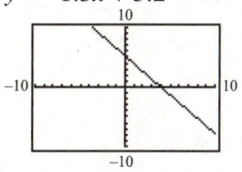

5. $y = -\frac{3}{10}x + \frac{32}{5}$

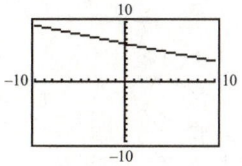

6. $y = \frac{2}{9}x - \frac{22}{3}$

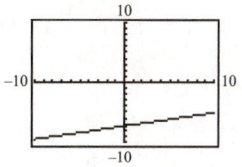

Exercise Set 6.2

1. $x - y = 6$

$y = 0$: $x - 0 = 6$
$x = 6$

$x = 4$: $4 - y = 6$
$-y = 2$
$y = -2$

$y = -1$: $x - (-1) = 6$
$x + 1 = 6$
$x = 5$

The ordered pairs are (6, 0), (4, –2), and (5, –1).

Copyright © 2012 Pearson Education, Inc. Publishing as Prentice Hall.

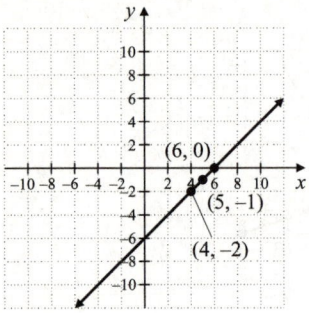

3. $y = -4x$

$x = 1$: $y = -4(1) = -4$

$x = 0$: $y = 4(0) = 0$

$x = -1$: $y = -4(-1) = 4$

The ordered pairs are (1, −4), (0, 0), and (−1, 4).

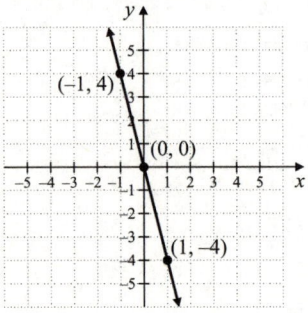

5. $y = \frac{1}{3}x$

$x = 0$: $y = \frac{1}{3}(0) = 0$

$x = 6$: $y = \frac{1}{3}(6) = 2$

$x = -3$: $y = \frac{1}{3}(-3) = -1$

The ordered pairs are (0, 0), (6, 2), and (−3, −1).

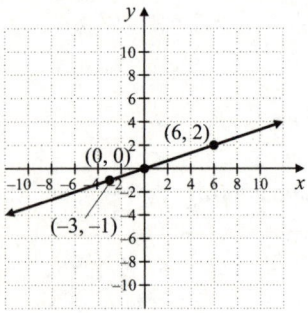

7. $y = -4x + 3$

$x = 0$: $y = -4(0) + 3 = 0 + 3 = 3$

$x = 1$: $y = -4(1) + 3 = -4 + 3 = -1$

$x = 2$: $y = -4(2) + 3 = -8 + 3 = -5$

The ordered pairs are (0, 3), (1, −1), and (2, −5).

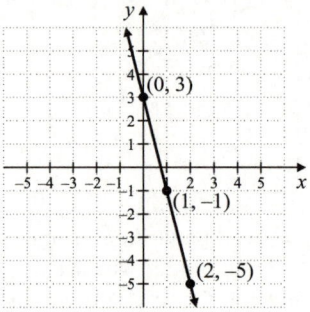

9. $x + y = 1$

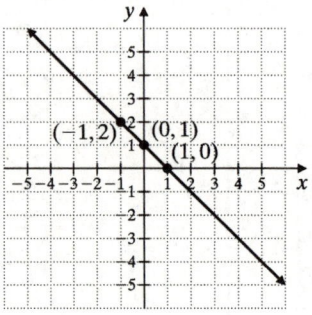

11. $x - y = -2$

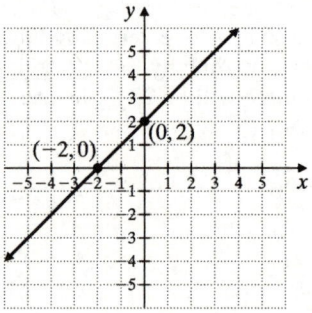

13. $x - 2y = 6$

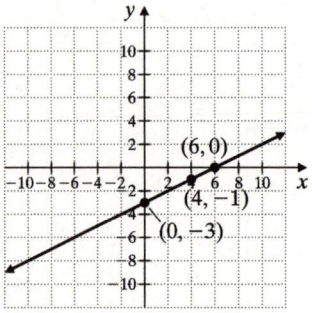

Copyright © 2012 Pearson Education, Inc. Publishing as Prentice Hall.

15. $y = 6x + 3$

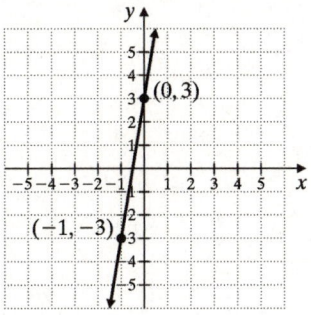

17. $x = -4$

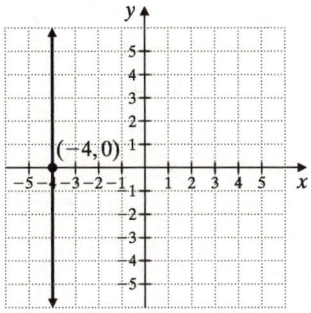

19. $y = 3$

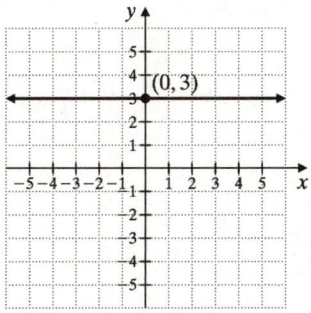

21. $y = x$

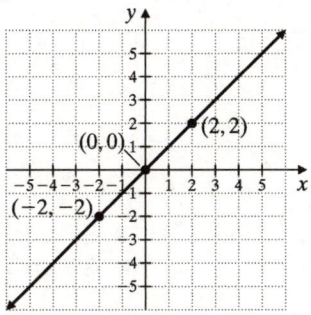

23. $x = -3y$

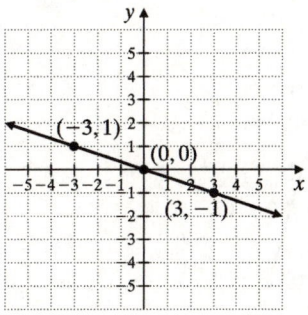

25. $x + 3y = 9$

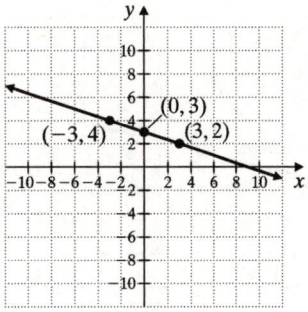

27. $y = \frac{1}{2}x + 2$

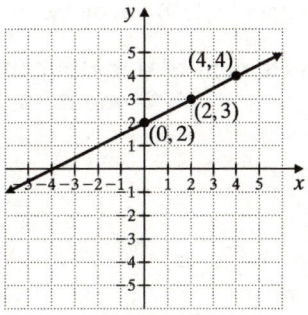

29. $3x - 2y = 12$

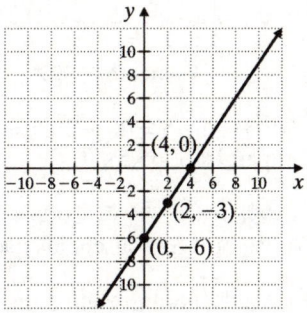

Copyright © 2012 Pearson Education, Inc. Publishing as Prentice Hall.

31. $y = -3.5x + 4$

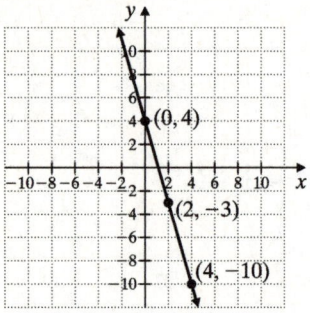

33. a.

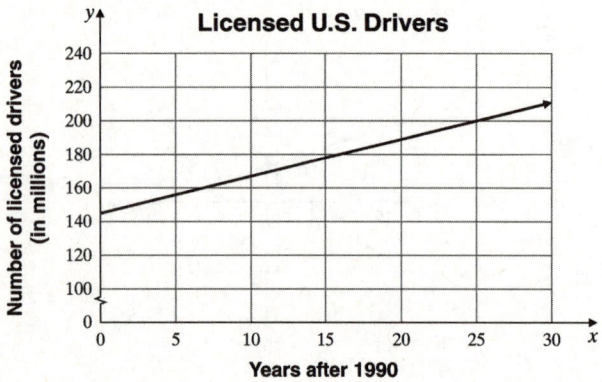

b. $x = 20$: $y = 2.2(20) + 145 = 44 + 145 = 189$
Yes, the point (20, 189) lies on the line. answers may vary

35. a.

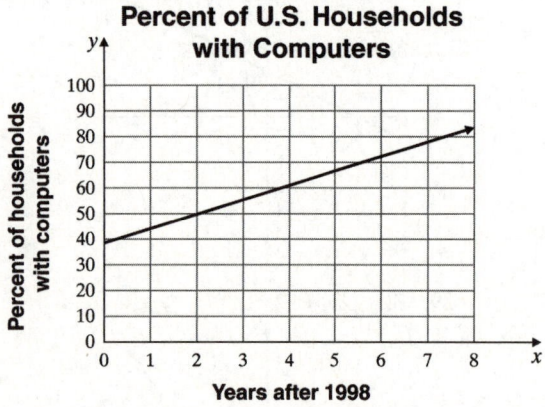

b. $x = 5$: $y = 5.6(5) + 38.5 = 28 + 38.5 = 66.5$
The ordered pair is (5, 66.5).

c. Year $= x + 1998 = 5 + 1998 = 2003$
In 2003, 66.5% of U.S. households had at least one computer.

37. The fourth vertex is the bottom-right corner of the rectangle. The x-coordinate must line the point up with the top-right corner, and the y-coordinate must line the point up with the bottom-left corner. The coordinates are (4, −1).

Copyright © 2012 Pearson Education, Inc. Publishing as Prentice Hall.

39. $x - y = -3$
$x = 0$: $0 - y = -3$
$y = 3$
$y = 0$: $x - 0 = -3$
$x = -3$

x	y
0	3
−3	0

41. $y = 2x$
$x = 0$: $y = 2(0) = 0$
$y = 0$: $0 = 2x$
$0 = x$

x	y
0	0
0	0

43. $y = 5x$
$y = 5x + 4$

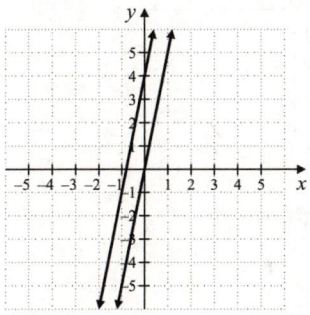

45. $y = -2x$
$y = -2x - 3$

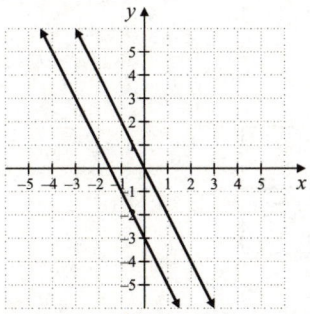

47. $y = x^2$

$x = 0$: $y = 0^2 = 0$

$x = 1$: $y = 1^2 = 1$

$x = -1$: $y = (-1)^2 = 1$

$x = 2$: $y = 2^2 = 4$

$x = -2$: $y = (-2)^2 = 4$

x	y
0	0
1	1
−1	1
2	4
−2	4

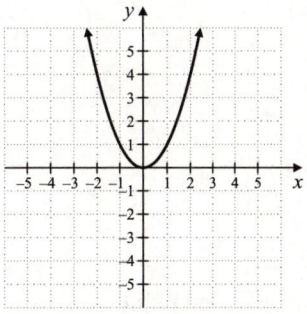

49. The perimeter is the distance around.
$x + 5 + y + 5 = 22$
$x + y + 10 = 22$
$x + y = 12$
$x = 3$: $3 + y = 12$
$y = 9$
If x is 3 centimeters, then y is 9 centimeters.

51. yes; answers may vary

Section 6.3 Practice

1. x-intercept: (2, 0)
y-intercept: (0, −4)

2. x-intercepts: (−4, 0), (2, 0)
y-intercept: (0, 2)

3. x-intercept and y-intercept: (0, 0)

4. $2x - y = 4$
$y = 0$: $2x - 0 = 4$
$2x = 4$
$x = 2$
x-intercept: (2, 0)

Copyright © 2012 Pearson Education, Inc. Publishing as Prentice Hall.

$x = 0$: $2(0) - y = 4$
$0 - y = 4$
$-y = 4$
$y = -4$

y-intercept: $(0, -4)$

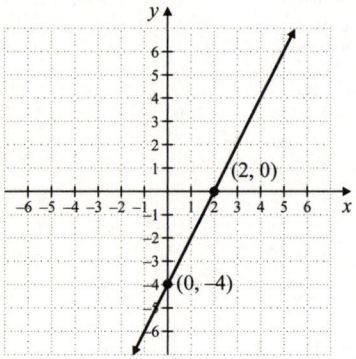

5. $y = 3x$
$y = 0$: $0 = 3x$
$0 = x$
x-intercept: $(0, 0)$
$x = 0$: $y = 3(0)$
$y = 0$
y-intercept: $(0, 0)$
Let $x = 1$ to find a second point.
$x = 1$: $y = 3$
$(1, 3)$ is another point on the line.

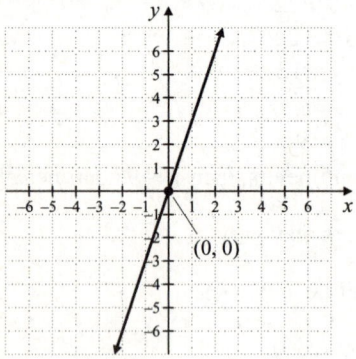

6. $x = -3$
The graph of $x = -3$ is a vertical line with x-intercept $(-3, 0)$.

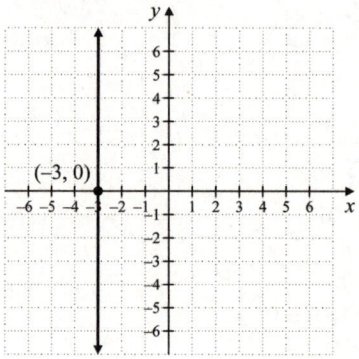

7. $y = 4$
The graph of $y = 4$ is a horizontal line with y-intercept $(0, 4)$.

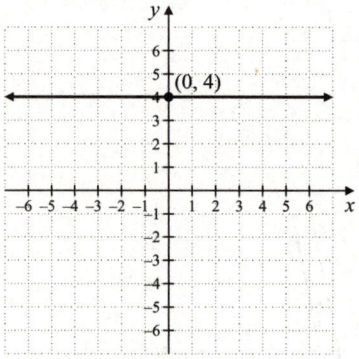

Calculator Explorations

1. $x = 3.78y$

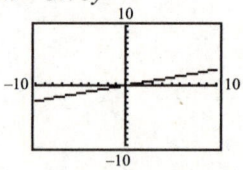

2. $-2.61y = x$

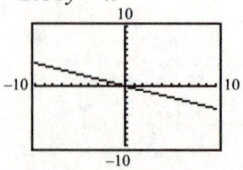

3. $3x + 7y = 21$

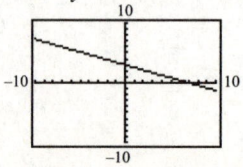

Copyright © 2012 Pearson Education, Inc. Publishing as Prentice Hall.

4. $-4x + 6y = 12$

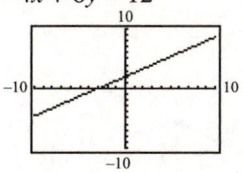

5. $-2.2x + 6.8y = 15.5$

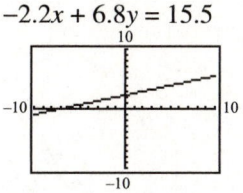

6. $5.9x - 0.8y = -10.4$

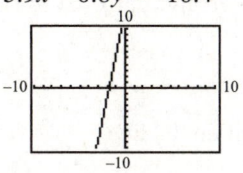

Vocabulary and Readiness Check

1. An equation that can be written in the form $Ax + By = C$ is called a <u>linear</u> equation in two variables.

2. The form $Ax + By = C$ is called <u>standard</u> form.

3. The graph of the equation $y = -1$ is a <u>horizontal</u> line.

4. The graph of the equation $x = 5$ is a <u>vertical</u> line.

5. A point where a graph crosses the y-axis is called a <u>y-intercept</u>.

6. A point where a graph crosses the x-axis is called an <u>x-intercept</u>.

7. Given an equation of a line, to find the x-intercept (if there is one), let <u>y</u> = 0 and solve for <u>x</u>.

8. Given an equation of a line, to find the y-intercept (if there is one), let <u>x</u> = 0 and solve for <u>y</u>.

9. False; a horizontal line (other than $y = 0$) has no x-intercept and a vertical line (other than $x = 0$) has no y-intercept.

10. True

11. True

12. False; $x = 5$: $y = 5(5) = 25$, so the point (5, 1) is not on the graph of $y = 5x$.

Exercise Set 6.3

1. x-intercept: (−1, 0)
y-intercept: (0, 1)

3. x-intercepts: (−2, 0), (2, 0)
y-intercept: (0, −2)

5. x-intercepts: (−2, 0), (1, 0), (3, 0)
y-intercept: (0, 3)

7. x-intercepts: (−1, 0), (1, 0)
y-intercepts: (0, 1), (0, −2)

9. $x - y = 3$
$y = 0$: $x - 0 = 0$
$x = 3$
x-intercept: (3, 0)
$x = 0$: $0 - y = 3$
$-y = 3$
$y = -3$
y-intercept: (0, −3)

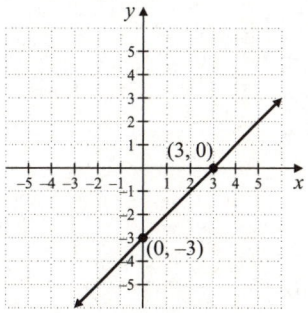

11. $x = 5y$
$y = 0$: $x = 5(0) = 0$
x-intercept: (0, 0)
$x = 0$: $0 = 5y$
$0 = y$
y-intercept: (0, 0)
Let $y = 1$ to find a second point.
$y = 1$: $x = 5$
(5, 1) is another point on the line.

Copyright © 2012 Pearson Education, Inc. Publishing as Prentice Hall.

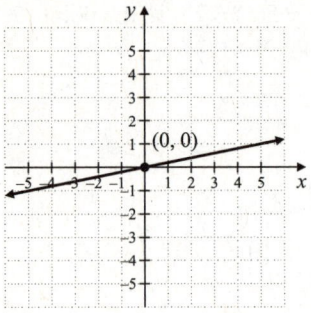

13. $-x + 2y = 6$

$y = 0$: $-x + 2(0) = 6$

$-x = 6$

$x = -6$

x-intercept: $(-6, 0)$

$x = 0$: $-0 + 2y = 6$

$2y = 6$

$y = 3$

y-intercept: $(0, 3)$

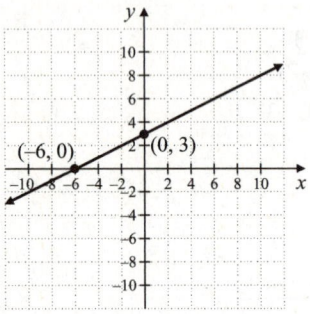

15. $2x - 4y = 8$

$y = 0$: $2x - 4(0) = 8$

$2x - 0 = 8$

$2x = 8$

$x = 4$

x-intercept: $(4, 0)$

$x = 0$: $2(0) - 4y = 8$

$0 - 4y = 8$

$-4y = 8$

$y = -2$

y-intercept: $(0, -2)$

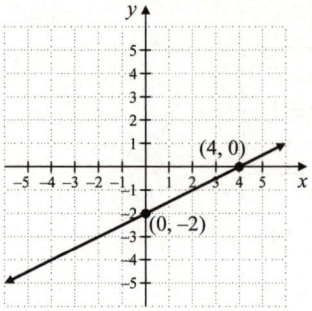

17. $y = 2x$

$y = 0$: $0 = 2x$

$0 = x$

x-intercept: $(0, 0)$

$(0, 0)$ is also the y-intercept. Let $x = 1$ to find a second point.

$x = 1$: $y = 2(1)$

$y = 2$

$(1, 2)$ is another point on the line.

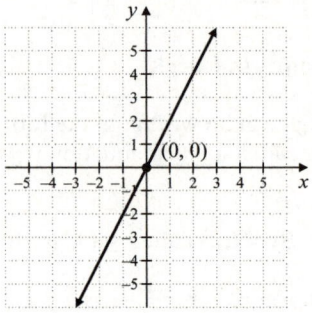

19. $y = 3x + 6$

$y = 0$: $0 = 3x + 6$

$-6 = 3x$

$-2 = x$

x-intercept: $(-2, 0)$

$x = 0$: $y = 3(0) + 6$

$y = 0 + 6$

$y = 6$

y-intercept: $(0, 6)$

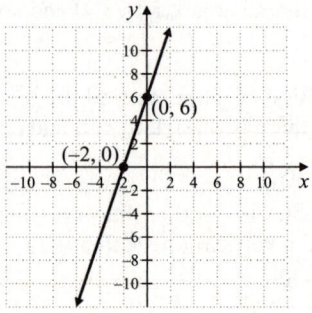

Copyright © 2012 Pearson Education, Inc. Publishing as Prentice Hall.

21. The graph of $x = -1$ is a vertical line with x-intercept $(-1, 0)$.

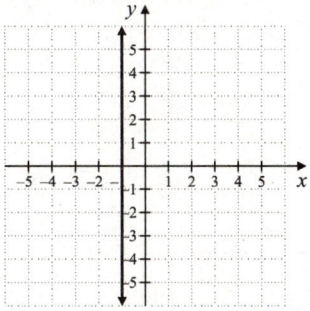

23. The graph of $y = 0$ is a horizontal line with y-intercept $(0, 0)$.

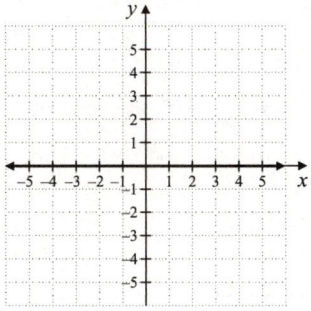

25. $y + 7 = 0$

$y = -7$

The graph of $y = -7$ is a horizontal line with y-intercept $(0, -7)$.

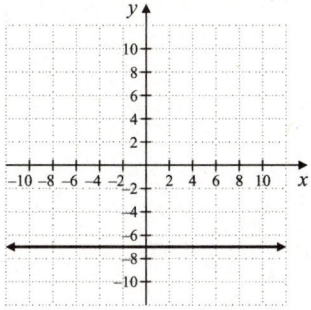

27. $x + 3 = 0$

$x = -3$

The graph of $x = -3$ is a vertical line with x-intercept $(0, -3)$.

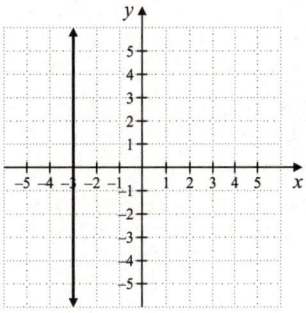

29. $x = y$

$y = 0$: $x = 0$

x-intercept: $(0, 0)$

$(0, 0)$ is also the y-intercept. Let $x = 3$ to find a second point.

$x = 3$: $3 = y$

$(3, 3)$ is another point on the line.

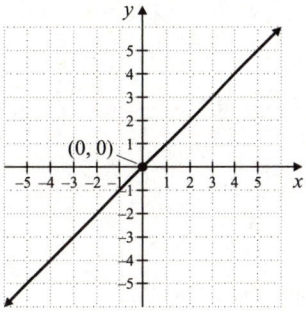

31. $x + 8y = 8$

$y = 0$: $x + 8(0) = 8$

$x + 0 = 8$

$x = 8$

x-intercept: $(8, 0)$

$x = 0$: $0 + 8y = 8$

$8y = 8$

$y = 1$

y-intercept: $(0, 1)$

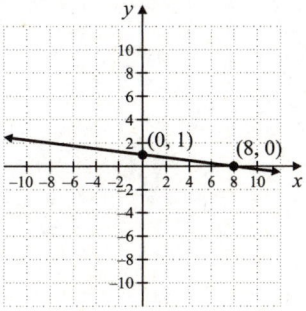

Copyright © 2012 Pearson Education, Inc. Publishing as Prentice Hall.

33. $5 = 6x - y$

$y = 0$: $5 = 6x - 0$

$5 = 6x$

$\frac{5}{6} = x$

x-intercept: $\left(\frac{5}{6}, 0\right)$

$x = 0$: $5 = 6(0) - y$

$5 = -y$

$-5 = y$

y-intercept: $(0, -5)$

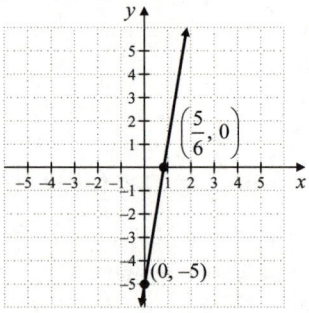

35. $-x + 10y = 11$

$y = 0$: $-x + 10(0) = 11$

$-x + 0 = 11$

$-x = 11$

$x = -11$

x-intercept: $(-11, 0)$

$x = 0$: $-0 + 10y = 11$

$10y = 11$

$y = \frac{11}{10}$

y-intercept: $\left(0, \frac{11}{10}\right)$

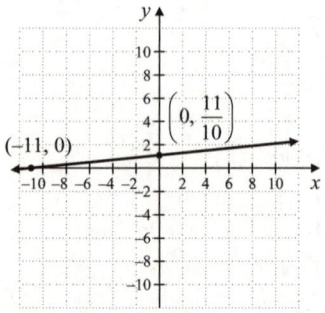

37. $x = -4\frac{1}{2}$

This is a vertical line with x-intercept $\left(-4\frac{1}{2}, 0\right)$.

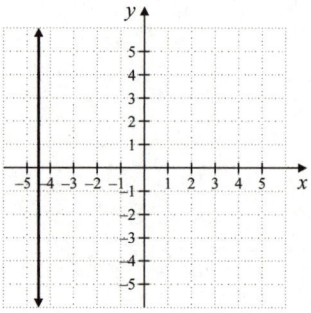

39. $y = 3\frac{1}{4}$

This is a horizontal line with y-intercept $\left(0, 3\frac{1}{4}\right)$.

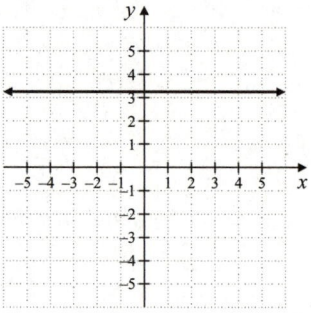

41. $y = -\frac{2}{3}x + 1$

$y = 0$: $0 = -\frac{2}{3}x + 1$

$\frac{2}{3}x = 1$

$x = \frac{3}{2}$

x-intercept: $\left(\frac{3}{2}, 0\right)$

$x = 0$: $y = -\frac{2}{3}(0) + 1$

$y = 1$

y-intercept: $(0, 1)$

Copyright © 2012 Pearson Education, Inc. Publishing as Prentice Hall.

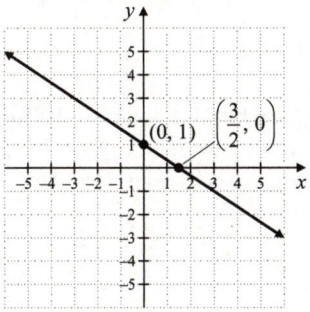

43. $4x - 6y + 2 = 0$

$y = 0$: $4x - 6(0) + 2 = 0$

$4x - 0 + 2 = 0$

$4x = -2$

$x = -\frac{1}{2}$

x-intercept: $\left(-\frac{1}{2}, 0\right)$

$x = 0$: $4(0) - 6y + 2 = 0$

$0 - 6y + 2 = 0$

$-6y = -2$

$y = \frac{1}{3}$

y-intercept: $\left(0, \frac{1}{3}\right)$

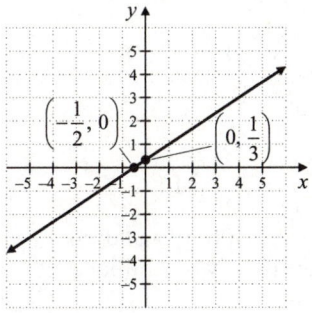

45. $\frac{-6-3}{2-8} = \frac{-9}{-6} = \frac{3}{2}$

47. $\frac{-8-(-2)}{-3-(-2)} = \frac{-8+2}{-3+2} = \frac{-6}{-1} = 6$

49. $\frac{0-6}{5-0} = \frac{-6}{5} = -\frac{6}{5}$

51. The graph of $y = 3$ is a horizontal line with y-intercept (0, 3). This is graph c.

53. The graph of $x = 3$ is a vertical line with x-intercept (3, 0). This is graph a.

55. The line can be on the x-axis or the y-axis so it can have infinitely many x- and y-intercepts.

57. A circle can have no x- and y-intercepts. That is, it does not have to intersect the axes.

59. answers may vary

61. $3x + 6y = 1200$

$x = 0$: $3(0) + 6y = 1200$

$6y = 1200$

$y = 200$

The ordered pair (0, 200) corresponds to manufacturing 0 chairs and 200 desks.

63. Manufacturing 50 desks corresponds to $y = 50$.

$3x + 6y = 1200$

$y = 50$: $3x + 6(50) = 1200$

$3x + 300 = 1200$

$3x = 900$

$x = 300$

When 50 desks are manufactured, 300 chairs can be manufactured.

65.

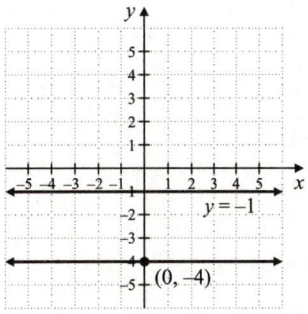

The equation of the line is $y = -4$.

67. a. $y = -1.9x + 59$

$y = 0$: $0 = -1.9x + 59$

$1.9x = 59$

$x \approx 31.1$

The x-intercept is (31.1, 0).

b. The x-intercept of (31.1, 0) means that 31.1 years after 2003, there may be no newspaper circulation.

Copyright © 2012 Pearson Education, Inc. Publishing as Prentice Hall.

Section 6.4 Practice

1. $m = \frac{y_2 - y_1}{x_2 - x_1} = \frac{-1-3}{4-(-2)} = \frac{-4}{6} = -\frac{2}{3}$

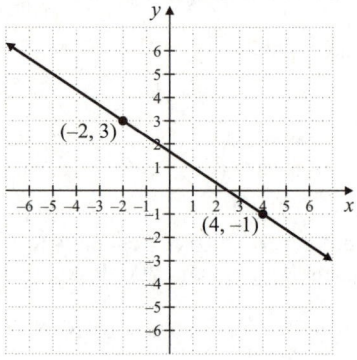

2. $m = \frac{y_2 - y_1}{x_2 - x_1} = \frac{5-1}{3-(-2)} = \frac{4}{5}$

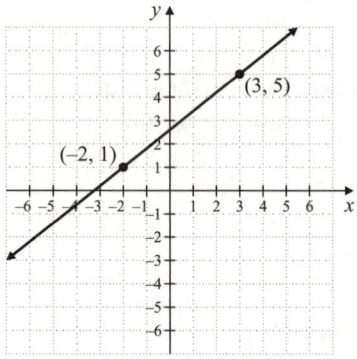

3. $5x + 4y = 10$

$4y = -5x + 10$

$y = -\frac{5}{4}x + \frac{10}{4}$

The slope is $-\frac{5}{4}$.

4. $-y = -2x + 7$

$\frac{-y}{-1} = \frac{-2x}{-1} + \frac{7}{-1}$

$y = 2x - 7$

The slope is 2.

5. $y = 3$ is a horizontal line. Horizontal lines have a slope of 0.

6. $x = -2$ is a vertical line. Vertical lines have undefined slopes.

7. a. $x + y = 5$ $\quad$ $2x + y = 5$

$y = -x + 5$ $\quad$ $y = -2x + 5$

slope = −1 $\quad$ slope = −2

The slopes are not the same, so the lines are not parallel. The product, (−1)(−2) = 2, is not −1, so the lines are not perpendicular.

b. $5y = 2x - 3$ $\quad$ $5x + 2y = 1$

$y = \frac{2}{5}x - \frac{3}{5}$ $\quad$ $2y = -5x + 1$

$y = -\frac{5}{2}x + \frac{1}{2}$

slope $= \frac{2}{5}$ $\quad$ slope $= -\frac{5}{2}$

The slopes are not the same, so the lines are not parallel. The product, $\left(\frac{2}{5}\right)\left(-\frac{5}{2}\right) = -1$, is −1, so the lines are perpendicular.

c. $y = 2x + 1$ $\quad$ $4x - 2y = 8$

$-2y = -4x + 8$

$y = \frac{-4x}{-2} + \frac{8}{-2}$

$y = 2x - 4$

slope = 2 $\quad$ slope = 2

The slopes are the same, so the lines are parallel.

8. grade $= \frac{\text{rise}}{\text{run}} = \frac{3}{20} = 0.15 = 15\%$

The grade is 15%.

9. $m = \frac{240-120}{1990-1980} = \frac{120}{10} = 12$

Each year the sales of food and drink from restaurants increases by \$12 billion.

Calculator Explorations

1.

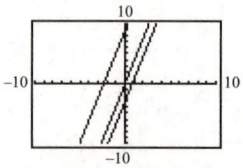

The lines are parallel since they all have a slope of 3.8. The graph of $y = 3.8x - 3$ is the graph of $y = 3.8x$ moved 3 units down with a y-intercept of −3. The graph of $y = 3.8x + 9$ is the graph of $y = 3.8x$ moved 9 units up with a y-intercept of 9.

Copyright © 2012 Pearson Education, Inc. Publishing as Prentice Hall.

2.

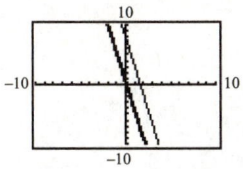

The lines are parallel since they all have a slope of −4.9. The graph of $y = -4.9x + 1$ is the graph of $y = -4.9x$ moved 1 unit up with a y-intercept of 1. The graph of $y = -4.9x + 8$ is the graph of $y = -4.9x$ moved 8 units up with a y-intercept of 8.

3.

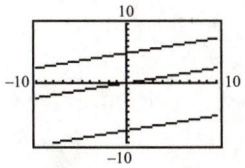

The lines are parallel since they all have a slope of $\frac{1}{4}$. The graph of $y = \frac{1}{4}x + 5$ is the graph of $y = \frac{1}{4}x$ moved 5 units up with a y-intercept of 5. The graph of $y = \frac{1}{4}x - 8$ is the graph of $y = \frac{1}{4}x$ moved 8 units down with a y-intercept of −8.

4.

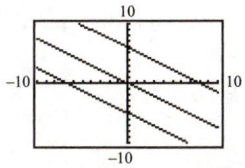

The lines are parallel since they all have a slope of $-\frac{3}{4}$. The graph of $y = -\frac{3}{4}x - 5$ is the graph of $y = -\frac{3}{4}x$ moved 5 units down with a y-intercept of −5. The graph of $y = -\frac{3}{4}x + 6$ is the graph of $y = -\frac{3}{4}x$ moved 6 units up with a y-intercept of 6.

Vocabulary and Readiness Check

1. The measure of the steepness or tilt of a line is called <u>slope</u>.
2. If an equation is written in the form $y = mx + b$, the value of the letter <u>m</u> is the value of the slope of the graph.
3. The slope of a horizontal line is <u>0</u>.
4. The slope of a vertical line is <u>undefined</u>.
5. If the graph of a line moves upward from left to right, the line has <u>positive</u> slope.
6. If the graph of a line moves downward from left to right, the line has <u>negative</u> slope.
7. Given two points of a line, slope $= \dfrac{\text{change in } \underline{y}}{\text{change in } \underline{x}}$.
8. The line slants downward, so its slope is negative.
9. The line slants upward, so its slope is positive.
10. The line is vertical, so its slope is undefined.
11. The line is horizontal, so its slope is 0.
12. Since $m = \frac{7}{6}$ is positive, the line slants upward.
13. Since $m = -3$ is negative, the line slants downward.
14. Since $m = 0$, the line is horizontal.
15. Since m is undefined, the line is vertical.

Exercise Set 6.4

1. $m = \dfrac{y_2 - y_1}{x_2 - x_1} = \dfrac{-2-5}{6-(-1)} = \dfrac{-7}{7} = -1$

3. $m = \dfrac{y_2 - y_1}{x_2 - x_1} = \dfrac{3-4}{5-1} = \dfrac{-1}{4} = -\dfrac{1}{4}$

5. $m = \dfrac{y_2 - y_1}{x_2 - x_1} = \dfrac{1-1}{-2-5} = \dfrac{0}{-7} = 0$

7. $m = \dfrac{y_2 - y_1}{x_2 - x_1} = \dfrac{5-3}{-4-(-4)} = \dfrac{5-3}{-4+4} = \dfrac{2}{0}$

   The slope is undefined.

Copyright © 2012 Pearson Education, Inc. Publishing as Prentice Hall.

9. $(x_1, y_1) = (-1, 2), (x_2, y_2) = (2, -2)$

$$m = \frac{y_2 - y_1}{x_2 - x_1} = \frac{-2-2}{2-(-1)} = \frac{-4}{3} = -\frac{4}{3}$$

11. $(x_1, y_1) = (1, -2), (x_2, y_2) = (3, 3)$

$$m = \frac{y_2 - y_1}{x_2 - x_1} = \frac{3-(-2)}{3-1} = \frac{3+2}{3-1} = \frac{5}{2}$$

13. Line 1 has a positive slope and line 2 has a negative slope, so line 1 has the greater slope.

15. Line 2 is steeper, so it has the greater slope.

17. $y = 5x - 2$
The slope is $m = 5$.

19. $y = -0.3x + 2.5$
The slope is $m = -0.3$.

21. $2x + y = 7$
$y = -2x + 7$
The slope is $m = -2$.

23. The line is vertical, so it has an undefined slope.

25. $2x - 3y = 10$
$-3y = -2x + 10$
$y = \frac{2}{3}x - \frac{10}{3}$
The slope is $m = \frac{2}{3}$.

27. $x = 1$ is a vertical line, so its slope is undefined.

29. $x = 2y$
$\frac{1}{2}x = y$
$y = \frac{1}{2}x$
The slope is $m = \frac{1}{2}$.

31. $y = -3$ is a horizontal line, so its slope is 0.

33. $-3x - 4y = 6$
$-4y = 3x + 6$
$y = -\frac{3}{4}x - \frac{3}{2}$
The slope is $m = -\frac{3}{4}$.

35. $20x - 5y = 1.2$
$-5y = -20x + 1.2$
$y = \frac{-20x}{-5} + \frac{1.2}{-5}$
$y = 4x - 0.24$
The slope is $m = 4$.

37. $y = \frac{2}{9}x + 3$
$y = -\frac{2}{9}x$
$\frac{2}{9} \neq -\frac{2}{9}$, so the lines are not parallel.
$\left(\frac{2}{9}\right)\left(-\frac{2}{9}\right) = -\frac{4}{81} \neq -1$, so the lines are not perpendicular.
The lines are neither parallel nor perpendicular.

39. $x - 3y = -6$
$-3y = -x - 6$
$y = \frac{1}{3}x + 2$
$y = 3x - 9$
$\frac{1}{3} \neq 3$, so the lines are not parallel.
$\left(\frac{1}{3}\right)(3) = 1 \neq -1$, so the lines are not perpendicular. The lines are neither parallel nor perpendicular.

41. $6x = 5y + 1$
$-5y = -6x + 1$
$y = \frac{6}{5}x - \frac{1}{5}$
$-12x + 10y = 1$
$10y = 12x + 1$
$y = \frac{6}{5}x + \frac{1}{10}$
Both lines have slope $\frac{6}{5}$ and the y-intercepts are different, so they are parallel.

Copyright © 2012 Pearson Education, Inc. Publishing as Prentice Hall.

43. $6 + 4x = 3y$

$2+\frac{4}{3}x = y$ or $y = \frac{4}{3}x+2$

$3x+4y=8$

$4y=-3x+8$

$y=-\frac{3}{4}x+2$

$\left(\frac{4}{3}\right)\left(-\frac{3}{4}\right)=-1$, so the lines are perpendicular.

45. $m=\frac{y_2-y_1}{x_2-x_1}=\frac{0-(-3)}{0-(-3)}=\frac{3}{3}=1$

a. The slope of a parallel line is 1.

b. The slope of a perpendicular line is $-\frac{1}{1}=-1$.

47. $m=\frac{y_2-y_1}{x_2-x_1}=\frac{5-(-4)}{3-(-8)}=\frac{5+4}{3+8}=\frac{9}{11}$

a. The slope of a parallel line is $\frac{9}{11}$.

b. The slope of a perpendicular line is $-\frac{1}{\frac{9}{11}}=-\frac{11}{9}$.

49. $\text{slope}=\frac{\text{rise}}{\text{run}}=\frac{6\text{ feet}}{10\text{ feet}}=\frac{3}{5}$

The pitch of the roof is $\frac{3}{5}$.

51. $\text{slope}=\frac{\text{rise}}{\text{run}}=\frac{2}{16}=0.125=12.5\%$

The grade of the road is 12.5%

53. $\text{grade}=\frac{\text{rise}}{\text{run}}=\frac{2580\text{ meters}}{6450\text{ meters}}=0.40=40\%$

The grade of the track is 40%.

55. Canton Avenue:

$\text{grade}=\frac{\text{rise}}{\text{run}}=\frac{11\text{ meters}}{30\text{ meters}}\approx 0.37=37\%$

The grade of Canton Avenue is 37%.

Baldwin Street:

$\text{grade}=\frac{\text{rise}}{\text{run}}=\frac{1\text{ meter}}{2.86\text{ meters}}\approx 0.35=35\%$

The grade of Baldwin Street is 35%.

57. $m=\frac{y_2-y_1}{x_2-x_1}=\frac{112-109.5}{2007-2005}=\frac{2.5}{2}=\frac{5}{4}$

Every 4 years there are/should be 5 million more U.S. households with televisions.

59. $m=\frac{y_2-y_1}{x_2-x_1}=\frac{9.4-8.8}{2008-2004}=\frac{0.6}{4}=0.15$

Every year, the median age of automobiles in the United States increases by 0.15 year.

61. $y-(-6)=2(x-4)$

$y+6=2x-8$

$y=2x-14$

63. $y-1=-6(x-(-2))$

$y-1=-6(x+2)$

$y-1=-6x-12$

$y=-6x-11$

65. $(x_1, y_1)=(0, 0)$, $(x_2, y_2)=(1, 1)$

$m=\frac{y_2-y_1}{x_2-x_1}=\frac{1-0}{1-0}=\frac{1}{1}=1$

The slope is $m = 1$; d.

67. The line is vertical, so its slope is undefined; b.

69. $(x_1, y_1)=(2, 0)$, $(x_2, y_2)=(4, -1)$

$m=\frac{y_2-y_1}{x_2-x_1}=\frac{-1-0}{4-2}=\frac{-1}{2}=-\frac{1}{2}$

The slope is $m=-\frac{1}{2}$; e.

71. $m=\frac{y_2-y_1}{x_2-x_1}=\frac{0-1}{0-2}=\frac{-1}{-2}=\frac{1}{2}$

$\frac{-1-1}{-2-2}=\frac{-2}{-4}=\frac{1}{2}$; $\frac{-2-1}{-4-2}=\frac{-3}{-6}=\frac{1}{2}$

$\frac{-1-0}{-2-0}=\frac{-1}{-2}=\frac{1}{2}$; $\frac{-2-0}{-4-0}=\frac{-2}{-4}=\frac{1}{2}$

$\frac{-2-(-1)}{-4-(-2)}=\frac{-2+1}{-4+2}=\frac{-1}{-2}=\frac{1}{2}$

73. answers may vary

75. From the graph, for year 2004 the average fuel economy was 29.5 miles per gallon.

Copyright © 2012 Pearson Education, Inc. Publishing as Prentice Hall.

77. The lowest point on the graph corresponds to 1999. The average fuel economy for that year was 28.3 miles per gallon.

79. Of the line segments listed, the segment from 2006 to 2007 is the steepest and therefore has the greatest slope.

81. $\text{pitch} = \dfrac{\text{rise}}{\text{run}}$

$$\frac{2}{5} = \frac{4}{\frac{x}{2}}$$

$$2 \cdot \frac{x}{2} = 5 \cdot 4$$

$$x = 20$$

83. a. (2004, 2025) and (2007, 2208)

b. $m = \dfrac{y_2 - y_1}{x_2 - x_1}$

$$= \frac{2208 - 2025}{2007 - 2004}$$

$$= \frac{183}{3}$$

$$= 61$$

The slope is 61.

c. For the years 2004 through 2007, the number of heart transplants increased at a rate of 61 per year.

85. Slope through (1, 3) and (2, 1):

$$m = \frac{1-3}{2-1} = \frac{-2}{1} = -2$$

Slope through (–4, 0) and (–3, –2):

$$m = \frac{-2-0}{-3-(-4)} = \frac{-2}{-3+4} = \frac{-2}{1} = -2$$

Slope through (1, 3) and (–4, 0):

$$m = \frac{0-3}{-4-1} = \frac{-3}{-5} = \frac{3}{5}$$

Slope through (2, 1) and (–3, –2):

$$m = \frac{-2-1}{-3-2} = \frac{-3}{-5} = \frac{3}{5}$$

Opposite sides are parallel and their slopes are equal.

87. $m = \dfrac{y_2 - y_1}{x_2 - x_1} = \dfrac{4.5 - 1.2}{-2.2 - (-3.8)} = \dfrac{3.3}{1.6} = 2.0625$

89. $m = \dfrac{y_2 - y_1}{x_2 - x_1} = \dfrac{-2.9 - (-10.1)}{9.8 - 14.3} = \dfrac{7.2}{-4.5} = -1.6$

91.

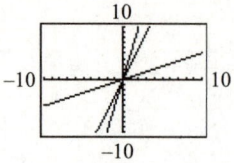

As the slope becomes larger, the line becomes steeper.

Section 6.5 Practice

1. From $y = \dfrac{2}{3}x - 4$, the y-intercept is (0, –4). The slope is $\dfrac{2}{3}$, so another point on the graph is (0 + 3, –4 + 2) or (3, –2).

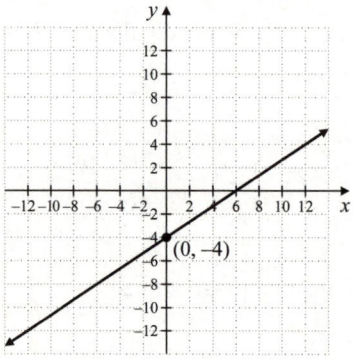

2. $3x + y = 2$

$$y = -3x + 2$$

The slope is $-3 = \dfrac{-3}{1}$ and the y-intercept is (0, 2). Another point on the graph is (0 + 1, 2 – 3) or (1, –1).

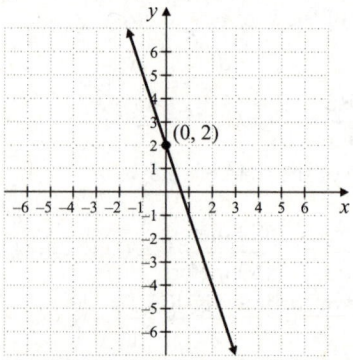

3. $y = mx + b$

$$y = \frac{1}{5}x + (-4)$$

$$y = \frac{1}{5}x - 4$$

Copyright © 2012 Pearson Education, Inc. Publishing as Prentice Hall.

4.
$$\begin{aligned} y - y_1 &= m(x - x_1) \\ y - (-4) &= -3(x - 2) \\ y + 4 &= -3(x - 2) \\ y + 4 &= -3x + 6 \\ y &= -3x + 2 \\ 3x + y &= 2 \end{aligned}$$

5. $m = \dfrac{-2-3}{5-1} = \dfrac{-5}{4}$
$$\begin{aligned} y - y_1 &= m(x - x_1) \\ y - 3 &= \frac{-5}{4}(x - 1) \\ 4(y - 3) &= 4\left[\frac{-5}{4}(x - 1)\right] \\ 4(y - 3) &= -5(x - 1) \\ 4y - 12 &= -5x + 5 \\ 5x + 4y &= 17 \end{aligned}$$

6. a. The ordered pairs are (10, 200), and (9, 250).
$$m = \frac{y_2 - y_1}{x_2 - x_1} = \frac{250 - 200}{9 - 10} = \frac{50}{-1} = -50$$
$$\begin{aligned} y - y_1 &= m(x - x_1) \\ y - 200 &= -50(x - 10) \\ y - 200 &= -50x + 500 \\ y &= -50x + 700 \end{aligned}$$

b. Let $x = 7.50$.
$y = -50x + 700$
$y = -50(7.50) + 700$
$y = -375 + 700$
$y = 325$
The predicted weekly sales is 325.

Calculator Explorations

1.

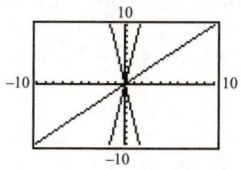

2.

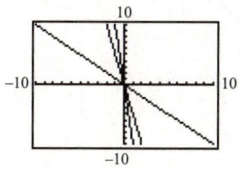

3.

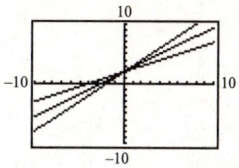

4.

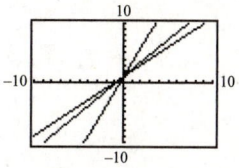

Vocabulary and Readiness Check

1. The form $y = mx + b$ is called <u>slope-intercept</u> form. When a linear equation in two variables is written in this form, <u>m</u> is the slope of its graph and (0, <u>b</u>) is its y-intercept.

2. The form $y - y_1 = m(x - x_1)$ is called <u>point-slope</u> form. When a linear equation in two variables is written in this form, <u>m</u> is the slope of its graph and <u>(x_1, y_1)</u> is a point on the graph.

3. $y - 7 = 4(x + 3)$; <u>point-slope</u> form

4. $5x - 9y = 11$; <u>standard</u> form

5. $y = \dfrac{3}{4}x - \dfrac{1}{3}$; <u>slope-intercept</u> form

6. $y + 2 = \dfrac{-1}{3}(x - 2)$; <u>point-slope form</u>

7. $y = \dfrac{1}{2}$; <u>horizontal</u> line

8. $x = -17$; <u>vertical</u> line

Exercise Set 6.5

1. From $y = 2x + 1$, the y-intercept is (0, 1). The slope is 2 or $\dfrac{2}{1}$, so another point on the graph is (0 + 1, 1 + 2) or (1, 3).

Copyright © 2012 Pearson Education, Inc. Publishing as Prentice Hall.

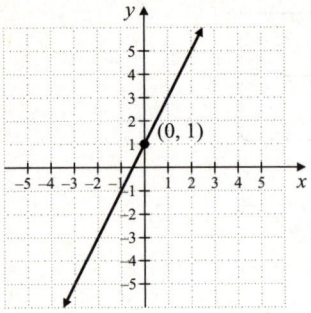

3. From $y = \frac{2}{3}x + 5$, the y-intercept is (0, 5). The slope is $\frac{2}{3}$, so another point on the graph is (0 + 3, 5 + 2) or (3, 7).

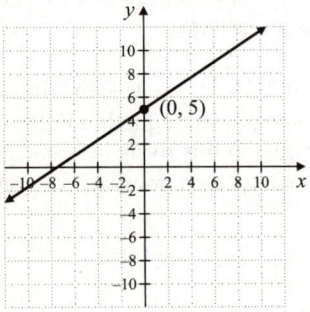

5. From $y = -5x$, the y-intercept is (0, 0). The slope is -5 or $\frac{-5}{1}$, so another point on the graph is (0 + 1, 0 + (–5)) or (1, –5).

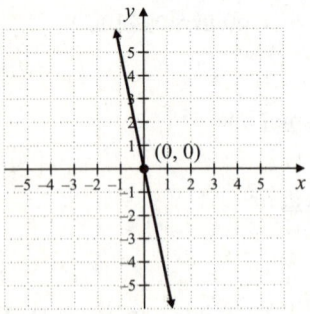

7. $4x + y = 6$

$y = -4x + 6$

The slope is $-4 = \frac{-4}{1}$ and the y-intercept is (0, 6). Another point on the graph is (0 + 1, 6 – 4) or (1, 2).

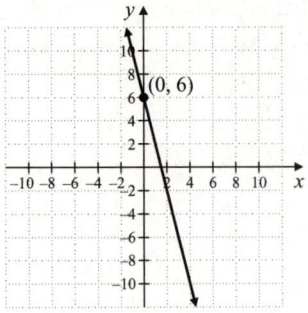

9. $4x - 7y = -14$

$-7y = -4x - 14$

$y = \frac{4}{7}x + 2$

The slope is $\frac{4}{7}$ and the y-intercept is (0, 2). Another point on the graph is (0 + 7, 2 + 4) or (7, 6).

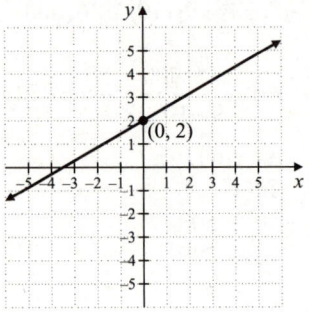

11. $x = \frac{5}{4}y$

$\frac{4}{5}x = y$

$y = \frac{4}{5}x + 0$

The slope is $\frac{4}{5}$ and the y-intercept is (0, 0). Another point on the graph is (0 + 5, 0 + 4) or (5, 4).

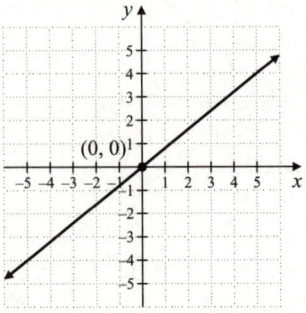

Copyright © 2012 Pearson Education, Inc. Publishing as Prentice Hall.

13. $y = mx + b$
$y = 5x + 3$

15. $y = mx + b$
$y = -4x + \left(-\frac{1}{6}\right)$
$y = -4x - \frac{1}{6}$

17. $y = mx + b$
$y = \frac{2}{3}x + 0$
$y = \frac{2}{3}x$

19. $y = mx + b$
$y = 0x + (-8)$
$y = -8$

21. $y = mx + b$
$y = -\frac{1}{5}x + \frac{1}{9}$

23. $y - y_1 = m(x - x_1)$
$y - 2 = 6(x - 2)$
$y - 2 = 6x - 12$
$y = 6x - 10$
$-6x + y = -10$

25. $y - y_1 = m(x - x_1)$
$y - (-5) = -8[x - (-1)]$
$y + 5 = -8(x + 1)$
$y + 5 = -8x - 8$
$y = -8x - 13$
$8x + y = -13$

27. $y - y_1 = m(x - x_1)$
$y - (-6) = \frac{3}{2}(x - 5)$
$y + 6 = \frac{3}{2}(x - 5)$
$2(y + 6) = 2\left[\frac{3}{2}(x - 5)\right]$
$2y + 12 = 3x - 15$
$2y = 3x - 27$
$-3x + 2y = -27$ or $3x - 2y = 27$

29. $y - y_1 = m(x - x_1)$
$y - 0 = -\frac{1}{2}[x - (-3)]$
$y = -\frac{1}{2}(x + 3)$
$2y = -1(x + 3)$
$2y = -x - 3$
$x + 2y = -3$

31. $m = \frac{y_2 - y_1}{x_2 - x_1} = \frac{6 - 2}{5 - 3} = \frac{4}{2} = 2$
$y - y_1 = m(x - x_1)$
$y - 2 = 2(x - 3)$
$y - 2 = 2x - 6$
$y + 4 = 2x$
$4 = 2x - y$
$2x - y = 4$

33. $m = \frac{y_2 - y_1}{x_2 - x_1} = \frac{-5 - 3}{-2 - (-1)} = \frac{-8}{-1} = 8$
$y - y_1 = m(x - x_1)$
$y - 3 = 8[x - (-1)]$
$y - 3 = 8(x + 1)$
$y - 3 = 8x + 8$
$y - 11 = 8x$
$-11 = 8x - y$
$8x - y = -11$

35. $m = \frac{y_2 - y_1}{x_2 - x_1} = \frac{-1 - 3}{-1 - 2} = \frac{-4}{-3} = \frac{4}{3}$
$y - y_1 = m(x - x_1)$
$y - 3 = \frac{4}{3}(x - 2)$
$3(y - 3) = 4(x - 2)$
$3y - 9 = 4x - 8$
$3y - 1 = 4x$
$-1 = 4x - 3y$
$4x - 3y = -1$

37. $m = \frac{y_2 - y_1}{x_2 - x_1} = \frac{\frac{1}{13} - 0}{-\frac{1}{8} - 0} = \frac{\frac{1}{13}}{-\frac{1}{8}} = \frac{1}{13}\left(-\frac{8}{1}\right) = -\frac{8}{13}$
$y - y_1 = m(x - x_1)$
$y - 0 = -\frac{8}{13}(x - 0)$
$y = -\frac{8}{13}x$
$13y = -8x$
$8x + 13y = 0$

Copyright © 2012 Pearson Education, Inc. Publishing as Prentice Hall.

39. $y = mx + b$

$y = -\frac{1}{2}x + \frac{5}{3}$

41. $m = \frac{y_2 - y_1}{x_2 - x_1} = \frac{10-7}{7-10} = \frac{3}{-3} = -1$

$y - y_1 = m(x - x_1)$

$y - 7 = -1(x - 10)$

$y - 7 = -x + 10$

$y = -x + 17$

43. A line with undefined slope is a vertical line. This one has an x-intercept of $\left(-\frac{3}{4}, 0\right)$. The equation is $x = -\frac{3}{4}$.

45. $y - y_1 = m(x - x_1)$

$y - 9 = 1[x - (-7)]$

$y - 9 = x + 7$

$y = x + 16$

47. $y = mx + b$

$y = -5x + 7$

49. A line parallel to $y = 5$ is a horizontal line.

$y = 2$

51. $m = \frac{y_2 - y_1}{x_2 - x_1} = \frac{3-0}{2-0} = \frac{3}{2}$

$y - y_1 = m(x - x_1)$

$y - 0 = \frac{3}{2}(x - 0)$

$y = \frac{3}{2}x$

53. A line perpendicular to the y-axis is a horizontal line.

$y = -3$

55. $y - y_1 = m(x - x_1)$

$y - (-2) = -\frac{4}{7}[x - (-1)]$

$y + 2 = -\frac{4}{7}(x + 1)$

$y + 2 = -\frac{4}{7}x - \frac{4}{7}$

$y = -\frac{4}{7}x - \frac{4}{7} - \frac{14}{7}$

$y = -\frac{4}{7}x - \frac{18}{7}$

57. a. The ordered pairs are (0, 302) and (4, 322).

b. $m = \frac{y_2 - y_1}{x_2 - x_1} = \frac{322-302}{4-0} = \frac{20}{4} = 5$

$y - y_1 = m(x - x_1)$

$y - 302 = 5(x - 0)$

$y - 302 = 5x$

$y = 5x + 302$

c. 2005 corresponds to $x = 2$.
$x = 2$: $y = 5(2) + 302 = 10 + 302 = 312$
There were approximately 312 million magazine subscriptions in 2005.

59. a. The ordered pairs are (1, 32) and (3, 96).

$m = \frac{s_2 - s_1}{t_2 - t_1} = \frac{96-32}{3-1} = \frac{64}{2} = 32$

$s - s_1 = m(t - t_1)$

$s - 32 = 32(t - 1)$

$s - 32 = 32t - 32$

$s = 32t$

b. $t = 4$: $s = 32(4)$

$s = 128$

The speed of the rock 4 seconds after it was dropped is 128 feet per second.

61. a. The ordered pairs are (3, 353,000) and (0, 83,000).

$m = \frac{y_2 - y_1}{x_2 - x_1} = \frac{83{,}000 - 353{,}000}{0-3}$

$= \frac{-270{,}000}{-3}$

$= 90{,}000$

$y - y_1 = m(x - x_1)$

$y - 353{,}000 = 90{,}000(x - 3)$

$y - 353{,}000 = 90{,}000x - 270{,}000$

$y = 90{,}000x + 83{,}000$

Copyright © 2012 Pearson Education, Inc. Publishing as Prentice Hall.

b. The year 2009 is 5 years past 2004.
$x = 2009 - 2004 = 5$
$y = 90{,}000(5) + 83{,}000$
$y = 450{,}000 + 83{,}000$
$y = 533{,}000$
533,000 vehicles are predicted for 2009.

63. a. The ordered pairs are (4, 5540) and (0, 5700).

$$m = \frac{y_2 - y_1}{x_2 - x_1} = \frac{5700 - 5540}{0 - 4} = \frac{160}{-4} = -40$$

$$\begin{aligned} y - y_1 &= m(x - x_1) \\ y - 5700 &= -40(x - 0) \\ y - 5700 &= -40x \\ y &= -40x + 5700 \end{aligned}$$

b. The year 2010 is 7 years past 2003, so it corresponds to $x = 7$.
$x = 7$: $y = -40(7) + 5700$
$y = -280 + 5700$
$y = 5420$
5420 cinema sites are predicted for 2010.

65. a. The ordered pairs are (3, 10,000) and (5, 8000).

$$\begin{aligned} m &= \frac{S_2 - S_1}{p_2 - p_1} \\ &= \frac{8000 - 10{,}000}{5 - 3} \\ &= \frac{-2000}{2} \\ &= -1000 \end{aligned}$$

$$\begin{aligned} S - S_1 &= m(p - p_1) \\ S - 10{,}000 &= -1000(p - 3) \\ S - 10{,}000 &= -1000p + 3000 \\ S &= -1000p + 13{,}000 \end{aligned}$$

b. $p = 3.50$: $S = -1000(3.50) + 13{,}000$
$S = -3500 + 13{,}000$
$S = 9500$
9500 Fun Noodles will be sold when the price is \$3.50 each.

67. $x = 2$: $x^2 - 3x + 1 = 2^2 - 3(2) + 1 = 4 - 6 + 1 = -1$

69. $x = -1$: $x^2 - 3x + 1 = (-1)^2 - 3(-1) + 1$
$= 1 + 3 + 1$
$= 5$

71. The graph of $y = 2x + 1$ has slope $m = 2$ and y-intercept (0, 1). This is graph b.

73. The graph of $y = -3x - 2$ has slope $m = -3$ and y-intercept (0, –2). This is graph d.

75. The slope of the line $y = 3x - 1$ is $m = 3$.

$$\begin{aligned} y - y_1 &= m(x - x_1) \\ y - 2 &= 3[x - (-1)] \\ y - 2 &= 3(x + 1) \\ y - 2 &= 3x + 3 \\ -5 &= 3x - y \\ 3x - y &= -5 \end{aligned}$$

77. A line perpendicular to the line $y = 3x - 1$ will have slope $-\frac{1}{3}$.

$$\begin{aligned} y - y_1 &= m(x - x_1) \\ y - 2 &= -\frac{1}{3}[x - (-1)] \\ y - 2 &= -\frac{1}{3}(x + 1) \\ 3y - 6 &= -1(x + 1) \\ 3y - 6 &= -x - 1 \\ 3y &= -x + 5 \\ x + 3y &= 5 \end{aligned}$$

Integrated Review

1. Select two points on the line, such as (0, 0) and (1, 2).

$$m = \frac{y_2 - y_1}{x_2 - x_1} = \frac{2 - 0}{1 - 0} = \frac{2}{1} = 2$$

2. Horizontal lines have slopes of $m = 0$.

3. Select two points on the line, such as (0, 1) and (–3, 3).

$$m = \frac{y_2 - y_1}{x_2 - x_1} = \frac{3 - 1}{-3 - 0} = \frac{2}{-3} = -\frac{2}{3}$$

4. Vertical lines have undefined slopes.

5. $y = -2x$
$y = 0$: $0 = -2x$
$0 = x$
The x-intercept is (0, 0).
$x = 0$: $y = -2(0) = 0$
The y-intercept is (0, 0).
Find another point, for example let $x = 1$.
$y = -2(1) = -2$
Another point on the line is (1, –2).

Copyright © 2012 Pearson Education, Inc. Publishing as Prentice Hall.

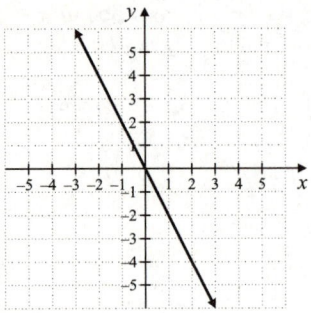

6. $x + y = 3$

$y = 0$: $x + 0 = 3$

$x = 3$

The x-intercept is (3, 0).

$x = 0$: $0 + y = 3$

$y = 3$

The y-intercept is (0, 3).

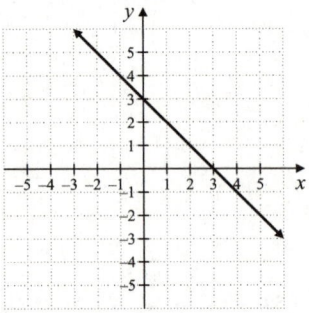

7. The graph of $x = -1$ is a vertical line with an x-intercept of (–1, 0).

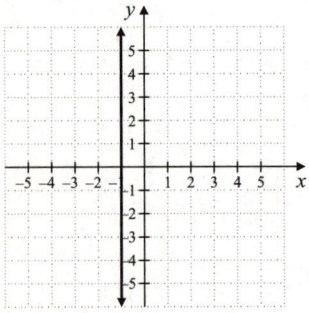

8. The graph of $y = 4$ is a horizontal line with a y-intercept of (0, 4).

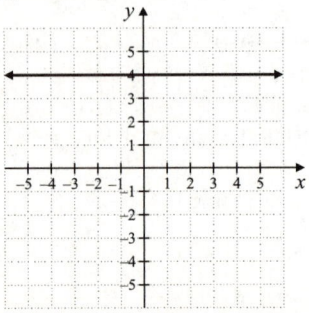

9. $x - 2y = 6$

$y = 0$: $x - 2(0) = 6$

$x - 0 = 6$

$x = 6$

The x-intercept is (6, 0).

$x = 0$: $0 - 2y = 6$

$-2y = 6$

$y = -3$

The y-intercept is (0, –3).

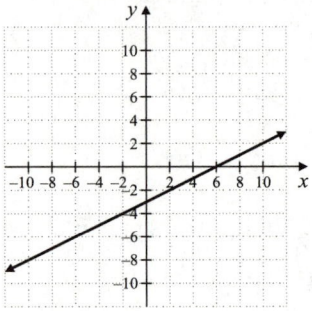

10. $y = 3x + 2$

$y = 0$: $0 = 3x + 2$

$-2 = 3x$

$-\frac{2}{3} = x$

The x-intercept is $\left(-\frac{2}{3}, 0\right)$.

$x = 0$: $y = 3(0) + 2$

$y = 0 + 2$

$y = 2$

The y-intercept is (0, 2).

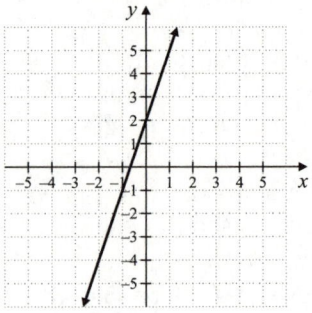

11. $y = -\frac{3}{4}x + 3$

$y = 0$: $0 = -\frac{3}{4}x + 3$

$-3 = -\frac{3}{4}x$

$4 = x$

The x-intercept is (4, 0).

Copyright © 2012 Pearson Education, Inc. Publishing as Prentice Hall.

$x = 0$: $y = -\frac{3}{4}(0) + 3$

$y = 0 + 3$

$y = 3$

The y-intercept is (0, 3).

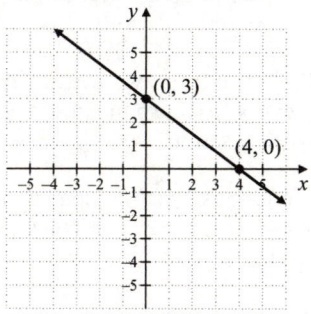

12. $5x - 2y = 8$

$y = 0$: $5x - 2(0) = 8$

$5x - 0 = 8$

$5x = 8$

$x = \frac{8}{5}$

The x-intercept is $\left(\frac{8}{5}, 0\right)$.

$x = 0$: $5(0) - 2y = 8$

$0 - 2y = 8$

$-2y = 8$

$y = -4$

The y-intercept is (0, –4).

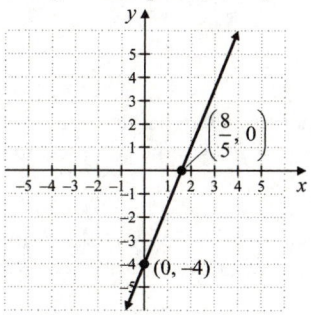

13. $y = mx + b$

$y = 3x - 1$

The slope is $m = 3$.

14. $y = mx + b$

$y = -6x + 2$

The slope is $m = -6$.

15. $y = mx + b$

$7x + 2y = 11$

$2y = -7x + 11$

$y = -\frac{7}{2}x + \frac{11}{2}$

The slope is $m = -\frac{7}{2}$.

16. $y = mx + b$

$2x - y = 0$

$-y = -2x$

$y = 2x$

$y = 2x + 0$

The slope is $m = 2$.

17. The graph of $x = 2$ is a vertical line. Vertical lines have undefined slopes.

18. The graph of $y = -4$ is a horizontal line. Horizontal lines have slopes of $m = 0$.

19. $y = mx + b$

$y = 2x + \left(-\frac{1}{3}\right)$

$y = 2x - \frac{1}{3}$

20. $y - y_1 = m(x - x_1)$

$y - 3 = -4[x - (-1)]$

$y - 3 = -4(x + 1)$

$y - 3 = -4x - 4$

$y = -4x - 1$

21. $m = \frac{y_2 - y_1}{x_2 - x_1} = \frac{-3 - 0}{-1 - 2} = \frac{-3}{-3} = 1$

$y - y_1 = m(x - x_1)$

$y - 0 = 1(x - 2)$

$y = x - 2$

$-x + y = -2$ or $x - y = 2$

22. $6x - y = 7$

$-y = -6x + 7$

$y = 6x - 7$

$m = 6$

$2x + 3y = 4$

$3y = -2x + 4$

$y = -\frac{2}{3}x + \frac{4}{3}$

$m = -\frac{2}{3}$

Since $6 \neq -\frac{2}{3}$, the lines are not parallel. Since

Copyright © 2012 Pearson Education, Inc. Publishing as Prentice Hall.

$6\left(-\frac{2}{3}\right) = -4 \neq -1$, the lines are not perpendicular. The lines are neither parallel nor perpendicular.

23. $3x - 6y = 4$ $\quad$ $y = -2x$

$-6y = -3x + 4$ $\quad$ $m = -2$

$y = \frac{1}{2}x - \frac{2}{3}$

$m = \frac{1}{2}$

Since $\left(\frac{1}{2}\right)(-2) = -1$, the lines are perpendicular.

24. a. The ordered pairs are (2002, 2133) and (2007, 3478).

b. $m = \frac{y_2 - y_1}{x_2 - x_1} = \frac{3478 - 2133}{2007 - 2002} = \frac{1345}{5} = 269$

c. For the years 2002 through 2007, the amount of yogurt produced increased at a rate of 269 million pounds per year.

Section 6.6 Practice

1. The domain is the set of x-coordinates: $\{-3, 4, 7\}$.
The range is the set of y-coordinates: $\{0, 1, 5, 6\}$.

2. a. Each x-value is only assigned to one y-value, so the relation is a function.

b. The x-value 1 is paired with two y-values, 4 and -3, so this set of ordered pairs is not a function.

3. a. This is the graph of the relation $\{(-3, -2), (-1, -1), (0, 0), (1, 1)\}$.
Each x-coordinate has exactly one y-value, so this is the graph of a function.

b. This is the graph of the relation $\{(-1, -1), (-1, 2), (1, 0), (3, 1)\}$.
The x-value -1 is paired with two y-values, -1 and 2, so this is not the graph of a function.

4. a. No vertical line will intersect the graph more than once, so the graph is the graph of a function.

b. No vertical line will intersect the graph more than once, so the graph is the graph of a function.

c. Vertical lines can be drawn that intersect the graph in two points, so the graph is not the graph of a function.

d. A vertical line can be drawn that intersects this line at every point, so the graph is not the graph of a function.

5. a, b, and **c** are functions because their graphs are nonvertical lines. **d** is not a function because its graph is a vertical line.

6. a. According to the graph, the time of sunrise on March 1st is 6:30 A.M.

b. According to the graph, the sun rises at 6 A.M. in the middle of March and the middle of September.

7. $f(x) = x^2 + 1$

a. $f(1) = 1^2 + 1 = 1 + 1 = 2$
Ordered pair: (1, 2)

b. $f(-3) = (-3)^2 + 1 = 9 + 1 = 10$
$(-3, 10)$

c. $f(0) = 0^2 + 1 = 0 + 1 = 1$
(0, 1)

8. a. In this function, x can be any real number. The domain of $h(x)$ is the set of all real numbers.

b. Since we cannot divide by 0, the domain of $f(x)$ is the set of all real numbers except 0.

9. a. The x-values go from -4 to 6, so the domain is $-4 \le x \le 6$.
The y-values go from -2 to 3, so the range is $-2 \le y \le 3$.

b. There are no restrictions on the x-values, so the domain is all real numbers.
The y-values are all less than or equal to 3, so the range is $y \le 3$.

Vocabulary and Readiness Check

1. A set of ordered pairs is called a <u>relation</u>.

Copyright © 2012 Pearson Education, Inc. Publishing as Prentice Hall.

2. A set of ordered pairs that assigns to each x-value exactly one y-value is called a function.

3. The set of all y-coordinates of a relation is called the range.

4. The set of all x-coordinates of a relation is called the domain.

5. All linear equations are functions except those whose graphs are vertical lines.

6. All linear equations are functions except those whose equations are of the form $x = c$.

7. If $f(3) = 7$, the corresponding ordered pair is $(3, 7)$.

8. The domain of $f(x) = x + 5$ is all real numbers.

9. For the function $y = mx + b$, the dependent variable is y and the independent variable is x.

Exercise Set 6.6

1. The domain is the set of x-coordinates: $\{-7, 0, 2, 10\}$.
The range is the set of y-coordinates: $\{-7, 0, 4, 10\}$.

3. The domain is the set of x-coordinates: $\{0, 1, 5\}$
The range is the set of y-coordinates: $\{-2\}$

5. Each x-value is only assigned to one y-value, so the relation is a function.

7. The x-value -1 is paired with more than one y-value, 0, 6, and 8, so the relation is not a function.

9. The vertical line $x = 1$ will intersect the graph in two points, so the graph is not the graph of a function.

11. No vertical line will intersect the graph more than once, so the graph is the graph of a function.

13. No vertical line will intersect the graph more than once, so the graph is the graph of a function.

15. Vertical lines can be drawn that intersect the graph in two points, so the graph is not the graph of a function.

17. If $x = -1$, the relation is not also a function; a.

19. $y - x = 7$
The graph of this linear equation is not a vertical line, so the equation describes a function.

21. $y = 6$
The graph of this linear equation is not a vertical line, so the equation describes a function.

23. $x = -2$
The graph of this linear equation is a vertical line, so the equation does not describe a function.

25. $x = y^2$
$y = 1$: $x = (1)^2 = 1$
$y = -1$: $x = (-1)^2 = 1$
Since there is an x-value that is paired with two y-values, the equation does not describe a function.

27. On June 1, the graph shows sunset to be at approximately 9:30 P.M.

29. At 3 P.M., the graph shows this happens on January 1 and December 1.

31. The graph passes the vertical line test, so it is the graph of a function.

33. Before October 1996, the graph shows the minimum wage was \$4.25 per hour.

35. According to the graph, the minimum wage will increase to over \$7.00 in 2009.

37. yes; answers may vary

39. According to the graph, the postage would be \$1.50.

41. From the graph, it would cost \$1 to mail a large envelope that weighs more than 1 ounce and less than or equal to 2 ounces.

43. yes; answers may vary

45. $f(x) = 2x - 5$
$f(-2) = 2(-2) - 5 = -4 - 5 = -9$
$f(0) = 2(0) - 5 = 0 - 5 = -5$
$f(3) = 2(3) - 5 = 6 - 5 = 1$

Copyright © 2012 Pearson Education, Inc. Publishing as Prentice Hall.

47. $f(x) = x^2 + 2$

$f(-2) = (-2)^2 + 2 = 4 + 2 = 6$

$f(0) = 0^2 + 2 = 0 + 2 = 2$

$f(3) = 3^2 + 2 = 9 + 2 = 11$

49. $f(x) = 3x$
$f(-2) = 3(-2) = -6$
$f(0) = 3(0) = 0$
$f(3) = 3(3) = 9$

51. $f(x) = |x|$
$f(-2) = |-2| = 2$
$f(0) = |0| = 0$
$f(3) = |3| = 3$

53. $h(x) = -5x$
$h(-1) = -5(-1) = 5$
$h(0) = -5(0) = 0$
$h(4) = -5(4) = -20$

55. $h(x) = 2x^2 + 3$

$h(-1) = 2(-1)^2 + 3 = 2(1) + 3 = 2 + 3 = 5$

$h(0) = 2(0)^2 + 3 = 2(0) + 3 = 0 + 3 = 3$

$h(4) = 2(4)^2 + 3 = 2(16) + 3 = 32 + 3 = 35$

57. The ordered-pair solution corresponding to $f(3) = 6$ is $(3, 6)$.

59. The ordered-pair solution corresponding to $g(0) = -\frac{1}{2}$ is $\left(0, -\frac{1}{2}\right)$.

61. The ordered-pair solution corresponding to $h(-2) = 9$ is $(-2, 9)$.

63. The domain of $f(x)$ is all real numbers.

65. $x + 5$ cannot be 0.
$x + 5 = 0$
$x = -5$
The domain of $f(x)$ is all real numbers except -5.

67. The domain is all real numbers. The range is $y \geq -4$.

69. The domain is all real numbers. The range is all real numbers.

71. The domain is all real numbers. The range is $\{2\}$.

73. When $x = 0$, $y = -1$, so the ordered-pair solution is $(0, -1)$.

75. When $x = 0$, $y = -1$, so $f(0) = -1$.

77. When $y = 0$, $x = -1$ and $x = 5$.

79. $2x + 5 < 7$
$2x < 2$
$x < 1$

81. $-x + 6 \leq 9$
$-x \leq 3$
$x \geq -3$

83. $\frac{3}{x} + \frac{3}{2x} + \frac{5}{x} = \frac{3 \cdot 2}{x \cdot 2} + \frac{3}{2x} + \frac{5 \cdot 2}{x \cdot 2}$

$= \frac{6}{2x} + \frac{3}{2x} + \frac{10}{2x}$

$= \frac{6 + 3 + 10}{2x}$

$= \frac{19}{2x}$

The perimeter is $\frac{19}{2x}$ meters.

85. A function f evaluated at -5 as 12 is written as $f(-5) = 12$.

87. answers may vary

89. $y = x + 7$ written in function notation is $f(x) = x + 7$.

91. $f(x) = \frac{136}{25}x$

a. $f(35) = \frac{136}{25}(35) = \frac{4760}{25} = \frac{952}{5} = 190.4$

The proper dosage for a 35-pound dog is 190.4 milligrams.

b. $f(70) = \frac{136}{25}(70) = \frac{9520}{25} = \frac{1904}{5} = 380.8$

The proper dosage for a 70-pound dog is 380.8 milligrams.

Copyright © 2012 Pearson Education, Inc. Publishing as Prentice Hall.

Section 6.7 Practice

1. $x - 4y > 8$

a. $(-3, 2)$: $-3 - 4(2) > 8$
$-3 - 8 > 8$
$-11 > 8$ False
$(-3, 2)$ is not a solution of the inequality.

b. $(9, 0)$: $9 - 4(0) > 8$
$9 - 0 > 8$
$9 > 8$ True
$(9, 0)$ is a solution of the inequality.

2. Graph the boundary line, $x - y = 3$, with a dashed line.
Test $(0, 0)$: $x - y > 3$
$0 - 0 > 3$
$0 > 3$ False
Shade the half-plane not containing $(0, 0)$.

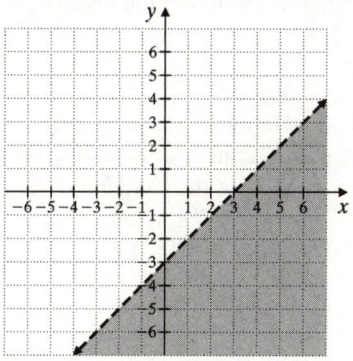

3. Graph the boundary line, $x - 4y = 4$, with a solid line.
Test $(0, 0)$: $x - 4y \le 4$
$0 - 4(0) \le 4$
$0 - 0 \le 4$
$0 \le 4$ True
Shade the half-plane containing $(0, 0)$.

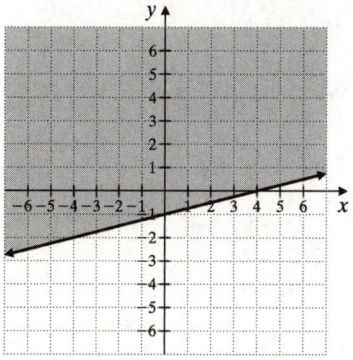

4. Graph the boundary line, $y = 3x$, with a dashed line.
Test $(1, 1)$: $y < 3x$
$1 < 3(1)$
$1 < 3$ True
Shade the half-plane containing $(1, 1)$.

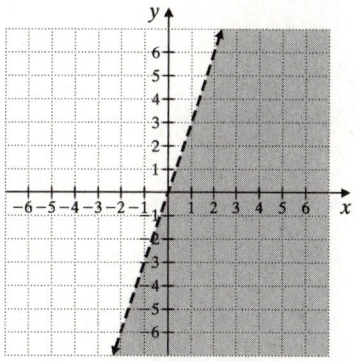

5. Graph the boundary line, $3x + 2y = 12$, with a solid line.
Test $(0, 0)$: $3x + 2y \ge 12$
$3(0) + 2(0) \ge 12$
$0 + 0 \ge 12$
$0 \ge 12$ False
Shade the half-plane not containing $(0, 0)$.

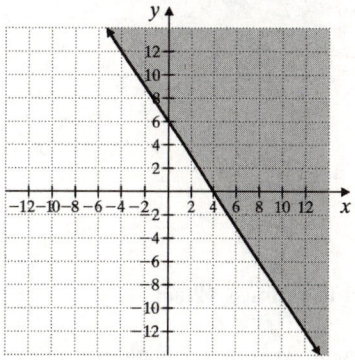

6. Graph the boundary line, $x = 2$, with a dashed line.
Test $(0, 0)$: $x < 2$
$0 < 2$ True
Shade the half-plane containing $(0, 0)$.

Copyright © 2012 Pearson Education, Inc. Publishing as Prentice Hall.

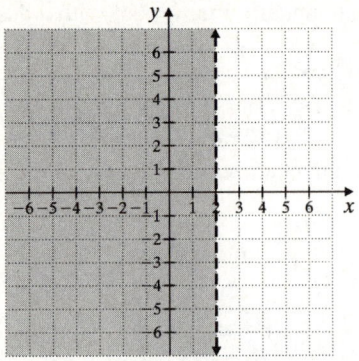

7. Graph the boundary line, $y = \frac{1}{4}x + 3$, with a solid line.

Test (0, 0): $y \geq \frac{1}{4}x + 3$

$0 \geq \frac{1}{4}(0) + 3$

$0 \geq 0 + 3$

$0 \geq 3$ False

Shade the half-plane not containing (0, 0).

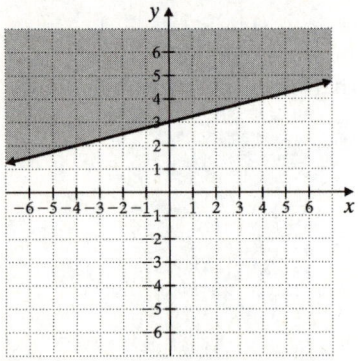

Vocabulary and Readiness Check

1. The statement $5x - 6y < 7$ is an example of a <u>linear inequality in two variables</u>.

2. A boundary line divides a plane into two regions called <u>half-planes</u>.

3. The graph of $5x - 6y < 7$ include its corresponding boundary line. <u>false</u>

4. When graphing a linear inequality, to determine which side of the boundary line to shade, choose a point *not* on the boundary line. <u>true</u>

5. The boundary line for the inequality $5x - 6y < 7$ is the graph of $5x - 6y = 7$. <u>true</u>

6. The graph shown is <u>$y > 2$</u>.

Exercise Set 6.7

1. $x - y > 3$

(0, 3): $0 - 3 > 3$

$-3 > 3$ False

(0, 3) is not a solution of the inequality.

(2, −1): $2 - (-1) > 3$

$2 + 1 > 3$

$3 > 3$ False

(2, −1) is not a solution of the inequality.

3. $3x - 5y \leq -4$

(2, 3): $3(2) - 5(3) \leq -4$

$6 - 15 \leq -4$

$-9 \leq -4$ True

(2, 3) is a solution of the inequality.

(−1, −1): $3(-1) - 5(-1) \leq -4$

$-3 + 5 \leq -4$

$2 \leq -4$ False

(−1, −1) is not a solution of the inequality.

5. $x < -y$

(0, 2): $0 < -2$ False

(0, 2) is not a solution of the inequality.

(−5, 1): $-5 < -1$ True

(−5, 1) is a solution of the inequality.

7. Graph the boundary line, $x + y = 1$, with a solid line.

Test (0, 0): $x + y \leq 1$

$0 + 0 \leq 1$

$0 \leq 1$ True

Shade the half-plane containing (0, 0).

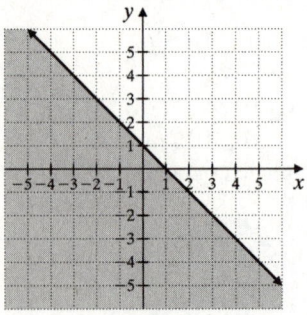

9. Graph the boundary line, $2x - y = -4$, with a dashed line.

Test (0, 0): $2x - y > -4$

$2(0) - 0 > -4$

$0 - 0 > -4$

$0 > -4$ True

Shade the half-plane containing (0, 0).

Copyright © 2012 Pearson Education, Inc. Publishing as Prentice Hall.

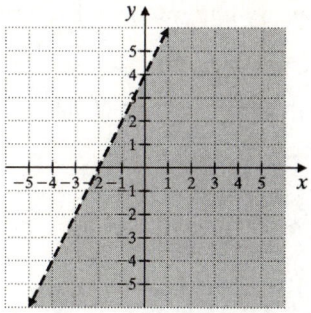

11. Graph the boundary line, $y = 2x$, with a solid line.

Test (1, 1): $y \geq 2x$

$1 \geq 2(1)$

$1 \geq 2$ False

Shade the half-plane not containing (1, 1).

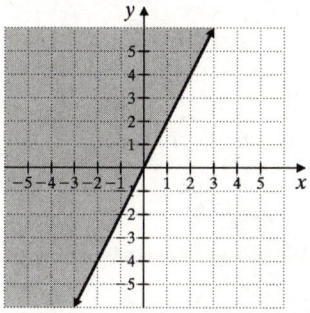

13. Graph the boundary line, $x = -3y$, with a dashed line.

Test (1, 1): $x < -3y$

$1 < -3(1)$

$1 < -3$ False

Shade the half-plane not containing (1, 1).

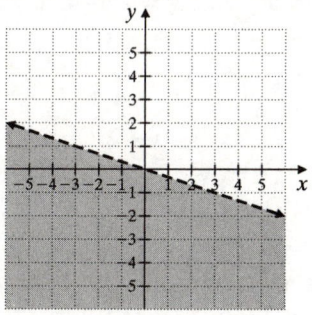

15. Graph the boundary line, $y = x + 5$, with a solid line.

Test (0, 0): $y \geq x + 5$

$0 \geq 0 + 5$

$0 \geq 5$ False

Shade the half-plane not containing (0, 0).

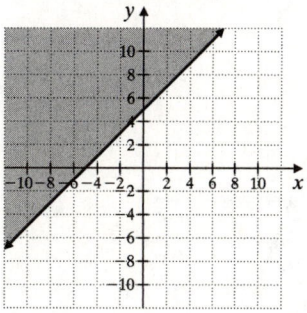

17. Graph the boundary line, $y = 4$, with a dashed line.

Test (0, 0): $y < 4$

$0 < 4$ True

Shade the half-plane containing (0, 0).

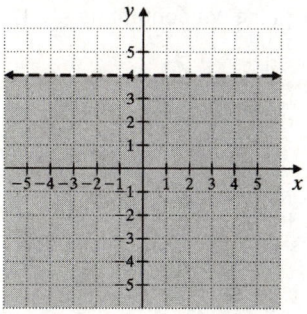

19. Graph the boundary line, $x = -3$, with a solid line.

Test (0, 0): $x \geq -3$

$0 \geq -3$ True

Shade the half-plane containing (0, 0).

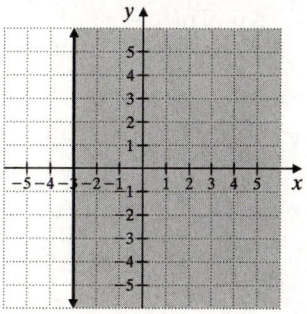

Copyright © 2012 Pearson Education, Inc. Publishing as Prentice Hall.

21. Graph the boundary line, $5x + 2y = 10$, with a solid line.

Test (0, 0): $5x + 2y \le 10$

$5(0) + 2(0) \le 10$

$0 + 0 \le 10$

$0 \le 10$ True

Shade the half-plane containing (0, 0).

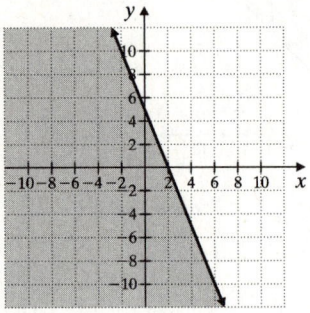

23. Graph the boundary line, $x = y$, with a dashed line.

Test (1, 4): $x > y$

$1 > 4$ False

Shade the half-plane not containing (1, 4).

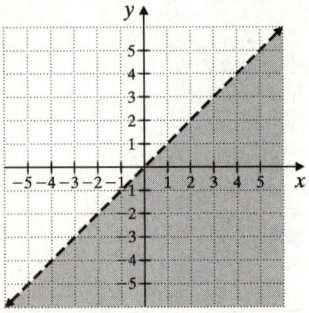

25. Graph the boundary line, $x - y = 6$, with a solid line.

Test (0, 0): $x - y \le 6$

$0 - 0 \le 6$

$0 \le 6$ True

Shade the half-plane containing (0, 0).

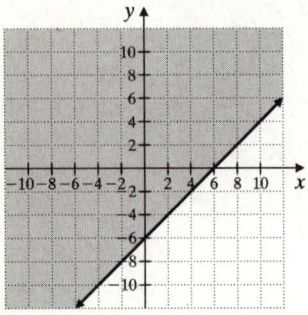

27. Graph the boundary line, $x = 0$, with a solid line.

Test (1, 1): $x \ge 0$

$1 \ge 0$ True

Shade the half-plane containing (1, 1).

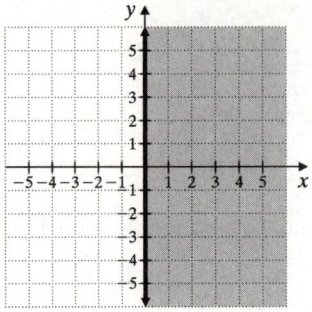

29. Shade the boundary line, $2x + 7y = 5$, with a dashed line.

Test (0, 0): $2x + 7y > 5$

$2(0) + 7(0) > 5$

$0 + 0 > 5$

$0 > 5$ False

Shade the half-plane not containing (0, 0).

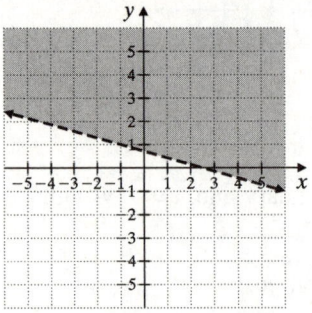

31. Graph the boundary line, $y = \frac{1}{2}x - 4$, with a solid line.

Test (0, 0): $y \ge \frac{1}{2}x - 4$

$0 \ge \frac{1}{2}(0) - 4$

$0 \ge 0 - 4$

$0 \ge -4$ True

Shade the half-plane containing (0, 0).

Copyright © 2012 Pearson Education, Inc. Publishing as Prentice Hall.

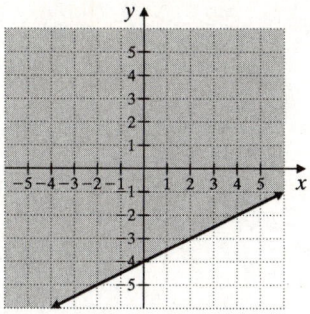

33. The point of intersection appears to be (−2, 1).

35. The point of intersection appears to be (−3, −1).

37. The graph is the half-plane with the dashed boundary line $x = 2$. The choice is a.

39. The graph is the half-plane with the dashed boundary line $y = 2$. The choice is b.

41. answers may vary

43. Test (1, 1):
$$\begin{aligned} 3x+4y &< 8 \\ 3(1)+4(1) &< 8 \\ 3+4 &< 8 \\ 7 &< 8 \quad \text{True} \end{aligned}$$
(1, 1) is included in the graph of $3x + 4y < 8$.

45. Test (1, 1):
$$\begin{aligned} y &\geq -\frac{1}{2}x \\ 1 &\geq -\frac{1}{2}(1) \\ 1 &\geq -\frac{1}{2} \quad \text{True} \end{aligned}$$
(1, 1) is included in the graph of $y \geq -\frac{1}{2}x$.

47. a. The sum of the number of days, x, times \$30, and the number of miles, y, times \$0.15, must be at most \$500.
$30x + 0.15y \leq 500$

b.

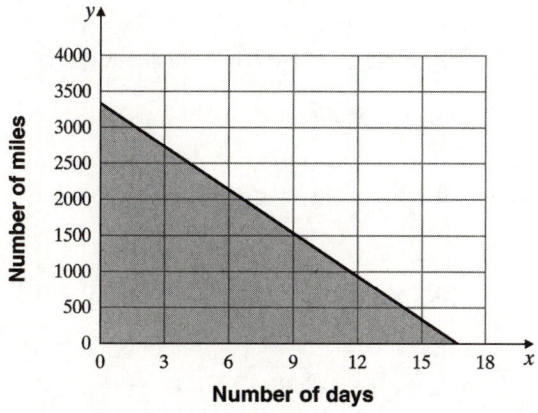

c. answers may vary

Copyright © 2012 Pearson Education, Inc. Publishing as Prentice Hall.

Section 6.8 Practice

1. Use (4, 8).

$y = kx$

$8 = k \cdot 4$

$\frac{8}{4} = \frac{k \cdot 4}{4}$

$2 = k$

Since $k = 2$, the equation is $y = 2x$.

2. Let $y = 15$ and $x = 45$.

$y = kx$

$15 = k(45)$

$\frac{15}{45} = \frac{k(45)}{45}$

$\frac{1}{3} = k$

The equation is $y = \frac{1}{3}x$.

Let $x = 3$.

$y = \frac{1}{3}x$

$y = \frac{1}{3} \cdot 3$

$y = 1$

Thus, when x is 3, y is 1.

3. Use (–1, –2) and (0, 0).

$\text{slope} = \frac{0-(-2)}{0-(-1)} = \frac{2}{1} = 2$

Thus, $k = 2$ and the variation equation is $y = 2x$.

4. Use (4, 5).

$y = \frac{k}{x}$

$5 = \frac{k}{4}$

$4 \cdot 5 = 4 \cdot \frac{k}{4}$

$20 = k$

Since $k = 20$, the equation is $y = \frac{20}{x}$.

5. Let $y = 4$ and $x = 0.8$.

$y = \frac{k}{x}$

$4 = \frac{k}{0.8}$

$0.8(4) = 0.8\left(\frac{k}{0.8}\right)$

$3.2 = k$

The equation is $y = \frac{3.2}{x}$.

Let $x = 20$.

$y = \frac{3.2}{20}$

$y = 0.16$

Thus, when x is 20, y is 0.16.

6. $A = kr^2$

$49\pi = k(7)^2$

$49\pi = 49k$

$\pi = k$

The formula for the area of a circle is $A = \pi r^2$.

Let $r = 4$.

$A = \pi r^2$

$A = \pi \cdot 4^2$

$A = 16\pi$

The area is 16π square feet.

7. $d = kt^2$

$144 = k(3)^2$

$144 = 9k$

$16 = k$

The equation is $d = 16t^2$.

Let $t = 5$.

$d = 16t^2$

$d = 16 \cdot 5^2$

$d = 16 \cdot 25$

$d = 400$

The object will fall 400 feet in 5 seconds.

Vocabulary and Readiness Check

1. $y = \frac{k}{x}$, where k is a constant. <u>inverse</u>

2. $y = kx$, where k is a constant. <u>direct</u>

3. $y = 5x$ <u>direct</u>

4. $y = \frac{5}{x}$ <u>inverse</u>

5. $y = \frac{7}{x^2}$ <u>inverse</u>

6. $y = 6.5x^4$ <u>direct</u>

Copyright © 2012 Pearson Education, Inc. Publishing as Prentice Hall.

7. $y = \frac{11}{x}$ <u>inverse</u>

8. $y = 18x$ <u>direct</u>

9. $y = 12x^2$ <u>direct</u>

10. $y = \frac{20}{x^3}$ <u>inverse</u>

Exercise Set 6.8

1. $y = kx$
$3 = k(6)$
$\frac{1}{2} = k$
$y = \frac{1}{2}x$

3. $y = kx$
$12 = k(2)$
$6 = k$
$y = 6x$

5. $m = \frac{y_2 - y_1}{x_2 - x_1} = \frac{3-0}{1-0} = \frac{3}{1} = 3$
$y = 3x$

7. $m = \frac{y_2 - y_1}{x_2 - x_1} = \frac{2-0}{3-0} = \frac{2}{3}$
$y = \frac{2}{3}x$

9. $y = \frac{k}{x}$
$7 = \frac{k}{1}$
$7 = k$
$y = \frac{7}{x}$

11. $y = \frac{k}{x}$
$0.05 = \frac{k}{10}$
$0.5 = k$
$y = \frac{0.5}{x}$

13. y varies directly as x is written as $y = kx$.

15. h varies inversely as t is written as $h = \frac{k}{t}$.

17. z varies directly as x^2 is written as $z = kx^2$.

19. y varies inversely as z^3 is written as $y = \frac{k}{z^3}$.

21. x varies inversely as $\sqrt{y}$ is written as $x = \frac{k}{\sqrt{y}}$.

23. $y = kx$
$y = 20$ when $x = 5$: $20 = k(5)$
$4 = k$
$y = 4x$
$x = 10$: $y = 4(10) = 40$
$y = 40$ when $x = 10$.

25. $y = \frac{k}{x}$
$y = 5$ when $x = 60$: $5 = \frac{k}{60}$
$300 = k$
$y = \frac{300}{x}$
$x = 100$: $y = \frac{300}{100} = 3$
$y = 3$ when $x = 100$.

27. $z = kx^2$
$z = 96$ when $x = 4$: $96 = k(4)^2$
$96 = 16k$
$6 = k$
$z = 6x^2$
$x = 3$: $z = 6(3)^2 = 6(9) = 54$
$z = 54$ when $x = 3$.

29. $a = \frac{k}{b^3}$
$a = \frac{3}{2}$ when $b = 2$: $\frac{3}{2} = \frac{k}{2^3}$
$\frac{3}{2} = \frac{k}{8}$
$12 = k$
$a = \frac{12}{b^3}$

Copyright © 2012 Pearson Education, Inc. Publishing as Prentice Hall.

$b = 3$: $a = \frac{12}{3^3} = \frac{12}{27} = \frac{4}{9}$

$a = \frac{4}{9}$ when $b = 3$.

31. Let p be the paycheck amount when h hours are worked.
$p = kh$
$p = 166.50$ when $h = 18$: $166.50 = k(18)$
$9.25 = k$
$p = 9.25h$
$h = 10$: $p = 9.25(10) = 92.50$
The pay is \$92.50 for 10 hours.

33. Let c be the cost per headphone when h headphones are manufactured.

$c = \frac{k}{h}$

$c = 9$ when $h = 5000$: $9 = \frac{k}{5000}$
$45{,}000 = k$

$c = \frac{45{,}000}{h}$

$h = 7500$: $c = \frac{45{,}000}{7500} = 6$

The cost to manufacture 7500 headphones is \$6 per headphone.

35. Let d be the distance when a weight of w is attached.
$d = kw$
$d = 4$ when $w = 60$: $4 = k(60)$
$\frac{1}{15} = k$

$d = \frac{1}{15}w$

$w = 80$: $d = \frac{1}{15}(80) = 5\frac{1}{3}$

The spring stretches $5\frac{1}{3}$ inches when 80 pounds is attached to the spring.

37. Let w be the weight of an object when it is d miles from the center of the Earth.

$w = \frac{k}{d^2}$
$w = 180$ when $d = 4000$:

$180 = \frac{k}{4000^2}$

$180 = \frac{k}{16{,}000{,}000}$

$2{,}880{,}000{,}000 = k$

$w = \frac{2{,}880{,}000{,}000}{d^2}$

$w = \frac{2{,}880{,}000{,}000}{4010^2}$

$d = 4010$: $= \frac{2{,}880{,}000{,}000}{16{,}080{,}100}$
≈ 179.1

The man will weigh about 179.1 pounds when he is 10 miles above the surface of the Earth.

39. $d = kt^2$
$d = 64$ when $t = 2$: $64 = k(2)^2$
$64 = 4k$
$16 = k$

$d = 16t^2$

$t = 10$: $d = 16(10)^2 = 16(100) = 1600$
He will fall 1600 feet in 10 seconds.

41.
$$\begin{array}{r} -3x + 4y = 7 \\ 3x - 2y = 9 \\ \hline 2y = 16 \end{array}$$

43.
$$\begin{array}{r} 5x - 0.4y = 0.7 \\ -9x + 0.4y = -0.2 \\ \hline -4x \qquad = 0.5 \end{array}$$

45. If y varies directly as x, then $y = kx$. If x is tripled, to become $3x$, then $y = k(3x) = 3(kx)$, and y is multiplied by 3.

47. If p varies directly with the square root of l, then $p = k\sqrt{l}$. If l is quadrupled, to become $4l$, then $k\sqrt{4l} = 2\left(k\sqrt{l}\right)$, and p is doubled.

Chapter 6 Vocabulary Check

1. An ordered pair is a <u>solution</u> of an equation in two variables if replacing the variables by the coordinates of the ordered pair results in a true statement.

2. The vertical number line in the rectangular coordinate system is called the <u>y-axis</u>.

Copyright © 2012 Pearson Education, Inc. Publishing as Prentice Hall.

3. A linear equation can be written in the form $Ax + By = C$.

4. A(n) x-intercept is a point of the graph where the graph crosses the x-axis.

5. The form $Ax + By = C$ is called standard form.

6. A(n) y-intercept is a point of the graph where the graph crosses the y-axis.

7. A set of ordered pairs that assigns to each x-value exactly one y-value is called a function.

8. The equation $y = 7x - 5$ is written in slope-intercept form.

9. The set of all x-coordinates of a relation is called the domain of the relation.

10. The set of all y-coordinates of a relation is called the range of the relation.

11. The set of ordered pairs is called a relation.

12. The equation $y + 1 = 7(x - 2)$ is written in point-slope form.

13. To find an x-intercept of a graph, let $y = 0$.

14. The horizontal number line in the rectangular coordinate system is called the x-axis.

15. To find a y-intercept of a graph, let $x = 0$.

16. The slope of a line measures the steepness or tilt of a line.

17. The equation $y = kx$ is an example of direct variation.

18. The equation $y = \frac{k}{x}$ is an example of inverse variation.

Chapter 6 Review

1–6.

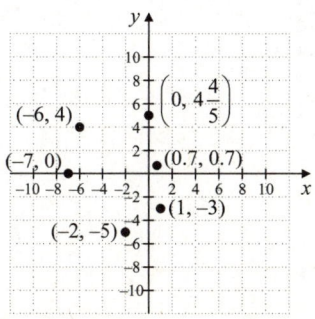

7. $-2 + y = 6x$

In (7,), the x-coordinate is 7.

$$\begin{aligned} -2 + y &= 6(7) \\ -2 + y &= 42 \\ y &= 44 \end{aligned}$$

The ordered-pair solution is (7, 44).

8. $y = 3x + 5$

In (, −8), the y-coordinate is −8.

$$\begin{aligned} -8 &= 3x + 5 \\ -13 &= 3x \\ -\frac{13}{3} &= x \end{aligned}$$

The ordered-pair solution is $\left(-\frac{13}{3}, -8\right)$.

9. $9 = -3x + 4y$

$y = 0$:
$$\begin{aligned} 9 &= -3x + 4(0) \\ 9 &= -3x + 0 \\ 9 &= -3x \\ -3 &= x \end{aligned}$$

$y = 3$:
$$\begin{aligned} 9 &= -3x + 4(3) \\ 9 &= -3x + 12 \\ -3 &= -3x \\ 1 &= x \end{aligned}$$

$x = 9$:
$$\begin{aligned} 9 &= -3(9) + 4y \\ 9 &= -27 + 4y \\ 36 &= 4y \\ 9 &= y \end{aligned}$$

x	y
−3	0
1	3
9	9

10. $y = 5$ for each value of x.

x	y
7	5
−7	5
0	5

Copyright © 2012 Pearson Education, Inc. Publishing as Prentice Hall.

11. $x = 2y$

$y = 0$: $x = 2(0)$
$x = 0$

$y = 5$: $x = 2(5)$
$x = 10$

$y = -5$: $x = 2(-5)$
$x = -10$

x	y
0	0
10	5
−10	−5

12. a. $y = 5x + 2000$

$x = 1$: $y = 5(1) + 2000 = 5 + 2000 = 2005$

$x = 100$: $y = 5(100) + 2000$
$= 500 + 2000$
$= 2500$

$x = 1000$: $y = 5(1000) + 2000$
$= 5000 + 2000$
$= 7000$

x	1	100	1000
y	2005	2500	7000

b. Let $y = 6430$ and solve for x.
$6430 = 5x + 2000$
$4430 = 5x$
$886 = x$
886 compact disc holders can be produced for \$6430.

13. $x - y = 1$

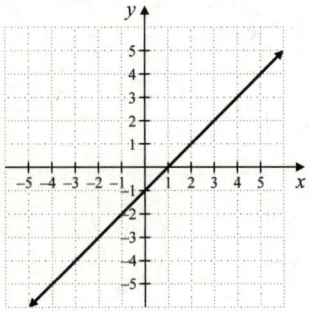

14. $x + y = 6$

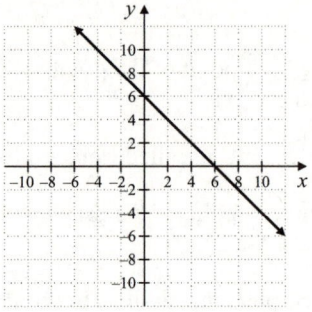

15. $x - 3y = 12$

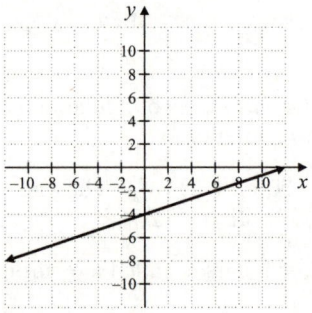

16. $5x - y = -8$

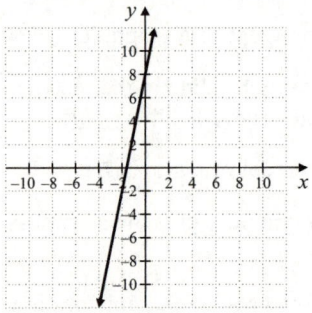

17. $x = 3y$

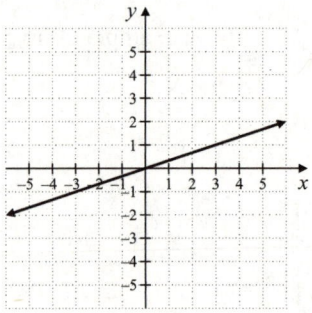

Copyright © 2012 Pearson Education, Inc. Publishing as Prentice Hall.

18. $y = -2x$

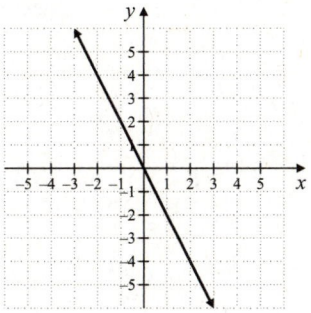

19. The x-intercept is (4, 0).
The y-intercept is (0, –2).

20. The x-intercepts are (–2, 0) and (2, 0).
The y-intercepts are (0, 2) and (0, –2).

21. $y = -3$ is a horizontal line with y-intercept (0, –3).

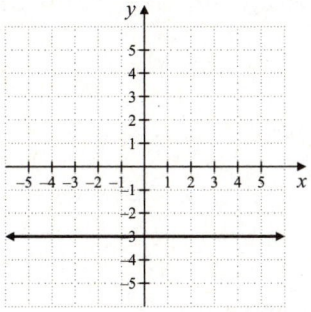

22. $x = 5$ is a vertical line with x-intercept (5, 0).

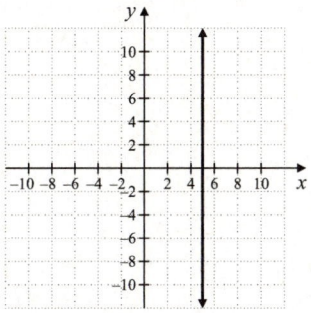

23. $x - 3y = 12$

$y = 0$: $x - 3(0) = 12$
$x - 0 = 12$
$x = 12$

x-intercept: (12, 0)

$x = 0$: $0 - 3y = 12$
$-3y = 12$
$y = -4$

y-intercept: (0, –4)

24. $-4x + y = 8$

$y = 0$: $-4x + 0 = 8$
$-4x = 8$
$x = -2$

x-intercept: (–2, 0)

$x = 0$: $-4(0) + y = 8$
$0 + y = 8$
$y = 8$

y-intercept: (0, 8)

25. $(x_1, y_1) = (-1, 2)$, $(x_2, y_2) = (3, -1)$

$$m = \frac{y_2 - y_1}{x_2 - x_1} = \frac{-1-2}{3-(-1)} = \frac{-3}{3+1} = -\frac{3}{4}$$

26. $(x_1, y_1) = (-2, -2)$, $(x_2, y_2) = (3, -1)$

$$m = \frac{y_2 - y_1}{x_2 - x_1} = \frac{-1-(-2)}{3-(-2)} = \frac{-1+2}{3+2} = \frac{1}{5}$$

27. When $m = 0$, the line is horizontal. The choice is d.

28. The slope is $m = -1$. The choice is b.

29. When the slope is undefined, the line is vertical. The choice is c.

30. The slope is $m = 4$. The choice is a.

31. $m = \dfrac{y_2 - y_1}{x_2 - x_1} = \dfrac{8-5}{6-2} = \dfrac{3}{4}$

32. $m = \dfrac{y_2 - y_1}{x_2 - x_1} = \dfrac{2-7}{1-4} = \dfrac{-5}{-3} = \dfrac{5}{3}$

33. $m = \dfrac{y_2 - y_1}{x_2 - x_1} = \dfrac{-9-3}{-2-1} = \dfrac{-12}{-3} = 4$

34. $m = \dfrac{y_2 - y_1}{x_2 - x_1} = \dfrac{-6-1}{3-(-4)} = \dfrac{-6-1}{3+4} = \dfrac{-7}{7} = -1$

35. $y = mx + b$
$y = 3x + 7$
The slope is $m = 3$.

36. $y = mx + b$

$x - 2y = 4$
$-2y = -x + 4$
$y = \frac{1}{2}x - 2$

The slope is $m = \frac{1}{2}$.

Copyright © 2012 Pearson Education, Inc. Publishing as Prentice Hall.

37. $y = mx + b$
$y = -2$
$y = 0x - 2$
The slope is $m = 0$.

38. $x = 0$ is a vertical line. The slope is undefined.

39. $x - y = -6$
$-y = -x - 6$
$y = x + 6$
$m = 1$

$x + y = 3$
$y = -x + 3$
$m = -1$

Since $(1)(-1) = -1$, the lines are perpendicular.

40. $3x + y = 7$
$y = -3x + 7$
$m = -3$

$-3x - y = 10$
$-y = 3x + 10$
$y = -3x - 10$
$m = -3$

Since the slopes are equal, the lines are parallel.

41. $y = 4x + \frac{1}{2}$
$m = 4$

$4x + 2y = 1$
$2y = -4x + 1$
$y = -2x + \frac{1}{2}$
$m = -2$

Since $4 \neq -2$ and $(4)(-2) = -8 \neq -1$, the lines are neither parallel nor perpendicular.

42. $y = 6x - \frac{1}{3}$
$m = 6$

$x + 6y = 6$
$6y = -x + 6$
$y = -\frac{1}{6}x + 1$
$m = -\frac{1}{6}$

Since $(6)\left(-\frac{1}{6}\right) = -1$, the lines are perpendicular.

43. $m = \frac{y_2 - y_1}{x_2 - x_1} = \frac{1.5 - 1.4}{2008 - 2004} = \frac{0.1}{4} = 0.025$
Every 1 year, 0.025 million (25,000) more students graduate with a bachelor's degree.

44. $m = \frac{y_2 - y_1}{x_2 - x_1} = \frac{16{,}600 - 14{,}800}{2007 - 2004} = \frac{1800}{3} = 600$
Every 1 year, 600 more people get kidney transplants.

45. $y = mx + b$
$x - 6y = -1$
$-6y = -x - 1$
$y = \frac{1}{6}x + \frac{1}{6}$
$m = \frac{1}{6}$; y-intercept $\left(0, \frac{1}{6}\right)$

46. $y = mx + b$
$3x + y = 7$
$y = -3x + 7$
$m = -3$; y-intercept $(0, 7)$

47. $y = mx + b$
$y = -5x + \frac{1}{2}$

48. $y = mx + b$
$y = \frac{2}{3}x + 6$

49. $y = mx + b$
$y = 2x + 1$
$m = 2$, y-intercept $(0, 1)$
The choice is d.

50. $y = mx + b$
$y = -4x$
$y = -4x + 0$
$m = -4$; y-intercept $(0, 0)$
The choice is c.

51. $y = mx + b$
$y = 2x$
$y = 2x + 0$
$m = 2$; y-intercept $(0, 0)$
The choice is a.

52. $y = mx + b$
$y = 2x - 1$
$m = 2$; y-intercept $(0, -1)$
The choice is b.

53. $y - y_1 = m(x - x_1)$
$y - 0 = 4(x - 2)$
$y = 4x - 8$
$-4x + y = -8$

54. $y - y_1 = m(x - x_1)$
$y - (-5) = -3(x - 0)$
$y + 5 = -3x$
$y = -3x - 5$
$3x + y = -5$

Copyright © 2012 Pearson Education, Inc. Publishing as Prentice Hall.

55.
$$\begin{aligned} y-y_1 &= m(x-x_1) \\ y-4 &= \frac{3}{5}(x-1) \\ 5(y-4) &= 5\cdot\frac{3}{5}(x-1) \\ 5y-20 &= 3(x-1) \\ 5y-20 &= 3x-3 \\ 5y &= 3x+17 \\ -3x+5y &= 17 \end{aligned}$$

56.
$$\begin{aligned} y-y_1 &= m(x-x_1) \\ y-3 &= -\frac{1}{3}[x-(-3)] \\ 3y-9 &= -1[x-(-3)] \\ 3y-9 &= -(x+3) \\ 3y-9 &= -x-3 \\ 3y &= -x+6 \\ x+3y &= 6 \end{aligned}$$

57. $m=\dfrac{y_2-y_1}{x_2-x_1}=\dfrac{-7-7}{2-1}=\dfrac{-14}{1}=-14$
$$\begin{aligned} y-y_1 &= m(x-x_1) \\ y-7 &= -14(x-1) \\ y-7 &= -14x+14 \\ y &= -14x+21 \end{aligned}$$

58. $m=\dfrac{y_2-y_1}{x_2-x_1}=\dfrac{6-5}{-4-(-2)}=\dfrac{6-5}{-4+2}=\dfrac{1}{-2}=-\dfrac{1}{2}$
$$\begin{aligned} y-y_1 &= m(x-x_1) \\ y-5 &= -\frac{1}{2}[x-(-2)] \\ y-5 &= -\frac{1}{2}(x+2) \\ 2(y-5) &= 2\left(-\frac{1}{2}\right)(x+2) \\ 2y-10 &= -1(x+2) \\ 2y-10 &= -x-2 \\ 2y &= -x+8 \\ y &= -\frac{1}{2}x+4 \end{aligned}$$

59. The x-value 7 is paired with two y-values, 1 and 5, so the relation is not a function.

60. Each x-value is only assigned to one y-value, so the relation is a function.

61. No vertical line will intersect the graph more than once, so the graph is the graph of a function.

62. No vertical line will intersect the graph more than once, so the graph is the graph of a function.

63. The vertical line $x = 3$ will intersect the graph at more than one point, so the graph is not the graph of a function.

64. No vertical line will intersect the graph more than once, so the graph is the graph of a function.

65. $f(x) = -2x + 6$
$f(0) = -2(0) + 6 = 0 + 6 = 6$

66. $f(x) = -2x + 6$
$f(-2) = -2(-2) + 6 = 4 + 6 = 10$

67. $f(x) = -2x + 6$
$$f\left(\frac{1}{2}\right)=-2\left(\frac{1}{2}\right)+6=-1+6=5$$

68. $f(x) = -2x + 6$
$$f\left(-\frac{1}{2}\right)=-2\left(-\frac{1}{2}\right)+6=1+6=7$$

69. Graph the boundary line, $x + 6y = 6$, with a dashed line.
Test (0, 0):
$$\begin{aligned} x+6y &< 6 \\ 0+6(0) &< 6 \\ 0+0 &< 6 \\ 0 &< 6 \quad \text{True} \end{aligned}$$
Shade the half-plane containing (0, 0).

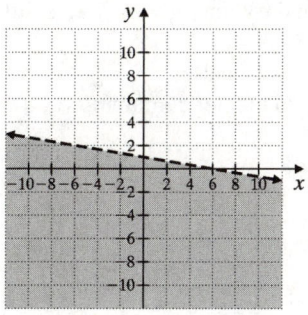

Copyright © 2012 Pearson Education, Inc. Publishing as Prentice Hall.

70. Graph the boundary line, $x + y = -2$, with a dashed line.
Test (0, 0): $x + y > -2$
$0 + 0 > -2$
$0 > -2$ True
Shade the half-plane containing (0, 0).

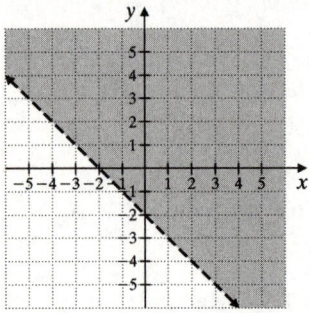

71. Graph the boundary line, $y = -7$, with a solid line.
Test (0, 0): $y \geq -7$
$0 \geq -7$ True
Shade the half-plane containing (0, 0).

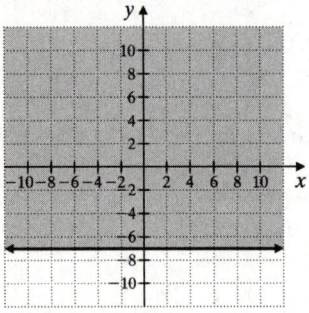

72. Graph the boundary line, $y = -4$, as a solid line.
Test (0, 0): $y \leq -4$
$0 \leq -4$ False
Shade the half-plane not containing (0, 0).

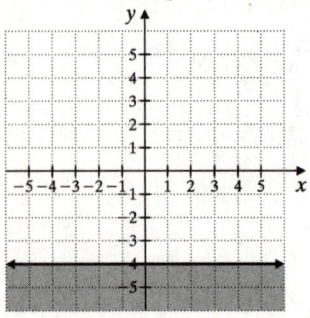

73. Graph the boundary line, $-x = y$, as a solid line.
Test (1, 1): $-x \leq y$
$-1 \leq 1$ True
Shade the half-plane containing (1, 1).

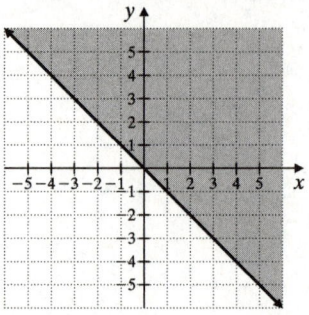

74. Graph the boundary line, $x = -y$, as a solid line.
Test (1, 1): $x \geq -y$
$1 \geq -1$ True
Shade the half-plane containing (1, 1).

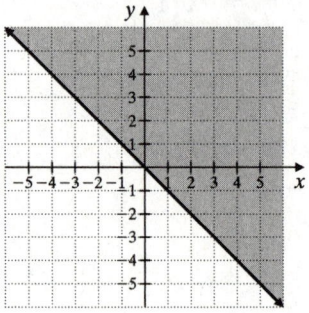

75. $y = kx$
$y = 40$ when $x = 4$: $40 = k(4)$
$10 = k$
$y = 10x$
$x = 11$: $y = 10(11) = 110$
$y = 110$ when $x = 11$.

76. $y = \dfrac{k}{x}$

$y = 4$ when $x = 6$: $4 = \dfrac{k}{6}$
$24 = k$

$y = \dfrac{24}{x}$

$x = 48$: $y = \dfrac{24}{48} = \dfrac{1}{2}$

$y = \dfrac{1}{2}$ when $x = 48$.

Copyright © 2012 Pearson Education, Inc. Publishing as Prentice Hall.

77. $y = \frac{k}{x^3}$

$y = 12.5$ when $x = 2$: $12.5 = \frac{k}{2^3}$

$12.5 = \frac{k}{8}$

$100 = k$

$y = \frac{100}{x^3}$

$x = 3$: $y = \frac{100}{3^3} = \frac{100}{27}$

$y = \frac{100}{27}$ when $x = 3$.

78. $y = kx^2$

$y = 175$ when $x = 5$: $175 = k(5)^2$

$175 = 25k$

$7 = k$

$y = 7x^2$

$x = 10$: $y = 7(10)^2 = 7(100) = 700$

$y = 700$ when $x = 10$.

79. Let c be the cost for manufacturing m milliliters.

$c = \frac{k}{m}$

$c = 6600$ when $m = 3000$: $6600 = \frac{k}{3000}$

$19{,}800{,}000 = k$

$c = \frac{19{,}800{,}000}{m}$

Let $m = 5000$: $c = \frac{19{,}800{,}000}{5000} = 3960$

It costs \$3960 to manufacture 5000 milliliters.

80. Let d be the distance when a weight of w is attached.

$d = kw$

$d = 8$ when $w = 150$: $8 = k(150)$

$\frac{4}{75} = k$

$d = \frac{4}{75}w$

Let $w = 90$: $d = \frac{4}{75} \cdot 90 = 4\frac{4}{5}$

The spring stretches $4\frac{4}{5}$ inches when 90 pounds is attached.

81. $2x - 5y = 9$

$y = 1$: $2x - 5(1) = 9$

$2x - 5 = 9$

$2x = 14$

$x = 7$

$x = 2$: $2(2) - 5y = 9$

$4 - 5y = 9$

$-5y = 5$

$y = -1$

$y = -3$: $2x - 5(-3) = 9$

$2x + 15 = 9$

$2x = -6$

$x = -3$

x	y
7	1
2	−1
−3	−3

82. $x = -3y$

$x = 0$: $0 = -3y$

$0 = y$

$y = 1$: $x = -3(1)$

$x = -3$

$x = 6$: $6 = -3y$

$-2 = y$

x	y
0	0
−3	1
6	−2

83. $2x - 3y = 6$

$y = 0$: $2x - 3(0) = 6$

$2x - 0 = 6$

$2x = 6$

$x = 3$

x-intercept: (3, 0)

$x = 0$: $2(0) - 3y = 6$

$0 - 3y = 6$

$-3y = 6$

$y = -2$

y-intercept: (0, −2)

Copyright © 2012 Pearson Education, Inc. Publishing as Prentice Hall.

84. $-5x + y = 10$

$y = 0$: $-5x + 0 = 10$

$-5x = 10$

$x = -2$

x-intercept: $(-2, 0)$

$x = 0$: $-5(0) + y = 10$

$0 + y = 10$

$y = 10$

y-intercept: $(0, 10)$

85. $x - 5y = 10$

$y = 0$: $x - 5(0) = 10$

$x - 0 = 10$

$x = 10$

x-intercept: $(10, 0)$

$x = 0$: $0 - 5y = 10$

$-5y = 10$

$y = -2$

y-intercept: $(0, -2)$

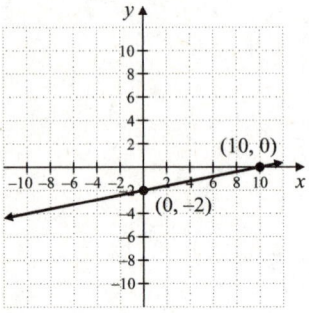

86. $x + y = 4$

$y = 0$: $x + 0 = 4$

$x = 4$

x-intercept: $(4, 0)$

$x = 0$: $0 + y = 4$

$y = 4$

y-intercept: $(0, 4)$

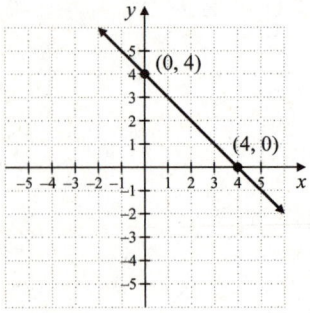

87. $y = -4x$

$y = 0$: $0 = -4x$

$0 = x$

x-intercept: $(0, 0)$

The y-intercept is also $(0, 0)$. Find another point.

Let $x = 1$.

$y = -4(1) = -4$

Another point is $(1, -4)$.

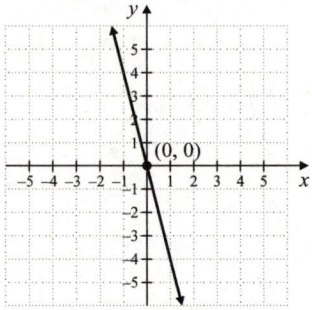

88. $2x + 3y = -6$

$y = 0$: $2x + 3(0) = -6$

$2x + 0 = -6$

$2x = -6$

$x = -3$

x-intercept: $(-3, 0)$

$x = 0$: $2(0) + 3y = -6$

$0 + 3y = -6$

$y = -2$

y-intercept: $(0, -2)$

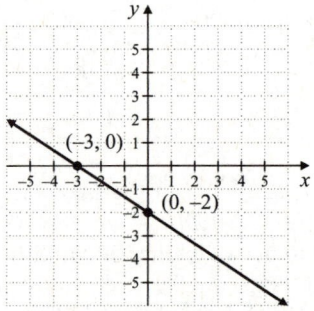

89. $x = 3$ is a vertical line with x-intercept $(3, 0)$.

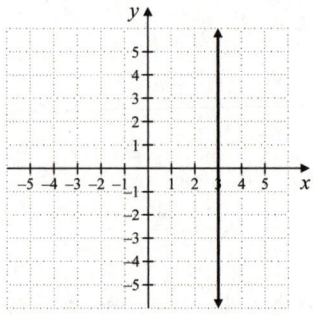

Copyright © 2012 Pearson Education, Inc. Publishing as Prentice Hall.

90. $y = -2$ is a horizontal line with y-intercept $(0, -2)$.

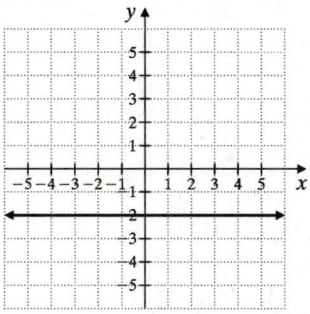

91. $m = \dfrac{y_2 - y_1}{x_2 - x_1} = \dfrac{2-(-5)}{-4-3} = \dfrac{2+5}{-4-3} = \dfrac{7}{-7} = -1$

92. $m = \dfrac{y_2 - y_1}{x_2 - x_1} = \dfrac{-8-3}{-6-1} = \dfrac{-11}{-7} = \dfrac{11}{7}$

93. $(x_1, y_1) = (0, -4), (x_2, y_2) = (2, 0)$

$m = \dfrac{y_2 - y_1}{x_2 - x_1} = \dfrac{0-(-4)}{2-0} = \dfrac{4}{2} = 2$

94. $(x_1, y_1) = (0, 2), (x_2, y_2) = (6, 0)$

$m = \dfrac{y_2 - y_1}{x_2 - x_1} = \dfrac{0-2}{6-0} = \dfrac{-2}{6} = -\dfrac{1}{3}$

95. $y = mx + b$

$$\begin{aligned} -2x + 3y &= -15 \\ 3y &= 2x - 15 \\ y &= \frac{2}{3}x - 5 \end{aligned}$$

$m = \dfrac{2}{3}$; y-intercept: $(0, -5)$

96. $y = mx + b$

$$\begin{aligned} 6x + y - 2 &= 0 \\ 6x + y &= 2 \\ y &= -6x + 2 \end{aligned}$$

$m = -6$; y-intercept: $(0, 2)$

97.

$$\begin{aligned} y - y_1 &= m(x - x_1) \\ y - (-7) &= -5(x - 3) \\ y + 7 &= -5x + 15 \\ y &= -5x + 8 \\ 5x + y &= 8 \end{aligned}$$

98.

$$\begin{aligned} y - y_1 &= m(x - x_1) \\ y - 6 &= 3(x - 0) \\ y - 6 &= 3x \\ -6 &= 3x - y \\ 3x - y &= -6 \end{aligned}$$

99. $m = \dfrac{y_2 - y_1}{x_2 - x_1} = \dfrac{5-9}{-2-(-3)} = \dfrac{5-9}{-2+3} = \dfrac{-4}{1} = -4$

$$\begin{aligned} y - y_1 &= m(x - x_1) \\ y - 9 &= -4[x - (-3)] \\ y - 9 &= -4(x + 3) \\ y - 9 &= -4x - 12 \\ y &= -4x - 3 \\ 4x + y &= -3 \end{aligned}$$

100. $m = \dfrac{y_2 - y_1}{x_2 - x_1} = \dfrac{-9-1}{5-3} = \dfrac{-10}{2} = -5$

$$\begin{aligned} y - y_1 &= m(x - x_1) \\ y - 1 &= -5(x - 3) \\ y - 1 &= -5x + 15 \\ y &= -5x + 16 \\ 5x + y &= 16 \end{aligned}$$

Chapter 6 Test

1. $12y - 7x = 5$

$$\begin{aligned} x = 1:\ 12y - 7(1) &= 5 \\ 12y - 7 &= 5 \\ 12y &= 12 \\ y &= 1 \end{aligned}$$

The ordered pair is $(1, 1)$.

2. $y = 17$ for each value of x. The ordered pair is $(-4, 17)$.

3. $(x_1, y_1) = (-1, -1), (x_2, y_2) = (4, 1)$

$m = \dfrac{y_2 - y_1}{x_2 - x_1} = \dfrac{1-(-1)}{4-(-1)} = \dfrac{1+1}{4+1} = \dfrac{2}{5}$

4. The slope of a horizontal line is $m = 0$.

5. $m = \dfrac{y_2 - y_1}{x_2 - x_1} = \dfrac{2-(-5)}{-1-6} = \dfrac{7}{-7} = -1$

6. $m = \dfrac{y_2 - y_1}{x_2 - x_1} = \dfrac{-1-(-8)}{-1-0} = \dfrac{-1+8}{-1} = \dfrac{7}{-1} = -7$

Copyright © 2012 Pearson Education, Inc. Publishing as Prentice Hall.

7. $y = mx + b$
$-3x + y = 5$
$y = 3x + 5$
$m = 3$

8. $x = 6$ is a vertical line. The slope of a vertical line is undefined.

9. $2x + y = 8$
$y = 0$: $2x + 0 = 8$
$2x = 8$
$x = 4$
x-intercept: (4, 0)
$x = 0$: $2(0) + y = 8$
$y = 8$
y-intercept: (0, 8)

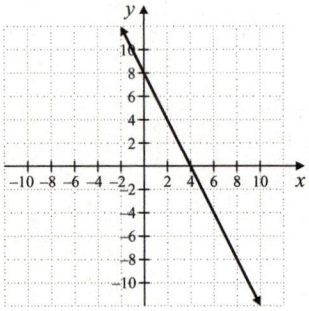

10. $-x + 4y = 5$
$y = 0$: $-x + 4(0) = 5$
$-x + 0 = 5$
$-x = 5$
$x = -5$
x-intercept: (–5, 0)
$x = 0$: $-0 + 4y = 5$
$4y = 5$
$y = \frac{5}{4}$
y-intercept: $\left(0, \frac{5}{4}\right)$

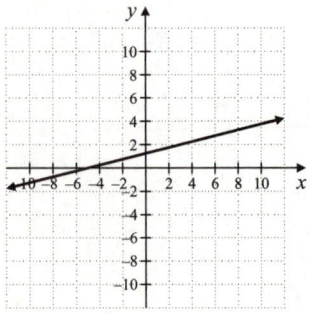

11. Graph the boundary line, $x - y = -2$, with a solid line.
Test (0, 0): $x - y \geq -2$
$0 - 0 \geq -2$
$0 \geq -2$ True
Shade the half-plane containing (0, 0).

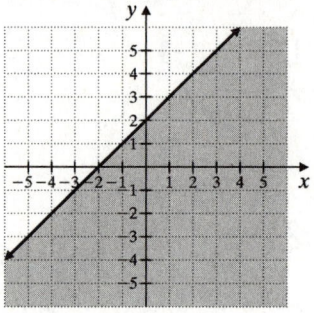

12. Graph the boundary line, $y = -4x$, with a solid line.
Test (1, 1): $y \geq -4x$
$1 \geq -4(1)$
$1 \geq -4$ True
Shade the half-plane containing (1, 1).

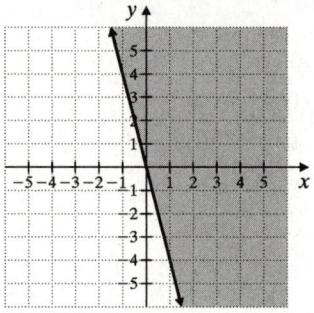

13. $5x - 7y = 10$
$y = 0$: $5x - 7(0) = 10$
$5x - 0 = 10$
$5x = 10$
$x = 2$
x-intercept: (2, 0)
$x = 0$: $5(0) - 7y = 10$
$0 - 7y = 10$
$-7y = 10$
$y = -\frac{10}{7}$
y-intercept: $\left(0, -\frac{10}{7}\right)$

Copyright © 2012 Pearson Education, Inc. Publishing as Prentice Hall.

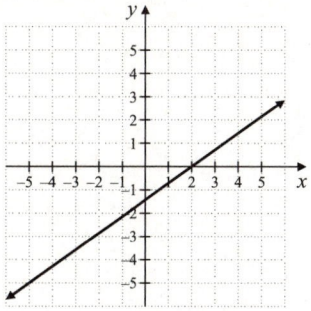

14. Graph the boundary line, $2x - 3y = -6$, with a dashed line.

Test (0, 0): $2x - 3y > -6$

$2(0) - 3(0) > -6$

$0 - 0 > -6$

$0 > -6$ True

Shade the half-plane containing (0, 0).

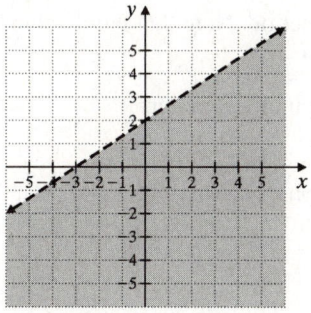

15. Graph the boundary line, $6x + y = -1$, with a dashed line.

Test (0, 0): $6x + y > -1$

$6(0) + 0 > -1$

$0 + 0 > -1$

$0 > -1$ True

Shade the half-plane containing (0, 0).

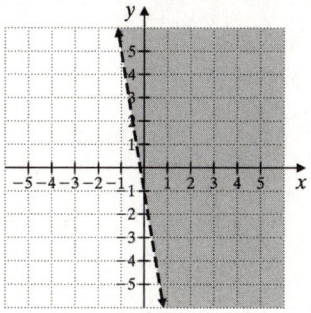

16. The graph of $y = -1$ is a horizontal line with a y-intercept of $(0, -1)$.

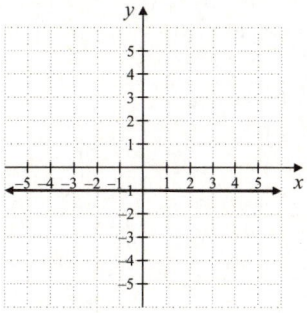

17. $y = 2x - 6$

The slope is $m = 2$ and the y-intercept is $(0, -6)$.

$-4x = 2y$

$-2x = y$

$y = -2x + 0$

The slope is $m = -2$ and the y-intercept is $(0, 0)$. Since the slopes are different, the lines are not parallel. Since $2(-2) = -4 \neq -1$, the lines are not perpendicular. The lines are neither parallel nor perpendicular.

18.

$$y - y_1 = m(x - x_1)$$

$$y - 2 = -\frac{1}{4}(x - 2)$$

$$4(y - 2) = 4\left(-\frac{1}{4}\right)(x - 2)$$

$$4y - 8 = -1(x - 2)$$

$$4y - 8 = -x + 2$$

$$4y = -x + 10$$

$$x + 4y = 10$$

19.

$$m = \frac{y_2 - y_1}{x_2 - x_1} = \frac{-7 - 0}{6 - 0} = \frac{-7}{6} = -\frac{7}{6}$$

$$y - y_1 = m(x - x_1)$$

$$y - 0 = -\frac{7}{6}(x - 0)$$

$$y = -\frac{7}{6}x$$

$$6y = 6\left(-\frac{7}{6}x\right)$$

$$6y = -7x$$

$$7x + 6y = 0$$

Copyright © 2012 Pearson Education, Inc. Publishing as Prentice Hall.

20. $m = \frac{y_2 - y_1}{x_2 - x_1} = \frac{3-(-5)}{1-2} = \frac{3+5}{-1} = -8$

$$y - y_1 = m(x - x_1)$$
$$y - (-5) = -8(x - 2)$$
$$y + 5 = -8x + 16$$
$$y = -8x + 11$$
$$8x + y = 11$$

21. $m = \frac{1}{8}$; $b = 12$

$$y = \frac{1}{8}x + 12$$
$$8y = x + 8(12)$$
$$8y = x + 96$$
$$x - 8y = -96$$

22. Each x-value is only assigned to one y-value, so the relation is a function.

23. The x-value -3 is assigned to two y-values, -3 and 2, so the relation is not a function. Note that the x-value 0 is also assigned to two y-values, 5 and 0.

24. No vertical line will intersect the graph more than once, so the graph is the graph of a function. From the graph, the domain is all real numbers and the range is all real numbers

25. No vertical line will intersect the graph more than once, so the graph is the graph of a function. From the graph, the domain is all real numbers and the range is $\{2\}$.

26. $f(x) = 2x - 4$

a. $f(-2) = 2(-2) - 4 = -4 - 4 = -8$

b. $f(0.2) = 2(0.2) - 4 = 0.4 - 4 = -3.6$

c. $f(0) = 2(0) - 4 = 0 - 4 = -4$

27. $f(x) = x^3 - x$

a. $f(-1) = (-1)^3 - (-1) = -1 + 1 = 0$

b. $f(0) = 0^3 - 0 = 0 - 0 = 0$

c. $f(4) = 4^3 - 4 = 64 - 4 = 60$

28. $2x + 2(2y) = 42$

$$2(x + 2y) = 2(21)$$
$$x + 2y = 21$$

Let $y = 8$: $x + 2(8) = 21$

$$x + 16 = 21$$
$$x = 5$$

When $y = 8$ meters, $x = 5$ meters.

Copyright © 2012 Pearson Education, Inc. Publishing as Prentice Hall.

29. a. The ordered pairs are (2003, 66.0), (2004, 65.4), (2005, 65.4), (2006, 65.6), (2007, 64.9), (2008, 63.7), (2009, 62.1).

b.

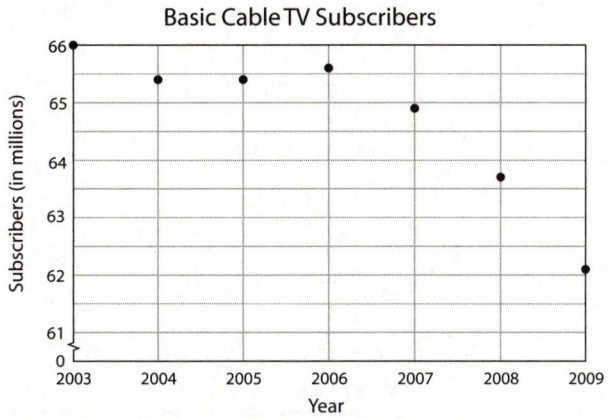

30. $m = \dfrac{y_2 - y_1}{x_2 - x_1} = \dfrac{1400 - 1484}{2007 - 2004} = \dfrac{-84}{3} = -28$

Every 1 year, 28 million fewer movie tickets are sold.

31. $y = kx$

$y = 10$ when $x = 15$: $10 = k(15)$

$$\frac{2}{3} = k$$

$y = \dfrac{2}{3}x$

Let $x = 42$: $y = \dfrac{2}{3}(42) = 28$

When $x = 42$, $y = 28$.

32. $y = \dfrac{k}{x^2}$

$y = 8$ when $x = 5$: $8 = \dfrac{k}{5^2}$

$$8 = \frac{k}{25}$$

$$200 = k$$

$y = \dfrac{200}{x^2}$

Let $x = 15$: $y = \dfrac{200}{15^2} = \dfrac{200}{225} = \dfrac{8}{9}$

When $x = 15$, $y = \dfrac{8}{9}$.

Cumulative Review Chapters 1–6

1. $6 \div 3 + 5^2 = 6 \div 3 + 25 = 2 + 25 = 27$

Copyright © 2012 Pearson Education, Inc. Publishing as Prentice Hall.

2. $\frac{10}{3}+\frac{5}{21}=\frac{10}{3}\cdot\frac{7}{7}+\frac{5}{21}=\frac{70}{21}+\frac{5}{21}=\frac{75}{21}=\frac{25}{7}$

3. $1+2[5(2\cdot 3+1)-10]=1+2[5(6+1)-10]$
$=1+2[5(7)-10]$
$=1+2(35-10)$
$=1+2(25)$
$=1+50$
$=51$

4. $16-3\cdot 3+2^4=16-3\cdot 3+16$
$=16-9+16$
$=7+16$
$=23$

5. 20,320 – (–282) = 20,320 + 282 = 20,602
The difference in elevation is 20,602 feet.

6. $1.7x-11-0.9x-25=(1.7x-0.9x)+(-11-25)$
$=0.8x+(-36)$
$=0.8x-36$

7. Twice a number, plus 6 is written as $2x + 6$.

8. The product of –15 and the sum of a number and $\frac{2}{3}$ is written as $-15\left(x+\frac{2}{3}\right)$.

9. The difference of a number and 4, divided by 7 is written as $(x - 4) \div 7$ or $\frac{x-4}{7}$.

10. The quotient of –9 and twice a number is written as $\frac{-9}{2x}$.

11. Five plus the sum of a number and 1 is written as $5 + (x + 1) = 5 + x + 1 = 6 + x$.

12. A number subtracted from –86 is written as $-86 - x$.

13. $\frac{5}{2}x=15$
$\frac{2}{5}\cdot\frac{5}{2}x=\frac{2}{5}\cdot 15$
$x=6$

14. $\frac{x}{4}-1=-7$
$\frac{x}{4}-1+1=-7+1$
$\frac{x}{4}=-6$
$4\cdot\frac{x}{4}=4(-6)$
$x=-24$

15. $2x<-4$
$\frac{2x}{2}<\frac{-4}{2}$
$x<-2$
$\{x|x<-2\}$

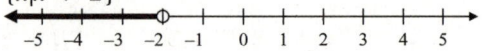

16. $5(x+4)\ge 4(2x+3)$
$5x+20\ge 8x+12$
$5x+20-20\ge 8x+12-20$
$5x\ge 8x-8$
$5x-8x\ge 8x-8-8x$
$-3x\ge -8$
$\frac{-3x}{-3}\le\frac{-8}{-3}$
$x\le\frac{8}{3}$
$\left\{x\middle|x\le\frac{8}{3}\right\}$

17. a. The degree of the trinomial $-2t^2+3t+6$ is 2, the greatest degree of any of its terms.

b. The degree of the binomial $15x - 10$ or $15x^1-10$ is 1.

c. The degree of the polynomial $7x+3x^3+2x^2-1$ is 3. It is not a monomial, binomial, nor a trinomial, so the answer is none of these.

18. $x+2y=6$
$x+2y-x=6-x$
$2y=6-x$
$\frac{2y}{2}=\frac{6-x}{2}$
$y=\frac{6-x}{2}$

Copyright © 2012 Pearson Education, Inc. Publishing as Prentice Hall.

19. $(-2x^2+5x-1)+(-2x^2+x+3)$
$=-2x^2+5x-1-2x^2+x+3$
$=(-2x^2-2x^2)+(5x+x)+(-1+3)$
$=-4x^2+6x+2$

20. $(-2x^2+5x-1)-(-2x^2+x+3)$
$=-2x^2+5x-1+2x^2-x-3$
$=(-2x^2+2x^2)+(5x-x)+(-1-3)$
$=0x^2+4x+(-4)$
$=4x-4$

21. $(3y+1)^2=(3y+1)(3y+1)$
$=(3y)(3y)+(3y)(1)+1(3y)+1(1)$
$=9y^2+3y+3y+1$
$=9y^2+6y+1$

22. $(x-12)^2=x^2-2(x)(12)+(12)^2$
$=x^2-24x+144$

23. $-9a^5+18a^2-3a=3a(-3a^4)+3a(6a)+3a(-1)$
$=3a(-3a^4+6a-1)$

24. $4x^2-36=4(x^2-9)$
$=4(x^2-3^2)$
$=4(x+3)(x-3)$

25. $x^2+4x-12$
Look for two numbers whose product is -12 and whose sum is 4.
$x^2+4x-12=(x-2)(x+6)$

26. $3x^2-20xy-7y^2=(3x+y)(x-7y)$

27. Factors of $8x^2$: $8x^2=8x\cdot x$, $8x^2=4x\cdot 2x$
Factors of 5: $5=-1\cdot -5$
$8x^2-22x+5=(4x-1)(2x-5)$

28. Factors of $18x^2$: $18x^2=18x\cdot x$, $18x^2=9x\cdot 2x$,
$18x^2=6x\cdot 3x$
Factors of -2: $-2=-1\cdot 2$, $-2=1\cdot -2$
$18x^2+35x-2=(18x-1)(x+2)$

29. $x^2-9x-22=0$
$(x-11)(x+2)=0$
$x-11=0$ or $x+2=0$
$x=11$ $\quad$ $x=-2$
The solutions are 11 and -2.

30. $x^2=x$
$x^2-x=0$
$x(x-1)=0$
$x=0$ or $x-1=0$
$x=1$
The solutions are 0 and 1.

31. $\dfrac{2x^2-11x+5}{5x-25}\div\dfrac{4x-2}{10}=\dfrac{2x^2-11x+5}{5x-25}\cdot\dfrac{10}{4x-2}$
$=\dfrac{(2x-1)(x-5)\cdot 2\cdot 5}{5(x-5)\cdot 2(2x-1)}$
$=\dfrac{1}{1}$ or 1

32. $\dfrac{2x^2-50}{4x^4-20x^3}=\dfrac{2(x^2-25)}{4x^3(x-5)}$
$=\dfrac{2(x-5)(x+5)}{2\cdot 2x^3(x-5)}$
$=\dfrac{x+5}{2x^3}$

33. $\dfrac{4b}{9a}=\dfrac{4b}{9a}\cdot 1=\dfrac{4b}{9a}\cdot\dfrac{3ab}{3ab}=\dfrac{4b(3ab)}{9a(3ab)}=\dfrac{12ab^2}{27a^2b}$

34. $\dfrac{1}{2x}=\dfrac{1}{2x}\cdot 1=\dfrac{1}{2x}\cdot\dfrac{7x^2}{7x^2}=\dfrac{1(7x^2)}{2x(7x^2)}=\dfrac{7x^2}{14x^3}$

35. $1+\dfrac{m}{m+1}=\dfrac{1}{1}+\dfrac{m}{m+1}$
$=\dfrac{1(m+1)}{1(m+1)}+\dfrac{m}{m+1}$
$=\dfrac{m+1+m}{m+1}$
$=\dfrac{2m+1}{m+1}$

Copyright © 2012 Pearson Education, Inc. Publishing as Prentice Hall.

36. $\dfrac{2x+1}{x-6}-\dfrac{x-4}{x-6}=\dfrac{(2x+1)-(x-4)}{x-6}$

$=\dfrac{2x+1-x+4}{x-6}$

$=\dfrac{x+5}{x-6}$

37. $3-\dfrac{6}{x}=x+8$

$x\left(3-\dfrac{6}{x}\right)=x(x+8)$

$3x-6=x^2+8x$

$0=x^2+5x+6$

$0=(x+3)(x+2)$

$x+3=0$ or $x+2=0$

$x=-3$ $\quad$ $x=-2$

The solutions are -3 and -2.

38. $3x^2+5x=2$

$3x^2+5x-2=0$

$(3x-1)(x+2)=0$

$3x-1=0$ or $x+2=0$

$x=\dfrac{1}{3}$ $\quad$ $x=-2$

The solutions are -2 and $\dfrac{1}{3}$.

39. $\dfrac{\frac{x+1}{y}}{\frac{x}{y}+2}=\dfrac{y\left(\frac{x+1}{y}\right)}{y\left(\frac{x}{y}+2\right)}=\dfrac{y\left(\frac{x+1}{y}\right)}{y\left(\frac{x}{y}\right)+y(2)}=\dfrac{x+1}{x+2y}$

40. $\dfrac{\frac{x}{2}-\frac{y}{6}}{\frac{x}{12}-\frac{y}{3}}=\dfrac{12\left(\frac{x}{2}-\frac{y}{6}\right)}{12\left(\frac{x}{12}-\frac{y}{3}\right)}$

$=\dfrac{12\left(\frac{x}{2}\right)-12\left(\frac{y}{6}\right)}{12\left(\frac{x}{12}\right)-12\left(\frac{y}{3}\right)}$

$=\dfrac{6x-2y}{x-4y}$ or $\dfrac{2(3x-y)}{x-4y}$

41. $3x+y=12$

a. In $(0,\ \ )$, the x-coordinate is 0.

$x=0$: $3(0)+y=12$

$0+y=12$

$y=12$

The ordered-pair solution is $(0, 12)$.

b. In $(\ \ ,6)$, the y-coordinate is 6.

$y=6$: $3x+6=12$

$3x=6$

$x=2$

The ordered-pair solution is $(2, 6)$.

c. In $(-1,\ \ )$, the x-coordinate is -1.

$x=-1$: $3(-1)+y=12$

$-3+y=12$

$y=15$

The ordered-pair solution is $(-1, 15)$.

42. $y=-5x$

$x=0$: $y=-5(0)=0$

$x=-1$: $y=-5(-1)=5$

$y=-10$: $-10=-5x$

$2=x$

x	y
0	0
-1	5
2	-10

43. $2x+y=5$

$y=0$: $2x+0=5$

$2x=5$

$x=\dfrac{5}{2}$

The x-intercept is $\left(\dfrac{5}{2}, 0\right)$.

$x=0$: $2(0)+y=5$

$0+y=5$

$y=5$

The y-intercept is $(0, 5)$.

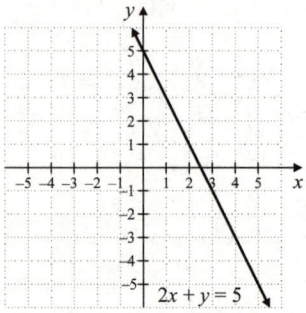

44. $(x_1, y_1)=(0, 5)$, $(x_2, y_2)=(-5, 4)$

$m=\dfrac{y_2-y_1}{x_2-x_1}=\dfrac{4-5}{-5-0}=\dfrac{-1}{-5}=\dfrac{1}{5}$

Copyright © 2012 Pearson Education, Inc. Publishing as Prentice Hall.

45. $y = mx + b$

$$-2x + 3y = 11$$
$$3y = 2x + 11$$
$$y = \frac{2}{3}x + \frac{11}{3}$$

The slope is $m = \frac{2}{3}$.

46. The graph of $x = -10$ is a vertical line. Slopes of vertical lines are undefined.

47. $y - y_1 = m(x - x_1)$

$$y - 5 = -2[x - (-1)]$$
$$y - 5 = -2(x + 1)$$
$$y - 5 = -2x - 2$$
$$y = -2x + 3$$
$$2x + y = 3$$

48. $y = mx + b$

$$2x - 5y = 10$$
$$-5y = -2x + 10$$
$$y = \frac{2}{5}x - 2$$

The slope is $m = \frac{2}{5}$ and the y-intercept is $(0, -2)$.

49. $g(x) = x^2 - 3$

a. $g(2) = 2^2 - 3 = 4 - 3 = 1$

The ordered pair is (2, 1).

b. $g(-2) = (-2)^2 - 3 = 4 - 3 = 1$

The ordered pair is (–2, 1).

c. $g(0) = 0^2 - 3 = 0 - 3 = -3$

The ordered pair is (0, –3).

50. $(x_1, y_1) = (2, 3)$, $(x_2, y_2) = (0, 0)$

$$m = \frac{y_2 - y_1}{x_2 - x_1} = \frac{0 - 3}{0 - 2} = \frac{-3}{-2} = \frac{3}{2}$$
$$y - y_1 = m(x - x_1)$$
$$y - 3 = \frac{3}{2}(x - 2)$$
$$2(y - 3) = 2\left(\frac{3}{2}\right)(x - 2)$$
$$2(y - 3) = 3(x - 2)$$
$$2y - 6 = 3x - 6$$
$$2y = 3x$$
$$0 = 3x - 2y$$
$$3x - 2y = 0$$

Copyright © 2012 Pearson Education, Inc. Publishing as Prentice Hall.

Chapter 7

Section 7.1 Practice

1.

$$\begin{aligned} 5x-2y&=-3 \\ 5(3)-2(9)&=-3 \\ 15-18&=-3 \\ -3&=-3 \end{aligned} \qquad \begin{aligned} y&=3x \\ 9&=3(3) \\ 9&=9 \end{aligned}$$

(3, 9) is a solution of the system.

2.

$$\begin{aligned} 2x-y&=8 \\ 2(3)-(-2)&=8 \\ 6+2&=8 \\ 8&=8 \end{aligned} \qquad \begin{aligned} x+3y&=4 \\ 3+3(-2)&=4 \\ 3+-6&=4 \\ -3&\neq 4 \end{aligned}$$

(3, –2) is not a solution of the system.

3.

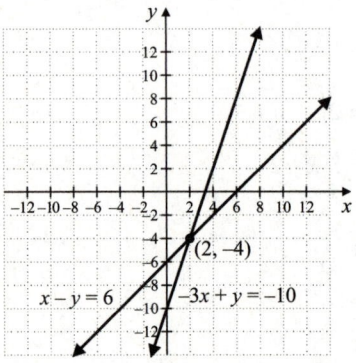

(2, –4) is a solution of the system.

4.

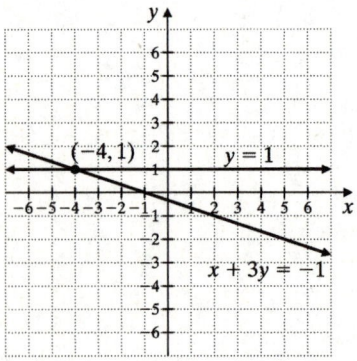

(–4, 1) is a solution of the system.

5.

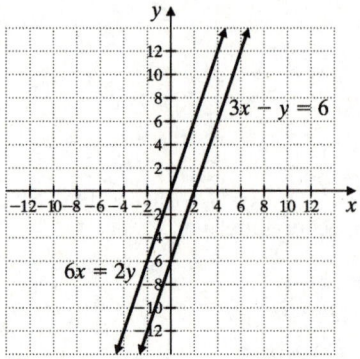

The system has no solution because the lines are parallel.

6.

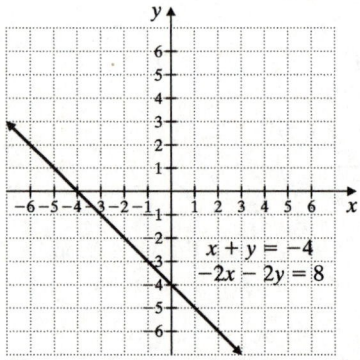

The system has an infinite number of solutions because the graphs of the equations are the same line.

7.

$$\begin{aligned} 5x+4y&=6 \\ 4y&=-5x+6 \\ y&=-\frac{5}{4}x+\frac{3}{2} \end{aligned} \qquad \begin{aligned} x-y&=3 \\ -y&=-x+3 \\ y&=x-3 \end{aligned}$$

The slope of the first line is $-\frac{5}{4}$, while the slope of the second line is 1. Since the slopes are not equal, the system has one solution.

8.

$$\begin{aligned} -\frac{2}{3}x+y&=6 \\ y&=\frac{2}{3}x+6 \end{aligned} \qquad \begin{aligned} 3y&=2x+5 \\ y&=\frac{2}{3}x+\frac{5}{3} \end{aligned}$$

Both lines have slope $\frac{2}{3}$, but the y-intercepts are different. The lines are parallel, so the system has no solution.

Copyright © 2012 Pearson Education, Inc. Publishing as Prentice Hall.

Calculator Explorations

1. $\begin{cases} y = -2.68x + 1.21 \\ y = 5.22x - 1.68 \end{cases}$

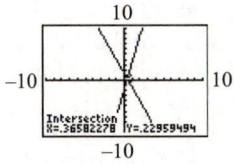

(0.37, 0.23) is the approximate point of intersection.

2. $\begin{cases} y = 4.25x + 3.89 \\ y = -1.88x + 3.21 \end{cases}$

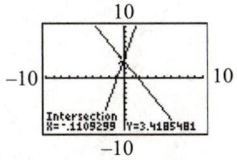

(–0.11, 3.42) is the approximate point of intersection.

3. $\begin{cases} 4.3x - 2.9y = 5.6 \rightarrow y = -(5.6 - 4.3x)/2.9 \\ 8.1x + 7.6y = -14.1 \rightarrow y = (-14.1 - 8.1x)/7.6 \end{cases}$

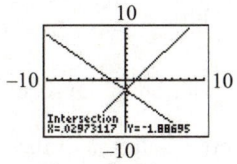

(0.03, –1.89) is the approximate point of intersection.

4. $\begin{cases} -3.6x - 8.6y = 10 \rightarrow y = -(10 + 3.6x)/8.6 \\ -4.5x + 9.6y = -7.7 \rightarrow y = (-7.7 + 4.5x)/9.6 \end{cases}$

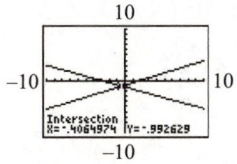

(–0.41, –0.99) is the approximate point of intersection.

Vocabulary and Readiness Check

1. In a system of linear equations in two variables, if the graphs of the equations are the same, the equations are dependent equations.

2. Two or more linear equations are called a system of linear equations.

3. A system of equations that has at least one solution is called a consistent system.

4. A solution of a system of two equations in two variables is an ordered pair of numbers that is a solution of both equations in the system.

5. A system of equations that has no solution is called an inconsistent system.

6. In a system of linear equations in two variables, if the graphs of the equations are different, the equations are independent equations.

7. The lines intersect at (–1, 3); therefore there is one solution.

8. Since the lines are parallel and do not intersect, there is no solution.

9. Since the lines are the same, there is an infinite number of solutions.

10. The lines intersect at (3, 4); therefore there is one solution.

Exercise Set 7.1

1. a. First equation:
$x + y = 8$
$2 + 4 \stackrel{?}{=} 8$
$6 = 8$ False
(2, 4) is not a solution of the first equation, so it is not a solution of the system.

   b. First equation:
$x + y = 8$
$5 + 3 \stackrel{?}{=} 8$
$8 = 8$ True
Second equation:
$3x + 2y = 21$
$3(5) + 2(3) \stackrel{?}{=} 21$
$15 + 6 \stackrel{?}{=} 21$
$21 = 21$ True
Since (5, 3) is a solution of both equations, it is a solution of the system.

3. a. First equation:
$3x - y = 5$
$3(3) - 4 \stackrel{?}{=} 5$
$9 - 4 \stackrel{?}{=} 5$
$5 = 5$ True

Copyright © 2012 Pearson Education, Inc. Publishing as Prentice Hall.

Second equation:

$x+2y=11$

$3+2(4) \stackrel{?}{=} 11$

$3+8 \stackrel{?}{=} 11$

$11=11$ True

Since (3, 4) is a solution of both equations, it is a solution of the system.

b. First equation:

$3x-y=5$

$3(0)-(-5) \stackrel{?}{=} 5$

$0+5 \stackrel{?}{=} 5$

$5=5$ True

Second equation:

$x+2y=11$

$0+2(-5) \stackrel{?}{=} 11$

$0+(-10) \stackrel{?}{=} 11$

$-10=11$ False

(0, −5) is not a solution of the second equation, so it is not a solution of the system.

5. a. First equation:

$2y=4x+6$

$2(-3) \stackrel{?}{=} 4(-3)+6$

$-6 \stackrel{?}{=} -12+6$

$-6=-6$ True

Second equation:

$2x-y=-3$

$2(-3)-(-3) \stackrel{?}{=} -3$

$-6+3 \stackrel{?}{=} -3$

$-3=-3$ True

(−3, −3) is a solution of both equations, it is a solution of the system.

b. First equation:

$2y=4x+6$

$2(3) \stackrel{?}{=} 4(0)+6$

$6 \stackrel{?}{=} 0+6$

$6=6$ True

Second equation:

$2x-y=-3$

$2(0)-3 \stackrel{?}{=} -3$

$0-3 \stackrel{?}{=} -3$

$-3=-3$ True

Since (0, 3) is a solution of both equations, it is a solution of the system.

7. a. First equation:

$-2=x-7y$

$-2 \stackrel{?}{=} -2-7(0)$

$-2 \stackrel{?}{=} -2-0$

$-2=-2$ True

Second equation:

$6x-y=13$

$6(-2)-0 \stackrel{?}{=} 13$

$-12-0 \stackrel{?}{=} 13$

$-12=13$ False

(−2, 0) is not a solution of the second equation, so it is not a solution of the system.

b. First equation:

$-2=x-7y$

$-2 \stackrel{?}{=} \frac{1}{2}-7\left(\frac{5}{14}\right)$

$-2 \stackrel{?}{=} \frac{1}{2}-\frac{5}{2}$

$-2=-\frac{4}{2}$

$-2=-2$ True

Second equation:

$6x-y=13$

$6\left(\frac{1}{2}\right)-\frac{5}{14} \stackrel{?}{=} 13$

$\frac{42}{14}-\frac{5}{14} \stackrel{?}{=} 13$

$\frac{37}{14}=13$ False

$\left(\frac{1}{2}, \frac{5}{14}\right)$ is not a solution of the second equation, so it is not a solution of the system.

9.

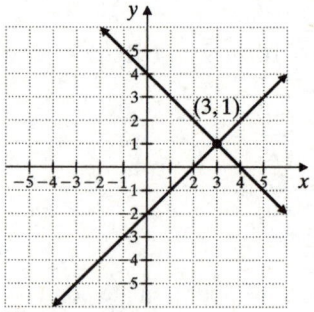

First equation:

$x+y=4$

$3+1 \stackrel{?}{=} 4$

$4=4$ True

Second equation:

$x-y=2$

$3-1 \stackrel{?}{=} 2$

$2=2$ True

The solution of the system is (3, 1).

Copyright © 2012 Pearson Education, Inc. Publishing as Prentice Hall.

11.

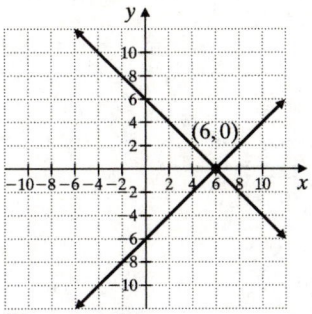

First equation:

$x + y = 6$

$6 + 0 \stackrel{?}{=} 6$

$6 = 6$ True

Second equation:

$-x + y = -6$

$-6 + 0 = -6$

$-6 = -6$ True

The solution of the system is (6, 0).

13.

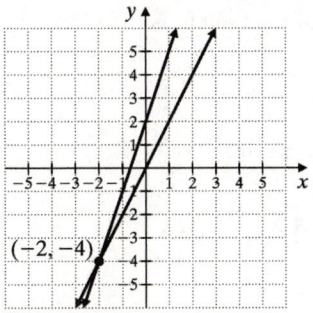

First equation:

$y = 2x$

$-4 \stackrel{?}{=} 2(-2)$

$-4 = -4$ True

Second equation:

$3x - y = -2$

$3(-2) - (-4) \stackrel{?}{=} -2$

$-6 + 4 \stackrel{?}{=} -2$

$-2 = -2$ True

The solution of the system is (−2, −4).

15.

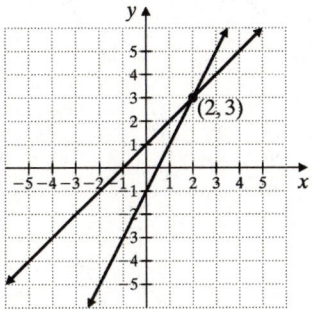

First equation:

$y = x + 1$

$3 \stackrel{?}{=} 2 + 1$

$3 = 3$ True

Second equation:

$y = 2x - 1$

$3 \stackrel{?}{=} 2(2) - 1$

$3 \stackrel{?}{=} 4 - 1$

$3 = 3$ True

The solution of the system is (2, 3).

17.

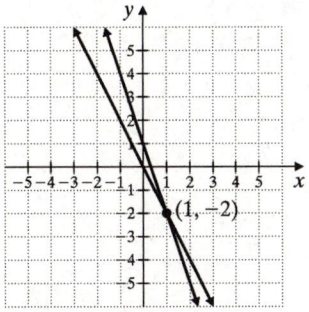

First equation:

$2x + y = 0$

$2(1) + (-2) \stackrel{?}{=} 0$

$2 - 2 \stackrel{?}{=} 0$

$0 = 0$ True

Second equation:

$3x + y = 1$

$3(1) + (-2) \stackrel{?}{=} 1$

$3 - 2 \stackrel{?}{=} 1$

$1 = 1$ True

The solution of the system is (1, −2).

19.

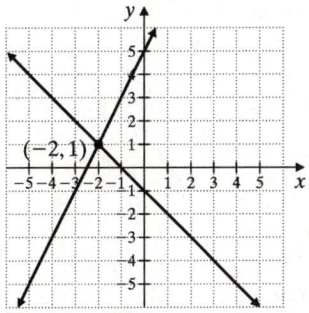

First equation:

$y = -x - 1$

$1 \stackrel{?}{=} -(-2) - 1$

$1 \stackrel{?}{=} 2 - 1$

$1 = 1$ True

Copyright © 2012 Pearson Education, Inc. Publishing as Prentice Hall.

Second equation:
$y = 2x + 5$
$1 \stackrel{?}{=} 2(-2) + 5$
$1 \stackrel{?}{=} -4 + 5$
$1 = 1$ True
The solution of the system is (–2, 1).

21.

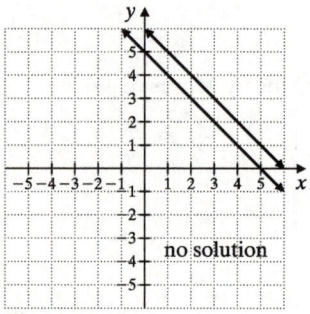

First equation:
$x + y = 5$
$y = -x + 5$
Second equation:
$x + y = 6$
$y = -x + 6$
The lines have the same slope, but different y-intercepts, so they are parallel. The system has no solution.

23.

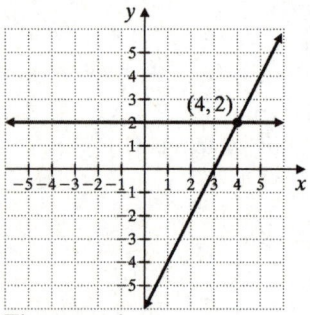

First equation:
$2x - y = 6$
$2(4) - 2 \stackrel{?}{=} 6$
$8 - 2 \stackrel{?}{=} 6$
$6 = 6$ True
Second equation:
$y = 2$
$2 = 2$ True
The solution of the system is (4, 2).

25.

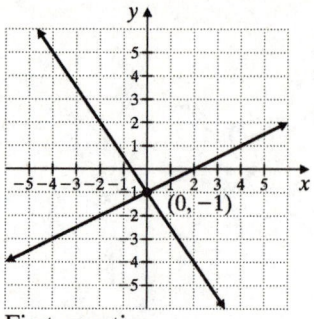

First equation:
$x - 2y = 2$
$0 - 2(-1) \stackrel{?}{=} 2$
$0 + 2 \stackrel{?}{=} 2$
$2 = 2$ True
Second equation:
$3x + 2y = -2$
$3(0) + 2(-1) \stackrel{?}{=} -2$
$0 - 2 \stackrel{?}{=} -2$
$-2 = -2$ True
The solution of the system is (0, –1).

27.

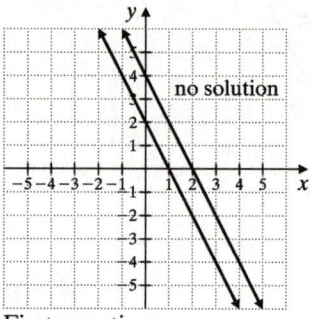

First equation:
$2x + y = 4$
$y = -2x + 4$
Second equation:
$6x = -3y + 6$
$3y = -6x + 6$
$y = -2x + 2$
The lines have the same slope, but different y-intercepts, so they are parallel. The system has no solution.

Copyright © 2012 Pearson Education, Inc. Publishing as Prentice Hall.

29.

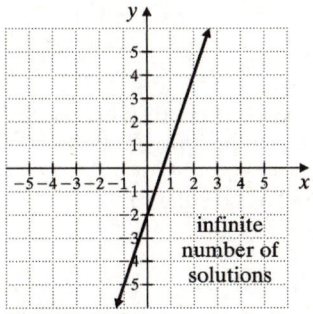

First equation:

$y - 3x = -2$

$y = 3x - 2$

Second equation:

$6x - 2y = 4$

$-2y = -6x + 4$

$y = 3x - 2$

The graphs of the equations are the same line, so the system has an infinite number of solutions.

31.

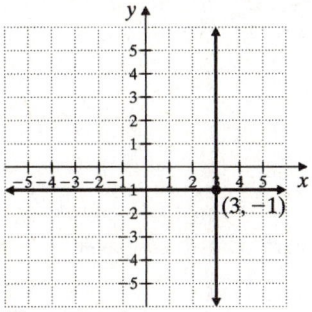

First equation:

$x = 3$

$3 = 3$ True

Second equation:

$y = -1$

$-1 = -1$ True

The solution of the system is (3, −1).

33.

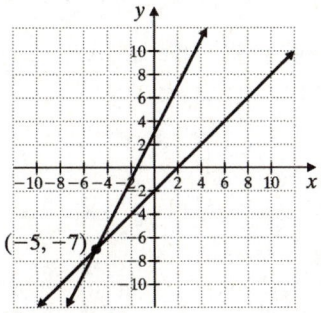

First equation:

$y = x - 2$

$-7 \stackrel{?}{=} -5 - 2$

$-7 = -7$ True

Second equation:

$y = 2x + 3$

$-7 \stackrel{?}{=} 2(-5) + 3$

$-7 \stackrel{?}{=} -10 + 3$

$-7 = -7$ True

The solution of the system is (−5, −7).

35.

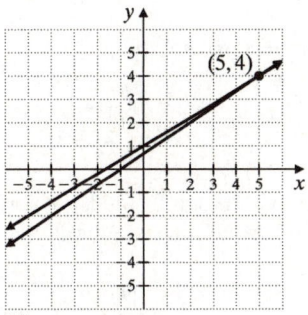

First equation:

$2x - 3y = -2$

$2(5) - 3(4) \stackrel{?}{=} -2$

$10 - 12 \stackrel{?}{=} -2$

$-2 = -2$ True

Second equation:

$-3x + 5y = 5$

$-3(5) + 5(4) \stackrel{?}{=} 5$

$-15 + 20 \stackrel{?}{=} 5$

$5 = 5$ True

The solution of the system is (5, 4).

37.

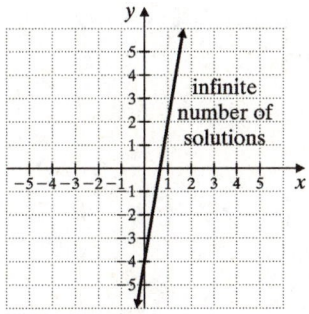

First equation:

$6x - y = 4$

$-y = -6x + 4$

$y = 6x - 4$

Second equation:

$\frac{1}{2}y = -2 + 3x$

$y = -4 + 6x$

$y = 6x - 4$

The graphs of the equations are the same line, so the system has an infinite number of solutions.

Copyright © 2012 Pearson Education, Inc. Publishing as Prentice Hall.

39. $4x+y=24$ $\quad$ $x+2y=2$

$y=-4x+24$ $\quad$ $2y=-x+2$

$y=-\frac{1}{2}x+1$

a. The slopes of the lines are different, so the lines intersect.

b. The system has one solution.

41. $2x+y=0$ $\quad$ $2y=6-4x$

$y=-2x$ $\quad$ $y=-2x+3$

a. The lines have the same slope, but different y-intercepts, so the lines are parallel.

b. The system has no solution.

43. $4x-y=6$ $\quad$ $\frac{1}{2}y=-3+2x$

$4x-6=y$ $\quad$ $y=-6+4x$

a. The lines are identical.

b. The system has an infinite number of solutions.

45. $x=5$ $\quad$ $y=-2$

vertical line $\quad$ horizontal line

a. The slopes of the lines are different, so the lines intersect.

b. The system has one solution.

47. $3y-2x=3$ $\quad$ $x+2y=9$

$3y=2x+3$ $\quad$ $2y=-x+9$

$y=\frac{2}{3}x+1$ $\quad$ $y=-\frac{1}{2}x+\frac{9}{2}$

a. The slopes of the lines are different, so the lines intersect.

b. The system has one solution.

49. $6y+4x=6$ $\quad$ $3y-3=-2x$

$6y=-4x+6$ $\quad$ $3y=-2x+3$

$y=-\frac{2}{3}x+1$ $\quad$ $y=-\frac{2}{3}x+1$

a. The lines are identical.

b. The system has an infinite number of solutions.

51. $x+y=4$ $\quad$ $x+y=3$

$y=-x+4$ $\quad$ $y=-x+3$

a. The lines have the same slope, but different y-intercepts, so the lines are parallel.

b. The system has no solution.

53. $5(x-3)+3x=1$

$5x-15+3x=1$

$8x=16$

$x=2$

55. $4\left(\frac{y+1}{2}\right)+3y=0$

$2(y+1)+3y=0$

$2y+2+3y=0$

$5y+2=0$

$5y=-2$

$y=-\frac{2}{5}$

57. $8a-2(3a-1)=6$

$8a-6a+2=6$

$2a=4$

$a=2$

59. Answers may vary. Possible answer:

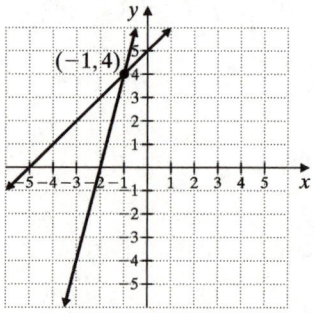

61. Answers may vary. Possible answer:

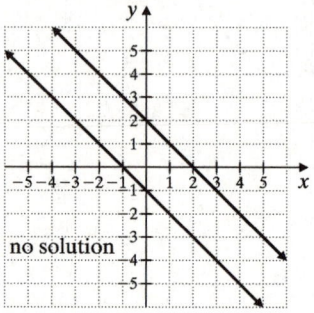

Copyright © 2012 Pearson Education, Inc. Publishing as Prentice Hall.

63. The lines cross at a point between 1988 and 1989, and again between 2001 and 2002. The number of pounds of imported fishery products was equal to the domestic catch between 1988 and 1989, and also between 2001 and 2002.

65. The average attendance per game for the Texas Rangers was greater than the average attendance per game for the Minnesota Twins in 2003, 2004, 2005, 2006, and 2007.

67. answers may vary

69. answers may vary

71. a. (4, 9) appears in both tables, so it is a solution of the system.

b.

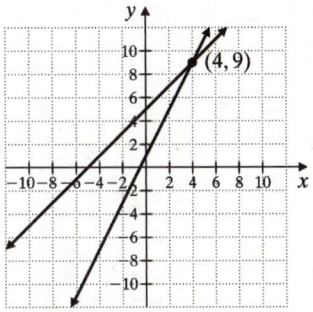

c. Yes; the two lines intersect at (4, 9).

Section 7.2 Practice

1. $\begin{cases} 2x+3y=13 \\ x=y+4 \end{cases}$

Substitute $y + 4$ for x in the first equation and solve for y.

$$\begin{aligned} 2(y+4)+3y&=13 \\ 2y+8+3y&=13 \\ 5y&=5 \\ y&=1 \end{aligned}$$

Solve for x.

$x = y + 4 = 1 + 4 = 5$

The solution of the system is (5, 1).

2. $\begin{cases} 4x-y=2 \\ y=5x \end{cases}$

Substitute $5x$ for y in the first equation and solve for x.

$$\begin{aligned} 4x-5x&=2 \\ -x&=2 \\ x&=-2 \end{aligned}$$

Solve for y.

$y = 5x = 5(-2) = -10$

The solution of the system is (−2, −10).

3. $\begin{cases} 3x+y=5 \\ 3x-2y=-7 \end{cases}$

Solve the first equation for y.

$$\begin{aligned} 3x+y&=5 \\ y&=-3x+5 \end{aligned}$$

Substitute $-3x + 5$ for y in the second equation.

$$\begin{aligned} 3x-2y&=-7 \\ 3x-2(-3x+5)&=-7 \\ 3x+6x-10&=-7 \\ 9x&=3 \\ x&=\frac{3}{9}=\frac{1}{3} \end{aligned}$$

Solve for y.

$$y=-3x+5=-3\left(\frac{1}{3}\right)+5=-1+5=4$$

The solution of the system is $\left(\frac{1}{3}, 4\right)$.

4. $\begin{cases} 5x-2y=6 \\ -3x+y=-3 \end{cases}$

Solve the second equation for y.

$$\begin{aligned} -3x+y&=-3 \\ y&=3x-3 \end{aligned}$$

Substitute $3x - 3$ for y in the first equation.

$$\begin{aligned} 5x-2y&=6 \\ 5x-2(3x-3)&=6 \\ 5x-6x+6&=6 \\ -x&=0 \\ x&=0 \end{aligned}$$

Solve for y.

$y = 3x - 3 = 3(0) - 3 = 0 - 3 = -3$

The solution of the system is (0, −3).

5. $\begin{cases} -x+3y=6 \\ y=\frac{1}{3}x+2 \end{cases}$

Substitute $\frac{1}{3}x+2$ for y in the first equation.

$$\begin{aligned} -x+3y&=6 \\ -x+3\left(\frac{1}{3}x+2\right)&=6 \\ -x+x+6&=6 \\ 0x&=0 \\ 0&=0 \end{aligned}$$

The statement $0 = 0$ indicates that this system has an infinite number of solutions. It is a graph of one line.

Copyright © 2012 Pearson Education, Inc. Publishing as Prentice Hall.

6. $\begin{cases} 2x-3y=6 \\ -4x+6y=12 \end{cases}$

$$2x-3y=6$$
$$-3y=-2x+6$$
$$y=\frac{2}{3}x-2$$

Substitute $\frac{2}{3}x-2$ for y in the second equation.

$$-4x+6y=12$$
$$-4x+6\left(\frac{2}{3}x-2\right)=12$$
$$-4x+4x-12=12$$
$$-12=12$$

The statement $-12 = 12$ indicates that this system has no solution. It is a graph of parallel lines.

Vocabulary and Readiness Check

1. $x = 1$: $y = 4(1) = 4$
The solution is (1, 4).

2. $0 = 34$ is a false statement. The system has no solution.

3. $0 = 0$ is a true statement. The system has an infinite number of solutions.

4. $y = 0$: $x = 0 + 5 = 5$
The solution is (5, 0).

5. $x = 0$: $0+y=0$
$$y=0$$
The solution is (0, 0).

6. $0 = 0$ is a true statement. The system has an infinite number of solutions.

Exercise Set 7.2

1. $\begin{cases} x+y=3 \\ x=2y \end{cases}$

Substitute $2y$ for x in the first equation and solve for y.

$$x+y=3$$
$$2y+y=3$$
$$3y=3$$
$$y=1$$

Now solve for x.
$x = 2y = 2(1) = 2$
The solution of the system is (2, 1).

3. $\begin{cases} x+y=6 \\ y=-3x \end{cases}$

Substitute $-3x$ for y in the first equation and solve for x.

$$x+y=6$$
$$x+(-3x)=6$$
$$-2x=6$$
$$x=-3$$

Now solve for y.
$y = -3x = -3(-3) = 9$
The solution of the system is (–3, 9).

5. $\begin{cases} y=3x+1 \\ 4y-8x=12 \end{cases}$

Substitute $3x + 1$ for y in the second equation and solve for x.

$$4y-8x=12$$
$$4(3x+1)-8x=12$$
$$12x+4-8x=12$$
$$4x+4=12$$
$$4x=8$$
$$x=2$$

Now solve for y.
$y = 3x + 1 = 3(2) + 1 = 6 + 1 = 7$
The solution of the system is (2, 7).

7. $\begin{cases} y=2x+9 \\ y=7x+10 \end{cases}$

Substitute $2x + 9$ for y in the second equation and solve for x.

$$y=7x+10$$
$$2x+9=7x+10$$
$$-5x=1$$
$$y=-\frac{1}{5}$$

Now solve for y.

$$y=2x+9=2\left(-\frac{1}{5}\right)+9=\frac{-2}{5}+\frac{45}{5}=\frac{43}{5}$$

The solution of the system is $\left(-\frac{1}{5}, \frac{43}{5}\right)$.

9. $\begin{cases} 3x-4y=10 \\ y=x-3 \end{cases}$

Substitute $x - 3$ for y in the first equation and solve for x.

Copyright © 2012 Pearson Education, Inc. Publishing as Prentice Hall.

$$3x-4y=10$$
$$3x-4(x-3)=10$$
$$3x-4x+12=10$$
$$-x+12=10$$
$$-x=-2$$
$$x=2$$

Now solve for y.
$y = x - 3 = 2 - 3 = -1$
The solution of the system is $(2, -1)$.

11. $\begin{cases} x+2y=6 \\ 2x+3y=8 \end{cases}$

Solve the first equation for x.
$$x+2y=6$$
$$x=-2y+6$$

Substitute $-2y + 6$ for x in the second equation and solve for y.
$$2x+3y=8$$
$$2(-2y+6)+3y=8$$
$$-4y+12+3y=8$$
$$-y=-4$$
$$y=4$$

Now solve for x.
$x = -2y + 6 = -2(4) + 6 = -8 + 6 = -2$
The solution of the system is $(-2, 4)$.

13. $\begin{cases} 3x+2y=16 \\ x=3y-2 \end{cases}$

Substitute $3y - 2$ for x in the first equation and solve for y.
$$3x+2y=16$$
$$3(3y-2)+2y=16$$
$$9y-6+2y=16$$
$$11y-6=16$$
$$11y=22$$
$$y=2$$

Now solve for x.
$x = 3y - 2 = 3(2) - 2 = 6 - 2 = 4$
The solution of the system is $(4, 2)$.

15. $\begin{cases} 2x-5y=1 \\ 3x+y=-7 \end{cases}$

Solve the second equation for y.
$$3x+y=-7$$
$$y=-3x-7$$

Substitute $-3x - 7$ for y in the first equation and solve for y.
$$2x-5y=1$$
$$2x-5(-3x-7)=1$$
$$2x+15x+35=1$$
$$17x=-34$$
$$x=-2$$

Now solve for y.
$y = -3x - 7 = -3(-2) - 7 = 6 - 7 = -1$
The solution of the system is $(-2, -1)$.

17. $\begin{cases} 4x+2y=5 \\ -2x=y+4 \end{cases}$

Solve the second equation for x.
$$-2x=y+4$$
$$x=-\frac{1}{2}y-2$$

Substitute $-\frac{1}{2}y-2$ for x in the first equation and solve for y.
$$4x+2y=5$$
$$4\left(-\frac{1}{2}y-2\right)+2y=5$$
$$-2y-8+2y=5$$
$$-8=5 \quad \text{False}$$

Since the statement $-8 = 5$ is false, the system has no solution.

19. $\begin{cases} 4x+y=11 \\ 2x+5y=1 \end{cases}$

Solve the first equation for y.
$$4x+y=11$$
$$y=-4x+11$$

Substitute $-4x + 11$ for y in the second equation and solve for y.
$$2x+5y=1$$
$$2x+5(-4x+11)=1$$
$$2x+(-20x)+55=1$$
$$-18x=-54$$
$$x=3$$

Now solve for y.
$y = -4x + 11 = -4(3) + 11 = -12 + 11 = -1$
The solution of the system is $(3, -1)$.

21. $\begin{cases} x+2y+5=-4+5y-x \\ 2x+x=y+4 \end{cases}$

Simplify each equation.
$\begin{cases} 2x+9=3y \\ 3x=y+4 \end{cases}$

Solve the second simplified equation for y.
$$3x=y+4$$
$$3x-4=y$$

Copyright © 2012 Pearson Education, Inc. Publishing as Prentice Hall.

Substitute $3x - 4$ for y in the first simplified equation and solve for x.

$$\begin{aligned} 2x+9&=3y \\ 2x+9&=3(3x-4) \\ 2x+9&=9x-12 \\ 2x+21&=9x \\ 21&=7x \\ 3&=x \end{aligned}$$

Now solve for y.

$y = 3x - 4 = 3(3) - 4 = 9 - 4 = 5$

The solution of the system is (3, 5).

23. $\begin{cases} 6x-3y=5 \\ x+2y=0 \end{cases}$

Solve the second equation for x.

$$\begin{aligned} x+2y&=0 \\ x&=-2y \end{aligned}$$

Substitute $-2y$ for x in the first equation and solve for y.

$$\begin{aligned} 6x-3y&=5 \\ 6(-2y)-3y&=5 \\ -12y-3y&=5 \\ -15y&=5 \\ y&=-\frac{5}{15}=-\frac{1}{3} \end{aligned}$$

Now solve for x.

$$x=-2y=-2\left(-\frac{1}{3}\right)=\frac{2}{3}$$

The solution of the system is $\left(\frac{2}{3}, -\frac{1}{3}\right)$.

25. $\begin{cases} 3x-y=1 \\ 2x-3y=10 \end{cases}$

Solve the first equation for y.

$$\begin{aligned} 3x-y&=1 \\ -y&=-3x+1 \\ y&=3x-1 \end{aligned}$$

Substitute $3x - 1$ for y in the second equation and solve for x.

$$\begin{aligned} 2x-3y&=10 \\ 2x-3(3x-1)&=10 \\ 2x-9x+3&=10 \\ -7x+3&=10 \\ -7x&=7 \\ x&=-1 \end{aligned}$$

Now solve for y.

$y = 3x - 1 = 3(-1) - 1 = -3 - 1 = -4$

The solution of the system is $(-1, -4)$.

27. $\begin{cases} -x+2y=10 \\ -2x+3y=18 \end{cases}$

Solve the first equation for x.

$$\begin{aligned} -x+2y&=10 \\ 2y-10&=x \end{aligned}$$

Substitute $2y - 10$ for x in the second equation and solve for y.

$$\begin{aligned} -2x+3y&=18 \\ -2(2y-10)+3y&=18 \\ -4y+20+3y&=18 \\ -y+20&=18 \\ 2&=y \end{aligned}$$

Now solve for x.

$x = 2y - 10 = 2(2) - 10 = 4 - 10 = -6$

The solution of the system is $(-6, 2)$.

29. $\begin{cases} 5x+10y=20 \\ 2x+6y=10 \end{cases}$

Solve the first equation for x. (Note that the second equation could also be easily solved for x.)

$$\begin{aligned} 5x+10y&=20 \\ 5x&=-10y+20 \\ x&=-2y+4 \end{aligned}$$

Substitute $-2y + 4$ for x in the second equation and solve for y.

$$\begin{aligned} 2x+6y&=10 \\ 2(-2y+4)+6y&=10 \\ -4y+8+6y&=10 \\ 2y+8&=10 \\ 2y&=2 \\ y&=1 \end{aligned}$$

Now solve for x.

$x = -2y + 4 = -2(1) + 4 = -2 + 4 = 2$

The solution of the system is (2, 1).

31. $\begin{cases} 3x+6y=9 \\ 4x+8y=16 \end{cases}$

Solve the first equation for x.

$$\begin{aligned} 3x+6y&=9 \\ 3x&=-6y+9 \\ x&=-2y+3 \end{aligned}$$

Substitute $-2y + 3$ for x in the second equation and solve for y.

$$\begin{aligned} 4x+8y&=16 \\ 4(-2y+3)+8y&=16 \\ -8y+12+8y&=16 \\ 0&=4 \quad \text{False} \end{aligned}$$

Since the statement $0 = 4$ is false, the system has no solution.

Copyright © 2012 Pearson Education, Inc. Publishing as Prentice Hall.

33. $\begin{cases} \frac{1}{3}x - y = 2 \\ x - 3y = 6 \end{cases}$

Solve the second equation for x.

$$x - 3y = 6$$
$$x = 3y + 6$$

Substitute $3y + 6$ for x in the first equation and solve for y.

$$\frac{1}{3}x - y = 2$$
$$\frac{1}{3}(3y+6) - y = 2$$
$$y + 2 - y = 2$$
$$2 = 2$$

Since $2 = 2$ is a true statement, the two equations in the original system are equivalent. The system has an infinite number of solutions.

35. $\begin{cases} x = \frac{3}{4}y - 1 \\ 8x - 5y = -6 \end{cases}$

Substitute $\frac{3}{4}y - 1$ for x in the second equation and solve for y.

$$8x - 5y = -6$$
$$8\left(\frac{3}{4}y - 1\right) - 5y = -6$$
$$6y - 8 - 5y = -6$$
$$y = 2$$

Now solve for x.

$$x = \frac{3}{4}y - 1 = \frac{3}{4}(2) - 1 = \frac{3}{2} - 1 = \frac{1}{2}$$

The solution of the system is $\left(\frac{1}{2}, 2\right)$.

37.
$$3x + 2y = 6$$
$$-2(3x + 2y) = -2(6)$$
$$-6x - 4y = -12$$

39.
$$-4x + y = 3$$
$$3(-4x + y) = 3(3)$$
$$-12x + 3y = 9$$

41.
$$\begin{array}{r} 3n + 6m \\ \underline{2n - 6m} \\ 5n \end{array}$$

43.
$$\begin{array}{r} -5a - 7b \\ \underline{5a - 8b} \\ -15b \end{array}$$

45. $\begin{cases} -5y + 6y = 3x + 2(x-5) - 3x + 5 \\ 4(x+y) - x + y = -12 \end{cases}$

Simplify each equation.

$\begin{cases} y = 2x - 5 \\ 3x + 5y = -12 \end{cases}$

Substitute $2x - 5$ for y in the second simplified equation and solve for x.

$$3x + 5y = -12$$
$$3x + 5(2x - 5) = -12$$
$$3x + 10x - 25 = -12$$
$$13x - 25 = -12$$
$$13x = 13$$
$$x = 1$$

Now solve for y.

$y = 2x - 5 = 2(1) - 5 = 2 - 5 = -3$

The solution of the system is $(1, -3)$.

47. answers may vary

49. no; answers may vary

51. c; answers may vary

53. Using a graphing calculator, the solution of the system is $(-2.6, 1.3)$.

55. Using a graphing calculator, the solution of the system is $(3.28, 2.1)$.

57. a. $\begin{cases} y = -1.1x + 7.1 \\ y = 1.9x + 4.7 \end{cases}$

Substitute $-1.1x + 7.1$ for y in the second equation and solve for x.

$$y = 1.9x + 4.7$$
$$-1.1x + 7.1 = 1.9x + 4.7$$
$$-1.1x + 2.4 = 1.9x$$
$$2.4 = 3x$$
$$0.8 = x$$

Now solve for y.

$$\begin{aligned} y &= -1.1x + 7.1 \\ &= -1.1(0.8) + 7.1 \\ &= -0.88 + 7.1 \\ &= 6.22 \\ &\approx 6 \end{aligned}$$

The rounded solution is $(0.8, 6)$.

b. In about 0.8 year after 2000, the sales of DVD and VHS units were both approximately \$6 billion.

Copyright © 2012 Pearson Education, Inc. Publishing as Prentice Hall.

c.

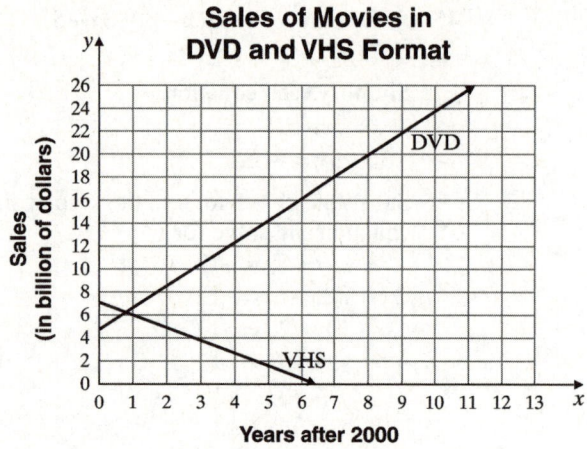

(Note: Not enough BluRay years of data to include in graph.)
answers may vary

d. 2007 corresponds to $x = 7$.
$x = 7$: $y = -1.1(7) + 7.1 = -7.7 + 7.1 = -0.6$
answers may vary

Section 7.3 Practice

1. $\begin{cases} x + y = 13 \\ x - y = 5 \end{cases}$

Add the equations to eliminate y; then solve for x.

$$\begin{array}{rl} x + y &= 13 \\ x - y &= 5 \\ \hline 2x \quad &= 18 \\ x \quad &= 9 \end{array}$$

Now solve for y.

$$\begin{aligned} x + y &= 13 \\ 9 + y &= 13 \\ y &= 4 \end{aligned}$$

The solution of the system is (9, 4).

2. $\begin{cases} 2x - y = -6 \\ -x + 4y = 17 \end{cases}$

Multiply the first equation by 4.

$$\begin{cases} 4(2x - y) = 4(-6) \\ -x + 4y = 17 \end{cases} \Rightarrow \begin{cases} 8x - 4y = -24 \\ -x + 4y = 17 \end{cases}$$

Add the equations to eliminate y; then solve for x.

$$\begin{array}{rl} 8x - 4y &= -24 \\ -x + 4y &= 17 \\ \hline 7x \quad &= -7 \\ x &= -1 \end{array}$$

Now solve for y.

Copyright © 2012 Pearson Education, Inc. Publishing as Prentice Hall.

$$\begin{aligned} 2x - y &= -6 \\ 2(-1) - y &= -6 \\ -2 - y &= -6 \\ -y &= -4 \\ y &= 4 \end{aligned}$$

The solution of the system is (−1, 4).

3. $\begin{cases} x - 3y = -2 \\ -3x + 9y = 5 \end{cases}$

Multiply the first equation by 3.

$$\begin{cases} 3(x-3y) = 3(-2) \\ -3x + 9y = 5 \end{cases} \Rightarrow \begin{cases} 3x - 9y = -6 \\ -3x + 9y = 5 \end{cases}$$

Add the equations to eliminate y; then solve for x.

$$\begin{array}{r} 3x - 9y = -6 \\ -3x + 9y = 5 \\ \hline 0 = -1 \end{array}$$

Since the statement 0 = −1 is false, there is no solution to the system.

4. $\begin{cases} 2x + 5y = 1 \\ -4x - 10y = -2 \end{cases}$

Multiply the first equation by 2.

$$\begin{cases} 2(2x+5y) = 2(1) \\ -4x - 10y = -2 \end{cases} \Rightarrow \begin{cases} 4x + 10y = 2 \\ -4x - 10y = -2 \end{cases}$$

Add the equations to eliminate y; then solve for x.

$$\begin{array}{r} 4x + 10y = 2 \\ -4x - 10y = -2 \\ \hline 0 = 0 \end{array}$$

Since the statement 0 = 0 is true, there are an infinite number of solutions.

5. $\begin{cases} 4x + 5y = 14 \\ 3x - 2y = -1 \end{cases}$

Multiply the first equation by 2 and multiply the second equation by 5.

$$\begin{cases} 2(4x+5y) = 2(14) \\ 5(3x-2y) = 5(-1) \end{cases} \Rightarrow \begin{cases} 8x + 10y = 28 \\ 15x - 10y = -5 \end{cases}$$

Add the equations to eliminate y; then solve for x.

$$\begin{array}{r} 8x + 10y = 28 \\ 15x - 10y = -5 \\ \hline 23x \qquad = 23 \\ x = 1 \end{array}$$

Now solve for y.

$$\begin{aligned} 4x + 5y &= 14 \\ 4(1) + 5y &= 14 \\ 5y &= 10 \\ y &= 2 \end{aligned}$$

The solution of the system is (1, 2).

Copyright © 2012 Pearson Education, Inc. Publishing as Prentice Hall.

6. $\begin{cases} -\frac{x}{3}+y=\frac{4}{3} \\ \frac{x}{2}-\frac{5}{2}y=-\frac{1}{2} \end{cases}$

Multiply the first equation by 3 and the second equation by 2.

$$\begin{cases} 3\left(-\frac{x}{3}+y\right)=3\left(\frac{4}{3}\right) \\ 2\left(\frac{x}{2}-\frac{5}{2}y\right)=2\left(-\frac{1}{2}\right) \end{cases} \Rightarrow \begin{cases} -x+3y=4 \\ x-5y=-1 \end{cases}$$

Add the equations to eliminate x; then solve for y.

$$\begin{array}{r} -x+3y=4 \\ x-5y=-1 \\ \hline -2y=3 \\ y=-\frac{3}{2} \end{array}$$

Now solve for y through elimination. Multiply the first equation by 15 and the second equation by 6.

$$\begin{cases} 15\left(-\frac{x}{3}+y\right)=15\left(\frac{4}{3}\right) \\ 6\left(\frac{x}{2}-\frac{5}{2}y\right)=6\left(-\frac{1}{2}\right) \end{cases} \Rightarrow \begin{cases} -5x+15y=20 \\ 3x-15y=-3 \end{cases}$$

Add the equations to eliminate y; then solve for x.

$$\begin{array}{r} -5x+15y=20 \\ 3x-15y=-3 \\ \hline -2x \quad =17 \\ x=-\frac{17}{2} \end{array}$$

The solution of the system is $\left(-\frac{17}{2}, -\frac{3}{2}\right)$.

Vocabulary and Readiness Check

1. $\begin{cases} 3x-2y=-9 \\ x+5y=14 \end{cases}$

Multiply the second equation by -3, then add the resulting equations.

$$\begin{array}{r} 3x-2y=-9 \\ -3x-15y=-42 \\ \hline -17y=-51 \end{array}$$

The y's are not eliminated; the statement is false.

2. $\begin{cases} 3x-2y=-9 \\ x+5y=14 \end{cases}$

Multiply the second equation by -3, then add the resulting equations.

$$\begin{array}{r} 3x-2y=-9 \\ -3x-15y=-42 \\ \hline -17y=-51 \end{array}$$

The statement is true.

Copyright © 2012 Pearson Education, Inc. Publishing as Prentice Hall.

3. $\begin{cases} 3x-2y=-9 \\ x+5y=14 \end{cases}$

Multiply the first equation by 5 and the second equation by 2, then add the two new equations.

$$\begin{array}{r} 15x-10y=-45 \\ 2x+10y=\ \ 28 \\ \hline 17x\qquad\ =-17 \end{array}$$

The statement is true.

4. $\begin{cases} 3x-2y=-9 \\ x+5y=14 \end{cases}$

Multiply the first equation by 5 and the second equation by –2, then add the two new equations.

$$\begin{array}{r} 15x-10y=-45 \\ -2x-10y=-28 \\ \hline 13x-20y=-73 \end{array}$$

The y's are not eliminated; the statement is false.

Exercise Set 7.3

1. $\begin{cases} 3x+y=5 \\ 6x-y=4 \end{cases}$

Add the equations to eliminate y; then solve for x.

$$\begin{array}{r} 3x+y=5 \\ 6x-y=4 \\ \hline 9x\quad\ =9 \\ x=1 \end{array}$$

Now solve for y.

$$\begin{aligned} 3x+y&=5 \\ 3(1)+y&=5 \\ 3+y&=5 \\ y&=2 \end{aligned}$$

The solution of the system is (1, 2).

3. $\begin{cases} x-2y=8 \\ -x+5y=-17 \end{cases}$

Add the equations to eliminate x; then solve for y.

$$\begin{array}{r} x-2y=8 \\ -x+5y=-17 \\ \hline 3y=-9 \\ y=-3 \end{array}$$

Now solve for x.

$$\begin{aligned} x-2y&=8 \\ x-2(-3)&=8 \\ x+6&=8 \\ x&=2 \end{aligned}$$

The solution of the system is (2, –3).

5. $\begin{cases} 3x+y=-11 \\ 6x-2y=-2 \end{cases}$

Multiply the first equation by 2.

$$\begin{cases} 2(3x+y)=2(-11) \\ 6x-2y=-2 \end{cases} \rightarrow \begin{cases} 6x+2y=-22 \\ 6x-2y=-2 \end{cases}$$

Add the equations to eliminate y; then solve for x.

$$\begin{array}{r} 6x+2y=-22 \\ 6x-2y=-2 \\ \hline 12x\qquad=-24 \\ x=-2 \end{array}$$

Now solve for y.

$$\begin{aligned} 3x+y&=-11 \\ 3(-2)+y&=-11 \\ -6+y&=-11 \\ y&=-5 \end{aligned}$$

The solution of the system is (–2, –5).

7. $\begin{cases} 3x+2y=11 \\ 5x-2y=29 \end{cases}$

Add the equations to eliminate y; then solve for x.

$$\begin{array}{r} 3x+2y=11 \\ 5x-2y=29 \\ \hline 8x=40 \\ x=5 \end{array}$$

Now solve for y.

$$\begin{aligned} 3x+2y&=11 \\ 3(5)+2y&=11 \\ 15+2y&=11 \\ 2y&=-4 \\ y&=-2 \end{aligned}$$

The solution of the system is (5, –2).

9. $\begin{cases} x+5y=18 \\ 3x+2y=-11 \end{cases}$

Multiply the first equation by –3.

$$\begin{cases} -3(x+5y)=-3(18) \\ 3x+2y=-11 \end{cases} \rightarrow \begin{cases} -3x-15y=-54 \\ 3x+2y=-11 \end{cases}$$

Add the equations to eliminate x; then solve for y.

$$\begin{array}{r} -3x-15y=-54 \\ 3x+2y=-11 \\ \hline -13y=-65 \\ y=5 \end{array}$$

Now solve for x.

Copyright © 2012 Pearson Education, Inc. Publishing as Prentice Hall.

$$\begin{aligned} x+5y&=18\\ x+5(5)&=18\\ x+25&=18\\ x&=-7 \end{aligned}$$

The solution of the system is (–7, 5).

11. $\begin{cases} x+y=6\\ x-y=6 \end{cases}$

Add the equations to eliminate y; then solve for x.

$$\begin{array}{r} x+y=6\\ x-y=6\\ \hline 2x\quad=12\\ x=6 \end{array}$$

Now solve for y.

$$\begin{aligned} x+y&=6\\ 6+y&=6\\ y&=0 \end{aligned}$$

The solution of the system is (6, 0).

13. $\begin{cases} 2x+3y=0\\ 4x+6y=3 \end{cases}$

Multiply the first equation by –2.

$$\begin{cases} -2(2x+3y)=-2(0)\\ 4x+6y=3 \end{cases} \rightarrow \begin{cases} -4x-6y=0\\ 4x+6y=3 \end{cases}$$

Add the equations to eliminate x.

$$\begin{array}{r} -4x-6y=0\\ 4x+6y=3\\ \hline 0=3 \end{array}$$

Since the statement 0 = 3 is false, the system has no solution.

15. $\begin{cases} -x+5y=-1\\ 3x-15y=3 \end{cases}$

Multiply the first equation by 3.

$$\begin{cases} 3(-x+5y)=3(-1)\\ 3x-15y=3 \end{cases} \rightarrow \begin{cases} -3x+15y=-3\\ 3x-15y=3 \end{cases}$$

Add the equations to eliminate x.

$$\begin{array}{r} -3x+15y=-3\\ 3x-15y=3\\ \hline 0=0 \end{array}$$

Since the statement 0 = 0 is true, the system has an infinite number of solutions.

17. $\begin{cases} 3x-2y=7\\ 5x+4y=8 \end{cases}$

Multiply the first equation by 2.

$$\begin{cases} 2(3x-2y)=2(7)\\ 5x+4y=8 \end{cases} \rightarrow \begin{cases} 6x-4y=14\\ 5x+4y=8 \end{cases}$$

Add the equations to eliminate y; then solve for x.

$$\begin{array}{r} 6x-4y=14\\ 5x+4y=8\\ \hline 11x\quad=22\\ x=2 \end{array}$$

Now solve for y.

$$\begin{aligned} 3x-2y&=7\\ 3(2)-2y&=7\\ 6-2y&=7\\ -2y&=1\\ y&=-\frac{1}{2} \end{aligned}$$

The solution of the system is $\left(2, -\frac{1}{2}\right)$.

19. $\begin{cases} 8x=-11y-16\\ 2x+3y=-4 \end{cases} \rightarrow \begin{cases} 8x+11y=-16\\ 2x+3y=-4 \end{cases}$

Multiply the second equation by –4.

$$\begin{cases} 8x+11y=-16\\ -4(2x+3y)=-4(-4) \end{cases} \rightarrow \begin{cases} 8x+11y=-16\\ -8x-12y=16 \end{cases}$$

Add the equations to eliminate x; then solve for y.

$$\begin{array}{r} 8x+11y=-16\\ -8x-12y=16\\ \hline -y=0\\ y=0 \end{array}$$

Now solve for x.

$$\begin{aligned} 8x+11y&=-16\\ 8x+11(0)&=-16\\ 8x&=-16\\ x&=-2 \end{aligned}$$

The solution of the system is (–2, 0).

21. $\begin{cases} 4x-3y=7\\ 7x+5y=2 \end{cases}$

Multiply the first equation by 5 and the second equation by 3.

$$\begin{cases} 5(4x-3y)=5(7)\\ 3(7x+5y)=3(2) \end{cases} \rightarrow \begin{cases} 20x-15y=35\\ 21x+15y=6 \end{cases}$$

Add the equations to eliminate y; then solve for x.

$$\begin{array}{r} 20x-15y=35\\ 21x+15y=6\\ \hline 41x\quad=41\\ x=1 \end{array}$$

Now solve for y.

Copyright © 2012 Pearson Education, Inc. Publishing as Prentice Hall.

$$\begin{aligned} 4x-3y&=7\\ 4(1)-3y&=7\\ 4-3y&=7\\ -3y&=3\\ y&=-1 \end{aligned}$$

The solution of the system is (1, –1).

23. $\begin{cases} 4x-6y=8\\ 6x-9y=12 \end{cases}$

Multiply the first equation by –3 and the second equation by 2.

$$\begin{cases} -3(4x-6y)=-3(8)\\ 2(6x-9y)=2(12) \end{cases} \rightarrow \begin{cases} -12x+18y=-24\\ 12x-18y=24 \end{cases}$$

Add the equations to eliminate x.

$$\begin{array}{r} -12x+18y=-24\\ 12x-18y=24\\ \hline 0=0 \end{array}$$

Since the statement 0 = 0 is true, the system has an infinite number of solutions.

25. $\begin{cases} 2x-5y=4\\ 3x-2y=4 \end{cases}$

Multiply the first equation by –3 and the second equation by 2.

$$\begin{cases} -3(2x-5y)=-3(4)\\ 2(3x-2y)=2(4) \end{cases} \rightarrow \begin{cases} -6x+15y=-12\\ 6x-4y=8 \end{cases}$$

Add the equations to eliminate x; then solve for y.

$$\begin{array}{r} -6x+15y=-12\\ 6x-4y=8\\ \hline 11y=-4\\ y=-\dfrac{4}{11} \end{array}$$

Multiply the first original equation by –2 and the second original equation by 5.

$$\begin{cases} -2(2x-5y)=-2(4)\\ 5(3x-2y)=5(4) \end{cases} \rightarrow \begin{cases} -4x+10y=-8\\ 15x-10y=20 \end{cases}$$

Add the equations to eliminate y; then solve for x.

$$\begin{array}{r} -4x+10y=-8\\ 15x-10y=20\\ \hline 11x\quad\;\; =12\\ x=\dfrac{12}{11} \end{array}$$

The solution of the system is $\left(\dfrac{12}{11}, -\dfrac{4}{11}\right)$.

27. $\begin{cases} \dfrac{x}{3}+\dfrac{y}{6}=1\\ \dfrac{x}{2}-\dfrac{y}{4}=0 \end{cases}$

Multiply the first equation by 6 and the second equation by 4.

$$\begin{cases} 6\left(\dfrac{x}{3}+\dfrac{y}{6}\right)=6(1)\\ 4\left(\dfrac{x}{2}-\dfrac{y}{4}\right)=4(0) \end{cases} \rightarrow \begin{cases} 2x+y=6\\ 2x-y=0 \end{cases}$$

Add the equations to eliminate y; then solve for x.

$$\begin{array}{r} 2x+y=6\\ 2x-y=0\\ \hline 4x=6\\ x=\dfrac{6}{4}=\dfrac{3}{2} \end{array}$$

Multiply the original first equation by 6 and the second original equation by –4.

$$\begin{cases} 6\left(\dfrac{x}{3}+\dfrac{y}{6}\right)=6(1)\\ -4\left(\dfrac{x}{2}-\dfrac{y}{4}\right)=-4(0) \end{cases} \rightarrow \begin{cases} 2x+y=6\\ -2x+y=0 \end{cases}$$

Add the equations to eliminate x; then solve for y.

$$\begin{array}{r} 2x+y=6\\ -2x+y=0\\ \hline 2y=6\\ y=3 \end{array}$$

The solution of the system is $\left(\dfrac{3}{2}, 3\right)$.

29. $\begin{cases} \dfrac{10}{3}x+4y=-4\\ 5x+6y=-6 \end{cases}$

Multiply the first equation by 3 and the second equation by –2.

$$\begin{cases} 3\left(\dfrac{10}{3}x+4y\right)=3(-4)\\ -2(5x+6y)=-2(-6) \end{cases} \rightarrow \begin{cases} 10x+12y=-12\\ -10x-12y=12 \end{cases}$$

Add the equations to eliminate x.

$$\begin{array}{r} 10x+12y=-12\\ -10x-12y=12\\ \hline 0=0 \end{array}$$

Since the statement 0 = 0 is true, the system has an infinite number of solutions.

Copyright © 2012 Pearson Education, Inc. Publishing as Prentice Hall.

31. $\begin{cases} x-\frac{y}{3}=-1 \\ -\frac{x}{2}+\frac{y}{8}=\frac{1}{4} \end{cases}$

Multiply the first equation by 3 and the second equation by 8.

$$\begin{cases} 3\left(x-\frac{y}{3}\right)=3(-1) \\ 8\left(-\frac{x}{2}+\frac{y}{8}\right)=8\left(\frac{1}{4}\right) \end{cases} \rightarrow \begin{cases} 3x-y=-3 \\ -4x+y=2 \end{cases}$$

Add the equations to eliminate y; then solve for x.

$$\begin{array}{r} 3x-y=-3 \\ -4x+y=2 \\ \hline -x \quad =-1 \\ x=1 \end{array}$$

Now solve for y.

$$x-\frac{y}{3}=-1$$
$$1-\frac{y}{3}=-1$$
$$-\frac{y}{3}=-2$$
$$y=6$$

The solution of the system is (1, 6).

33. $\begin{cases} -4(x+2)=3y \\ 2x-2y=3 \end{cases}$

Rewrite the first equation.

$$-4(x+2)=3y$$
$$-4x-8=3y$$
$$-4x=3y+8$$
$$-4x-3y=8$$

$$\begin{cases} -4x-3y=8 \\ 2x-2y=3 \end{cases}$$

Multiply the second equation by 2.

$$\begin{cases} -4x-3y=8 \\ 2(2x-2y)=2(3) \end{cases} \rightarrow \begin{cases} -4x-3y=8 \\ 4x-4y=6 \end{cases}$$

Add the equations to eliminate x; then solve for y.

$$\begin{array}{r} -4x-3y=8 \\ 4x-4y=6 \\ \hline -7y=14 \\ y=-2 \end{array}$$

Now solve for x.

$$2x-2y=3$$
$$2x-2(-2)=3$$
$$2x+4=3$$
$$2x=-1$$
$$x=-\frac{1}{2}$$

The solution of the system is $\left(-\frac{1}{2}, -2\right)$.

35. $\begin{cases} \frac{x}{3}-y=2 \\ -\frac{x}{2}+\frac{3y}{2}=-3 \end{cases}$

Multiply the first equation by 3 and the second by 2.

$$\begin{cases} 3\left(\frac{x}{3}-y\right)=3(2) \\ 2\left(-\frac{x}{2}+\frac{3y}{2}\right)=2(-3) \end{cases} \rightarrow \begin{cases} x-3y=6 \\ -x+3y=-6 \end{cases}$$

Add the equations to eliminate y.

$$\begin{array}{r} x-3y=6 \\ -x+3y=-6 \\ \hline 0=0 \end{array}$$

Since the statement $0 = 0$ is true, the system has an infinite number of solutions.

37. $\begin{cases} \frac{3}{5}x-y=-\frac{4}{5} \\ 3x+\frac{y}{2}=-\frac{9}{5} \end{cases}$

Multiply the first equation by 5 and the second equation by 10 to eliminate fractions.

$$\begin{cases} 5\left(\frac{3}{5}x-y\right)=5\left(-\frac{4}{5}\right) \\ 10\left(3x+\frac{y}{2}\right)=10\left(-\frac{9}{5}\right) \end{cases} \rightarrow \begin{cases} 3x-5y=-4 \\ 30x+5y=-18 \end{cases}$$

Add the equations to eliminate y; then solve for x.

$$\begin{array}{r} 3x-5y=-4 \\ 30x+5y=-18 \\ \hline 33x \quad =-22 \end{array}$$

$$x=-\frac{22}{33}=-\frac{2}{3}$$

Now use the equation $3x - 5y = -4$ to solve for y.

Copyright © 2012 Pearson Education, Inc. Publishing as Prentice Hall.

$$3x-5y=-4$$
$$3\left(-\frac{2}{3}\right)-5y=-4$$
$$-2-5y=-4$$
$$-5y=-2$$
$$y=\frac{2}{5}$$

The solution of the system is $\left(-\frac{2}{3},\frac{2}{5}\right)$.

39. $\begin{cases}3.5x+2.5y=17\\-1.5x-7.5y=-33\end{cases}$

Multiply the first equation by 30 and the second equation by 10.

$$\begin{cases}30(3.5x+2.5y)=30(17)\\10(-1.5x-7.5y)=10(-33)\end{cases}$$
$$\rightarrow\begin{cases}105x+75y=510\\-15x-75y=-330\end{cases}$$

Add the equations to eliminate y; then solve for x.

$$\begin{array}{r}105x+75y=510\\-15x-75y=-330\\\hline 90x\qquad=180\\x=2\end{array}$$

Now solve for y.

$$3.5x+2.5y=17$$
$$3.5(2)+2.5y=17$$
$$7+2.5y=17$$
$$2.5y=10$$
$$y=4$$

The solution of the system is (2, 4).

41. $\begin{cases}0.02x+0.04y=0.09\\-0.1x+0.3y=0.8\end{cases}$

Multiply the first equation by 100 and the second equation by 10 to eliminate decimals.

$$\begin{cases}100(0.02x+0.04y)=100(0.09)\\10(-0.1x+0.3y)=10(0.8)\end{cases}$$
$$\rightarrow\begin{cases}2x+4y=9\\-x+3y=8\end{cases}$$

Multiply the second equation by 2.

$$\begin{cases}2x+4y=9\\2(-x+3y)=2(8)\end{cases}\rightarrow\begin{cases}2x+4y=9\\-2x+6y=16\end{cases}$$

Add the equations to eliminate x; then solve for y.

$$\begin{array}{r}2x+4y=9\\-2x+6y=16\\\hline 10y=25\\y=2.5\end{array}$$

Use the equation $-x+3y=8$ to solve for x.

$$-x+3y=8$$
$$-x+3(2.5)=8$$
$$-x+7.5=8$$
$$-x=0.5$$
$$x=-0.5$$

The solution of the system is (–0.5, 2.5).

43. Twice a number, added to 6, is 3 less than the number is written as $2x+6=x-3$.

45. Three times a number, subtracted from 20, is 2 is written as $20-3x=2$.

47. The product of 4 and the sum of a number and 6 is twice the number is written as $4(x+6)=2x$.

49. To eliminate the variable y, multiply the second equation by 2.

$$3x-y=-12$$
$$2(3x-y)=2(-12)$$
$$6x-2y=-24$$

51. b; answers may vary

53. answers may vary

55. **a.** When $b=15$, the system has an infinite number of solutions.

b. When b is any real number except 15, the system has no solutions.

57. $\begin{cases}2x+3y=14\\3x-4y=-69.1\end{cases}$

Multiply the first equation by –3 and the second equation by 2.

$$\begin{cases}-3(2x+3y)=-3(14)\\2(3x-4y)=2(-69.1)\end{cases}$$
$$\rightarrow\begin{cases}-6x-9y=-42\\6x-8y=-138.2\end{cases}$$

Add the equations to eliminate x; then solve for y.

$$\begin{array}{r}-6x-9y=-42\\6x-8y=-138.2\\\hline -17y=-180.2\\y=10.6\end{array}$$

Now solve for x.

Copyright © 2012 Pearson Education, Inc. Publishing as Prentice Hall.

$$2x+3y=14$$
$$2x+3(10.6)=14$$
$$2x+31.8=14$$
$$2x=-17.8$$
$$x=-8.9$$

The solution of the system is (–8.9, 10.6).

59. a. $\begin{cases} 9.1x-y=-295 \\ 14x-y=-262 \end{cases}$

Multiply the second equation by –1.

$$\begin{cases} 9.1x-y=-295 \\ -1(14x-y)=-1(-262) \end{cases}$$
$$\rightarrow \begin{cases} 9.1x-y=-295 \\ -14x+y=262 \end{cases}$$

Add the equations to eliminate y; then solve for x.

$$\begin{array}{r} 9.1x-y=-295 \\ -14x+y=262 \\ \hline -4.9x \quad =-33 \end{array}$$
$$x \approx 6.7 \quad \text{or } 7$$

Now solve for y.

$$9.1x-y=-295$$
$$9.1(7)-y=-295$$
$$63.7-y=-295$$
$$y=358.7 \text{ or } 359$$

The solution is (7, 359) or (7, 360). (Note that (7, 360) is the result if you use the second equation when solving for y.)

b. The solution (7, 359) means that in about 2013 (2006 + 7) the number of pharmacy technician jobs will be approximately equal to the number of network system analyst jobs.

c. There will be about 359 thousand (or 360 thousand) such jobs.

Integrated Review

1. $\begin{cases} 2x-3y=-11 \\ y=4x-3 \end{cases}$

Substitute $4x - 3$ for y in the first equation and solve for x.

$$2x-3y=-11$$
$$2x-3(4x-3)=-11$$
$$2x-12x+9=-11$$
$$-10x=-20$$
$$x=2$$

Now solve for y.

$y = 4x - 3 = 4(2) - 3 = 8 - 3 = 5$

The solution of the system is (2, 5).

2. $\begin{cases} 4x-5y=6 \\ y=3x-10 \end{cases}$

Substitute $3x - 10$ for y in the first equation and solve for x.

$$4x-5y=6$$
$$4x-5(3x-10)=6$$
$$4x-15x+50=6$$
$$-11x=-44$$
$$x=4$$

Now solve for y.

$y = 3x - 10 = 3(4) - 10 = 12 - 10 = 2$

The solution of the system is (4, 2).

3. $\begin{cases} x+y=3 \\ x-y=7 \end{cases}$

Add the equations to eliminate y; then solve for x.

$$\begin{array}{r} x+y=3 \\ x-y=7 \\ \hline 2x \quad =10 \end{array}$$
$$x=5$$

Now solve for y.

$$x+y=3$$
$$5+y=3$$
$$y=-2$$

The solution of the system is (5, –2).

4. $\begin{cases} x-y=20 \\ x+y=-8 \end{cases}$

Add the equations to eliminate y; then solve for x.

$$\begin{array}{r} x-y=20 \\ x+y=-8 \\ \hline 2x=12 \end{array}$$
$$x=6$$

Now solve for y.

$$x+y=-8$$
$$6+y=-8$$
$$y=-14$$

The solution of the system is (6, –14).

5. $\begin{cases} x+2y=1 \\ 3x+4y=-1 \end{cases}$

Solve the first equation for x.

$$x+2y=1$$
$$x=-2y+1$$

Substitute $-2y + 1$ for x in the second equation and solve for y.

Copyright © 2012 Pearson Education, Inc. Publishing as Prentice Hall.

$$\begin{aligned} 3x+4y&=-1\\ 3(-2y+1)+4y&=-1\\ -6y+3+4y&=-1\\ -2y&=-4\\ y&=2 \end{aligned}$$

Now solve for x.

$x = -2y + 1 = -2(2) + 1 = -4 + 1 = -3$

The solution of the system is $(-3, 2)$.

6. $\begin{cases} x+3y=5\\ 5x+6y=-2 \end{cases}$

Solve the first equation for x.

$$\begin{aligned} x+3y&=5\\ x&=-3y+5 \end{aligned}$$

Substitute $-3y + 5$ for x in the second equation and solve for y.

$$\begin{aligned} 5x+6y&=-2\\ 5(-3y+5)+6y&=-2\\ -15y+25+6y&=-2\\ -9y&=-27\\ y&=3 \end{aligned}$$

Now solve for x.

$x = -3y + 5 = -3(3) + 5 = -9 + 5 = -4$

The solution of the system is $(-4, 3)$.

7. $\begin{aligned} y&=x+3\\ 3x&=2y-6 \end{aligned}$

Substitute $x + 3$ for y in the second equation and solve for x.

$$\begin{aligned} 3x&=2y-6\\ 3x&=2(x+3)-6\\ 3x&=2x+6-6\\ x&=0 \end{aligned}$$

Now solve for y.

$y = x + 3 = 0 + 3 = 3$

The solution of the system is $(0, 3)$.

8. $\begin{cases} y=-2x\\ 2x-3y=-16 \end{cases}$

Substitute $-2x$ for y in the second equation and solve for x.

$$\begin{aligned} 2x-3y&=-16\\ 2x-3(-2x)&=-16\\ 2x+6x&=-16\\ 8x&=-16\\ x&=-2 \end{aligned}$$

Now solve for y.

$y = -2x = -2(-2) = 4$

The solution of the system is $(-2, 4)$.

9. $\begin{cases} y=2x-3\\ y=5x-18 \end{cases}$

Substitute $2x - 3$ for y in the second equation and solve for x.

$$\begin{aligned} y&=5x-18\\ 2x-3&=5x-18\\ 15&=3x\\ 5&=x \end{aligned}$$

Now solve for y.

$y = 2x - 3 = 2(5) - 3 = 10 - 3 = 7$

The solution of the system is $(5, 7)$.

10. $\begin{cases} y=6x-5\\ y=4x-11 \end{cases}$

Substitute $6x - 5$ for y in the second equation and solve for x.

$$\begin{aligned} y&=4x-11\\ 6x-5&=4x-11\\ 2x&=-6\\ x&=-3 \end{aligned}$$

Now solve for y.

$y = 6x - 5 = 6(-3) - 5 = -18 - 5 = -23$

The solution of the system is $(-3, -23)$.

11. $\begin{cases} x+\frac{1}{6}y=\frac{1}{2}\\ 3x+2y=3 \end{cases}$

Multiply the first equation by 6 to eliminate fractions.

$$\begin{aligned} 6\left(x+\frac{1}{6}y\right)&=6\left(\frac{1}{2}\right)\\ 6x+y&=3 \end{aligned}$$

Now solve for y.

$y = -6x + 3$

Substitute $-6x + 3$ for y in the second equation and solve for x.

$$\begin{aligned} 3x+2y&=3\\ 3x+2(-6x+3)&=3\\ 3x-12x+6&=3\\ -9x&=-3\\ x&=\frac{3}{9}=\frac{1}{3} \end{aligned}$$

Now solve for y.

$$y=-6x+3=-6\left(\frac{1}{3}\right)+3=-2+3=1$$

The solution of the system is $\left(\frac{1}{3}, 1\right)$.

Copyright © 2012 Pearson Education, Inc. Publishing as Prentice Hall.

12. $x+\frac{1}{3}y=\frac{5}{12}$
$8x+3y=4$

Multiply the first equation by 12 to eliminate fractions.

$$12\left(x+\frac{1}{3}y\right)=12\left(\frac{5}{12}\right)$$
$$12x+4y=5$$

Multiply the revised first equation by –3 and the second original equation by 4.

$$\begin{cases}-3(12x+4y)=-3(5)\\4(8x+3y)=4(4)\end{cases}\rightarrow\begin{cases}-36x-12y=-15\\32x+12y=16\end{cases}$$

Add the equations to eliminate y; then solve for x.

$$\begin{array}{r}-36x-12y=-15\\32x+12y=16\\\hline -4x\qquad\quad=1\end{array}$$
$$x=-\frac{1}{4}$$

Now solve for y.

$$8x+3y=4$$
$$8\left(-\frac{1}{4}\right)+3y=4$$
$$-2+3y=4$$
$$3y=6$$
$$y=2$$

The solution of the system is $\left(-\frac{1}{4}, 2\right)$.

13. $\begin{cases}x-5y=1\\-2x+10y=3\end{cases}$

Solve the first equation for x.

$$x-5y=1$$
$$x=5y+1$$

Substitute $5y + 1$ for x in the second equation and then solve for y.

$$-2x+10y=3$$
$$-2(5y+1)+10y=3$$
$$-10y-2+10y=3$$
$$-2=3$$

Since the statement is false, the system has no solution.

14. $\begin{cases}-x+2y=3\\3x-6y=-9\end{cases}$

Solve the first equation for x.

$$-x+2y=3$$
$$x=2y-3$$

Substitute $2y - 3$ for x in the second equation and solve for y.

$$3x-6y=-9$$
$$3(2y-3)-6y=-9$$
$$6y-9-6y=-9$$
$$0=0$$

Since the statement 0 = 0 is true, the system has an infinite number of solutions.

15. $\begin{cases}0.2x-0.3y=-0.95\\0.4x+0.1y=0.55\end{cases}$

Multiply both equations by 100 to eliminate decimals.

$$\begin{cases}100(0.2x-0.3y)=100(-0.95)\\100(0.4x+0.1y)=100(0.55)\end{cases}$$
$$\rightarrow\begin{cases}20x-30y=-95\\40x+10y=55\end{cases}$$

Multiply the second revised equation by 3.

$$\begin{cases}20x-30y=-95\\3(40x+10y)=3(55)\end{cases}\rightarrow\begin{cases}20x-30y=-95\\120x+30y=165\end{cases}$$

Add the equations to eliminate y; then solve for x.

$$\begin{array}{r}20x-30y=-95\\120x+30y=165\\\hline 140x\qquad\quad=70\end{array}$$
$$x=0.5$$

Now solve for y.

$$0.4x+0.1y=0.55$$
$$0.4(0.5)+0.1y=0.55$$
$$0.2+0.1y=0.55$$
$$0.1y=0.35$$
$$y=3.5$$

The solution of the system is (0.5, 3.5).

16. $\begin{cases}0.08x-0.04y=-0.11\\0.02x-0.06y=-0.09\end{cases}$

Multiply both equations by 100 to eliminate decimals.

$$\begin{cases}100(0.08x-0.04y)=100(-0.11)\\100(0.02x-0.06y=100(-0.09)\end{cases}$$
$$\rightarrow\begin{cases}8x-4y=-11\\2x-6y=-9\end{cases}$$

Multiply the second revised equation by –4.

$$\begin{cases}8x-4y=-11\\-4(2x-6y)=-4(-9)\end{cases}\rightarrow\begin{cases}8x-4y=-11\\-8x+24y=36\end{cases}$$

Add the equations to eliminate x; then solve for y.

$$\begin{array}{r}8x-4y=-11\\-8x+24y=36\\\hline 20y=25\end{array}$$
$$y=1.25$$

Now solve for x.

Copyright © 2012 Pearson Education, Inc. Publishing as Prentice Hall.

$$\begin{aligned} 0.08x - 0.04y &= -0.11 \\ 0.08x - 0.04(1.25) &= -0.11 \\ 0.08x - 0.05 &= -0.11 \\ 0.08x &= -0.06 \\ x &= -0.75 \end{aligned}$$

The solution of the system is (–0.75, 1.25).

17. $\begin{cases} x = 3y - 7 \\ 2x - 6y = -14 \end{cases}$

Substitute $3y - 7$ for x in the second equation and solve for y.

$$\begin{aligned} 2x - 6y &= -14 \\ 2(3y-7) - 6y &= -14 \\ 6y - 14 - 6y &= -14 \\ 0 &= 0 \end{aligned}$$

Since the statement 0 = 0 is true, the system has an infinite number of solutions.

18. $\begin{cases} y = \dfrac{x}{2} - 3 \\ 2x - 4y = 0 \end{cases}$

Substitute $\dfrac{x}{2} - 3$ for y in the second equation and solve for x.

$$\begin{aligned} 2x - 4y &= 0 \\ 2x - 4\left(\frac{x}{2} - 3\right) &= 0 \\ 2x - 2x + 12 &= 0 \\ 12 &= 0 \end{aligned}$$

Since the statement 12 = 0 is false, the system has no solution.

19. $\begin{cases} 2x + 5y = -1 \\ 3x - 4y = 33 \end{cases}$

Multiply the first equation by 3 and the second equation by –2.

$$\begin{cases} 3(2x+5y) = 3(-1) \\ -2(3x-4y) = -2(33) \end{cases} \rightarrow \begin{cases} 6x + 15y = -3 \\ -6x + 8y = -66 \end{cases}$$

Add the equations to eliminate x; then solve for y.

$$\begin{array}{r} 6x + 15y = -3 \\ -6x + 8y = -66 \\ \hline 23y = -69 \\ y = -3 \end{array}$$

Now solve for x.

$$\begin{aligned} 2x + 5y &= -1 \\ 2x + 5(-3) &= -1 \\ 2x - 15 &= -1 \\ 2x &= 14 \\ x &= 7 \end{aligned}$$

The solution of the system is (7, –3).

20. $\begin{cases} 7x - 3y = 2 \\ 6x + 5y = -21 \end{cases}$

Multiply the first equation by 5 and the second equation by 3.

$$\begin{cases} 5(7x-3y) = 5(2) \\ 3(6x+5y) = 3(-21) \end{cases} \rightarrow \begin{cases} 35x - 15y = 10 \\ 18x + 15y = -63 \end{cases}$$

Add the equations to eliminate y; then solve for x.

$$\begin{array}{r} 35x - 15y = 10 \\ 18x + 15y = -63 \\ \hline 53x \quad = -53 \\ x = -1 \end{array}$$

Now solve for y.

$$\begin{aligned} 7x - 3y &= 2 \\ 7(-1) - 3y &= 2 \\ -7 - 3y &= 2 \\ -3y &= 9 \\ y &= -3 \end{aligned}$$

The solution of the system is (–1, –3).

21. answers may vary

22. answers may vary

Section 7.4 Practice

1. $\begin{cases} x + y = 50 \\ x - y = 22 \end{cases}$

Add the equations to eliminate y; then solve for x.

$$\begin{array}{r} x + y = 50 \\ x - y = 22 \\ \hline 2x \quad = 72 \\ x = 36 \end{array}$$

Now solve for y.

$$\begin{aligned} x + y &= 50 \\ 36 + y &= 50 \\ y &= 14 \end{aligned}$$

The two numbers are 36 and 14.

2. $\begin{cases} A + C = 587 \\ 7A + 5C = 3379 \end{cases}$

Multiply the first equation by –5.

$$\begin{cases} -5(A+C) = -5(587) \\ 7A + 5C = 3379 \end{cases} \rightarrow \begin{cases} -5A - 5C = -2935 \\ 7A + 5C = 3379 \end{cases}$$

Add the equations to eliminate C; then solve for A.

$$\begin{array}{r} -5A - 5C = -2935 \\ 7A + 5C = 3379 \\ \hline 2A \quad = 444 \\ A = 222 \end{array}$$

Copyright © 2012 Pearson Education, Inc. Publishing as Prentice Hall.

Now solve for C.

$$A + C = 587$$
$$222 + C = 587$$
$$C = 365$$

There were 222 adults and 365 children.

3.

	r	$\cdot\ t =$	d
Faster car	x	3	$3x$
Slower car	y	3	$3y$

$$\begin{cases} 3x + 3y = 440 \\ x = y + 10 \end{cases}$$

Substitute $y + 1$ for x in the first equation and solve for y.

$$3x + 3y = 440$$
$$3(y + 10) + 3y = 440$$
$$3y + 30 + 3y = 440$$
$$6y = 410$$
$$y = 68\frac{1}{3}$$

Now solve for x.

$$x = y + 10 = 68\frac{1}{3} + 10 = 78\frac{1}{3}$$

One car's speed is $68\frac{1}{3}$ mph and the other car's speed is $78\frac{1}{3}$ mph.

4. Let x be the liters of 20% solution.
Let y be the liters of 70% solution.

$$\begin{cases} x + y = 50 \\ 0.2x + 0.7y = 0.6(50) \end{cases}$$

Multiply the first equation by -2 and the second equation by 10.

$$\begin{cases} -2(x + y) = -2(50) \\ 10(0.2x + 0.7y) = 10(30) \end{cases}$$

$$\rightarrow \begin{cases} -2x - 2y = -100 \\ 2x + 7y = 300 \end{cases}$$

Add the equations to eliminate x; then solve for y.

$$\begin{array}{r} -2x - 2y = -100 \\ 2x + 7y = 300 \\ \hline 5y = 200 \\ y = 40 \end{array}$$

Now solve for x.

$$x + y = 50$$
$$x + 40 = 50$$
$$x = 10$$

10 liters of the 20% alcohol solution and 40 liters of the 70% alcohol solution make 50 liters of the 60% alcohol solution.

Exercise Set 7.4

1. In choice b, the length is not 3 feet longer than the width. In choice a, the perimeter is $2(8 + 5) = 2(13) = 26$ feet, not 30 feet.
Choice c gives the solution, since $9 = 6 + 3$ and $2(9 + 6) = 2(15) = 30$.

3. In choice a, the total cost is
$2(3) + 3(4) = 6 + 12 = \$18$, not \$17.
In choice c, the total cost is
$2(2) + 3(5) = 4 + 15 = \$19$, not \$17.
Choice b gives the solution, since
$2(4) + 3(3) = 8 + 9 = \$17$ and
$5(4) + 4(3) = 20 + 12 = \$32$.

5. In choice b, the total number of coins is $20 + 44 = 64$, not 100.
In choice c, the total value of the coins is $60(0.10) + 40(0.25) = 6.00 + 10.00 = \$16.00$, not \$13.00.
Choice a gives the solution, since $80 + 20 = 100$ and $80(0.10) + 20(0.25) = 8.00 + 5.00 = \$13.00$.

7. Let x be the first number and y the second.

$$\begin{cases} x + y = 15 \\ x - y = 7 \end{cases}$$

9. Let x be the amount in the larger account and y be the amount in the smaller account.

$$\begin{cases} x + y = 6500 \\ x = y + 800 \end{cases}$$

11. Let x be the first number and y be the second.

$$\begin{cases} x + y = 83 \\ x - y = 17 \end{cases}$$

Add the equations to eliminate y then solve for x.

$$\begin{array}{r} x + y = 83 \\ x - y = 17 \\ \hline 2x \quad = 100 \\ x = 50 \end{array}$$

Now solve for y.

$$x + y = 83$$
$$50 + y = 83$$
$$y = 33$$

The numbers are 50 and 33.

Copyright © 2012 Pearson Education, Inc. Publishing as Prentice Hall.

13. Let x be the first number and y the second.

$$\begin{cases} x+2y=8 \\ 2x+y=25 \end{cases}$$

Solve the first equation for x.

$$\begin{aligned} x+2y&=8 \\ x&=8-2y \end{aligned}$$

Substitute $8-2y$ for x in the second equation and solve for y.

$$\begin{aligned} 2x+y&=25 \\ 2(8-2y)+y&=25 \\ 16-4y+y&=25 \\ 16-3y&=25 \\ -3y&=9 \\ y&=-3 \end{aligned}$$

Now solve for x.

$x = 8 - 2y = 8 - 2(-3) = 8 + 6 = 14$

The numbers are 14 and -3.

15. Let x be the number of runs that Ryan Howard batted in and y be the number that Josh Hamilton batted in.

$$\begin{cases} y=x-16 \\ x+y=276 \end{cases}$$

Substitute $x-16$ for y in the second equation and solve for x.

$$\begin{aligned} x+y&=276 \\ x+x-16&=276 \\ 2x&=292 \\ x&=146 \end{aligned}$$

Now solve for y.

$y = x - 16 = 146 - 16 = 130$

Ryan Howard batted in 146 runs and Josh Hamilton batted in 130 runs.

17. Let a be the price of an adult's ticket and c be the price of a child's ticket.

$$\begin{cases} 3a+4c=159 \\ 2a+3c=112 \end{cases}$$

Multiply the first equation by -2 and the second equation by 3.

$$\begin{cases} -2(3a+4c)=-2(159) \\ 3(2a+3c)=3(112) \end{cases} \rightarrow \begin{cases} -6a-8c=-318 \\ 6a+9c=336 \end{cases}$$

Add the equations to eliminate a and solve for c.

$$\begin{aligned} -6a-8c&=-318 \\ 6a+9c&=336 \\ \hline c&=18 \end{aligned}$$

Now solve for a.

$$\begin{aligned} 2a+3c&=112 \\ 2a+3(18)&=112 \\ 2a+54&=112 \\ 2a&=58 \\ a&=29 \end{aligned}$$

The price of an adult's ticket is \$29 and the price of a child's ticket is \$18.

19. Let x be quarters and y be nickels.

$$\begin{cases} x+y=80 \\ 0.25x+0.05y=14.60 \end{cases}$$

Solve the first equation in terms of y.

$$\begin{aligned} x+y&=80 \\ y&=-x+80 \end{aligned}$$

Substitute $-x+80$ for y in the second equation and solve for x.

$$\begin{aligned} 0.25x+0.05y&=14.60 \\ 0.25x+0.05(-x+80)&=14.60 \\ 0.25x-0.05x+4&=14.60 \\ 0.20x&=10.60 \\ x&=53 \end{aligned}$$

Now solve for y.

$y = -x + 80 = -53 + 80 = 27$

There are 53 quarters and 27 nickels.

21. Let x be the value of one McDonald's share and let y be the value of one Ohio Art Company share.

$$\begin{cases} 35x+69y=2360 \\ x=y+60 \end{cases}$$

Substitute $y+60$ for x in the first equation and solve for y.

$$\begin{aligned} 35x+69y&=2360 \\ 35(y+60)+69y&=2360 \\ 35y+2100+69y&=2360 \\ 2100+104y&=2360 \\ 104y&=260 \\ y&=2.5 \end{aligned}$$

Now solve for x.

$x = y + 60 = 2.5 + 60 = 62.5$

On that day, the closing price of the McDonald's stock was \$62.5 per share and the closing price of The Ohio Art Company stock was \$2.50 per share.

23. Let x be the daily fee and y be the mileage charge.

$$\begin{cases} 4x+450y=240.50 \\ 3x+200y=146.00 \end{cases}$$

Multiply the first equation by 3 and the second by -4.

Copyright © 2012 Pearson Education, Inc. Publishing as Prentice Hall.

$$\begin{cases} 3(4x+450y)=3(240.50) \\ -4(3x+200y)=-4(146.00) \end{cases}$$

$$\rightarrow \begin{cases} 12x+1350y=721.5 \\ -12x-800y=-584 \end{cases}$$

Add the equations to eliminate x and solve for y.

$$\begin{aligned} 12x+1350y &= 721.5 \\ -12x-800y &= -584 \\ \hline 550y &= 137.5 \\ y &= 0.25 \end{aligned}$$

Now solve for x.

$$\begin{aligned} 3x+200y &= 146 \\ 3x+200(0.25) &= 146 \\ 3x+50 &= 146 \\ 3x &= 96 \\ x &= 32 \end{aligned}$$

There is a \$32 daily fee and a \$0.25 per mile mileage charge.

25. $$\begin{cases} 18=2(x+y) \\ 18=\frac{9}{2}(x-y) \end{cases}$$

Multiply the first equation by $\frac{1}{2}$ and the second equation by $\frac{2}{9}$.

$$\begin{cases} \frac{1}{2}(18)=\frac{1}{2}[2(x+y)] \\ \frac{2}{9}(18)=\frac{2}{9}\left[\frac{9}{2}(x-y)\right] \end{cases} \rightarrow \begin{cases} 9=x+y \\ 4=x-y \end{cases}$$

Add the equations to eliminate y; then solve for x.

$$\begin{aligned} 9 &= x+y \\ 4 &= x-y \\ \hline 13 &= 2x \\ 6.5 &= x \end{aligned}$$

Now solve for y.

$$\begin{aligned} 9 &= x+y \\ 9 &= 6.5+y \\ 2.5 &= y \end{aligned}$$

The rate that Pratap can row in still water is 6.5 miles per hour and the rate of the current is 2.5 miles per hour.

27. Let x = rate of flight in still wind. Then y = rate of wind.

$$\begin{cases} 780=\frac{3}{2}(x+y) \\ 780=2(x-y) \end{cases}$$

Multiply the first equation by $\frac{2}{3}$ and the second equation by $\frac{1}{2}$.

$$\begin{cases} \frac{2}{3}(780)=\frac{2}{3}\left[\frac{3}{2}(x+y)\right] \\ \frac{1}{2}(780)=\frac{1}{2}[2(x-y)] \end{cases} \rightarrow \begin{cases} 520=x+y \\ 390=x-y \end{cases}$$

Add the equations to eliminate y; then solve for x.

$$\begin{aligned} 520 &= x+y \\ 390 &= x-y \\ \hline 910 &= 2x \\ 455 &= x \end{aligned}$$

Now solve for y.

$$\begin{aligned} 780 &= 2(x-y) \\ 780 &= 2(455)-2y \\ 780 &= 910-2y \\ -130 &= -2y \\ 65 &= y \end{aligned}$$

The speed of the plane in still air is 455 mph and the speed of the wind is 65 mph.

29. Let x be the number of hours that Kevin spent on his bicycle and y be the number of hours he spent walking. Then the distance he rode was $40x$ miles and the distance he walked was $4y$ miles.

$$\begin{cases} x+y=6 \\ 40x+4y=186 \end{cases}$$

Solve the first equation for y.

$$\begin{aligned} x+y &= 6 \\ y &= 6-x \end{aligned}$$

Substitute $6-x$ for y in the second equation and solve for x.

$$\begin{aligned} 40x+4y &= 186 \\ 40x+4(6-x) &= 186 \\ 40x+24-4x &= 186 \\ 36x+24 &= 186 \\ 36x &= 162 \\ x &= 4.5 \end{aligned}$$

Kevin spent 4.5 hours on his bike.

31. Let x be ounces of 4% solution, and let y be ounces of 12% solution.

$$\begin{cases} x+y=12 \\ 0.04x+0.12y=0.09(12) \end{cases}$$

Solve the first equation in terms of x.

$$\begin{aligned} x+y &= 12 \\ x &= -y+12 \end{aligned}$$

Substitute $-y+12$ for x in the second equation and solve for y.

Copyright © 2012 Pearson Education, Inc. Publishing as Prentice Hall.

$$\begin{aligned} 0.04x+0.12y&=1.08\\ 0.04(-y+12)+0.12y&=1.08\\ -0.04y+0.48+0.12y&=1.08\\ 0.08y&=0.60\\ y&=7.5 \end{aligned}$$

Now solve for x.
$x = -y + 12 = -7.5 + 12 = 4.5$
Darren will need 4.5 ounces of the 4% solution and 7.5 ounces of the 12% solution to make 12 ounces of the 9% solution.

33. Let x be the number of pounds of high-quality coffee, and let y be the number of pounds of the cheaper coffee.

$$\begin{cases} x+y=200\\ 4.95x+2.65y=200(3.95) \end{cases}$$

Solve the first equation for x.

$$\begin{aligned} x+y&=200\\ x&=200-y \end{aligned}$$

Substitute $200 - y$ for x in the second equation and solve for y.

$$\begin{aligned} 4.95x+2.65y&=200(3.95)\\ 4.95(200-y)+2.65y&=790\\ 990-4.95y+2.65y&=790\\ 990-2.30y&=790\\ -2.30y&=-200\\ y&=\frac{200}{2.30}\\ y&\approx 87 \end{aligned}$$

$x = 200 - y \approx 200 - 87 = 113$
Wayne should blend 113 pounds of the coffee that sells for \$4.95 per pound with 87 pounds of the cheaper coffee.

35. Let x be one angle, and let y be the other angle.

$$\begin{cases} x+y=90\\ y=2x \end{cases}$$

Substitute $2x$ for y in the first equation and solve for x.

$$\begin{aligned} x+y&=90\\ x+2x&=90\\ 3x&=90\\ x&=30 \end{aligned}$$

Now solve for y.
$y = 2x = 2(30) = 60$
One angle is 30° and the other is 60°.

37. Let x be the measure of one angle and y be the measure of the other.

$$\begin{cases} x+y=90\\ x=10+3y \end{cases}$$

Substitute $10 + 3y$ for x in the first equation and solve for y.

$$\begin{aligned} x+y&=90\\ 10+3y+y&=90\\ 10+4y&=90\\ 4y&=80\\ y&=20 \end{aligned}$$

$x = 10 + 3y = 10 + 3(20) = 10 + 60 = 70$
The angles measure 20° and 70°.

39. Let x be the number of pieces sold at the original price, and let y be the number of pieces sold at the discounted price.

$$\begin{cases} x+y=90\\ 9.5x+7.5y=721 \end{cases}$$

Solve the first equation in terms of x.

$$\begin{aligned} x+y&=90\\ x&=-y+90 \end{aligned}$$

Substitute $-y + 90$ for x in the second equation and solve for y.

$$\begin{aligned} 9.5x+7.5y&=721\\ 9.5(-y+90)+7.5y&=721\\ -9.5y+855+7.5y&=721\\ -2.0y&=-134\\ y&=67 \end{aligned}$$

Now solve for x.
$x = -y + 90 = -67 + 90 = 23$
They sold 23 pieces at \$9.50 each and 67 pieces at \$7.50 each.

41. Let x be the rate of the faster group and y be the rate of the slower group.

$$\begin{cases} 240x+240y=1200\\ y=x-\frac{1}{2} \end{cases}$$

Substitute $x-\frac{1}{2}$ for y in the fist equation and solve for x.

$$\begin{aligned} 240x+240y&=1200\\ 240x+240\left(x-\frac{1}{2}\right)&=1200\\ 240x+240x-120&=1200\\ 480x-120&=1200\\ 480x&=1320\\ x&=2.75 \end{aligned}$$

$$y=x-\frac{1}{2}=x-0.5=2.75-0.5=2.25$$

The hiking rates are $2.75 = 2\frac{3}{4}$ miles per hour and $2.25 = 2\frac{1}{4}$ miles per hour.

Copyright © 2012 Pearson Education, Inc. Publishing as Prentice Hall.

43. Let x be the number of gallons of 30% solution and y be the number of gallons of 60% solution.

$$\begin{cases} x+y=150 \\ 0.3x+0.6y=0.5(150) \end{cases}$$

Solve the first equation in terms of x.

$x = 150 - y$

Substitute $150 - y$ for x in the second equation and solve for y.

$$\begin{aligned} 0.3x+0.6y &= 0.5(150) \\ 0.3x+0.6y &= 75 \\ 0.3(150-y)+0.6y &= 75 \\ 45-0.3y+0.6y &= 75 \\ 0.3y &= 30 \\ y &= 100 \end{aligned}$$

Now solve for x.

$x = 150 - y = 150 - 100 = 50$

Combining 50 gallons of the 30% solution and 100 gallons of the 60% solution is necessary to create 150 gallons of a 50% solution.

45. Let x be the length and y the width.

$$\begin{cases} 2(x+y)=144 \\ x=y+12 \end{cases}$$

Substitute $y + 12$ for x in the first equation and solve for y.

$$\begin{aligned} 2(x+y) &= 144 \\ 2(y+12+y) &= 144 \\ 2(2y+12) &= 144 \\ 4y+24 &= 144 \\ 4y &= 120 \\ y &= 30 \end{aligned}$$

$x = y + 12 = 30 + 12 = 42$

The length is 42 inches and the width is 30 inches.

47. $4^2 = 4 \cdot 4 = 16$

49. $(6x)^2 = (6x)(6x) = 36x^2$

51. $(10y^3)^2 = (10y^3)(10y^3) = 100y^6$

53. The price of the result must be between \$0.49 and \$0.65, so choice a is the only possibility.

55. $$\begin{cases} y+2x=33 \\ y=2x-3 \end{cases}$$

Substitute $2x - 3$ for y in the first equation and solve for y.

$$\begin{aligned} y+2x &= 33 \\ 2x-3+2x &= 33 \\ 4x &= 36 \\ x &= 9 \end{aligned}$$

Now solve for y.

$y = 2x - 3 = 2(9) - 3 = 18 - 3 = 15$

The width is 9 feet and the length is 15 feet.

Chapter 7 Vocabulary Check

1. In a system of linear equations in two variables, if the graphs of the equations are the same, the equations are <u>dependent</u> equations.
2. Two or more linear equations are called a <u>system of linear equations</u>.
3. A system of equations that has at least one solution is called a <u>consistent</u> system.
4. A <u>solution</u> of a system of two equations in two variables is an ordered pair of numbers that is a solution of both equations in the system.
5. Two algebraic methods for solving systems of equations are <u>addition</u> and <u>substitution</u>.
6. A system of equations that has no solution is called an <u>inconsistent</u> system.
7. In a system of linear equations in two variables, if the graphs of the equations are different, the equations are <u>independent</u> equations.

Chapter 7 Review

1. a. $$\begin{cases} 2x-3y=12 \\ 3x+4y=1 \end{cases}$$

First equation:

$$\begin{aligned} 2x-3y &= 12 \\ 2(12)-3(4) &\stackrel{?}{=} 12 \\ 24-12 &\stackrel{?}{=} 12 \\ 12 &= 12 \quad \text{True} \end{aligned}$$

Second equation:

$$\begin{aligned} 3x+4y &= 1 \\ 3(12)-4(4) &\stackrel{?}{=} 1 \\ 36-16 &\stackrel{?}{=} 1 \\ 20 &= 1 \quad \text{False} \end{aligned}$$

(12, 4) is not a solution of the system.

b. First equation:

$$\begin{aligned} 2x-3y &= 12 \\ 2(3)-3(-2) &\stackrel{?}{=} 12 \\ 6+6 &\stackrel{?}{=} 12 \\ 12 &= 12 \quad \text{True} \end{aligned}$$

Copyright © 2012 Pearson Education, Inc. Publishing as Prentice Hall.

Second equation:

$$3x+4y=1$$
$$3(3)+4(-2) \stackrel{?}{=} 1$$
$$9-8 \stackrel{?}{=} 1$$
$$1=1 \quad \text{True}$$

(3, –2) is a solution of the system.

2. a. $\begin{cases} 2x+3y=1 \\ 3y-x=4 \end{cases}$

First equation:

$$2x+3y=1$$
$$2(2)+3(2) \stackrel{?}{=} 1$$
$$4+6 \stackrel{?}{=} 1$$
$$10=1 \quad \text{False}$$

(2, 2) is not a solution of the system.

b. First equation:

$$2x+3y=1$$
$$2(-1)+3(1) \stackrel{?}{=} 1$$
$$-2+3 \stackrel{?}{=} 1$$
$$1=1 \quad \text{True}$$

Second equation:

$$3y-x=4$$
$$3(1)-(-1) \stackrel{?}{=} 4$$
$$3+1 \stackrel{?}{=} 4$$
$$4=4 \quad \text{True}$$

(–1, 1) is a solution of the system.

3. a. $\begin{cases} 5x-6y=18 \\ 2y-x=-4 \end{cases}$

First equation:

$$5x-6y=18$$
$$5(-6)-6(-8) \stackrel{?}{=} 18$$
$$-30+48 \stackrel{?}{=} 18$$
$$18=18 \quad \text{True}$$

Second equation:

$$2y-x=-4$$
$$2(-8)-(-6) \stackrel{?}{=} -4$$
$$-16+6 \stackrel{?}{=} -4$$
$$-10=-4 \quad \text{False}$$

(–6, –8) is not a solution of the system.

b. First equation:

$$5x-6y=18$$
$$5(3)-6\left(\frac{5}{2}\right) \stackrel{?}{=} 18$$
$$15-15 \stackrel{?}{=} 18$$
$$0=18 \quad \text{False}$$

$\left(3, \frac{5}{2}\right)$ is not a solution of the system.

4. a. $\begin{cases} 4x+y=0 \\ -8x-5y=9 \end{cases}$

First equation:

$$4x+y=0$$
$$4\left(\frac{3}{4}\right)+(-3) \stackrel{?}{=} 0$$
$$3-3 \stackrel{?}{=} 0$$
$$0=0 \quad \text{True}$$

Second equation:

$$-8x-5y=9$$
$$-8\left(\frac{3}{4}\right)-5(-3) \stackrel{?}{=} 9$$
$$-6+15 \stackrel{?}{=} 9$$
$$9=9 \quad \text{True}$$

$\left(\frac{3}{4}, -3\right)$ is a solution of the system.

b. First equation:

$$4x+y=0$$
$$4(-2)+8 \stackrel{?}{=} 0$$
$$-8+8 \stackrel{?}{=} 0$$
$$0=0 \quad \text{True}$$

Second equation:

$$-8x-5y=9$$
$$-8(-2)-5(8) \stackrel{?}{=} 9$$
$$16-40 \stackrel{?}{=} 9$$
$$-24=9 \quad \text{False}$$

(–2, 8) is not a solution of the system.

5. $\begin{cases} x+y=5 \\ x-y=1 \end{cases}$

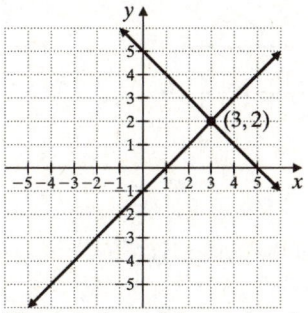

The solution of the system is (3, 2).

Copyright © 2012 Pearson Education, Inc. Publishing as Prentice Hall.

6. $\begin{cases} x+y=3 \\ x-y=-1 \end{cases}$

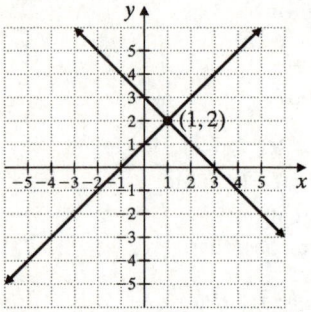

The solution of the system is (1, 2).

7. $\begin{cases} x=5 \\ y=-1 \end{cases}$

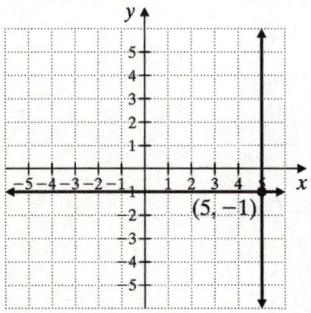

The solution of the system is (5, −1).

8. $\begin{cases} x=-3 \\ y=2 \end{cases}$

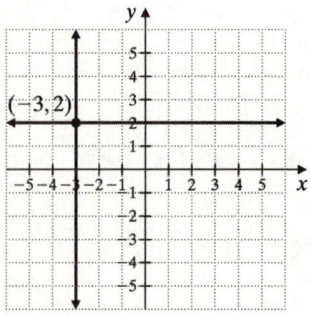

The solution of the system is (−3, 2).

9. $\begin{cases} 2x+y=5 \\ x=-3y \end{cases}$

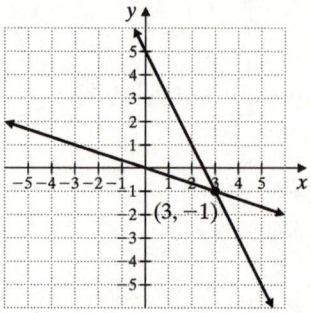

The solution of the system is (3, −1).

10. $\begin{cases} 3x+y=-2 \\ y=-5x \end{cases}$

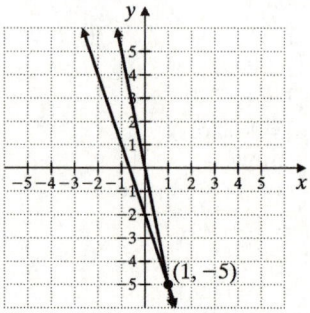

The solution of the system is (1, −5).

11. $\begin{cases} y=3x \\ -6x+2y=6 \end{cases}$

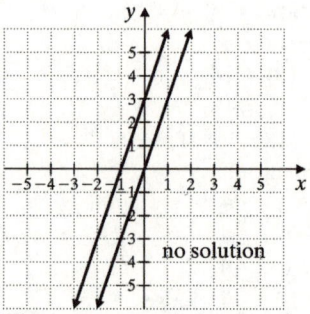

There are no solutions to the system because the lines are parallel.

Copyright © 2012 Pearson Education, Inc. Publishing as Prentice Hall.

12. $\begin{cases} x-2y=2 \\ -2x+4y=-4 \end{cases}$

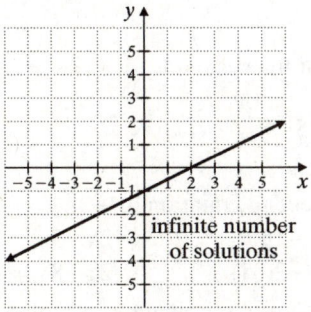

Since they are the same line, the system has an infinite number of solutions.

13. $\begin{cases} y=2x+6 \\ 3x-2y=-11 \end{cases}$

Substitute $2x + 6$ for y in the second equation and solve for x.

$$\begin{aligned} 3x-2y&=-11 \\ 3x-2(2x+6)&=-11 \\ 3x-4x-12&=-11 \\ -x&=1 \\ x&=-1 \end{aligned}$$

Now solve for y.

$y = 2x + 6 = 2(-1) + 6 = -2 + 6 = 4$

The solution of the system is $(-1, 4)$.

14. $\begin{cases} y=3x-7 \\ 2x-3y=7 \end{cases}$

Substitute $3x - 7$ for y in the second equation and solve for x.

$$\begin{aligned} 2x-3y&=7 \\ 2x-3(3x-7)&=7 \\ 2x-9x+21&=7 \\ -7x&=-14 \\ x&=2 \end{aligned}$$

Now solve for y.

$y = 3x - 7 = 3(2) - 7 = 6 - 7 = -1$

The solution of the system is $(2, -1)$.

15. $\begin{cases} x+3y=-3 \\ 2x+y=4 \end{cases}$

Solve the second equation for y.

$$\begin{aligned} 2x+y&=4 \\ y&=-2x+4 \end{aligned}$$

Substitute $-2x + 4$ for y in the first equation and solve for x.

$$\begin{aligned} x+3y&=-3 \\ x+3(-2x+4)&=-3 \\ x-6x+12&=-3 \\ -5x&=-15 \\ x&=3 \end{aligned}$$

Now solve for y.

$y = -2x + 4 = -2(3) + 4 = -6 + 4 = -2$

The solution of the system is $(3, -2)$.

16. $\begin{cases} 3x+y=11 \\ x+2y=12 \end{cases}$

Solve the first equation for y.

$$\begin{aligned} 3x+y&=11 \\ y&=-3x+11 \end{aligned}$$

Substitute $-3x + 11$ for y in the second equation and solve for x.

$$\begin{aligned} x+2y&=12 \\ x+2(-3x+11)&=12 \\ x-6x+22&=12 \\ -5x&=-10 \\ x&=2 \end{aligned}$$

Now solve for y.

$y = -3x + 11 = -3(2) + 11 = -6 + 11 = 5$

The solution of the system is $(2, 5)$.

17. $\begin{cases} 4y=2x+6 \\ x-2y=-3 \end{cases}$

Solve the second equation for x.

$$\begin{aligned} x-2y&=-3 \\ x&=2y-3 \end{aligned}$$

Substitute $2y - 3$ for x in the first equation and solve for y.

$$\begin{aligned} 4y&=2x+6 \\ 4y&=2(2y-3)+6 \\ 4y&=4y-6+6 \\ 0&=0 \end{aligned}$$

Since the statement $0 = 0$ is true, the system has an infinite number of solutions.

18. $\begin{cases} 9x=6y+3 \\ 6x-4y=2 \end{cases}$

Solve the first equation for x.

$$\begin{aligned} 9x&=6y+3 \\ x&=\frac{6}{9}y+\frac{3}{9} \\ x&=\frac{2}{3}y+\frac{1}{3} \end{aligned}$$

Substitute $\frac{2}{3}y+\frac{1}{3}$ for x in the second equation and solve for y.

Copyright © 2012 Pearson Education, Inc. Publishing as Prentice Hall.

$$6x-4y=2$$
$$6\left(\frac{2}{3}y+\frac{1}{3}\right)-4y=2$$
$$4y+2-4y=2$$
$$0=0$$

Since the statement 0 = 0 is true, the system has an infinite number of solutions.

19. $\begin{cases} x+y=6 \\ y=-x-4 \end{cases}$

Substitute $-x-4$ for y in the first equation and solve for x.

$$x+y=6$$
$$x+(-x-4)=6$$
$$0=10 \quad \text{False}$$

Since the statement 0 = 10 is false, the system has no solution.

20. $\begin{cases} -3x+y=6 \\ y=3x+2 \end{cases}$

Substitute $3x+2$ for y in the first equation and solve for x.

$$-3x+y=6$$
$$-3x+3x+2=6$$
$$2=6 \quad \text{False}$$

Since the statement 2 = 6 is false, the system has no solution.

21. $\begin{cases} 2x+3y=-6 \\ x-3y=-12 \end{cases}$

Add the equations to eliminate y and solve for x.

$$\begin{array}{r} 2x+3y=-6 \\ x-3y=-12 \\ \hline 3x \quad\quad =-18 \\ x=-6 \end{array}$$

Now solve for y.

$$x-3y=-12$$
$$-6-3y=-12$$
$$-3y=-6$$
$$y=2$$

The solution of the system is (–6, 2).

22. $\begin{cases} 4x+y=15 \\ -4x+3y=-19 \end{cases}$

Add the equations to eliminate x and solve for y.

$$\begin{array}{r} 4x+y=15 \\ -4x+3y=-19 \\ \hline 4y=-4 \\ y=-1 \end{array}$$

Now solve for x.

$$4x+y=15$$
$$4x-1=15$$
$$4x=16$$
$$x=4$$

The solution of the system is (4, –1).

23. $\begin{cases} 2x-3y=-15 \\ x+4y=31 \end{cases}$

Multiply the second equation by –2.

$$\begin{cases} 2x-3y=-15 \\ -2(x+4y)=-2(31) \end{cases} \rightarrow \begin{cases} 2x-3y=-15 \\ -2x-8y=-62 \end{cases}$$

Add the equations to eliminate x and solve for y.

$$\begin{array}{r} 2x-3y=-15 \\ -2x-8y=-62 \\ \hline -11y=-77 \\ y=7 \end{array}$$

Now solve for x.

$$x+4y=31$$
$$x+4(7)=31$$
$$x+28=31$$
$$x=3$$

The solution of the system is (3, 7).

24. $\begin{cases} x-5y=-22 \\ 4x+3y=4 \end{cases}$

Multiply the first equation by –4.

$$\begin{cases} -4(x-5y)=-4(-22) \\ 4x+3y=4 \end{cases} \rightarrow \begin{cases} -4x+20y=88 \\ 4x+3y=4 \end{cases}$$

Add the equations to eliminate x and solve for y.

$$\begin{array}{r} -4x+20y=88 \\ 4x+3y=4 \\ \hline 23y=92 \\ y=4 \end{array}$$

Now solve for x.

$$x-5y=-22$$
$$x-5(4)=-22$$
$$x-20=-22$$
$$x=-2$$

The solution of the system is (–2, 4).

25. $\begin{cases} 2x-6y=-1 \\ -x+3y=\frac{1}{2} \end{cases}$

Multiply the second equation by 2.

$$\begin{cases} 2x-6y=-1 \\ 2(-x+3y)=2\left(\frac{1}{2}\right) \end{cases} \rightarrow \begin{cases} 2x-6y=-1 \\ -2x+6y=1 \end{cases}$$

Add the equations to eliminate x.

Copyright © 2012 Pearson Education, Inc. Publishing as Prentice Hall.

$$\begin{array}{r} 2x-6y=-1 \\ -2x+6y=1 \\ \hline 0=0 \end{array}$$

Since the statement 0 = 0 is true, the system has an infinite number of solutions.

26. $\begin{cases} 0.6x-0.3y=-1.5 \\ 0.04x-0.02y=-0.1 \end{cases}$

Multiply the first equation by 10 and the second by 100 to eliminate decimals.

$$\begin{cases} 10(0.6x-0.3y)=10(-1.5) \\ 100(0.04x-0.02y)=100(-0.1) \end{cases}$$

$$\rightarrow \begin{cases} 6x-3y=-15 \\ 4x-2y=-10 \end{cases}$$

Multiply the first revised equation by 2 and the second revised equation by –3.

$$\begin{cases} 2(6x-3y)=2(-15) \\ -3(4x-2y)=-3(-10) \end{cases} \rightarrow \begin{cases} 12x-6y=-30 \\ -12x+6y=30 \end{cases}$$

Add the equations to eliminate y.

$$\begin{array}{r} 12x-6y=-30 \\ -12x+6y=30 \\ \hline 0=0 \end{array}$$

Since the statement 0 = 0 is true, the system has an infinite number of solutions.

27. $\begin{cases} \frac{3}{4}x+\frac{2}{3}y=2 \\ x+\frac{y}{3}=6 \end{cases}$

Multiply the first equation by 12 and the second by 3 to eliminate fractions.

$$\begin{cases} 12\left(\frac{3}{4}x+\frac{2}{3}y\right)=12(2) \\ 3\left(x+\frac{y}{3}\right)=3(6) \end{cases} \rightarrow \begin{cases} 9x+8y=24 \\ 3x+y=18 \end{cases}$$

Multiply the second revised equation by –3.

$$\begin{cases} 9x+8y=24 \\ -3(3x+y)=-3(18) \end{cases} \rightarrow \begin{cases} 9x+8y=24 \\ -9x-3y=-54 \end{cases}$$

Add the equations to eliminate x and solve for y.

$$\begin{array}{r} 9x+8y=24 \\ -9x-3y=-54 \\ \hline 5y=-30 \\ y=-6 \end{array}$$

Now solve for x.

$$x+\frac{y}{3}=6$$
$$x+\left(\frac{-6}{3}\right)=6$$
$$x+(-2)=6$$
$$x=8$$

The solution of the system is (8, –6).

28. $\begin{cases} 10x+2y=0 \\ 3x+5y=33 \end{cases}$

Multiply the first equation by 5 and the second by –2.

$$\begin{cases} 5(10x+2y)=5(0) \\ -2(3x+5y)=-2(33) \end{cases} \rightarrow \begin{cases} 50x+10y=0 \\ -6x-10y=-66 \end{cases}$$

Add the equations to eliminate y and solve for x.

$$\begin{array}{r} 50x+10y=0 \\ -6x-10y=-66 \\ \hline 44x \quad =-66 \end{array}$$

$$x=\frac{-66}{44}=-\frac{3}{2}$$

Now solve for y.

$$10x+2y=0$$
$$10\left(-\frac{3}{2}\right)+2y=0$$
$$-15+2y=0$$
$$2y=15$$
$$y=\frac{15}{2}$$

The solution of the system is $\left(-\frac{3}{2}, \frac{15}{2}\right)$.

29. Let x be the smaller number and y be the larger.

$$\begin{cases} x+y=16 \\ 3y-x=72 \end{cases}$$

Solve the first equation for y.

$$x+y=16$$
$$y=-x+16$$

Substitute $-x + 16$ for y in the second equation and solve for x.

$$3y-x=72$$
$$3(-x+16)-x=72$$
$$-3x+48-x=72$$
$$-4x=24$$
$$x=-6$$

Now solve for y.

$y=-x+16=-(-6)+16=6+16=22$

The two numbers are –6 and 22.

Copyright © 2012 Pearson Education, Inc. Publishing as Prentice Hall.

30. Let x be the number of orchestra seats and y be the number of balcony seats.

$$\begin{cases} x+y=360 \\ 45x+35y=15{,}150 \end{cases}$$

Solve the first equation for x.

$x=-y+360$

Substitute $-y+360$ for x in the second equation and solve for y.

$$\begin{aligned} 45x+35y&=15{,}150 \\ 45(-y+360)+35y&=15{,}150 \\ -45y+16{,}200+35y&=15{,}150 \\ -10y&=-1050 \\ y&=105 \end{aligned}$$

Now solve for x.

$$\begin{aligned} x+y&=360 \\ x+105&=360 \\ x&=255 \end{aligned}$$

There are 255 orchestra seats and 105 balcony seats.

31. Let x be the rate of the current and y be the speed in still water.

$$\begin{cases} 19(x-y)=340 \\ 14(x+y)=340 \end{cases}$$

Multiply the first equation by $\frac{1}{19}$ and the second by $\frac{1}{14}$.

$$\begin{cases} \frac{1}{19}[19(x-y)]=\frac{1}{19}(340) \\ \frac{1}{14}[14(x+y)]=\frac{1}{14}(340) \end{cases} \rightarrow \begin{cases} x-y=17.9 \\ x+y=24.3 \end{cases}$$

Add the equations to eliminate y and solve for x.

$$\begin{array}{r} x-y=17.9 \\ x+y=24.3 \\ \hline 2x\quad=42.2 \\ x=21.1 \end{array}$$

Now solve for x.

$$\begin{aligned} x+y&=24.3 \\ 21.1+y&=24.3 \\ y&=3.2 \end{aligned}$$

The speed in still water is 21.1 mph and the current of the river is 3.2 mph.

32. Let x = number of cc of 6% acid solution and y = number of cc of 14% acid solution.

$$\begin{cases} x+y=50 \\ 0.06x+0.14y=0.12(50) \rightarrow 0.06x+0.14y=6 \end{cases}$$

Solve the first equation for x.

$$\begin{aligned} x+y&=50 \\ x&=-y+50 \end{aligned}$$

Substitute $-y+50$ for x in the second equation and solve for y.

$$\begin{aligned} 0.06x+0.14y&=6 \\ 0.06(-y+50)+0.14y&=6 \\ -0.06y+3+0.14y&=6 \\ 0.08y&=3 \\ y&=37.5 \end{aligned}$$

Now solve for x.

$x=-y+50=-37.5+50=12.5$

12.5 cc of the 6% solution and 37.5 cc of the 14% solution are needed to make 50 cc of the 12% solution.

33. Let x be the cost of an egg and y be the cost of a strip of bacon.

$$\begin{cases} 3x+4y=3.80 \\ 2x+3y=2.75 \end{cases}$$

Multiply the first equation by 2 and the second by -3.

$$\begin{cases} 2(3x+4y)=2(3.80) \\ -3(2x+3y)=-3(2.75) \end{cases} \rightarrow \begin{cases} 6x+8y=7.60 \\ -6x-9y=-8.25 \end{cases}$$

Add the equations to eliminate x and solve for y.

$$\begin{array}{r} 6x+8y=7.60 \\ -6x-9y=-8.25 \\ \hline -y=-0.65 \\ y=0.65 \end{array}$$

Now solve for x.

$$\begin{aligned} 2x+3y&=2.75 \\ 2x+3(0.65)&=2.75 \\ 2x+1.95&=2.75 \\ 2x&=0.80 \\ x&=0.40 \end{aligned}$$

Each egg costs \$0.40 and each strip of bacon costs \$0.65.

34. Let x be time spent jogging, and let y be the time spent walking.

$$\begin{cases} x+y=3 \\ 7.5x+4y=15 \end{cases}$$

Solve the first equation for x.

$$\begin{aligned} x+y&=3 \\ x&=-y+3 \end{aligned}$$

Substitute $-y+3$ for x in the second equation and solve for y.

$$\begin{aligned} 7.5x+4y&=15 \\ 7.5(-y+3)+4y&=15 \\ -7.5y+22.5+4y&=15 \\ -3.5y&=-7.5 \\ y&=2.14 \end{aligned}$$

Copyright © 2012 Pearson Education, Inc. Publishing as Prentice Hall.

Now solve for x.

$$x + y = 3$$
$$x + 2.14 = 3$$
$$x = 0.86$$

He spent 0.86 hour jogging and 2.14 hours walking.

35. $\begin{cases} x - 2y = 1 \\ 2x + 3y = -12 \end{cases}$

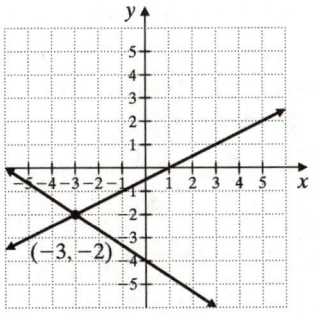

The solution of the system is (−3, −2).

36. $\begin{cases} 3x - y = -4 \\ 6x - 2y = -8 \end{cases}$

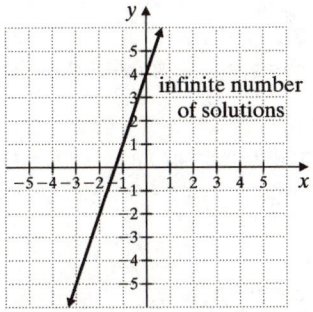

The system has an infinite number of solutions because it is one line.

37. $\begin{cases} x + 4y = 11 \\ 5x - 9y = -3 \end{cases}$

Solve the first equation for x.

$$x + 4y = 11$$
$$x = -4y + 11$$

Substitute $-4y + 11$ for x in the second equation and solve for y.

$$5x - 9y = -3$$
$$5(-4y + 11) - 9y = -3$$
$$-20y + 55 - 9y = -3$$
$$-29y = -58$$
$$y = 2$$

Now solve for x.

$x = -4y + 11 = -4(2) + 11 = -8 + 11 = 3$

The solution of the system is (3, 2).

38. $\begin{cases} x + 9y = 16 \\ 3x - 8y = 13 \end{cases}$

Solve the first equation for x.

$$x + 9y = 16$$
$$x = -9y + 16$$

Substitute $-9y + 16$ for x in the second equation and solve for y.

$$3x - 8y = 13$$
$$3(-9y + 16) - 8y = 13$$
$$-27y + 48 - 8y = 13$$
$$-35y = -35$$
$$y = 1$$

Now solve for x.

$x = -9y + 16 = -9(1) + 16 = -9 + 16 = 7$

The solution of the system is (7, 1).

39. $y = -2x$

$4x + 7y = -15$

Substitute $-2x$ for y in the second equation and solve for x.

$$4x + 7y = -15$$
$$4x + 7(-2x) = -15$$
$$4x - 14x = -15$$
$$-10x = -15$$
$$x = \frac{-15}{-10} = \frac{3}{2}$$

Now solve for y.

$$y = -2x = -2\left(\frac{3}{2}\right) = -3$$

The solution of the system is $\left(\frac{3}{2}, -3\right)$.

40. $\begin{cases} 3y = 2x + 15 \\ -2x + 3y = 21 \end{cases} \rightarrow \begin{cases} 2x - 3y = -15 \\ -2x + 3y = 21 \end{cases}$

Add the equations to eliminate x.

$$\begin{array}{r} 2x - 3y = -15 \\ -2x + 3y = 21 \\ \hline 0 = 6 \end{array} \quad \text{False}$$

Since the statement 0 = 6 in false, the system has no solution.

41. $\begin{cases} 3x - y = 4 \\ 4y = 12x - 16 \end{cases}$

Solve the first equation for y.

$$3x - y = 4$$
$$3x - 4 = y$$

Substitute $3x - 4$ for y in the second equation and solve for x.

Copyright © 2012 Pearson Education, Inc. Publishing as Prentice Hall.

$$4y = 12x - 16$$
$$4(3x - 4) = 12x - 16$$
$$12x - 16 = 12x - 16$$
$$0 = 0$$

Since the statement 0 = 0 is true, the system has an infinite number of solutions.

42. $\begin{cases} x + y = 19 \\ x - y = -3 \end{cases}$

Add the equations to eliminate y and solve for x.

$$\begin{array}{r} x + y = 19 \\ x - y = -3 \\ \hline 2x \quad = 16 \\ x = 8 \end{array}$$

Now solve for y.

$$x + y = 19$$
$$8 + y = 19$$
$$y = 11$$

The solution of the system is (8, 11).

43. $\begin{cases} x - 3y = -11 \\ 4x + 5y = -10 \end{cases}$

Solve the first equation for x.

$$x - 3y = -11$$
$$x = 3y - 11$$

Substitute $3y - 11$ for x in the second equation and solve for y.

$$4x + 5y = -10$$
$$4(3y - 11) + 5y = -10$$
$$12y - 44 + 5y = -10$$
$$17y = 34$$
$$y = 2$$

Now solve for x.

$x = 3y - 11 = 3(2) - 11 = 6 - 11 = -5$

The solution of the system is (–5, 2).

44. $\begin{cases} -x - 15y = 44 \\ 2x + 3y = 20 \end{cases}$

Solve the first equation for x.

$$-x - 15y = 44$$
$$-15y - 44 = x$$

Substitute $-15y - 44$ for x in the second equation and solve for y.

$$2x + 3y = 20$$
$$2(-15y - 44) + 3y = 20$$
$$-30y - 88 + 3y = 20$$
$$-27y = 108$$
$$y = -4$$

Now solve for x.

$x = -15y - 44 = -15(-4) - 44 = 60 - 44 = 16$

The solution of the system is (16, –4).

45. $\begin{cases} 2x + y = 3 \\ 6x + 3y = 9 \end{cases}$

Solve the first equation for y.

$$2x + y = 3$$
$$y = -2x + 3$$

Substitute $-2x + 3$ for y in the second equation and solve for x.

$$6x + 3y = 9$$
$$6x + 3(-2x + 3) = 9$$
$$6x - 6x + 9 = 9$$
$$0 = 0$$

Since the statement 0 = 0 is true, the system has an infinite number of solutions.

46. $\begin{cases} -3x + y = 5 \\ -3x + y = -2 \end{cases}$

Solve the first equation for y.

$$-3x + y = 5$$
$$y = 3x + 5$$

Substitute $3x + 5$ for y in the second equation and solve for x.

$$-3x + y = -2$$
$$-3x + 3x + 5 = -2$$
$$5 = -2 \quad \text{False}$$

Since the statement 5 = –2 is false, the system has no solution.

47. Let x be the smaller number and y be the larger number.

$\begin{cases} x + y = 12 \\ 3x + y = 20 \end{cases}$

Solve the first equation for y.

$$x + y = 12$$
$$y = -x + 12$$

Substitute $-x + 12$ for y in the second equation and solve for x.

$$3x + y = 20$$
$$3x + (-x + 12) = 20$$
$$2x + 12 = 20$$
$$2x = 8$$
$$x = 4$$

Now solve for y.

$y = -x + 12 = -4 + 12 = 8$

The two numbers are 4 and 8.

48. Let x be the smaller number and y be the larger number.

$\begin{cases} x - y = -18 \\ 2x - y = -23 \end{cases}$

Solve the first equation for x.

Copyright © 2012 Pearson Education, Inc. Publishing as Prentice Hall.

$x - y = -18$
$x = y - 18$

Substitute $y - 18$ for x in the second equation and solve for y.

$$\begin{aligned} 2x - y &= -23 \\ 2(y-18) - y &= -23 \\ 2y - 36 - y &= -23 \\ y &= 13 \end{aligned}$$

Now solve for x.
$x = y - 18 = 13 - 18 = -5$
The two numbers are -5 and 13.

49. Let x be nickels and y be dimes.

$$\begin{cases} x + y = 65 \\ 0.05x + 0.1y = 5.30 \end{cases}$$

Solve the first equation for x.
$x + y = 65$
$x = -y + 65$

Substitute $-y + 65$ for x in the second equation and solve for y.

$$\begin{aligned} 0.05x + 0.1y &= 5.30 \\ 0.05(-y+65) + 0.1y &= 5.30 \\ -0.05y + 3.25 + 0.1y &= 5.30 \\ 0.05y &= 2.05 \\ y &= 41 \end{aligned}$$

Now solve for x.
$x = -y + 65 = -41 + 65 = 24$.
There are 24 nickels and 41 dimes.

50. Let x be the number of 13¢ stamps and y be the number of 22¢ stamps.

$$\begin{cases} x + y = 26 \\ 0.13x + 0.22y = 4.19 \end{cases}$$

Solve the first equation for x.
$x + y = 26$
$x = -y + 26$

Substitute $-y + 26$ for x in the second equation and solve for y.

$$\begin{aligned} 0.13x + 0.22y &= 4.19 \\ 0.13(-y+26) + 0.22y &= 4.19 \\ -0.13y + 3.38 + 0.22y &= 4.19 \\ 0.09y &= 0.81 \\ y &= 9 \end{aligned}$$

Now solve for x.
$x = -y + 26 = -9 + 26 = 17$
They purchased 17 13¢ stamps and 9 22¢ stamps.

Chapter 7 Test

1. False; a system of two linear equations can have no solutions, exactly one solution, or infinitely many solutions.

2. False; a solution has to be a solution to both equations to be a solution of the system.

3. True; when the resulting statement is false the system has no solution.

4. False; when $3x = 0 \rightarrow x = 0$; the system does have a solution.

5. First equation:

$$\begin{aligned} 2x - 3y &= 5 \\ 2(1) - 3(-1) &\stackrel{?}{=} 5 \\ 2 + 3 &\stackrel{?}{=} 5 \\ 5 &= 5 \quad \text{True} \end{aligned}$$

Second equation:

$$\begin{aligned} 6x + y &= 1 \\ 6(1) + (-1) &\stackrel{?}{=} 1 \\ 6 - 1 &\stackrel{?}{=} 1 \\ 5 &= 1 \quad \text{False} \end{aligned}$$

Since the statement $5 = 1$ is false, $(1, -1)$ is not a solution of the system.

6. $\begin{cases} 4x - 3y = 24 \\ 4x + 5y = -8 \end{cases}$

First equation:

$$\begin{aligned} 4x - 3y &= 24 \\ 4(3) - 3(-4) &\stackrel{?}{=} 24 \\ 12 + 12 &\stackrel{?}{=} 24 \\ 24 &= 24 \quad \text{True} \end{aligned}$$

Second equation:

$$\begin{aligned} 4x + 5y &= -8 \\ 4(3) + 5(-4) &\stackrel{?}{=} -8 \\ 12 - 20 &\stackrel{?}{=} -8 \\ -8 &= -8 \quad \text{True} \end{aligned}$$

$(3, -4)$ is a solution of the system.

Copyright © 2012 Pearson Education, Inc. Publishing as Prentice Hall.

7. $\begin{cases} x-y=2 \\ 3x-y=-2 \end{cases}$

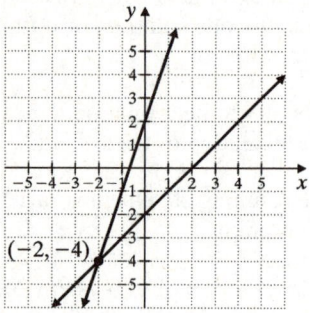

8. $\begin{cases} y=-3x \\ 3x+y=6 \end{cases}$

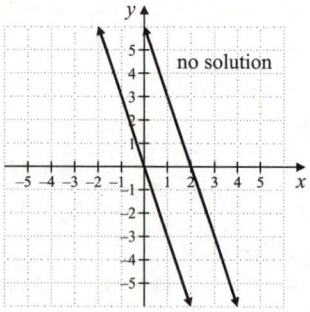

9. $\begin{cases} 3x-2y=-14 \\ y=x+5 \end{cases}$

Substitute $x + 5$ for y in the first equation and solve for x.

$$\begin{aligned} 3x-2y&=-14 \\ 3x-2(x+5)&=-14 \\ 3x-2x-10&=-14 \\ x-10&=-14 \\ x&=-4 \end{aligned}$$

$y = x + 5 = -4 + 5 = 1$

The solution of the system is (–4, 1).

10. $\begin{cases} \frac{1}{2}x+2y=-\frac{15}{4} \\ 4x=-y \end{cases}$

Multiply the second equation by –1.

$(-1)4x=(-1)(-y)$

$-4x = y$

Substitute $-4x$ for y in the first equation.

$$\begin{aligned} \frac{1}{2}x+2y&=-\frac{15}{4} \\ \frac{1}{2}x+2(-4x)&=-\frac{15}{4} \\ \frac{1}{2}x-8x&=-\frac{15}{4} \\ 4\left(\frac{1}{2}x-8x\right)&=4\left(-\frac{15}{4}\right) \\ 2x-32x&=-15 \\ -30x&=-15 \\ x&=\frac{15}{30}=\frac{1}{2} \end{aligned}$$

Now solve for y.

$y=-4x=-4\left(\frac{1}{2}\right)=-2$

The solution of the system is $\left(\frac{1}{2}, -2\right)$.

11. $\begin{cases} x+y=28 \\ x-y=12 \end{cases}$

Add the equations to eliminate y.

$$\begin{array}{r} x+y=28 \\ x-y=12 \\ \hline 2x \quad =40 \\ x=20 \end{array}$$

Now solve for y.

$$\begin{aligned} x+y&=28 \\ 20+y&=28 \\ y&=8 \end{aligned}$$

The solution of the system is (20, 8).

12. $\begin{cases} 4x-6y=7 \\ -2x+3y=0 \end{cases}$

Multiply the second equation by 2.

$\begin{cases} 4x-6y=7 \\ 2(-2x+3y)=2(0) \end{cases} \rightarrow \begin{cases} 4x-6y=7 \\ -4x+6y=0 \end{cases}$

Add the equations to eliminate x.

$$\begin{array}{r} 4x-6y=7 \\ -4x+6y=0 \\ \hline 0=7 \end{array}$$

Since the statement 0 = 7 is false, the system has no solution.

13. $\begin{cases} 3x+y=7 \\ 4x+3y=1 \end{cases}$

Solve the first equation for y.

$$\begin{aligned} 3x+y&=7 \\ y&=7-3x \end{aligned}$$

Substitute $7 - 3x$ for y in the second equation

Copyright © 2012 Pearson Education, Inc. Publishing as Prentice Hall.

and solve for x.

$$\begin{aligned} 4x+3y&=1\\ 4x+3(7-3x)&=1\\ 4x+21-9x&=1\\ 21-5x&=1\\ -5x&=-20\\ x&=4 \end{aligned}$$

$y = 7 - 3x = 7 - 3(4) = 7 - 12 = -5$

The solution of the system is (4, –5).

14. $\begin{cases} 3(2x+y)=4x+20\\ x-2y=3 \end{cases}$

Simplify the first equation.

$$\begin{aligned} 3(2x+y)&=4x+20\\ 6x+3y&=4x+20\\ 2x+3y&=20 \end{aligned}$$

$\begin{cases} 2x+3y=20\\ x-2y=3 \end{cases}$

Multiply the second equation by –2.

$\begin{cases} 2x+3y=20\\ -2(x-2y)=-2(3) \end{cases} \rightarrow \begin{cases} 2x+3y=20\\ -2x+4y=-6 \end{cases}$

Add the equations to eliminate x; then solve for y.

$$\begin{array}{r} 2x+3y=20\\ -2x+4y=-6\\ \hline 7y=14\\ y=2 \end{array}$$

Now solve for x.

$$\begin{aligned} x-2y&=3\\ x-2(2)&=3\\ x-4&=3\\ x&=7 \end{aligned}$$

The solution of the system is (7, 2).

15. $\begin{cases} \dfrac{x-3}{2}=\dfrac{2-y}{4}\\ \dfrac{7-2x}{3}=\dfrac{y}{2} \end{cases}$

Multiply the first equation by 4 and the second equation by 6 to eliminate fractions and simplify.

$\begin{cases} 4\left(\dfrac{x-3}{2}\right)=4\left(\dfrac{2-y}{4}\right)\\ 6\left(\dfrac{7-2x}{3}\right)=6\left(\dfrac{y}{2}\right) \end{cases} \rightarrow \begin{cases} 2x-6=2-y\\ 14-4x=3y \end{cases}$

$\rightarrow \begin{cases} 2x+y=8\\ 4x+3y=14 \end{cases}$

Multiply the first revised equation by –3.

$\begin{cases} -3(2x+y)=-3(8)\\ 4x+3y=14 \end{cases} \rightarrow \begin{cases} -6x-3y=-24\\ 4x+3y=14 \end{cases}$

Add the equations to eliminate y.

$$\begin{array}{r} -6x-3y=-24\\ 4x+3y=14\\ \hline -2x\qquad=-10\\ x=5 \end{array}$$

Now solve for y.

$$\begin{aligned} 2x+y&=8\\ 2(5)+y&=8\\ 10+y&=8\\ y&=-2 \end{aligned}$$

The solution of the system is (5, –2).

16. $\begin{cases} 8x-4y=12\\ y=2x-3 \end{cases}$

Substitute $2x - 3$ for y in the first equation and solve for x.

$$\begin{aligned} 8x-4y&=12\\ 8x-4(2x-3)&=12\\ 8x-8x+12&=12\\ 0&=0 \end{aligned}$$

Since the statement 0 = 0 is true, the system has an infinite number of solutions.

17. $\begin{cases} 0.01x-0.06y=-0.23\\ 0.2x+0.4y=0.2 \end{cases}$

Multiply the first equation by 100 and the second equation by 10 to eliminate decimals.

$\begin{cases} 100(0.01x-0.06y)=100(-0.23)\\ 10(0.2x+0.4y)=10(0.2) \end{cases}$

$\rightarrow \begin{cases} x-6y=-23\\ 2x+4y=2 \end{cases}$

Multiply the first equation by –2.

$\begin{cases} -2(x-6y)=-2(-23)\\ 2x+4y=2 \end{cases} \rightarrow \begin{cases} -2x+12y=46\\ 2x+4y=2 \end{cases}$

Add the equations to eliminate x; then solve for y.

$$\begin{array}{r} -2x+12y=46\\ 2x+4y=2\\ \hline 16y=48\\ y=3 \end{array}$$

Now solve for x.

$$\begin{aligned} x-6y&=-23\\ x-6(3)&=-23\\ x-18&=-23\\ x&=-5 \end{aligned}$$

The solution of the system is (–5, 3).

Copyright © 2012 Pearson Education, Inc. Publishing as Prentice Hall.

18. $\begin{cases} x-\frac{2}{3}y=3 \\ -2x+3y=10 \end{cases}$

Solve the first equation for x.

$$x-\frac{2}{3}y=3$$
$$x=\frac{2}{3}y+3$$

Substitute $\frac{2}{3}y+3$ for x in the second equation and solve for y.

$$-2x+3y=10$$
$$-2\left(\frac{2}{3}y+3\right)+3y=10$$
$$-\frac{4}{3}y-6+3y=10$$
$$-4y-18+9y=30$$
$$5y=48$$
$$y=\frac{48}{5}$$

Now solve for x.

$$x=\frac{2}{3}y+3=\frac{2}{3}\left(\frac{48}{5}\right)+3=\frac{32}{5}+\frac{15}{5}=\frac{47}{5}$$

The solution of the system is $\left(\frac{47}{5},\frac{48}{5}\right)$.

19. Let x be the first number and the y be the second.

$\begin{cases} x+y=124 \\ x-y=32 \end{cases}$

Add the equations to eliminate y then solve for x.

$$\begin{array}{r} x+y=124 \\ x-y=32 \\ \hline 2x \quad =156 \\ x=78 \end{array}$$

Now solve for y.

$$x+y=124$$
$$78+y=124$$
$$y=46$$

The numbers are 78 and 46.

20. Let x = number of cc of 12% solution and y = number of cc of 16% solution.

$\begin{cases} x+80=y \\ 0.12x+0.22(80)=0.16y \end{cases}$

Substitute $x + 80$ for y in the second equation; then solve for x.

$$0.12x+17.6=0.16y$$
$$0.12x+17.6=0.16(x+80)$$
$$0.12x+17.6=0.16x+12.8$$
$$4.8=0.04x$$
$$120=x$$

120 cc of the 12% saline is needed to make the 16% solution.

21. Let t be the number of farms in Texas and let m be the number of farms in Missouri.

$\begin{cases} t+m=336 \\ t=m+116 \end{cases}$

Substitute $m + 116$ for t in the first equation and solve for m.

$$t+m=336$$
$$m+116+m=336$$
$$2m+116=336$$
$$2m=220$$
$$m=110$$

$t = m + 116 = 110 + 116 = 226$

There are 226 thousand farms in Texas and 110 thousand in Missouri.

22. Let x be speed of one hiker and y be the speed of the other hiker.

$\begin{cases} 4x+4y=36 \\ y=2x \end{cases}$

Substitute $2x$ for y in the first equation; then solve for x.

$$4x+4y=36$$
$$4x+4(2x)=36$$
$$4x+8x=36$$
$$12x=36$$
$$x=3$$

Now solve for y.

$y = 2x = 2(3) = 6$

One hiker hikes at 3 mph and the other hikes at 6 mph.

23. Purchases of country music were equal to purchases of rap/hip-hop music in 1999, between 2003 and 2004, between 2004 and 2005, and between 2005 and 2006.

24. There were more purchases of country music than of rap/hip-hop music in 1998, 2004, 2006, and 2007.

Cumulative Review Chapters 1–7

1. a. $-14 - 8 + 10 - (-6) = -6$

b. $1.6 - (-10.3) + (-5.6) = 6.3$

Copyright © 2012 Pearson Education, Inc. Publishing as Prentice Hall.

2. a. $5^2 = 25$

b. $2^5 = 32$

3. reciprocal of $22 = \frac{1}{22}$

4. opposite of $22 = -22$

5. reciprocal of $\frac{3}{16} = \frac{16}{3}$

6. opposite of $\frac{3}{16} = -\frac{3}{16}$

7. reciprocal of $-10 = -\frac{1}{10}$

8. opposite of $-10 = 10$

9. reciprocal of $-\frac{9}{13} = -\frac{13}{9}$

10. opposite of $-\frac{9}{13} = \frac{9}{13}$

11. reciprocal of $1.7 = \frac{1}{1.7}$

12. opposite of $1.7 = -1.7$

13. a.
$$\begin{aligned} x+y&=8\\ x+3&=8\\ x&=5 \end{aligned}$$

b.
$$\begin{aligned} x+y&=8\\ y&=8-x \end{aligned}$$

14.
$$\begin{aligned} 5(x-1)&=6x\\ 5x-5&=6x\\ -5&=x \end{aligned}$$

15.
$$\begin{aligned} -2(x-5)+10&=-3(x+2)+x\\ -2x+10+10&=-3x-6+x\\ -2x+20&=-2x-6\\ 0x&=-26\\ 0&=-26 \quad \text{False} \end{aligned}$$
Since the statement $0 = -26$ is false, there is no solution.

16.
$$\begin{aligned} 5(y-5)&=5y+10\\ 5y-25&=5y+10\\ 0y&=35\\ 0&=35 \quad \text{False} \end{aligned}$$
Since the statement $0 = 35$ is false, there is no solution.

17.
$$\begin{aligned} \frac{x}{2}-1&=\frac{2}{3}x-3\\ 6\left(\frac{x}{2}-1\right)&=6\left(\frac{2}{3}x-3\right)\\ 3x-6&=4x-18\\ 12&=x \end{aligned}$$

18.
$$\begin{aligned} 7(x-2)-6(x+1)&=20\\ 7x-14-6x-6&=20\\ x-20&=20\\ x&=40 \end{aligned}$$

19.
$$\begin{aligned} -5x+7&<2(x-3)\\ -5x+7&<2x-6\\ 13&<7x\\ \frac{13}{7}&<x \end{aligned}$$
$$\left\{x \,\middle|\, x>\frac{13}{7}\right\}$$

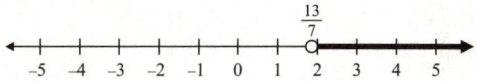

20. $P = a + b + c$ for b.
$b = P - a - c$

21. $\left(\frac{m}{n}\right)^7 = \frac{m^7}{n^7}$ where $n \neq 0$

22. $\frac{a^7b^{10}}{ab^{15}} = \frac{a^6}{b^5}$ where $b \neq 0$

23. $\left(\frac{2x^4}{3y^5}\right)^4 = \frac{2^4x^{4\cdot4}}{3^4y^{5\cdot4}} = \frac{16x^{16}}{81y^{20}}$ where $y \neq 0$

24. $(7a^2b^{-3})^2 = 7^2a^{2\cdot2}b^{-3\cdot2} = 49a^4b^{-6} = \frac{49a^4}{b^6}$

where $b \neq 0$.

25.
$$\begin{aligned} &(2x^3+8x^2-6x)-(2x^3-x^2+1)\\ &=2x^3+8x^2-6x-2x^3+x^2-1\\ &=9x^2-6x-1 \end{aligned}$$

Copyright © 2012 Pearson Education, Inc. Publishing as Prentice Hall.

26. $\left(5x^2+6x+\frac{1}{2}\right)+\left(x^2-\frac{4}{3}x-\frac{10}{21}\right)$
$=6x^2+\frac{14}{3}x+\frac{1}{42}$

27.
$$\begin{array}{r} 2x+4 \\ 3x-1\overline{\big)\,6x^2+10x-5} \\ \underline{6x^2-\ 2x} \\ 12x-5 \\ \underline{12x-4} \\ -1 \end{array}$$

$$\frac{6x^2+10x-1}{3x-1}=2x+4-\frac{1}{3x-1}$$

28. $9x^2=3\cdot3\cdot x\cdot x$
$6x^3=2\cdot3\cdot x\cdot x\cdot x$
$21x^5=3\cdot7\cdot x\cdot x\cdot x\cdot x\cdot x$
The common factors are 3 and $x\cdot x=x^2$, so the GCF is $3x^2$.

29.
$$\begin{aligned} x(2x-7)&=4 \\ 2x^2-7x&=4 \\ 2x^2-7x-4&=0 \\ (2x+1)(x-4)&=0 \end{aligned}$$
$2x+1=0$ or $x-4=0$
$2x=-1$ $\quad x=4$
$x=-\frac{1}{2}$

The solutions are $x=-\frac{1}{2}$ or $x=4$.

30. $x(2x-7)=0$
$x=0$ or $2x-7=0$
$2x=7$
$x=\frac{7}{2}$

The solutions are $x=0$ or $x=\frac{7}{2}$.

31. x = one leg
$x+2$ = other leg
$x+4$ = hypotenuse
$$\begin{aligned} x^2+(x+2)^2&=(x+4)^2 \\ x^2+x^2+4x+4&=x^2+8x+16 \\ x^2-4x-12&=0 \\ (x-6)(x+2)&=0 \end{aligned}$$
$x-6=0$ or $x+2=0$
$x=6$ $\quad x=-2$

Discard $x=-2$ because length cannot be negative.
one leg: $x=6$
other leg: $x+2=6+2=8$
hypotenuse: $x+4=6+4=10$
The lengths of the sides of the right triangle are 6, 8, and 10 units.

32. $A=bh$
Let x = base; then $h=3x+5$.
$$\begin{aligned} A&=x(3x+5)=182 \\ 3x^2+5x-182&=0 \\ (3x+26)(x-7)&=0 \end{aligned}$$
$3x+26=0$ or $x-7=0$
$3x=-26$ $\quad x=7$
$x=-\frac{26}{3}$

Discard $x=-\frac{26}{3}$ because length cannot be negative.
base = 7 ft
height = $3x+5=3(7)+5=21+5=26$ ft

33. $\frac{2y}{2y-7}-\frac{7}{2y-7}=\frac{2y-7}{2y-7}=1$

34.
$$\begin{aligned} \frac{2}{x-6}+\frac{3}{x+1}&=\frac{2(x+1)}{(x-6)(x+1)}+\frac{3(x-6)}{(x+1)(x-6)} \\ &=\frac{2x+2+3x-18}{(x-6)(x+1)} \\ &=\frac{5x-16}{(x-6)(x+1)} \end{aligned}$$

35. $\dfrac{\frac{x}{y}+\frac{3}{2x}}{\frac{x}{2}+y}=\dfrac{(2xy)\left(\frac{x}{y}+\frac{3}{2x}\right)}{(2xy)\left(\frac{x}{2}+y\right)}=\dfrac{2x^2+3y}{x^2y+2xy^2}$

36. $m=\frac{y_2-y_1}{x_2-x_1}=\frac{3-(-8)}{-1-2}=-\frac{11}{3}$

37. $y=mx+b$
If $y=-1$, then $m=0$.

38. Pick any 2 points on the line $x=2$.
Let $(x_1, y_1)=(2, 0)$ and $(x_2, y_2)=(2, 5)$.
$m=\frac{y_2-y_1}{x_2-x_1}=\frac{5-0}{2-2}=\frac{5}{0}=$ undefined

Copyright © 2012 Pearson Education, Inc. Publishing as Prentice Hall.

39. $m = \frac{y_2 - y_1}{x_2 - x_1} = \frac{4-5}{-3-2} = \frac{-1}{-5} = \frac{1}{5}$

$y - y_1 = m(x - x_1)$

$y - 5 = \frac{1}{5}(x - 2)$

$y - 5 = \frac{1}{5}x - \frac{2}{5}$

$5(y-5) = 5\left(\frac{1}{5}x - \frac{2}{5}\right)$

$5y - 25 = x - 2$

$-x + 5y = 23$

40. $y - y_1 = m(x - x_1)$

$y - 3 = -5(x - (-2))$

$y - 3 = -5x - 10$

$y = -5x - 7$

41. $\{(0, 2), (3, 3), (-1, 0), (3, -2)\}$

Domain: $\{-1, 0, 3\}$

Range: $\{-2, 0, 2, 3\}$

42. $f(x) = 5x^2 - 6$

$f(0) = 5(0)^2 - 6 = -6$

$f(-2) = 5(-2)^2 - 6 = 5(4) - 6 = 20 - 6 = 14$

43. $\begin{cases} 2x - 3y = 6 \\ x = 2y \end{cases}$

First equation:

$2x - 3y = 6$

$2(12) - 3(6) \stackrel{?}{=} 6$

$24 - 18 \stackrel{?}{=} 6$

$6 = 6$ True

Second equation:

$x = 2y$

$12 \stackrel{?}{=} 2(6)$

$12 = 12$ True

(12, 6) is a solution of the system.

44. a. $\begin{cases} 2x - y = 6 \\ 3x + 2y = -5 \end{cases}$

First equation:

$2(1) - (-4) \stackrel{?}{=} 6$

$2 + 4 \stackrel{?}{=} 6$

$6 = 6$ True

Second equation:

$3x + 2y = -5$

$3(1) + 2(-4) \stackrel{?}{=} -5$

$3 + (-8) \stackrel{?}{=} -5$

$-5 = -5$ True

(1, –4) is a solution of the system.

b. First equation:

$2x - y = 6$

$2(0) - (6) \stackrel{?}{=} 6$

$-6 = 6$ False

(0, 6) is not a solution of the system.

c. First equation:

$2x - y = 6$

$2(3) - 0 = 6$

$6 = 6$ True

Second equation:

$3x + 2y = -5$

$3(3) + 2(0) \stackrel{?}{=} -5$

$9 + 0 \stackrel{?}{=} -5$

$9 = -5$ False

(3, 0) is not a solution of the system.

45. $\begin{cases} x + 2y = 7 \\ 2x + 2y = 13 \end{cases}$

Multiply the first equation by –1.

$\begin{cases} -1(x + 2y) = -1(7) \\ 2x + 2y = 13 \end{cases} \rightarrow \begin{cases} -x - 2y = -7 \\ 2x + 2y = 13 \end{cases}$

Add the equations to eliminate *y*; then solve for *x*.

$$\begin{array}{r} -x - 2y = -7 \\ \underline{2x + 2y = 13} \\ x \quad\quad = 6 \end{array}$$

Now solve for *y*.

$x + 2y = 7$

$6 + 2y = 7$

$2y = 1$

$y = \frac{1}{2}$

The solution of the system is $\left(6, \frac{1}{2}\right)$.

Copyright © 2012 Pearson Education, Inc. Publishing as Prentice Hall.

46. $\begin{cases} 3x-4y=10 \\ y=2x \end{cases}$

Substitute $2x$ for y in the first equation and solve for x.

$$\begin{aligned} 3x-4y&=10 \\ 3x-4(2x)&=10 \\ 3x-8x&=10 \\ -5x&=10 \\ x&=-2 \end{aligned}$$

Now solve for y.

$y = 2x = 2(-2) = -4$

The solution of the system is $(-2, -4)$.

47. $\begin{cases} -x-\dfrac{y}{2}=\dfrac{5}{2} \\ \dfrac{x}{6}-\dfrac{y}{2}=0 \end{cases}$

Multiply the first equation by -6 and the second equation by 6.

$$\begin{cases} -6\left(-x-\dfrac{y}{2}\right)=-6\left(\dfrac{5}{2}\right) \\ 6\left(\dfrac{x}{6}-\dfrac{y}{2}\right)=6(0) \end{cases} \rightarrow \begin{cases} 6x+3y=-15 \\ x-3y=0 \end{cases}$$

Add the equations to eliminate y; then solve for x.

$$\begin{array}{r} 6x+3y=-15 \\ x-3y=0 \\ \hline 7x \quad =-15 \\ x=-\dfrac{15}{7} \end{array}$$

Now solve for y.

$$\begin{aligned} -x-\frac{y}{2}&=\frac{5}{2} \\ -\left(-\frac{15}{7}\right)-\frac{y}{2}&=\frac{5}{2} \\ 14\left(\frac{15}{7}-\frac{y}{2}\right)&=14\left(\frac{5}{2}\right) \\ 30-7y&=35 \\ -7y&=5 \\ y&=-\frac{5}{7} \end{aligned}$$

The solution of the system is $\left(-\dfrac{15}{7}, -\dfrac{5}{7}\right)$.

48. $\begin{cases} x=5y-3 \\ x=8y+4 \end{cases}$

Substitute $5y - 3$ for x in the second equation and solve for y.

$$\begin{aligned} x&=8y+4 \\ 5y-3&=8y+4 \\ -7&=3y \\ -\frac{7}{3}&=y \end{aligned}$$

Now solve for x.

$$x=5y-3=5\left(-\frac{7}{3}\right)-3=-\frac{35}{3}-\frac{9}{3}=-\frac{44}{3}$$

The solution of the system is $\left(-\dfrac{44}{3}, -\dfrac{7}{3}\right)$.

49. Let x be the first number and y be the second.

$\begin{cases} x+y=37 \\ x-y=21 \end{cases}$

Add the equations to eliminate y; then solve for x.

$$\begin{array}{r} x+y=37 \\ x-y=21 \\ \hline 2x \quad =58 \\ x=29 \end{array}$$

Now solve for y.

$$\begin{aligned} x+y&=37 \\ 29+y&=37 \\ y&=8 \end{aligned}$$

The two numbers are 29 and 8.

50. **a.** It is not a function because each x-coordinate (except 0) has more than one y-coordinate.

b. It is a function because each x-coordinate has exactly one y-coordinate.

c. It is not a function because all but two x-coordinates have more than one y-coordinate.

Copyright © 2012 Pearson Education, Inc. Publishing as Prentice Hall.

Chapter 8

Section 8.1 Practice

1. $\sqrt{100} = 10$, because $10^2 = 100$ and 10 is positive.

2. $-\sqrt{81} = -9$. The negative sign in front of the radical indicates the negative square root of 81.

3. $\sqrt{\frac{25}{81}} = \frac{5}{9}$, because $\left(\frac{5}{9}\right)^2 = \frac{25}{81}$ and $\frac{5}{9}$ is positive.

4. $\sqrt{1} = 1$, because $1^2 = 1$ and 1 is positive.

5. $\sqrt{0.81} = 0.9$ because $(0.9)^2 = 0.81$ and 0.9 is positive.

6. $\sqrt[3]{27} = 3$ because $3^3 = 27$.

7. $\sqrt[3]{-8} = -2$ because $(-2)^3 = -8$.

8. $\sqrt[3]{\frac{1}{64}} = \frac{1}{4}$ because $\left(\frac{1}{4}\right)^3 = \frac{1}{64}$.

9. $\sqrt[4]{-16}$ is not a real number since the index 4 is even and the radicand -16 is negative.

10. $\sqrt[5]{-1} = -1$ because $(-1)^5 = -1$.

11. $\sqrt[4]{256} = 4$ because $4^4 = 256$ and 4 is positive.

12. $\sqrt[6]{-1}$ is not a real number because the index 6 is even and the radicand -1 is negative.

13. To three decimal places, $\sqrt{22} \approx 4.690$.

14. $\sqrt{z^8} = z^4$ because $(z^4)^2 = z^8$.

15. $\sqrt{x^{20}} = x^{10}$ because $(x^{10})^2 = x^{20}$.

16. $\sqrt{4x^6} = 2x^3$ because $(2x^3)^2 = 4x^6$.

17. $\sqrt[3]{8y^{12}} = 2y^4$ because $(2y^4)^3 = 8y^{12}$.

18. $\sqrt[3]{-64x^9y^{24}} = -4x^3y^8$ because $(-4x^3y^8)^3 = -64x^9y^{24}$.

19. $\sqrt[3]{-64x^9y^{24}} = -4x^3y^8$ because $(-4x^3y^8)^3 = -64x^9y^{24}$.

Calculator Explorations

1. $\sqrt{6} \approx 2.449$; since 6 is between perfect squares 4 and 9, $\sqrt{6}$ is between $\sqrt{4} = 2$ and $\sqrt{9} = 3$.

2. $\sqrt{14} \approx 3.742$; since 14 is between perfect squares 9 and 16, $\sqrt{14}$ is between $\sqrt{9} = 3$ and $\sqrt{16} = 4$.

3. $\sqrt{11} \approx 3.317$; since 11 is between perfect squares 9 and 16, $\sqrt{11}$ is between $\sqrt{9} = 3$ and $\sqrt{16} = 4$.

4. $\sqrt{200} \approx 14.142$; since 200 is between perfect squares 196 and 225, $\sqrt{200}$ is between $\sqrt{196} = 14$ and $\sqrt{225} = 15$.

5. $\sqrt{82} \approx 9.055$; since 82 is between perfect squares 81 and 100, $\sqrt{82}$ is between $\sqrt{81} = 9$ and $\sqrt{100} = 10$.

6. $\sqrt{46} \approx 6.782$; since 46 is between perfect squares 36 and 49, $\sqrt{46}$ is between $\sqrt{36} = 6$ and $\sqrt{49} = 7$.

7. $\sqrt[3]{40} \approx 3.420$

8. $\sqrt[3]{71} \approx 4.141$

9. $\sqrt[4]{20} \approx 2.115$

10. $\sqrt[4]{15} \approx 1.968$

11. $\sqrt[5]{18} \approx 1.783$

12. $\sqrt[6]{2} \approx 1.122$

Copyright © 2012 Pearson Education, Inc. Publishing as Prentice Hall.

Vocabulary and Readiness Check

1. The symbol $\sqrt{}$ is used to denote the positive, or principal, square root.

2. In the expression $\sqrt[4]{16}$, the number 4 is called the index, the number 16 is called the radicand, and $\sqrt{}$ is called the radical sign.

3. The reverse operation of squaring a number is finding a square root of a number.

4. For a positive number a,
$-\sqrt{a}$ is the negative square root of a and
$\sqrt{a}$ is the positive square root of a.

5. An nth root of a number a is a number whose nth power is a.

6. $\sqrt{4} = 2$; the statement is false.

7. $\sqrt{-9}$ is not a real number; the statement is false.

8. $\sqrt{1000} \approx 31.623$; the statement is false.

9. True

10. True

Exercise Set 8.1

1. $\sqrt{16} = 4$, because $4^2 = 16$ and 4 is positive.

3. $\sqrt{\frac{1}{25}} = \frac{1}{5}$, because $\left(\frac{1}{5}\right)^2 = \frac{1}{25}$ and $\frac{1}{5}$ is positive.

5. $-\sqrt{100} = -10$. The negative sign indicates the negative square root of 100.

7. $\sqrt{-4}$ is not a real number, because there is no real number whose square is -4.

9. $-\sqrt{121} = -11$. The negative sign indicates the negative square root of 121.

11. $\sqrt{\frac{9}{25}} = \frac{3}{5}$, because $\left(\frac{3}{5}\right)^2 = \frac{9}{5}$ and $\frac{3}{5}$ is positive.

13. $\sqrt{900} = 30$, because $30^2 = 900$ and 30 is positive.

15. $\sqrt{144} = 12$, because $12^2 = 144$ and 12 is positive.

17. $\sqrt{\frac{1}{100}} = \frac{1}{10}$, because $\left(\frac{1}{10}\right)^2 = \frac{1}{100}$ and $\frac{1}{10}$ is positive.

19. $\sqrt{0.25} = 0.5$, because $0.5^2 = 0.25$ and 0.5 is positive.

21. $\sqrt[3]{125} = 5$, because $5^3 = 125$.

23. $\sqrt[3]{-64} = -4$, because $(-4)^3 = -64$.

25. $-\sqrt[3]{8} = -2$, because $2^3 = 8$.

27. $\sqrt[3]{\frac{1}{8}} = \frac{1}{2}$, because $\left(\frac{1}{2}\right)^3 = \frac{1}{8}$.

29. $\sqrt[3]{-125} = -5$, because $(-5)^3 = -125$.

31. $\sqrt[5]{32} = 2$, because $2^5 = 32$.

33. $\sqrt{81} = 9$, because $9^2 = 81$ and 9 is positive.

35. $\sqrt[4]{-16}$ is not a real number since the index 4 is even and the radicand -16 is negative.

37. $\sqrt[3]{-\frac{27}{64}} = -\frac{3}{4}$, because $\left(-\frac{3}{4}\right)^3 = -\frac{27}{64}$.

39. $-\sqrt[4]{625} = -5$, because $5^4 = 625$.

41. $\sqrt[6]{1} = 1$, because $1^6 = 1$ and 1 is positive.

43. $\sqrt{7} \approx 2.646$

45. $\sqrt{37} \approx 6.083$

47. $\sqrt{136} \approx 11.662$

49. $\sqrt{2} \approx 1.41$
$90\sqrt{2} \approx 90(1.41) = 126.90$
The distance from home plate to second base is approximately 126.90 feet.

Copyright © 2012 Pearson Education, Inc. Publishing as Prentice Hall.

51. $\sqrt{m^2} = m$, because $m^2 = m^2$.

53. $\sqrt{x^4} = x^2$, because $(x^2)^2 = x^4$.

55. $\sqrt{9x^8} = 3x^4$, because $(3x^4)^2 = 9x^8$.

57. $\sqrt{81x^2} = 9x$ because $(9x)^2 = 81x^2$.

59. $\sqrt{a^2b^4} = ab^2$, because $(ab^2)^2 = a^2b^4$.

61. $\sqrt{16a^6b^4} = 4a^3b^2$, because $(4a^3b^2)^2 = 16a^6b^4$

63. $\sqrt[3]{a^6b^{18}} = a^2b^6$, because $(a^2b^6)^3 = a^6b^{18}$.

65. $\sqrt[3]{-8x^3y^{27}} = -2xy^9$, because $(-2xy^9)^3 = -8x^3y^{27}$

67. $\sqrt{\dfrac{x^6}{36}} = \dfrac{x^3}{6}$, because $\left(\dfrac{x^3}{6}\right)^2 = \dfrac{x^6}{36}$.

69. $\sqrt{\dfrac{25y^2}{9}} = \dfrac{5y}{3}$, because $\left(\dfrac{5y}{3}\right)^2 = \dfrac{25y^2}{9}$.

71. $50 = 25 \cdot 2$

73. $32 = 16 \cdot 2$ or
$32 = 4 \cdot 8$

75. $28 = 4 \cdot 7$

77. $27 = 9 \cdot 3$

79. a. $\sqrt[7]{-1}$ is a real number because the index is odd.

b. $\sqrt[3]{-125}$ is a real number because the index is odd.

c. $\sqrt[6]{-128}$ is not a real number because the index is even and the radicand is negative.

d. $\sqrt[8]{-1}$ is not a real number because the index is even and the radicand is negative.

81. The length of the side is $\sqrt{49}$. Since $7^2 = 49$, $\sqrt{49} = 7$ and the sides of the square have length 7 miles.

83. The length of a side is $\sqrt{9.61}$ inches. Since $(3.1)^2 = 9.61$, $\sqrt{9.61} = 3.1$. The length of a side is 3.1 inches.

85. $\sqrt{\sqrt{81}} = \sqrt{9} = 3$, since $3^2 = 9$ and $9^2 = 81$.

87. $\sqrt{\sqrt{10{,}000}} = 10$ since $10^2 = 100$ and $100^2 = 10{,}000$.

89. Since $\sqrt{18}$ is between $\sqrt{16}$ and $\sqrt{25}$, then $\sqrt{18}$ is between 4 and 5.

91. Since $\sqrt{80}$ is between $\sqrt{64}$ and $\sqrt{81}$, then $\sqrt{80}$ is between 8 and 9.

93. $T = 2\pi\sqrt{\dfrac{L}{g}} = 2\pi\sqrt{\dfrac{30}{32}} \approx 2(3.14)(0.968) \approx 6.1$
The period of the pendulum is 6.1 seconds.

95. answers may vary

97.

x	$y = \sqrt{x}$
0	$\sqrt{0} = 0$
1	$\sqrt{1} = 1$
3	$\sqrt{3} \approx 1.7$
4	$\sqrt{4} = 2$
9	$\sqrt{9} = 3$

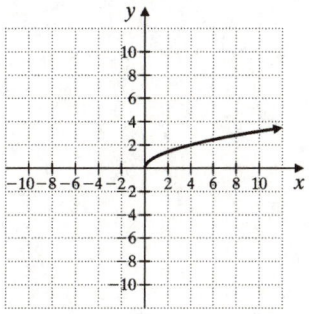

Copyright © 2012 Pearson Education, Inc. Publishing as Prentice Hall.

99. $\sqrt{x^2} = |x|$

101. $\sqrt{(x+2)^2} = |x+2|$

103. The graph of $y = \sqrt{x-2}$ 'starts' at (2, 0); answers may vary.

105. The graph of $y = \sqrt{x+4}$ 'starts' at (–4, 0); answers may vary.

Section 8.2 Practice

1. $\sqrt{40} = \sqrt{4 \cdot 10} = \sqrt{4} \cdot \sqrt{10} = 2\sqrt{10}$

2. $\sqrt{18} = \sqrt{9 \cdot 2} = \sqrt{9} \cdot \sqrt{2} = 3\sqrt{2}$

3. $\sqrt{500} = \sqrt{100 \cdot 5} = \sqrt{100} \cdot \sqrt{5} = 10\sqrt{5}$

4. $\sqrt{15}$
The radicand 15 contains no perfect square factors other than 1. Thus $\sqrt{15}$ is in simplest form.

5. $$\begin{aligned} 7\sqrt{75} &= 7 \cdot \sqrt{25 \cdot 3} \\ &= 7 \cdot \sqrt{25} \cdot \sqrt{3} \\ &= 7 \cdot 5 \cdot \sqrt{3} \\ &= 35\sqrt{3} \end{aligned}$$

6. $\sqrt{\frac{16}{81}} = \frac{\sqrt{16}}{\sqrt{81}} = \frac{4}{9}$

7. $\sqrt{\frac{2}{25}} = \frac{\sqrt{2}}{\sqrt{25}} = \frac{\sqrt{2}}{5}$

8. $\sqrt{\frac{45}{49}} = \frac{\sqrt{45}}{\sqrt{49}} = \frac{\sqrt{9} \cdot \sqrt{5}}{7} = \frac{3\sqrt{5}}{7}$

9. $\sqrt{x^{11}} = \sqrt{x^{10} \cdot x} = \sqrt{x^{10}} \cdot \sqrt{x} = x^5\sqrt{x}$

10. $$\begin{aligned} \sqrt{18x^4} &= \sqrt{9 \cdot 2 \cdot x^4} \\ &= \sqrt{9x^4 \cdot 2} \\ &= \sqrt{9x^4} \cdot \sqrt{2} \\ &= 3x^2\sqrt{2} \end{aligned}$$

11. $\sqrt{\frac{27}{x^8}} = \frac{\sqrt{27}}{\sqrt{x^8}} = \frac{\sqrt{9 \cdot 3}}{x^4} = \frac{\sqrt{9} \cdot \sqrt{3}}{x^4} = \frac{3\sqrt{3}}{x^4}$

12. $$\begin{aligned} \sqrt{\frac{7y^7}{25}} &= \frac{\sqrt{7y^7}}{\sqrt{25}} \\ &= \frac{\sqrt{y^6 \cdot 7y}}{5} \\ &= \frac{\sqrt{y^6} \cdot \sqrt{7}}{5} \\ &= \frac{y^3\sqrt{7y}}{5} \end{aligned}$$

13. $\sqrt[3]{88} = \sqrt[3]{8 \cdot 11} = \sqrt[3]{8} \cdot \sqrt[3]{11} = 2\sqrt[3]{11}$

14. $\sqrt[3]{50}$
The number 50 contains no perfect cube factors, so $\sqrt[3]{50}$ cannot be simplified further.

15. $\sqrt[3]{\frac{10}{27}} = \frac{\sqrt[3]{10}}{\sqrt[3]{27}} = \frac{\sqrt[3]{10}}{3}$

16. $\sqrt[3]{\frac{81}{8}} = \frac{\sqrt[3]{81}}{\sqrt[3]{8}} = \frac{\sqrt[3]{27 \cdot 3}}{2} = \frac{\sqrt[3]{27} \cdot \sqrt[3]{3}}{2} = \frac{3\sqrt[3]{3}}{2}$

Vocabulary and Readiness Check

1. If $\sqrt{a}$ and $\sqrt{b}$ are real numbers, then $\sqrt{a \cdot b} = \underline{\sqrt{a} \cdot \sqrt{b}}$.

2. If $\sqrt{a}$ and $\sqrt{b}$ are real numbers, then $\sqrt{\frac{a}{b}} = \underline{\frac{\sqrt{a}}{\sqrt{b}}}$.

3. $\sqrt{16 \cdot 25} = \sqrt{\underline{16}} \cdot \sqrt{\underline{25}} = \underline{4} \cdot \underline{5} = \underline{20}$

4. $\sqrt{36 \cdot 3} = \sqrt{\underline{36}} \cdot \sqrt{\underline{3}} = \underline{6} \cdot \sqrt{\underline{3}} = \underline{6\sqrt{3}}$

5. $$\begin{aligned} \sqrt{48} &= 2\sqrt{12} \\ &= 2\sqrt{4 \cdot 3} \\ &= 2\sqrt{4} \cdot \sqrt{3} \\ &= 2 \cdot 2\sqrt{3} \\ &= 4\sqrt{3} \end{aligned}$$
The statement is false.

6. True; 5 has no perfect cube factors.

Exercise Set 8.2

1. $\sqrt{20} = \sqrt{4 \cdot 5} = \sqrt{4}\sqrt{5} = 2\sqrt{5}$

Copyright © 2012 Pearson Education, Inc. Publishing as Prentice Hall.

3. $\sqrt{50} = \sqrt{25 \cdot 2} = \sqrt{25} \cdot \sqrt{2} = 5\sqrt{2}$

5. $\sqrt{33}$ is in simplest form.

7. $\sqrt{98} = \sqrt{49 \cdot 2} = \sqrt{49} \cdot \sqrt{2} = 7\sqrt{2}$

9. $\sqrt{60} = \sqrt{4 \cdot 15} = \sqrt{4} \cdot \sqrt{15} = 2\sqrt{15}$

11. $\sqrt{180} = \sqrt{36 \cdot 5} = \sqrt{36} \cdot \sqrt{5} = 6\sqrt{5}$

13. $\sqrt{52} = \sqrt{4 \cdot 13} = \sqrt{4} \cdot \sqrt{13} = 2\sqrt{13}$

15. $3\sqrt{25} = 3 \cdot 5 = 15$

17. $7\sqrt{63} = 7\sqrt{9 \cdot 7} = 7 \cdot \sqrt{9} \cdot \sqrt{7} = 7 \cdot 3 \cdot \sqrt{7} = 21\sqrt{7}$

19.
$$\begin{aligned} -5\sqrt{27} &= -5 \cdot \sqrt{9 \cdot 3} \\ &= -5 \cdot \sqrt{9} \cdot \sqrt{3} \\ &= -5 \cdot 3 \cdot \sqrt{3} \\ &= -15\sqrt{3} \end{aligned}$$

21. $\sqrt{\frac{8}{25}} = \frac{\sqrt{8}}{\sqrt{25}} = \frac{\sqrt{4 \cdot 2}}{5} = \frac{\sqrt{4}\sqrt{2}}{5} = \frac{2\sqrt{2}}{5}$

23. $\sqrt{\frac{27}{121}} = \frac{\sqrt{27}}{\sqrt{121}} = \frac{\sqrt{9 \cdot 3}}{11} = \frac{\sqrt{9} \cdot \sqrt{3}}{11} = \frac{3\sqrt{3}}{11}$

25. $\sqrt{\frac{9}{4}} = \frac{\sqrt{9}}{\sqrt{4}} = \frac{3}{2}$

27. $\sqrt{\frac{125}{9}} = \frac{\sqrt{125}}{\sqrt{9}} = \frac{\sqrt{25 \cdot 5}}{3} = \frac{\sqrt{25} \cdot \sqrt{5}}{3} = \frac{5\sqrt{5}}{3}$

29. $\sqrt{\frac{11}{36}} = \frac{\sqrt{11}}{\sqrt{36}} = \frac{\sqrt{11}}{6}$

31.
$$\begin{aligned} -\sqrt{\frac{27}{144}} &= -\frac{\sqrt{27}}{\sqrt{144}} \\ &= -\frac{\sqrt{9 \cdot 3}}{12} \\ &= -\frac{\sqrt{9} \cdot \sqrt{3}}{12} \\ &= -\frac{3\sqrt{3}}{12} \\ &= -\frac{\sqrt{3}}{4} \end{aligned}$$

33. $\sqrt{x^7} = \sqrt{x^6 \cdot x} = \sqrt{x^6}\sqrt{x} = x^3\sqrt{x}$

35. $\sqrt{x^{13}} = \sqrt{x^{12} \cdot x} = \sqrt{x^{12}} \cdot \sqrt{x} = x^6\sqrt{x}$

37. $\sqrt{36a^3} = \sqrt{36a^2 \cdot a} = \sqrt{36a^2}\sqrt{a} = 6a\sqrt{a}$

39. $\sqrt{96x^4} = \sqrt{16x^4 \cdot 6} = \sqrt{16x^4} \cdot \sqrt{6} = 4x^2\sqrt{6}$

41. $\sqrt{\frac{12}{m^2}} = \frac{\sqrt{12}}{\sqrt{m^2}} = \frac{\sqrt{4 \cdot 3}}{m} = \frac{\sqrt{4}\sqrt{3}}{m} = \frac{2\sqrt{3}}{m}$

43. $\sqrt{\frac{9x}{y^{10}}} = \frac{\sqrt{9x}}{\sqrt{y^{10}}} = \frac{\sqrt{9}\sqrt{x}}{y^5} = \frac{3\sqrt{x}}{y^5}$

45. $\sqrt{\frac{88}{x^{12}}} = \frac{\sqrt{88}}{\sqrt{x^{12}}} = \frac{\sqrt{4 \cdot 22}}{x^6} = \frac{\sqrt{4}\sqrt{22}}{x^6} = \frac{2\sqrt{22}}{x^6}$

47. $8\sqrt{4} = 8 \cdot 2 = 16$

49. $\sqrt{\frac{36}{121}} = \frac{\sqrt{36}}{\sqrt{121}} = \frac{6}{11}$

51. $\sqrt{175} = \sqrt{25 \cdot 7} = \sqrt{25} \cdot \sqrt{7} = 5\sqrt{7}$

53. $\sqrt{\frac{20}{9}} = \frac{\sqrt{20}}{\sqrt{9}} = \frac{\sqrt{4 \cdot 5}}{3} = \frac{\sqrt{4}\sqrt{5}}{3} = \frac{2\sqrt{5}}{3}$

55. $\sqrt{24m^7} = \sqrt{4m^6 \cdot 6m} = \sqrt{4m^6} \cdot \sqrt{6m} = 2m^3\sqrt{6m}$

57.
$$\begin{aligned} \sqrt{\frac{23y^3}{4x^6}} &= \frac{\sqrt{23y^3}}{\sqrt{4x^6}} \\ &= \frac{\sqrt{y^2 \cdot 23y}}{2x^3} \\ &= \frac{\sqrt{y^2}\sqrt{23y}}{2x^3} \\ &= \frac{y\sqrt{23y}}{2x^3} \end{aligned}$$

59. $\sqrt[3]{24} = \sqrt[3]{8 \cdot 3} = \sqrt[3]{8} \cdot \sqrt[3]{3} = 2\sqrt[3]{3}$

61. $\sqrt[3]{250} = \sqrt[3]{125 \cdot 2} = \sqrt[3]{125} \cdot \sqrt[3]{2} = 5\sqrt[3]{2}$

63. $\sqrt[3]{\frac{5}{64}} = \frac{\sqrt[3]{5}}{\sqrt[3]{64}} = \frac{\sqrt[3]{5}}{4}$

Copyright © 2012 Pearson Education, Inc. Publishing as Prentice Hall.

65. $\sqrt[3]{\frac{23}{8}} = \frac{\sqrt[3]{23}}{\sqrt[3]{8}} = \frac{\sqrt[3]{23}}{2}$

67. $\sqrt[3]{\frac{15}{64}} = \frac{\sqrt[3]{15}}{\sqrt[3]{64}} = \frac{\sqrt[3]{15}}{4}$

69. $\sqrt[3]{80} = \sqrt[3]{8 \cdot 10} = \sqrt[3]{8} \cdot \sqrt[3]{10} = 2\sqrt[3]{10}$

71. $6x + 8x = 14x$

73. $(2x+3)(x-5) = 2x^2 - 10x + 3x - 15$
$= 2x^2 - 7x - 15$

75. $9y^2 - 9y^2 = 0$

77. $\sqrt{x^6 y^3} = \sqrt{x^6 y^2 y} = \sqrt{x^6} \cdot \sqrt{y^2} \cdot \sqrt{y} = x^3 y\sqrt{y}$

79. $\sqrt{98x^5 y^4} = \sqrt{49x^4 y^4 \cdot 2x}$
$= \sqrt{49x^4 y^4} \cdot \sqrt{2x}$
$= 7x^2 y^2 \sqrt{2x}$

81. $\sqrt[3]{-8x^6} = \sqrt[3]{-8} \cdot \sqrt[3]{x^6} = -2x^3$

83. $\sqrt[3]{80} = \sqrt[3]{8 \cdot 10} = \sqrt[3]{8} \cdot \sqrt[3]{10} = 2\sqrt[3]{10}$
Each side length is $2\sqrt[3]{10}$ inches.

85. answers may vary; possible answer: let $a = 1$ and $b = 1$ so $\sqrt{a^2 + b^2} = \sqrt{2} \neq a + b = 2$.

87. $\frac{\sqrt{6A}}{6} = \frac{\sqrt{6 \cdot 120}}{6}$
$= \frac{\sqrt{720}}{6}$
$= \frac{\sqrt{144 \cdot 5}}{6}$
$= \frac{\sqrt{144} \cdot \sqrt{5}}{6}$
$= \frac{12\sqrt{5}}{6}$
$= 2\sqrt{5}$
The length of a side is $2\sqrt{5}$ inches.

89. $\sqrt{31{,}329} = 177$
The roof of the Water Cube is 177 meters by 177 meters.

91. Use $\frac{\sqrt{6A}}{6}$ with $A = 30.375$.
$\frac{\sqrt{6 \cdot 30.375}}{6} = \frac{\sqrt{182.25}}{6} = \frac{13.5}{6} = 2.25$

93. $C = 100\sqrt[3]{n} + 700$
$= 100\sqrt[3]{1000} + 700$
$= 100 \cdot 10 + 700$
$= 1000 + 700$
$= 1700$
The cost is \$1700.

95. $h = 169$ and $w = 64$.
$B = \sqrt{\frac{hw}{3600}}$
$= \sqrt{\frac{169 \cdot 64}{3600}}$
$= \frac{\sqrt{169 \cdot 64}}{\sqrt{3600}}$
$= \frac{\sqrt{169}\sqrt{64}}{60}$
$= \frac{13 \cdot 8}{60}$
$= \frac{104}{60}$
$= \frac{26}{15} \approx 1.7$
The body surface area is about 1.7 square meters.

Section 8.3 Practice

1. $6\sqrt{11} + 9\sqrt{11} = (6+9)\sqrt{11} = 15\sqrt{11}$

2. $\sqrt{7} - 3\sqrt{7} = 1\sqrt{7} - 3\sqrt{7} = (1-3)\sqrt{7} = -2\sqrt{7}$

3. $\sqrt{2} + \sqrt{2} - \sqrt{15} = 1\sqrt{2} + 1\sqrt{2} - \sqrt{15}$
$= (1+1)\sqrt{2} - \sqrt{15}$
$= 2\sqrt{2} - \sqrt{15}$

4. $3\sqrt{3} - 3\sqrt{2}$ cannot be simplified further since the radicands are not the same.

5. $\sqrt{27} + \sqrt{75} = \sqrt{9 \cdot 3} + \sqrt{25 \cdot 3}$
$= \sqrt{9} \cdot \sqrt{3} + \sqrt{25} \cdot \sqrt{3}$
$= 3\sqrt{3} + 5\sqrt{3}$
$= 8\sqrt{3}$

Copyright © 2012 Pearson Education, Inc. Publishing as Prentice Hall.

6. $3\sqrt{20}-7\sqrt{45}=3\sqrt{4\cdot 5}-7\sqrt{9\cdot 5}$
$=3\sqrt{4}\cdot\sqrt{5}-7\sqrt{9}\cdot\sqrt{5}$
$=3\cdot 2\sqrt{5}-7\cdot 3\sqrt{5}$
$=6\sqrt{5}-21\sqrt{5}$
$=-15\sqrt{5}$

7. $\sqrt{36}-\sqrt{48}-4\sqrt{3}-\sqrt{9}=6-\sqrt{16\cdot 3}-4\sqrt{3}-3$
$=6-\sqrt{16}\cdot\sqrt{3}-4\sqrt{3}-3$
$=6-4\sqrt{3}-4\sqrt{3}-3$
$=3-8\sqrt{3}$

8. $\sqrt{9x^4}-\sqrt{36x^3}+\sqrt{x^3}$
$=3x^2-\sqrt{36x^2\cdot x}+\sqrt{x^2\cdot x}$
$=3x^2-\sqrt{36x^2}\cdot\sqrt{x}+\sqrt{x^2}\cdot\sqrt{x}$
$=3x^2-6x\sqrt{x}+x\sqrt{x}$
$=3x^2-5x\sqrt{x}$

9. $10\sqrt[3]{81p^6}-\sqrt[3]{24p^6}=10\sqrt[3]{27p^6\cdot 3}-\sqrt[3]{8p^6\cdot 3}$
$=10\sqrt[3]{27p^6}\cdot\sqrt[3]{3}-\sqrt[3]{8p^6}\cdot\sqrt[3]{3}$
$=10\cdot 3p^2\sqrt[3]{3}-2p^2\sqrt[3]{3}$
$=30p^2\sqrt[3]{3}-2p^2\sqrt[3]{3}$
$=28p^2\sqrt[3]{3}$

Vocabulary and Readiness Check

1. Radicals that have the same index and same radicand are called <u>like radicals</u>.

2. The expressions $7\sqrt[3]{2x}$ and $-\sqrt[3]{2x}$ are called <u>like radicals</u>.

3. $11\sqrt{2}+6\sqrt{2}=\underline{17\sqrt{2}}$; b.

4. $\sqrt{5}$ is the same as $\underline{1\sqrt{5}}$; b.

5. $\sqrt{5}+\sqrt{5}=\underline{2\sqrt{5}}$; c.

6. $9\sqrt{7}-\sqrt{7}=\underline{8\sqrt{7}}$; a.

Exercise Set 8.3

1. $4\sqrt{3}-8\sqrt{3}=(4-8)\sqrt{3}=-4\sqrt{3}$

3. $3\sqrt{6}+8\sqrt{6}-2\sqrt{6}-5=(3+8-2)\sqrt{6}-5$
$=9\sqrt{6}-5$

5. $6\sqrt{5}-5\sqrt{5}+\sqrt{2}=(6-5)\sqrt{5}+\sqrt{2}=\sqrt{5}+\sqrt{2}$

7. $2\sqrt{3}+5\sqrt{3}-\sqrt{2}=(2+5)\sqrt{3}-\sqrt{2}=7\sqrt{3}-\sqrt{2}$

9. $2\sqrt{2}-7\sqrt{2}-6=(2-7)\sqrt{2}-6=-5\sqrt{2}-6$

11. $\sqrt{12}+\sqrt{27}=\sqrt{4\cdot 3}+\sqrt{9\cdot 3}$
$=\sqrt{4}\sqrt{3}+\sqrt{9}\sqrt{3}$
$=2\sqrt{3}+3\sqrt{3}$
$=(2+3)\sqrt{3}$
$=5\sqrt{3}$

13. $\sqrt{45}+3\sqrt{20}=\sqrt{9\cdot 5}+3\sqrt{4\cdot 5}$
$=\sqrt{9}\sqrt{5}+3\sqrt{4}\sqrt{5}$
$=3\sqrt{5}+3\cdot 2\sqrt{5}$
$=3\sqrt{5}+6\sqrt{5}$
$=(3+6)\sqrt{5}$
$=9\sqrt{5}$

15. $2\sqrt{54}-\sqrt{20}+\sqrt{45}-\sqrt{24}$
$=2\sqrt{9\cdot 6}-\sqrt{4\cdot 5}+\sqrt{9\cdot 5}-\sqrt{4\cdot 6}$
$=2\sqrt{9}\sqrt{6}-\sqrt{4}\sqrt{5}+\sqrt{9}\sqrt{5}-\sqrt{4}\sqrt{6}$
$=2\cdot 3\sqrt{6}-2\sqrt{5}+3\sqrt{5}-2\sqrt{6}$
$=6\sqrt{6}-2\sqrt{5}+3\sqrt{5}-2\sqrt{6}$
$=(6-2)\sqrt{6}+(3-2)\sqrt{5}$
$=4\sqrt{6}+1\sqrt{5}$
$=4\sqrt{6}+\sqrt{5}$

17. $4x-3\sqrt{x^2}+\sqrt{x}=4x-3x+\sqrt{x}=x+\sqrt{x}$

19. $\sqrt{25x}+\sqrt{36x}-11\sqrt{x}=\sqrt{25}\sqrt{x}+\sqrt{36}\sqrt{x}-11\sqrt{x}$
$=5\sqrt{x}+6\sqrt{x}-11\sqrt{x}$
$=(5+6-11)\sqrt{x}$
$=0$

21. $\sqrt{\frac{5}{9}}+\sqrt{\frac{5}{81}}=\frac{\sqrt{5}}{\sqrt{9}}+\frac{\sqrt{5}}{\sqrt{81}}$
$=\frac{\sqrt{5}}{3}+\frac{\sqrt{5}}{9}$
$=\frac{3\sqrt{5}}{9}+\frac{\sqrt{5}}{9}$
$=\left(\frac{3}{9}+\frac{1}{9}\right)\sqrt{5}$
$=\frac{4}{9}\sqrt{5}$
$=\frac{4\sqrt{5}}{9}$

Copyright © 2012 Pearson Education, Inc. Publishing as Prentice Hall.

23. $\sqrt{\frac{3}{4}}-\sqrt{\frac{3}{64}}=\frac{\sqrt{3}}{\sqrt{4}}-\frac{\sqrt{3}}{\sqrt{64}}$
$=\frac{\sqrt{3}}{2}-\frac{\sqrt{3}}{8}$
$=\frac{4\sqrt{3}}{8}-\frac{\sqrt{3}}{8}$
$=\left(\frac{4}{8}-\frac{1}{8}\right)\sqrt{3}$
$=\frac{3}{8}\sqrt{3}$
$=\frac{3\sqrt{3}}{8}$

25. $12\sqrt{5}-\sqrt{5}-4\sqrt{5}=(12-1-4)\sqrt{5}=7\sqrt{5}$

27. $\sqrt{75}+\sqrt{48}=\sqrt{25\cdot 3}+\sqrt{16\cdot 3}$
$=\sqrt{25}\sqrt{3}+\sqrt{16}\sqrt{3}$
$=5\sqrt{3}+4\sqrt{3}$
$=(5+4)\sqrt{3}$
$=9\sqrt{3}$

29. $\sqrt{5}+\sqrt{15}$ is in simplest form.

31. $3\sqrt{x^3}-x\sqrt{4x}=3\sqrt{x^2\cdot x}-x\sqrt{4\cdot x}$
$=3\sqrt{x^2}\sqrt{x}-x\sqrt{4}\sqrt{x}$
$=3x\sqrt{x}-2x\sqrt{x}$
$=(3x-2x)\sqrt{x}$
$=x\sqrt{x}$

33. $\sqrt{8}+\sqrt{9}+\sqrt{18}+\sqrt{81}=\sqrt{4\cdot 2}+3+\sqrt{9\cdot 2}+9$
$=\sqrt{4}\sqrt{2}+3+\sqrt{9}\sqrt{2}+9$
$=2\sqrt{2}+3+3\sqrt{2}+9$
$=(2+3)\sqrt{2}+3+9$
$=5\sqrt{2}+12$

35. $4+8\sqrt{2}-9=8\sqrt{2}-5$

37. $2\sqrt{45}-2\sqrt{20}=2\sqrt{9\cdot 5}-2\sqrt{4\cdot 5}$
$=2\sqrt{9}\sqrt{5}-2\sqrt{4}\sqrt{5}$
$=2\cdot 3\sqrt{5}-2\cdot 2\sqrt{5}$
$=6\sqrt{5}-4\sqrt{5}$
$=(6-4)\sqrt{5}$
$=2\sqrt{5}$

39. $\sqrt{35}-\sqrt{140}=\sqrt{35}-\sqrt{4\cdot 35}$
$=\sqrt{35}-\sqrt{4}\sqrt{35}$
$=\sqrt{35}-2\sqrt{35}$
$=(1-2)\sqrt{35}$
$=-1\sqrt{35}$
$=-\sqrt{35}$

41. $6-2\sqrt{3}-\sqrt{3}=6+(-2-1)\sqrt{3}=6-3\sqrt{3}$

43. $3\sqrt{9x}+2\sqrt{x}=3\sqrt{9}\sqrt{x}+2\sqrt{x}$
$=3\cdot 3\sqrt{x}+2\sqrt{x}$
$=9\sqrt{x}+2\sqrt{x}$
$=(9+2)\sqrt{x}$
$=11\sqrt{x}$

45. $\sqrt{9x^2}+\sqrt{81x^2}-11\sqrt{x}=3x+9x-11\sqrt{x}$
$=12x-11\sqrt{x}$

47. $\sqrt{3x^3}+3x\sqrt{x}=\sqrt{x^2\cdot 3x}+3x\sqrt{x}$
$=\sqrt{x^2}\sqrt{3x}+3x\sqrt{x}$
$=x\sqrt{3x}+3x\sqrt{x}$

49. $\sqrt{32x^2}+\sqrt{32x^2}+\sqrt{4x^2}$
$=\sqrt{16x^2\cdot 2}+\sqrt{16x^2\cdot 2}+2x$
$=\sqrt{16x^2}\sqrt{2}+\sqrt{16x^2}\sqrt{2}+2x$
$=4x\sqrt{2}+4x\sqrt{2}+2x$
$=(4x+4x)\sqrt{2}+2x$
$=8x\sqrt{2}+2x$

51. $\sqrt{40x}+\sqrt{40x^4}-2\sqrt{10x}-\sqrt{5x^4}$
$=\sqrt{4\cdot 10x}+\sqrt{4x^4\cdot 10}-2\sqrt{10x}-\sqrt{x^4\cdot 5}$
$=\sqrt{4}\sqrt{10x}+\sqrt{4x^4}\sqrt{10}-2\sqrt{10x}-\sqrt{x^4}\sqrt{5}$
$=2\sqrt{10x}+2x^2\sqrt{10}-2\sqrt{10x}-x^2\sqrt{5}$
$=(2-2)\sqrt{10x}+2x^2\sqrt{10}-x^2\sqrt{5}$
$=0\sqrt{10x}+2x^2\sqrt{10}-x^2\sqrt{5}$
$=2x^2\sqrt{10}-x^2\sqrt{5}$

53. $2\sqrt[3]{9}+5\sqrt[3]{9}-\sqrt[3]{25}=(2+5)\sqrt[3]{9}-\sqrt[3]{25}$
$=7\sqrt[3]{9}-\sqrt[3]{25}$

55. $2\sqrt[3]{2}-7\sqrt[3]{2}-6=(2-7)\sqrt[3]{2}-6=-5\sqrt[3]{2}-6$

Copyright © 2012 Pearson Education, Inc. Publishing as Prentice Hall.

57. $\sqrt[3]{81}+\sqrt[3]{24}=\sqrt[3]{27\cdot 3}+\sqrt[3]{8\cdot 3}$
$=\sqrt[3]{27}\sqrt[3]{3}+\sqrt[3]{8}\sqrt[3]{3}$
$=3\sqrt[3]{3}+2\sqrt[3]{3}$
$=(3+2)\sqrt[3]{3}$
$=5\sqrt[3]{3}$

59. $\sqrt[3]{8}+\sqrt[3]{54}-5=2+\sqrt[3]{27}\sqrt[3]{2}-5=-3+3\sqrt[3]{2}$

61. $2\sqrt[3]{8x^3}+2\sqrt[3]{16x^3}=2\cdot 2x+2\sqrt[3]{8x^3\cdot 2}$
$=4x+2\sqrt[3]{8x^3}\sqrt[3]{2}$
$=4x+2\cdot 2x\sqrt[3]{2}$
$=4x+4x\sqrt[3]{2}$

63. $12\sqrt[3]{y^7}-y^2\sqrt[3]{8y}=12\sqrt[3]{y^6\cdot y}-y^2\sqrt[3]{8}\sqrt[3]{y}$
$=12\sqrt[3]{y^6}\sqrt[3]{y}-2y^2\sqrt[3]{y}$
$=12y^2\sqrt[3]{y}-2y^2\sqrt[3]{y}$
$=(12y^2-2y^2)\sqrt[3]{y}$
$=10y^2\sqrt[3]{y}$

65. $\sqrt{40x}+x\sqrt[3]{40}-2\sqrt{10x}-x\sqrt[3]{5}$
$=\sqrt{4\cdot 10x}+x\sqrt[3]{8\cdot 5}-2\sqrt{10x}-x\sqrt[3]{5}$
$=\sqrt{4}\sqrt{10x}+x\sqrt[3]{8}\sqrt[3]{5}-2\sqrt{10x}-x\sqrt[3]{5}$
$=2\sqrt{10x}+2x\sqrt[3]{5}-2\sqrt{10x}-x\sqrt[3]{5}$
$=(2-2)\sqrt{10x}+(2x-x)\sqrt[3]{5}$
$=x\sqrt[3]{5}$

67. $(x+6)^2=(x)^2+2(x)(6)+(6)^2=x^2+12x+36$

69. $(2x-1)^2=(2x)^2-2(2x)(1)+(1)^2=4x^2-4x+1$

71. answers may vary

73. $P=2l+2w$
$=2\cdot 3\sqrt{5}+2\cdot\sqrt{5}$
$=6\sqrt{5}+2\sqrt{5}$
$=(6+2)\sqrt{5}$
$=8\sqrt{5}$
The perimeter is $8\sqrt{5}$ inches.

75. Two triangular end pieces and two rectangular side panels are needed. Each side panel has area $8\cdot 3=24$ square feet.
$2\cdot 24+2\cdot\frac{3\sqrt{27}}{4}=48+\frac{3\sqrt{9\cdot 3}}{2}$
$=48+\frac{3\sqrt{9}\sqrt{3}}{2}$
$=48+\frac{3\cdot 3\sqrt{3}}{2}$
$=48+\frac{9\sqrt{3}}{2}$
The total area of wood needed is $\left(48+\frac{9\sqrt{3}}{2}\right)$ square feet.

77. The expression can be simplified.
$4\sqrt{2}+3\sqrt{2}=(4+3)\sqrt{2}=7\sqrt{2}$

79. The expression $6+7\sqrt{6}$ cannot be simplified.

81. The expression can be simplified.
$\sqrt{7}+\sqrt{7}+\sqrt{7}=(1+1+1)\sqrt{7}=3\sqrt{7}$

83. $\sqrt{\frac{x^3}{16}}-x\sqrt{\frac{9x}{25}}+\frac{\sqrt{81x^3}}{2}$
$=\frac{\sqrt{x^3}}{\sqrt{16}}-x\frac{\sqrt{9x}}{\sqrt{25}}+\frac{\sqrt{81x^3}}{2}$
$=\frac{\sqrt{x^2\cdot x}}{4}-x\frac{\sqrt{9\cdot x}}{5}+\frac{\sqrt{81x^2\cdot x}}{2}$
$=\frac{\sqrt{x^2}\sqrt{x}}{4}-x\frac{\sqrt{9}\sqrt{x}}{5}+\frac{\sqrt{81x^2}\sqrt{x}}{2}$
$=\frac{x\sqrt{x}}{4}-\frac{3x\sqrt{x}}{5}+\frac{9x\sqrt{x}}{2}$
$=\left(\frac{1}{4}-\frac{3}{5}+\frac{9}{2}\right)x\sqrt{x}$
$=\left(\frac{5}{20}-\frac{12}{20}+\frac{90}{20}\right)x\sqrt{x}$
$=\frac{83}{20}x\sqrt{x}$
$=\frac{83x\sqrt{x}}{20}$

Section 8.4 Practice

1. $\sqrt{5}\cdot\sqrt{2}=\sqrt{5\cdot 2}=\sqrt{10}$

2. $\sqrt{7}\cdot\sqrt{7}=\sqrt{7\cdot 7}=\sqrt{49}=7$

Copyright © 2012 Pearson Education, Inc. Publishing as Prentice Hall.

3. $\sqrt{6}\cdot\sqrt{3}=\sqrt{18}=\sqrt{9\cdot 2}=\sqrt{9}\cdot\sqrt{2}=3\sqrt{2}$

4. $\sqrt{10x}\cdot\sqrt{2x}=\sqrt{10x\cdot 2x}$
$=\sqrt{20x^2}$
$=\sqrt{4x^2\cdot 5}$
$=\sqrt{4x^2}\cdot\sqrt{5}$
$=2x\sqrt{5}$

5. a. $\sqrt{7}\left(\sqrt{7}-\sqrt{3}\right)=\sqrt{7}\cdot\sqrt{7}-\sqrt{7}\cdot\sqrt{3}=7-\sqrt{21}$

b. $\sqrt{5x}\left(\sqrt{x}-3\sqrt{5}\right)=\sqrt{5x}\cdot\sqrt{x}-\sqrt{5x}\cdot 3\sqrt{5}$
$=\sqrt{5x\cdot x}-3\sqrt{5x\cdot 5}$
$=\sqrt{5\cdot x^2}-3\sqrt{25\cdot x}$
$=\sqrt{5}\cdot\sqrt{x^2}-3\cdot\sqrt{25}\cdot\sqrt{x}$
$=x\sqrt{5}-3\cdot 5\cdot\sqrt{x}$
$=x\sqrt{5}-15\sqrt{x}$

c. $\left(\sqrt{x}+\sqrt{5}\right)\left(\sqrt{x}-\sqrt{3}\right)$
$=\sqrt{x}\cdot\sqrt{x}-\sqrt{x}\cdot\sqrt{3}+\sqrt{5}\cdot\sqrt{x}-\sqrt{5}\cdot\sqrt{3}$
$=x-\sqrt{3x}+\sqrt{5x}-\sqrt{15}$

6. a. $\left(\sqrt{3}+8\right)\left(\sqrt{3}-8\right)=\left(\sqrt{3}\right)^2-8^2$
$=3-64$
$=-61$

b. $\left(\sqrt{5x}+4\right)^2=\left(\sqrt{5x}\right)^2+2\left(\sqrt{5x}\right)(4)+(4)^2$
$=5x+8\sqrt{5x}+16$

7. $\dfrac{\sqrt{15}}{\sqrt{3}}=\sqrt{\dfrac{15}{3}}=\sqrt{5}$

8. $\dfrac{\sqrt{90}}{\sqrt{2}}=\sqrt{\dfrac{90}{2}}=\sqrt{45}=\sqrt{9\cdot 5}=\sqrt{9}\cdot\sqrt{5}=3\sqrt{5}$

9. $\dfrac{\sqrt{125x^3}}{\sqrt{5x}}=\sqrt{\dfrac{125x^3}{5x}}=\sqrt{25x^2}=5x$

10. $\dfrac{5}{\sqrt{3}}=\dfrac{5}{\sqrt{3}}\cdot\dfrac{\sqrt{3}}{\sqrt{3}}=\dfrac{5\cdot\sqrt{3}}{\sqrt{3}\cdot\sqrt{3}}=\dfrac{5\sqrt{3}}{3}$

11. $\dfrac{\sqrt{7}}{\sqrt{20}}=\dfrac{\sqrt{7}}{\sqrt{4\cdot 5}}$
$=\dfrac{\sqrt{7}}{2\sqrt{5}}\cdot\dfrac{\sqrt{5}}{\sqrt{5}}$
$=\dfrac{\sqrt{7}\cdot\sqrt{5}}{2\sqrt{5}\cdot\sqrt{5}}$
$=\dfrac{\sqrt{35}}{2\cdot 5}$
$=\dfrac{\sqrt{35}}{10}$

12. $\sqrt{\dfrac{2}{45x}}=\dfrac{\sqrt{2}}{\sqrt{45x}}$
$=\dfrac{\sqrt{2}}{\sqrt{9}\cdot\sqrt{5x}}$
$=\dfrac{\sqrt{2}}{3\sqrt{5x}}\cdot\dfrac{\sqrt{5x}}{\sqrt{5x}}$
$=\dfrac{\sqrt{2}\cdot\sqrt{5x}}{3\sqrt{5x}\cdot\sqrt{5x}}$
$=\dfrac{\sqrt{10x}}{3\cdot 5x}$
$=\dfrac{\sqrt{10x}}{15x}$

13. $\dfrac{3}{2+\sqrt{7}}=\dfrac{3\left(2-\sqrt{7}\right)}{\left(2+\sqrt{7}\right)\left(2-\sqrt{7}\right)}$
$=\dfrac{3\left(2-\sqrt{7}\right)}{2^2-\left(\sqrt{7}\right)^2}$
$=\dfrac{3\left(2-\sqrt{7}\right)}{4-7}$
$=\dfrac{3\left(2-\sqrt{7}\right)}{-3}$
$=-\dfrac{3\left(2-\sqrt{7}\right)}{3}$
$=-1\left(2-\sqrt{7}\right)$
$=-2+\sqrt{7}$

Copyright © 2012 Pearson Education, Inc. Publishing as Prentice Hall.

14. $\dfrac{\sqrt{2}+5}{\sqrt{2}-1}=\dfrac{\left(\sqrt{2}+5\right)\left(\sqrt{2}+1\right)}{\left(\sqrt{2}-1\right)\left(\sqrt{2}+1\right)}$

$=\dfrac{2+\sqrt{2}+5\sqrt{2}+5}{2-1}$

$=\dfrac{7+6\sqrt{2}}{1}$

$=7+6\sqrt{2}$

15. $\dfrac{7}{2-\sqrt{x}}=\dfrac{7\left(2+\sqrt{x}\right)}{\left(2-\sqrt{x}\right)\left(2+\sqrt{x}\right)}=\dfrac{7\left(2+\sqrt{x}\right)}{4-x}$

Vocabulary and Readiness Check

1. $\sqrt{7}\cdot\sqrt{3}=\underline{\sqrt{21}}$

2. $\sqrt{10}\cdot\sqrt{10}=\underline{\sqrt{100}\text{ or }10}$

3. $\dfrac{\sqrt{15}}{\sqrt{3}}=\underline{\sqrt{\dfrac{15}{3}}\text{ or }\sqrt{5}}$

4. The process of eliminating the radical in the denominator of a radical expression is called <u>rationalizing the denominator</u>.

5. The conjugate of $2+\sqrt{3}$ is $\underline{2-\sqrt{3}}$.

Exercise Set 8.4

1. $\sqrt{8}\cdot\sqrt{2}=\sqrt{16}=4$

3. $\sqrt{10}\cdot\sqrt{5}=\sqrt{50}=\sqrt{25\cdot 2}=\sqrt{25}\sqrt{2}=5\sqrt{2}$

5. $\left(\sqrt{6}\right)^2=6$

7. $\sqrt{2x}\cdot\sqrt{2x}=\left(\sqrt{2x}\right)^2=2x$

9. $\left(2\sqrt{5}\right)^2=\left(2\sqrt{5}\right)\left(2\sqrt{5}\right)=4\left(\sqrt{5}\right)^2=4\cdot 5=20$

11. $\left(6\sqrt{x}\right)^2=\left(6\sqrt{x}\right)\left(6\sqrt{x}\right)=36\left(\sqrt{x}\right)^2=36x$

13. $\sqrt{3x^5}\cdot\sqrt{6x}=\sqrt{3x^5\cdot 6x}$

$=\sqrt{18x^6}$

$=\sqrt{9x^6\cdot 2}$

$=\sqrt{9x^6}\sqrt{2}$

$=3x^3\sqrt{2}$

15. $\sqrt{2xy^2}\cdot\sqrt{8xy}=\sqrt{2xy^2\cdot 8xy}$

$=\sqrt{16x^2y^3}$

$=\sqrt{16x^2y^2\cdot y}$

$=\sqrt{16x^2y^2}\sqrt{y}$

$=4xy\sqrt{y}$

17. $\sqrt{6}\left(\sqrt{5}+\sqrt{7}\right)=\sqrt{6}\cdot\sqrt{5}+\sqrt{6}\cdot\sqrt{7}=\sqrt{30}+\sqrt{42}$

19. $\sqrt{10}\left(\sqrt{2}+\sqrt{5}\right)=\sqrt{10}\cdot\sqrt{2}+\sqrt{10}\cdot\sqrt{5}$

$=\sqrt{20}+\sqrt{50}$

$=\sqrt{4\cdot 5}+\sqrt{25\cdot 2}$

$=\sqrt{4}\sqrt{5}+\sqrt{25}\sqrt{2}$

$=2\sqrt{5}+5\sqrt{2}$

21. $\sqrt{7y}\left(\sqrt{y}-2\sqrt{7}\right)=\sqrt{7y}\cdot\sqrt{y}-\sqrt{7y}\cdot 2\sqrt{7}$

$=\sqrt{7y\cdot y}-2\sqrt{7y\cdot 7}$

$=\sqrt{7y^2}-2\sqrt{49y}$

$=\sqrt{y^2\cdot 7}-2\sqrt{49\cdot y}$

$=\sqrt{y^2}\sqrt{7}-2\sqrt{49}\sqrt{y}$

$=y\sqrt{7}-2\cdot 7\sqrt{y}$

$=y\sqrt{7}-14\sqrt{y}$

23. $\left(\sqrt{3}+6\right)\left(\sqrt{3}-6\right)=\left(\sqrt{3}\right)^2-6^2=3-36=-33$

25. $\left(\sqrt{3}+\sqrt{5}\right)\left(\sqrt{2}-\sqrt{5}\right)$

$=\sqrt{3}\cdot\sqrt{2}-\sqrt{3}\sqrt{5}+\sqrt{5}\cdot\sqrt{2}-\sqrt{5}\cdot\sqrt{5}$

$=\sqrt{6}-\sqrt{15}+\sqrt{10}-\sqrt{25}$

$=\sqrt{6}-\sqrt{15}+\sqrt{10}-5$

Copyright © 2012 Pearson Education, Inc. Publishing as Prentice Hall.

27. $\left(2\sqrt{11}+1\right)\left(\sqrt{11}-6\right)$
$=2\sqrt{11}\cdot\sqrt{11}-2\sqrt{11}\cdot 6+1\cdot\sqrt{11}-1\cdot 6$
$=2\cdot 11-12\sqrt{11}+\sqrt{11}-6$
$=22-11\sqrt{11}-6$
$=16-11\sqrt{11}$

29. $\left(\sqrt{x}+6\right)\left(\sqrt{x}-6\right)=\left(\sqrt{x}\right)^2-(6)^2=x-36$

31. $\left(\sqrt{x}-7\right)^2=\left(\sqrt{x}\right)^2-2\left(\sqrt{x}\right)(7)+(7)^2$
$=x-14\sqrt{x}+49$

33. $\left(\sqrt{6y}+1\right)^2=\left(\sqrt{6y}\right)^2+2\left(\sqrt{6y}\right)(1)+(1)^2$
$=6y+2\sqrt{6y}+1$

35. $\dfrac{\sqrt{32}}{\sqrt{2}}=\sqrt{\dfrac{32}{2}}=\sqrt{16}=4$

37. $\dfrac{\sqrt{21}}{\sqrt{3}}=\sqrt{\dfrac{21}{3}}=\sqrt{7}$

39. $\dfrac{\sqrt{90}}{\sqrt{5}}=\sqrt{\dfrac{90}{5}}=\sqrt{18}=\sqrt{9\cdot 2}=\sqrt{9}\sqrt{2}=3\sqrt{2}$

41. $\dfrac{\sqrt{75y^5}}{\sqrt{3y}}=\sqrt{\dfrac{75y^5}{3y}}=\sqrt{25y^4}=5y^2$

43. $\dfrac{\sqrt{150}}{\sqrt{2}}=\sqrt{\dfrac{150}{2}}=\sqrt{75}=\sqrt{25\cdot 3}=\sqrt{25}\sqrt{3}=5\sqrt{3}$

45. $\dfrac{\sqrt{72y^5}}{\sqrt{3y^3}}=\sqrt{\dfrac{72y^5}{3y^3}}$
$=\sqrt{24y^2}$
$=\sqrt{4y^2\cdot 6}$
$=\sqrt{4y^2}\sqrt{6}$
$=2y\sqrt{6}$

47. $\dfrac{\sqrt{24x^3y^4}}{\sqrt{2xy}}=\sqrt{\dfrac{24x^3y^4}{2xy}}$
$=\sqrt{12x^2y^3}$
$=\sqrt{4x^2y^2\cdot 3y}$
$=\sqrt{4x^2y^2}\sqrt{3y}$
$=2xy\sqrt{3y}$

49. $\dfrac{\sqrt{3}}{\sqrt{5}}=\dfrac{\sqrt{3}}{\sqrt{5}}\cdot\dfrac{\sqrt{5}}{\sqrt{5}}=\dfrac{\sqrt{15}}{5}$

51. $\dfrac{7}{\sqrt{2}}=\dfrac{7}{\sqrt{2}}\cdot\dfrac{\sqrt{2}}{\sqrt{2}}=\dfrac{7\sqrt{2}}{2}$

53. $\dfrac{1}{\sqrt{6y}}=\dfrac{1}{\sqrt{6y}}\cdot\dfrac{\sqrt{6y}}{\sqrt{6y}}=\dfrac{\sqrt{6y}}{6y}$

55. $\sqrt{\dfrac{5}{18}}=\dfrac{\sqrt{5}}{\sqrt{18}}$
$=\dfrac{\sqrt{5}}{\sqrt{9\cdot 2}}$
$=\dfrac{\sqrt{5}}{3\sqrt{2}}\cdot\dfrac{\sqrt{2}}{\sqrt{2}}$
$=\dfrac{\sqrt{5}\sqrt{2}}{3\sqrt{2}\cdot\sqrt{2}}$
$=\dfrac{\sqrt{10}}{3\cdot 2}$
$=\dfrac{\sqrt{10}}{6}$

57. $\sqrt{\dfrac{3}{x}}=\dfrac{\sqrt{3}}{\sqrt{x}}=\dfrac{\sqrt{3}}{\sqrt{x}}\cdot\dfrac{\sqrt{x}}{\sqrt{x}}=\dfrac{\sqrt{3x}}{x}$

59. $\sqrt{\dfrac{1}{8}}=\dfrac{\sqrt{1}}{\sqrt{8}}$
$=\dfrac{1}{\sqrt{4\cdot 2}}$
$=\dfrac{1}{2\sqrt{2}}$
$=\dfrac{1}{2\sqrt{2}}\cdot\dfrac{\sqrt{2}}{\sqrt{2}}$
$=\dfrac{\sqrt{2}}{2\cdot 2}$
$=\dfrac{\sqrt{2}}{4}$

Copyright © 2012 Pearson Education, Inc. Publishing as Prentice Hall.

61. $\sqrt{\frac{2}{15}} = \frac{\sqrt{2}}{\sqrt{15}} = \frac{\sqrt{2}}{\sqrt{15}} \cdot \frac{\sqrt{15}}{\sqrt{15}} = \frac{\sqrt{30}}{15}$

63. $\sqrt{\frac{3}{20}} = \frac{\sqrt{3}}{\sqrt{20}}$
$= \frac{\sqrt{3}}{\sqrt{4 \cdot 5}}$
$= \frac{\sqrt{3}}{2\sqrt{5}}$
$= \frac{\sqrt{3}}{2\sqrt{5}} \cdot \frac{\sqrt{5}}{\sqrt{5}}$
$= \frac{\sqrt{15}}{2 \cdot 5}$
$= \frac{\sqrt{15}}{10}$

65. $\frac{3x}{\sqrt{2x}} = \frac{3x}{\sqrt{2x}} \cdot \frac{\sqrt{2x}}{\sqrt{2x}} = \frac{3x\sqrt{2x}}{2x} = \frac{3\sqrt{2x}}{2}$

67. $\frac{8y}{\sqrt{5}} = \frac{8y}{\sqrt{5}} \cdot \frac{\sqrt{5}}{\sqrt{5}} = \frac{8y\sqrt{5}}{5}$

69. $\sqrt{\frac{x}{36y}} = \frac{\sqrt{x}}{\sqrt{36y}}$
$= \frac{\sqrt{x}}{\sqrt{36}\sqrt{y}}$
$= \frac{\sqrt{x}}{6\sqrt{y}}$
$= \frac{\sqrt{x}}{6\sqrt{y}} \cdot \frac{\sqrt{y}}{\sqrt{y}}$
$= \frac{\sqrt{xy}}{6 \cdot y}$
$= \frac{\sqrt{xy}}{6y}$

71. $\sqrt{\frac{y}{12x}} = \frac{\sqrt{y}}{\sqrt{12x}}$
$= \frac{\sqrt{y}}{\sqrt{4}\sqrt{3x}}$
$= \frac{\sqrt{y}}{2\sqrt{3x}}$
$= \frac{\sqrt{y}}{2\sqrt{3x}} \cdot \frac{\sqrt{3x}}{\sqrt{3x}}$
$= \frac{\sqrt{3xy}}{2 \cdot 3x}$
$= \frac{\sqrt{3xy}}{6x}$

73. $\frac{3}{\sqrt{2}+1} = \frac{3}{\sqrt{2}+1} \cdot \frac{\sqrt{2}-1}{\sqrt{2}-1}$
$= \frac{3\left(\sqrt{2}-1\right)}{\left(\sqrt{2}\right)^2 - 1^2}$
$= \frac{3\sqrt{2}-3}{2-1}$
$= \frac{3\sqrt{2}-3}{1}$
$= 3\sqrt{2}-3$

75. $\frac{4}{2-\sqrt{5}} = \frac{4}{2-\sqrt{5}} \cdot \frac{2+\sqrt{5}}{2+\sqrt{5}}$
$= \frac{4\left(2+\sqrt{5}\right)}{2^2 - \left(\sqrt{5}\right)^2}$
$= \frac{8+4\sqrt{5}}{4-5}$
$= \frac{8+4\sqrt{5}}{-1}$
$= -8-4\sqrt{5}$

77. $\frac{\sqrt{5}+1}{\sqrt{6}-\sqrt{5}} = \frac{\sqrt{5}+1}{\sqrt{6}-\sqrt{5}} \cdot \frac{\sqrt{6}+\sqrt{5}}{\sqrt{6}+\sqrt{5}}$
$= \frac{\left(\sqrt{5}+1\right)\left(\sqrt{6}+\sqrt{5}\right)}{\left(\sqrt{6}\right)^2 - \left(\sqrt{5}\right)^2}$
$= \frac{\sqrt{5}\sqrt{6}+\sqrt{5}\sqrt{5}+1\cdot\sqrt{6}+1\cdot\sqrt{5}}{6-5}$
$= \frac{\sqrt{30}+5+\sqrt{6}+\sqrt{5}}{1}$
$= \sqrt{30}+5+\sqrt{6}+\sqrt{5}$

Copyright © 2012 Pearson Education, Inc. Publishing as Prentice Hall.

79. $\dfrac{\sqrt{3}+1}{\sqrt{2}-1}=\dfrac{\sqrt{3}+1}{\sqrt{2}-1}\cdot\dfrac{\sqrt{2}+1}{\sqrt{2}+1}$

$=\dfrac{(\sqrt{3}+1)(\sqrt{2}+1)}{(\sqrt{2})^2-1}$

$=\dfrac{\sqrt{3}\sqrt{2}+\sqrt{3}\cdot 1+1\cdot\sqrt{2}+1^2}{2-1}$

$=\dfrac{\sqrt{6}+\sqrt{3}+\sqrt{2}+1}{1}$

$=\sqrt{6}+\sqrt{3}+\sqrt{2}+1$

81. $\dfrac{5}{2+\sqrt{x}}=\dfrac{5}{2+\sqrt{x}}\cdot\dfrac{2-\sqrt{x}}{2-\sqrt{x}}$

$=\dfrac{5(2-\sqrt{x})}{2^2-(\sqrt{x})^2}$

$=\dfrac{10-5\sqrt{x}}{4-x}$

83. $\dfrac{3}{\sqrt{x}-4}=\dfrac{3}{\sqrt{x}-4}\cdot\dfrac{\sqrt{x}+4}{\sqrt{x}+4}$

$=\dfrac{3(\sqrt{x}+4)}{(\sqrt{x})^2-(4)^2}$

$=\dfrac{3\sqrt{x}+12}{x-16}$

85. $x+5=7^2$

$x=49-5$

$x=44$

87. $4z^2+6z-12=(2z)^2$

$4z^2+6z-12=4z^2$

$6z-12=0$

$6z=12$

$z=2$

89. $9x^2+5x+4=(3x+1)^2$

$9x^2+5x+4=9x^2+6x+1$

$5x+4=6x+1$

$4=x+1$

$3=x$

91. Area = (length)(width)

$13\sqrt{2}\cdot 5\sqrt{6}=13\cdot 5\cdot\sqrt{2}\cdot\sqrt{6}$

$=65\sqrt{12}$

$=65\sqrt{4}\sqrt{3}$

$=65\cdot 2\sqrt{3}$

$=130\sqrt{3}$

The area is $130\sqrt{3}$ square meters.

93. $\sqrt{\dfrac{A}{\pi}}=\dfrac{\sqrt{A}}{\sqrt{\pi}}=\dfrac{\sqrt{A}}{\sqrt{\pi}}\cdot\dfrac{\sqrt{\pi}}{\sqrt{\pi}}=\dfrac{\sqrt{A\pi}}{\pi}$

95. $\sqrt{5}\cdot\sqrt{5}=(\sqrt{5})^2=5$

The statement is true.

97. $\sqrt{3x}\cdot\sqrt{3x}=(\sqrt{3x})^2=3x$

The statement is false.

99. $\sqrt{11}+\sqrt{2}$ cannot be simplified because the radicands are different. The statement is false.

101. answers may vary

103. answers may vary

105. $\dfrac{\sqrt{3}+1}{\sqrt{2}-1}=\dfrac{\sqrt{3}+1}{\sqrt{2}-1}\cdot\dfrac{\sqrt{3}-1}{\sqrt{3}-1}$

$=\dfrac{(\sqrt{3})^2-1^2}{(\sqrt{2}-1)(\sqrt{3}-1)}$

$=\dfrac{3-1}{\sqrt{2}\sqrt{3}-1\cdot\sqrt{2}-1\cdot\sqrt{3}+1\cdot 1}$

$=\dfrac{2}{\sqrt{6}-\sqrt{2}-\sqrt{3}+1}$

Integrated Review

1. $\sqrt{36}=6$, because $6^2=36$ and 6 is positive.

2. $\sqrt{48}=\sqrt{16\cdot 3}=\sqrt{16}\cdot\sqrt{3}=4\sqrt{3}$

3. $\sqrt{x^4}=x^2$, because $(x^2)^2=x^4$.

4. $\sqrt{y^7}=\sqrt{y^6\cdot y}=\sqrt{y^6}\sqrt{y}=y^3\sqrt{y}$

5. $\sqrt{16x^2}=4x$, because $(4x)^2=16x^2$.

Copyright © 2012 Pearson Education, Inc. Publishing as Prentice Hall.

6. $\sqrt{18x^{11}} = \sqrt{9x^{10} \cdot 2x} = \sqrt{9x^{10}}\sqrt{2x} = 3x^5\sqrt{2x}$

7. $\sqrt[3]{8} = 2$, because $2^3 = 8$.

8. $\sqrt[4]{81} = 3$, because $3^4 = 81$.

9. $\sqrt[3]{-27} = -3$, because $(-3)^3 = -27$.

10. $\sqrt{-4}$ is not a real number.

11. $\sqrt{\frac{11}{9}} = \frac{\sqrt{11}}{\sqrt{9}} = \frac{\sqrt{11}}{3}$

12. $\sqrt[3]{\frac{7}{64}} = \frac{\sqrt[3]{7}}{\sqrt[3]{64}} = \frac{\sqrt[3]{7}}{4}$

13. $-\sqrt{16} = -4$. The negative sign indicates the negative square root of 16.

14. $-\sqrt{25} = -5$. The negative sign indicates the negative square root of 25.

15. $\sqrt{\frac{9}{49}} = \frac{\sqrt{9}}{\sqrt{49}} = \frac{3}{7}$

16. $\sqrt{\frac{1}{64}} = \frac{\sqrt{1}}{\sqrt{64}} = \frac{1}{8}$

17. $\sqrt{a^8a^2} = \sqrt{a^8}\sqrt{b^2} = a^4b$

18. $\sqrt{x^{10}y^{20}} = \sqrt{x^{10}}\sqrt{y^{20}} = x^5y^{10}$

19. $\sqrt{25m^6} = \sqrt{25}\sqrt{m^6} = 5m^3$

20. $\sqrt{9n^{16}} = \sqrt{9}\sqrt{n^{16}} = 3n^8$

21. $5\sqrt{7} + \sqrt{7} = (5+1)\sqrt{7} = 6\sqrt{7}$

22. $\sqrt{50} - \sqrt{8} = \sqrt{25 \cdot 2} - \sqrt{4 \cdot 2}$
$= \sqrt{25}\sqrt{2} - \sqrt{4}\sqrt{2}$
$= 5\sqrt{2} - 2\sqrt{2}$
$= (5-2)\sqrt{2}$
$= 3\sqrt{2}$

23. $5\sqrt{2} - 5\sqrt{3}$ cannot be simplified.

24. $2\sqrt{x} + \sqrt{25x} - \sqrt{36x} + 3x$
$= 2\sqrt{x} + \sqrt{25}\sqrt{x} - \sqrt{36}\sqrt{x} + 3x$
$= 2\sqrt{x} + 5\sqrt{x} - 6\sqrt{x} + 3x$
$= (2+5-6)\sqrt{x} + 3x$
$= \sqrt{x} + 3x$

25. $\sqrt{2} \cdot \sqrt{15} = \sqrt{2 \cdot 15} = \sqrt{30}$

26. $\sqrt{3} \cdot \sqrt{3} = \sqrt{3 \cdot 3} = \sqrt{9} = 3$

27. $\left(2\sqrt{7}\right)^2 = \left(2\sqrt{7}\right)\left(2\sqrt{7}\right) = 4\left(\sqrt{7}\right)^2 = 4 \cdot 7 = 28$

28. $\left(3\sqrt{5}\right)^2 = \left(3\sqrt{5}\right)\left(3\sqrt{5}\right) = 9\left(\sqrt{5}\right)^2 = 9 \cdot 5 = 45$

29. $\sqrt{3}\left(\sqrt{11}+1\right) = \sqrt{3} \cdot \sqrt{11} + \sqrt{3} \cdot 1 = \sqrt{33} + \sqrt{3}$

30. $\sqrt{6}\left(\sqrt{3}-2\right) = \sqrt{6} \cdot \sqrt{3} - \sqrt{6} \cdot 2$
$= \sqrt{18} - 2\sqrt{6}$
$= \sqrt{9 \cdot 2} - 2\sqrt{6}$
$= 3\sqrt{2} - 2\sqrt{6}$

31. $\sqrt{8y}\sqrt{2y} = \sqrt{8y \cdot 2y} = \sqrt{16y^2} = 4y$

32. $\sqrt{15x^2} \cdot \sqrt{3x^2} = \sqrt{15x^2 \cdot 3x^2}$
$= \sqrt{45x^4}$
$= \sqrt{9x^4 \cdot 5}$
$= 3x^2\sqrt{5}$

33. $\left(\sqrt{x}-5\right)\left(\sqrt{x}+2\right) = \sqrt{x} \cdot \sqrt{x} + 2\sqrt{x} - 5\sqrt{x} - 5 \cdot 2$
$= x - 3\sqrt{x} - 10$

34. $\left(3+\sqrt{2}\right)^2 = (3)^2 + 2(3)\left(\sqrt{2}\right) + \left(\sqrt{2}\right)^2$
$= 9 + 6\sqrt{2} + 2$
$= 11 + 6\sqrt{2}$

35. $\frac{\sqrt{8}}{\sqrt{2}} = \sqrt{\frac{8}{2}} = \sqrt{4} = 2$

36. $\frac{\sqrt{45}}{\sqrt{15}} = \sqrt{\frac{45}{15}} = \sqrt{3}$

Copyright © 2012 Pearson Education, Inc. Publishing as Prentice Hall.

37. $\dfrac{\sqrt{24x^5}}{\sqrt{2x}} = \sqrt{\dfrac{24x^5}{2x}} = \sqrt{12x^4} = \sqrt{4x^4 \cdot 3} = 2x^2\sqrt{3}$

38. $\dfrac{\sqrt{75a^4b^5}}{\sqrt{5ab}} = \sqrt{\dfrac{75a^4b^5}{5ab}}$

$= \sqrt{15a^3b^4}$

$= \sqrt{a^2b^4 \cdot 15a}$

$= ab^2\sqrt{15a}$

39. $\sqrt{\dfrac{1}{6}} = \dfrac{\sqrt{1}}{\sqrt{6}} = \dfrac{1}{\sqrt{6}} = \dfrac{1}{\sqrt{6}} \cdot \dfrac{\sqrt{6}}{\sqrt{6}} = \dfrac{\sqrt{6}}{6}$

40. $\dfrac{x}{\sqrt{20}} = \dfrac{x}{\sqrt{4 \cdot 5}} = \dfrac{x}{2\sqrt{5}} = \dfrac{x}{2\sqrt{5}} \cdot \dfrac{\sqrt{5}}{\sqrt{5}} = \dfrac{x\sqrt{5}}{2 \cdot 5} = \dfrac{x\sqrt{5}}{10}$

41. $\dfrac{4}{\sqrt{6}+1} = \dfrac{4}{\sqrt{6}+1} \cdot \dfrac{\sqrt{6}-1}{\sqrt{6}-1}$

$= \dfrac{4\left(\sqrt{6}-1\right)}{\left(\sqrt{6}\right)^2 - 1^2}$

$= \dfrac{4\sqrt{6}-4}{6-1}$

$= \dfrac{4\sqrt{6}-4}{5}$

42. $\dfrac{\sqrt{2}+1}{\sqrt{x}-5} = \dfrac{\sqrt{2}+1}{\sqrt{x}-5} \cdot \dfrac{\sqrt{x}+5}{\sqrt{x}+5}$

$= \dfrac{\left(\sqrt{2}+1\right)\left(\sqrt{x}+5\right)}{\left(\sqrt{x}\right)^2 - 5^2}$

$= \dfrac{\sqrt{2}\sqrt{x}+5\sqrt{2}+1\sqrt{x}+1 \cdot 5}{x-25}$

$= \dfrac{\sqrt{2x}+5\sqrt{2}+\sqrt{x}+5}{x-25}$

Section 8.5 Practice

1. $\sqrt{x-2} = 7$

$\left(\sqrt{x-2}\right)^2 = 7^2$

$x-2 = 49$

$x = 51$

2. $\sqrt{6x-1} = \sqrt{x}$

$\left(\sqrt{6x-1}\right)^2 = \left(\sqrt{x}\right)^2$

$6x-1 = x$

$5x-1 = 0$

$5x = 1$

$x = \dfrac{1}{5}$

3. $\sqrt{x}+9 = 2$

$\sqrt{x} = -7$

$\sqrt{x}$ cannot equal -7. Thus, the equation has no solution.

4. $\sqrt{9y^2+2y-10} = 3y$

$\left(\sqrt{9y^2+2y-10}\right)^2 = (3y)^2$

$9y^2+2y-10 = 9y^2$

$2y-10 = 0$

$2y = 10$

$y = 5$

5. $\sqrt{x+1}-x = -5$

$\sqrt{x+1} = x-5$

$\left(\sqrt{x+1}\right)^2 = (x-5)^2$

$x+1 = x^2-10x+25$

$0 = x^2-11x+24$

$0 = (x-8)(x-3)$

$0 = x-8$ or $0 = x-3$

$8 = x$ $\quad$ $3 = x$

Replacing x with 3 results in a false statement. 3 is an extraneous solution. The only solution is 8.

6. $\sqrt{x}+3 = \sqrt{x+15}$

$\left(\sqrt{x}+3\right)^2 = \left(\sqrt{x+15}\right)^2$

$x+6\sqrt{x}+9 = x+15$

$6\sqrt{x} = 6$

$\sqrt{x} = 1$

$x = 1$

Exercise Set 8.5

1. $\sqrt{x} = 9$

$\left(\sqrt{x}\right)^2 = 9^2$

$x = 81$

Copyright © 2012 Pearson Education, Inc. Publishing as Prentice Hall.

3. $\sqrt{x+5}=2$

$\left(\sqrt{x+5}\right)^2=2^2$

$x+5=4$

$x=-1$

5. $\sqrt{x}-2=5$

$\sqrt{x}=7$

$\left(\sqrt{x}\right)^2=7^2$

$x=49$

7. $3\sqrt{x}+5=2$

$3\sqrt{x}=-3$

$\sqrt{x}=-1$

$\sqrt{x}$ cannot equal -1. Thus, the equation has no solution.

9. $\sqrt{x}=\sqrt{3x-8}$

$\left(\sqrt{x}\right)^2=\left(\sqrt{3x-8}\right)^2$

$x=3x-8$

$-2x=-8$

$x=4$

11. $\sqrt{4x-3}=\sqrt{x+3}$

$\left(\sqrt{4x-3}\right)^2=\left(\sqrt{x+3}\right)^2$

$4x-3=x+3$

$3x-3=3$

$3x=6$

$x=2$

13. $\sqrt{9x^2+2x-4}=3x$

$\left(\sqrt{9x^2+2x-4}\right)^2=(3x)^2$

$9x^2+2x-4=9x^2$

$2x-4=0$

$2x=4$

$x=2$

15. $\sqrt{x}=x-6$

$\left(\sqrt{x}\right)^2=(x-6)^2$

$x=x^2-12x+36$

$0=x^2-13x+36$

$0=(x-9)(x-4)$

$0=x-9$ or $0=x-4$

$9=x$ $\quad$ $4=x$

$x=4$ does not check, so the solution is $x=9$.

17. $\sqrt{x+7}=x+5$

$\left(\sqrt{x+7}\right)^2=(x+5)^2$

$x+7=x^2+10x+25$

$0=x^2+9x+18$

$0=(x+6)(x+3)$

$x+6=0$ or $x+3=0$

$x=-6$ $\quad$ $x=-3$

$x=-6$ does not check, so the solution is $x=-3$.

19. $\sqrt{3x+7}-x=3$

$\sqrt{3x+7}=x+3$

$\left(\sqrt{3x+7}\right)^2=(x+3)^2$

$3x+7=x^2+6x+9$

$0=x^2+3x+2$

$0=(x+2)(x+1)$

$x+2=0$ or $x+1=0$

$x=-2$ $\quad$ $x=-1$

21. $\sqrt{16x^2+2x+2}=4x$

$\left(\sqrt{16x^2+2x+2}\right)^2=(4x)^2$

$16x^2+2x+2=16x^2$

$2x+2=0$

$2x=-2$

$x=-1$

$x=-1$ does not check, so the equation has no solution.

Copyright © 2012 Pearson Education, Inc. Publishing as Prentice Hall.

23. $\sqrt{2x^2+6x+9}=3$

$\left(\sqrt{2x^2+6x+9}\right)^2=3^2$

$2x^2+6x+9=9$

$2x^2+6x=0$

$2x(x+3)=0$

$2x=0$ or $x+3=0$

$x=0$ $\quad x=-3$

25. $\sqrt{x-7}=\sqrt{x}-1$

$\left(\sqrt{x-7}\right)^2=\left(\sqrt{x}-1\right)^2$

$x-7=x-2\sqrt{x}+1$

$2\sqrt{x}=8$

$\sqrt{x}=4$

$\left(\sqrt{x}\right)^2=4^2$

$x=16$

27. $\sqrt{x}+2=\sqrt{x+24}$

$\left(\sqrt{x}+2\right)^2=\left(\sqrt{x+24}\right)^2$

$x+4\sqrt{x}+4=x+24$

$4\sqrt{x}=20$

$\sqrt{x}=5$

$\left(\sqrt{x}\right)^2=5^2$

$x=25$

29. $\sqrt{x+8}=\sqrt{x}+2$

$\left(\sqrt{x+8}\right)^2=\left(\sqrt{x}+2\right)^2$

$x+8=x+4\sqrt{x}+4$

$4=4\sqrt{x}$

$1=\sqrt{x}$

$1^2=\left(\sqrt{x}\right)^2$

$1=x$

31. $\sqrt{2x+6}=4$

$\left(\sqrt{2x+6}\right)^2=4^2$

$2x+6=16$

$2x=10$

$x=5$

33. $\sqrt{x+6}+1=3$

$\sqrt{x+6}=2$

$\left(\sqrt{x+6}\right)^2=2^2$

$x+6=4$

$x=-2$

35. $\sqrt{x+6}+5=3$

$\sqrt{x+6}=-2$

$\sqrt{x+6}$ cannot equal -2. Thus, the equation has no solution.

37. $\sqrt{16x^2-3x+6}=4x$

$\left(\sqrt{16x^2-3x+6}\right)^2=(4x)^2$

$16x^2-3x+6=16x^2$

$-3x+6=0$

$-3x=-6$

$x=2$

39. $-\sqrt{x}=-6$

$\sqrt{x}=6$

$\left(\sqrt{x}\right)^2=6^2$

$x=36$

41. $\sqrt{x+9}=\sqrt{x}-3$

$\left(\sqrt{x+9}\right)^2=\left(\sqrt{x}-3\right)^2$

$x+9=x-6\sqrt{x}+9$

$0=-6\sqrt{x}$

$0=\sqrt{x}$

$0^2=\left(\sqrt{x}\right)^2$

$0=x$

$x=0$ does not check, so the equation has no solution.

43. $\sqrt{2x+1}+3=5$

$\sqrt{2x+1}=2$

$\left(\sqrt{2x+1}\right)^2=2^2$

$2x+1=4$

$2x=3$

$x=\frac{3}{2}$

Copyright © 2012 Pearson Education, Inc. Publishing as Prentice Hall.

45. $\sqrt{x}+3=7$
$\sqrt{x}=4$
$\left(\sqrt{x}\right)^2=4^2$
$x=16$

47. $\sqrt{4x}=\sqrt{2x+6}$
$\left(\sqrt{4x}\right)^2=\left(\sqrt{2x+6}\right)^2$
$4x=2x+6$
$2x=6$
$x=3$

49. $\sqrt{2x+1}=x-7$
$\left(\sqrt{2x+1}\right)^2=(x-7)^2$
$2x+1=x^2-14x+49$
$0=x^2-16x+48$
$0=(x-4)(x-12)$
$x-4=0$ or $x-12=0$
$x=4$ $\quad$ $x=12$
$x=4$ does not check, so the only solution is $x=12$.

51. $x=\sqrt{2x-2}+1$
$x-1=\sqrt{2x-2}$
$(x-1)^2=\left(\sqrt{2x-2}\right)^2$
$x^2-2x+1=2x-2$
$x^2-4x+3=0$
$(x-3)(x-1)=0$
$x-3=0$ or $x-1=0$
$x=3$ $\quad$ $x=1$

53. $\sqrt{1-8x}-x=4$
$\sqrt{1-8x}=x+4$
$\left(\sqrt{1-8x}\right)^2=(x+4)^2$
$1-8x=x^2+8x+16$
$0=x^2+16x+15$
$0=(x+15)(x+1)$
$x+15=0$ or $x+1=0$
$x=-15$ $\quad$ $x=-1$
$x=-15$ does not check, so the solution is $x=-1$.

55. $3x-8=19$
$3x=27$
$x=9$

57. Let x be the width of the rectangle, then the length is $2x$.
$2(2x)+2x=24$
$4x+2x=24$
$6x=24$
$x=4$
$2x=2(4)=8$
The length of the rectangle is 8 inches.

59. $\sqrt{x-3}+3=\sqrt{3x+4}$
$\left(\sqrt{x-3}+3\right)^2=\left(\sqrt{3x+4}\right)^2$
$x-3+6\sqrt{x-3}+9=3x+4$
$6\sqrt{x-3}=2x-2$
$\left(6\sqrt{x-3}\right)^2=(2x-2)^2$
$36(x-3)=4x^2-8x+4$
$36x-108=4x^2-8x+4$
$0=4x^2-44x+112$
$0=4(x-7)(x-4)$
$x-7=0$ or $x-4=0$
$x=7$ $\quad$ $x=4$

61. answers may vary

63. a. For $V=20$, $b=\sqrt{\dfrac{20}{2}}=\sqrt{10}\approx 3.2$.

For $V=200$, $b=\sqrt{\dfrac{200}{2}}=\sqrt{100}=10$.

For $V=2000$, $b=\sqrt{\dfrac{2000}{2}}=\sqrt{1000}\approx 31.6$.

V	20	200	2000
b	3.2	10	31.6

b. No, the volume increases by a factor of $\sqrt{10}$.

65.

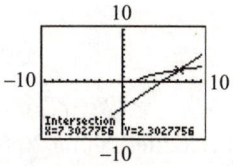

The solution of $\sqrt{x-2}=x-5$ is $x\approx 7.30$.

Copyright © 2012 Pearson Education, Inc. Publishing as Prentice Hall.

67.

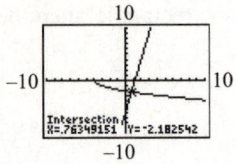

The solution of $-\sqrt{x+4}=5x-6$ is $x \approx 0.76$.

Section 8.6 Practice

1. $a^2+b^2=c^2$
$3^2+4^2=c^2$
$9+16=c^2$
$25=c^2$
$\sqrt{25}=c$
$5=c$

The hypotenuse has a length of 5 centimeters.

2. $a^2+b^2=c^2$
$5^2+b^2=10^2$
$25+b^2=100$
$b^2=75$
$b=\sqrt{75}=5\sqrt{3}\approx 8.66$

The length of the leg is exactly $5\sqrt{3}$ miles or approximately 8.66 miles.

3. The property owner is using a right triangle with one leg measuring 40 feet and hypotenuse measuring 65 feet to find the unknown distance.

$a^2+b^2=c^2$
$40^2+b^2=65^2$
$1600+b^2=4225$
$b^2=2625$
$b=\sqrt{2625}=5\sqrt{105}$

The distance across the pond is $5\sqrt{105}$ feet or approximately 51.2 feet.

4. $v=\sqrt{2gh}$
$=\sqrt{2\cdot 32\cdot 20}$
$=\sqrt{1280}$
$=16\sqrt{5}$

The velocity of the object after falling 20 feet is exactly $16\sqrt{5}$ feet per second or approximately 35.8 feet per second.

Exercise Set 8.6

1. $a^2+b^2=c^2$
$2^2+3^2=c^2$
$4+9=c^2$
$13=c^2$
$\sqrt{13}=\sqrt{c^2}$
$\sqrt{13}=c$
$c=\sqrt{13}\approx 3.61$

3. $a^2+b^2=c^2$
$3^2+b^2=6^2$
$9+b^2=36$
$b^2=27$
$\sqrt{b^2}=\sqrt{27}$
$b=3\sqrt{3}\approx 5.20$

5. $a^2+b^2=c^2$
$7^2+24^2=c^2$
$49+576=c^2$
$625=c^2$
$\sqrt{625}=\sqrt{c^2}$
$25=c$

7. $a^2+b^2=c^2$
$\left(\sqrt{3}\right)^2+b^2=5^2$
$3+b^2=25$
$b^2=22$
$\sqrt{b^2}=\sqrt{22}$
$b=\sqrt{22}\approx 4.69$

9. $a^2+b^2=c^2$
$4^2+b^2=13^2$
$16+b^2=169$
$b^2=153$
$\sqrt{b^2}=\sqrt{153}$
$b=3\sqrt{17}$
$b=3\sqrt{17}\approx 12.37$

Copyright © 2012 Pearson Education, Inc. Publishing as Prentice Hall.

11. $a^2+b^2=c^2$
$4^2+5^2=c^2$
$16+25=c^2$
$41=c^2$
$\sqrt{41}=\sqrt{c^2}$
$c=\sqrt{41}\approx 6.40$

13. $a^2+b^2=c^2$
$a^2+2^2=6^2$
$a^2+4=36$
$a^2=32$
$\sqrt{a^2}=\sqrt{32}$
$a=4\sqrt{2}$
$a=4\sqrt{2}\approx 5.66$

15. $a^2+b^2=c^2$
$\left(\sqrt{10}\right)^2+b^2=10^2$
$10+b^2=100$
$b^2=90$
$\sqrt{b^2}=\sqrt{90}$
$b=3\sqrt{10}\approx 9.49$

17. The pole, wire, and ground form a right triangle with legs of 5 feet and 20 feet.
$a^2+b^2=c^2$
$5^2+20^2=c^2$
$25+400=c^2$
$425=c^2$
$\sqrt{425}=\sqrt{c^2}$
$\sqrt{425}=c$
$c=\sqrt{425}\approx 20.6$
The length of the wire is 20.6 feet.

19. The diagonal brace is the hypotenuse of a right triangle with legs measuring 6 feet and 10 feet.
$a^2+b^2=c^2$
$6^2+10^2=c^2$
$36+100=c^2$
$136=c^2$
$\sqrt{136}=\sqrt{c^2}$
$c=\sqrt{136}\approx 11.7$
The brace needs to be 11.7 feet.

21. $b=\sqrt{\dfrac{3V}{h}}$
$6=\sqrt{\dfrac{3V}{2}}$
$6^2=\left(\sqrt{\dfrac{3V}{2}}\right)^2$
$36=\dfrac{3V}{2}$
$72=3V$
$24=V$
The volume is 24 cubic feet.

23. $s=\sqrt{30fd}$
$s=\sqrt{30\cdot 0.35\cdot 280}$
$s=\sqrt{2940}$
$s\approx 54$
The car was moving at a speed of 54 miles per hour.

25. $v=\sqrt{2.5r}$
$v=\sqrt{2.5(300)}$
$v=\sqrt{750}$
$v\approx 27.4$
The maximum safe speed is 27 miles per hour.

27. $d=3.5\sqrt{h}$
$d=3.5\sqrt{285.4}$
$d\approx 59.1$
You can see a distance of 59.1 kilometers.

29. $d=3.5\sqrt{h}$
$d=3.5\sqrt{295.7}$
$d\approx 60.2$
You can see a distance of 60.2 kilometers.

31. $\sqrt{9}=3$ and $-\sqrt{9}=-3$, so -3 and 3 are numbers whose square is 9.

33. $\sqrt{100}=10$ and $-\sqrt{100}=-10$, so -10 and 10 are numbers whose square is 100.

35. $\sqrt{64}=8$ and $-\sqrt{64}=-8$, so -8 and 8 are numbers whose square is 64.

Copyright © 2012 Pearson Education, Inc. Publishing as Prentice Hall.

37. First find y.

$$a^2+b^2=c^2$$
$$3^2+y^2=7^2$$
$$9+y^2=49$$
$$y^2=40$$
$$y=\sqrt{40}$$
$$y=\sqrt{4\cdot 10}$$
$$y=2\sqrt{10}$$

Let b be the second leg of the right triangle with hypotenuse 5.

$$a^2+b^2=c^2$$
$$3^2+b^2=5^2$$
$$9+b^2=25$$
$$b^2=16$$
$$\sqrt{b^2}=\sqrt{16}$$
$$b=4$$

Now find x.

$$x=y-4$$
$$x=2\sqrt{10}-4$$

39. The distance is the length of the hypotenuse of a right triangle. One leg has length $3\cdot 30=90$ miles, and the other leg has length $3\cdot 60=180$ miles.

$$a^2+b^2=c^2$$
$$90^2+180^2=c^2$$
$$8100+32{,}400=c^2$$
$$40{,}500=c^2$$
$$\sqrt{40{,}500}=\sqrt{c^2}$$
$$201\approx c$$

They are 201 miles apart.

41. answers may vary

Chapter 8 Vocabulary Check

1. The expressions $5\sqrt{x}$ and $7\sqrt{x}$ are examples of <u>like radicals</u>.

2. In the expression $\sqrt[3]{45}$ the number 3 is the <u>index</u>, the number 45 is the <u>radicand</u>, and $\sqrt{}$ is called the <u>radical</u> sign.

3. The <u>conjugate</u> of $a+b$ is $a-b$.

4. The <u>principal square root</u> of 25 is 5.

5. The process eliminating the radical in the denominator of a radical expression is called <u>rationalizing the denominator</u>.

6. The Pythagorean theorem states that for a right triangle, $(\text{leg})^2+(\text{leg})^2=(\underline{\text{hypotenuse}})^2$.

Chapter 8 Review

1. $\sqrt{81}=9$, because $9^2=81$ and 9 is positive.

2. $-\sqrt{49}=-7$. The negative indicates the negative square root of 49.

3. $\sqrt[3]{27}=3$, because $3^3=27$.

4. $\sqrt[4]{81}=3$, because $3^4=81$.

5. $-\sqrt{\dfrac{9}{64}}=-\dfrac{3}{8}$ because $\left(\dfrac{3}{8}\right)^2=\dfrac{9}{64}$.

6. $\sqrt{\dfrac{36}{81}}=\dfrac{6}{9}$ because $\left(\dfrac{6}{9}\right)^2=\dfrac{36}{81}$, and $\dfrac{6}{9}=\dfrac{2}{3}$.

7. $\sqrt[4]{16}=2$ because $2^4=16$.

8. $\sqrt[3]{-8}=-2$ because $(-2)^3=-8$.

9. c; $\sqrt{-4}$ is not a real number because the radicand is negative and the index is even.

10. a, c; $\sqrt{-5}$ and $\sqrt[4]{-5}$ are not real numbers because the radicands are negatives and the indexes are even.

11. $\sqrt{x^{12}}=x^6$, because $(x^6)^2=x^{12}$.

12. $\sqrt{x^8}=x^4$, because $(x^4)^2=x^8$.

13. $\sqrt{9y^2}=3y$, because $(3y)^2=9y^2$.

14. $\sqrt{25x^4}=5x^2$, because $(5x^2)^2=25x^4$.

15. $\sqrt{40}=\sqrt{4\cdot 10}=\sqrt{4}\sqrt{10}=2\sqrt{10}$

16. $\sqrt{24}=\sqrt{4\cdot 6}=\sqrt{4}\sqrt{6}=2\sqrt{6}$

Copyright © 2012 Pearson Education, Inc. Publishing as Prentice Hall.

17. $\sqrt{54}=\sqrt{9\cdot 6}=\sqrt{9}\sqrt{6}=3\sqrt{6}$

18. $\sqrt{88}=\sqrt{4\cdot 22}=\sqrt{4}\sqrt{22}=2\sqrt{22}$

19. $\sqrt{x^5}=\sqrt{x^4\cdot x}=\sqrt{x^4}\sqrt{x}=x^2\sqrt{x}$

20. $\sqrt{y^7}=\sqrt{y^6\cdot y}=\sqrt{y^6}\sqrt{y}=y^3\sqrt{y}$

21. $\sqrt{20x^2}=\sqrt{4x^2\cdot 5}=\sqrt{4x^2}\sqrt{5}=2x\sqrt{5}$

22. $\sqrt{50y^4}=\sqrt{25y^4\cdot 2}=\sqrt{25y^4}\sqrt{2}=5y^2\sqrt{2}$

23. $\sqrt[3]{54}=\sqrt[3]{27\cdot 2}=\sqrt[3]{27}\sqrt[3]{2}=3\sqrt[3]{2}$

24. $\sqrt[3]{88}=\sqrt[3]{8\cdot 11}=\sqrt[3]{8}\sqrt[3]{11}=2\sqrt[3]{11}$

25. $\sqrt{\frac{18}{25}}=\frac{\sqrt{18}}{\sqrt{25}}=\frac{\sqrt{9\cdot 2}}{5}=\frac{\sqrt{9}\sqrt{2}}{5}=\frac{3\sqrt{2}}{5}$

26. $\sqrt{\frac{75}{64}}=\frac{\sqrt{75}}{\sqrt{64}}=\frac{\sqrt{25\cdot 3}}{8}=\frac{\sqrt{25}\sqrt{3}}{8}=\frac{5\sqrt{3}}{8}$

27. $-\sqrt{\frac{50}{9}}=-\frac{\sqrt{50}}{\sqrt{9}}=-\frac{\sqrt{25\cdot 2}}{3}=-\frac{\sqrt{25}\sqrt{2}}{3}=-\frac{5\sqrt{2}}{3}$

28. $-\sqrt{\frac{12}{49}}=-\frac{\sqrt{12}}{\sqrt{49}}=-\frac{\sqrt{4\cdot 3}}{7}=-\frac{\sqrt{4}\sqrt{3}}{7}=-\frac{2\sqrt{3}}{7}$

29. $\sqrt{\frac{11}{x^2}}=\frac{\sqrt{11}}{\sqrt{x^2}}=\frac{\sqrt{11}}{x}$

30. $\sqrt{\frac{7}{y^4}}=\frac{\sqrt{7}}{\sqrt{y^4}}=\frac{\sqrt{7}}{y^2}$

31. $\sqrt{\frac{y^5}{100}}=\frac{\sqrt{y^5}}{\sqrt{100}}=\frac{\sqrt{y^4\cdot y}}{10}=\frac{\sqrt{y^4}\sqrt{y}}{10}=\frac{y^2\sqrt{y}}{10}$

32. $\sqrt{\frac{x^3}{81}}=\frac{\sqrt{x^3}}{\sqrt{81}}=\frac{\sqrt{x^2\cdot x}}{9}=\frac{\sqrt{x^2}\sqrt{x}}{9}=\frac{x\sqrt{x}}{9}$

33. $5\sqrt{8}-8\sqrt{2}=(5-8)\sqrt{2}=-3\sqrt{2}$

34. $\sqrt{3}-6\sqrt{3}=(1-6)\sqrt{3}=-5\sqrt{3}$

35. $6\sqrt{5}+3\sqrt{6}-2\sqrt{5}+\sqrt{6}=(6-2)\sqrt{5}+(3+1)\sqrt{6}$
$=4\sqrt{5}+4\sqrt{6}$

36. $-\sqrt{7}+8\sqrt{2}-\sqrt{7}-6\sqrt{2}=(-1-1)\sqrt{7}+(8-6)\sqrt{2}$
$=-2\sqrt{7}+2\sqrt{2}$

37. $\sqrt{28}+\sqrt{63}+\sqrt{56}=\sqrt{4\cdot 7}+\sqrt{9\cdot 7}+\sqrt{4\cdot 14}$
$=\sqrt{4}\sqrt{7}+\sqrt{9}\sqrt{7}+\sqrt{4}\sqrt{14}$
$=2\sqrt{7}+3\sqrt{7}+2\sqrt{14}$
$=(2+3)\sqrt{7}+2\sqrt{14}$
$=5\sqrt{7}+2\sqrt{14}$

38. $\sqrt{75}+\sqrt{48}-\sqrt{16}=\sqrt{25\cdot 3}+\sqrt{16\cdot 3}-4$
$=\sqrt{25}\sqrt{3}+\sqrt{16}\sqrt{3}-4$
$=5\sqrt{3}+4\sqrt{3}-4$
$=(5+4)\sqrt{3}-4$
$=9\sqrt{3}-4$

39. $\sqrt{\frac{5}{9}}-\sqrt{\frac{5}{36}}=\frac{\sqrt{5}}{\sqrt{9}}-\frac{\sqrt{5}}{\sqrt{36}}$
$=\frac{\sqrt{5}}{3}-\frac{\sqrt{5}}{6}$
$=\frac{2\sqrt{5}}{6}-\frac{\sqrt{5}}{6}$
$=\frac{\sqrt{5}}{6}$

40. $\sqrt{\frac{11}{25}}+\sqrt{\frac{11}{16}}=\frac{\sqrt{11}}{\sqrt{25}}+\frac{\sqrt{11}}{\sqrt{16}}$
$=\frac{\sqrt{11}}{5}+\frac{\sqrt{11}}{4}$
$=\frac{4\sqrt{11}}{20}+\frac{5\sqrt{11}}{20}$
$=\frac{9\sqrt{11}}{20}$

41. $\sqrt{45x^2}+3\sqrt{5x^2}-7x\sqrt{5}+10$
$=\sqrt{9x^2\cdot 5}+3\sqrt{x^2\cdot 5}-7x\sqrt{5}+10$
$=\sqrt{9x^2}\sqrt{5}+3\sqrt{x^2}\sqrt{5}-7x\sqrt{5}+10$
$=3x\sqrt{5}+3x\sqrt{5}-7x\sqrt{5}+10$
$=(3x+3x-7x)\sqrt{5}+10$
$=-x\sqrt{5}+10$ or $10-x\sqrt{5}$

Copyright © 2012 Pearson Education, Inc. Publishing as Prentice Hall.

42. $\sqrt{50x}-9\sqrt{2x}+\sqrt{72x}-\sqrt{3x}$
$=\sqrt{25\cdot 2x}-9\sqrt{2x}+\sqrt{36\cdot 2x}-\sqrt{3x}$
$=\sqrt{25}\sqrt{2x}-9\sqrt{2x}+\sqrt{36}\sqrt{2x}-\sqrt{3x}$
$=5\sqrt{2x}-9\sqrt{2x}+6\sqrt{2x}-\sqrt{3x}$
$=(5-9+6)\sqrt{2x}-\sqrt{3x}$
$=2\sqrt{2x}-\sqrt{3x}$

43. $\sqrt{3}\cdot\sqrt{6}=\sqrt{3\cdot 6}=\sqrt{18}=\sqrt{9\cdot 2}=\sqrt{9}\cdot\sqrt{2}=3\sqrt{2}$

44. $\sqrt{5}\cdot\sqrt{15}=\sqrt{5\cdot 15}$
$=\sqrt{75}$
$=\sqrt{25\cdot 3}$
$=\sqrt{25}\cdot\sqrt{3}$
$=5\sqrt{3}$

45. $\sqrt{2}\left(\sqrt{5}-\sqrt{7}\right)=\sqrt{2}\sqrt{5}-\sqrt{2}\sqrt{7}=\sqrt{10}-\sqrt{14}$

46. $\sqrt{5}\left(\sqrt{11}+\sqrt{3}\right)=\sqrt{5}\sqrt{11}+\sqrt{5}\sqrt{3}=\sqrt{55}+\sqrt{15}$

47. $\left(\sqrt{3}+2\right)\left(\sqrt{6}-5\right)=\sqrt{3}\sqrt{6}-5\sqrt{3}+2\sqrt{6}-2\cdot 5$
$=\sqrt{18}-5\sqrt{3}+2\sqrt{6}-10$
$=\sqrt{9\cdot 2}-5\sqrt{3}+2\sqrt{6}-10$
$=3\sqrt{2}-5\sqrt{3}+2\sqrt{6}-10$

48. $\left(\sqrt{5}+1\right)\left(\sqrt{5}-3\right)=\sqrt{5}\sqrt{5}-3\sqrt{5}+1\sqrt{5}-1\cdot 3$
$=5+(-3+1)\sqrt{5}-3$
$=2-2\sqrt{5}$

49. $\left(\sqrt{x}-2\right)^2=\left(\sqrt{x}\right)^2-2\left(\sqrt{x}\right)(2)+2^2$
$=x-4\sqrt{x}+4$

50. $\left(\sqrt{y}+4\right)^2=\left(\sqrt{y}\right)^2+2\left(\sqrt{y}\right)(4)+4^2$
$=y+8\sqrt{y}+16$

51. $\dfrac{\sqrt{27}}{\sqrt{3}}=\sqrt{\dfrac{27}{3}}=\sqrt{9}=3$

52. $\dfrac{\sqrt{20}}{\sqrt{5}}=\sqrt{\dfrac{20}{5}}=\sqrt{4}=2$

53. $\dfrac{\sqrt{160}}{\sqrt{8}}=\sqrt{\dfrac{160}{8}}=\sqrt{20}=\sqrt{4\cdot 5}=\sqrt{4}\sqrt{5}=2\sqrt{5}$

54. $\dfrac{\sqrt{96}}{\sqrt{3}}=\sqrt{\dfrac{96}{3}}=\sqrt{32}=\sqrt{16\cdot 2}=\sqrt{16}\sqrt{2}=4\sqrt{2}$

55. $\dfrac{\sqrt{30x^6}}{\sqrt{2x^3}}=\sqrt{\dfrac{30x^6}{2x^3}}$
$=\sqrt{15x^3}$
$=\sqrt{x^2\cdot 15x}$
$=\sqrt{x^2}\sqrt{15x}$
$=x\sqrt{15x}$

56. $\dfrac{\sqrt{54x^5y^2}}{\sqrt{3xy^2}}=\sqrt{\dfrac{54x^5y^2}{3xy^2}}$
$=\sqrt{18x^4}$
$=\sqrt{9x^4\cdot 2}$
$=\sqrt{9x^4}\sqrt{2}$
$=3x^2\sqrt{2}$

57. $\dfrac{\sqrt{2}}{\sqrt{11}}=\dfrac{\sqrt{2}}{\sqrt{11}}\cdot\dfrac{\sqrt{11}}{\sqrt{11}}=\dfrac{\sqrt{2\cdot 11}}{11}=\dfrac{\sqrt{22}}{11}$

58. $\dfrac{\sqrt{3}}{\sqrt{13}}=\dfrac{\sqrt{3}}{\sqrt{13}}\cdot\dfrac{\sqrt{13}}{\sqrt{13}}=\dfrac{\sqrt{3\cdot 13}}{13}=\dfrac{\sqrt{39}}{13}$

59. $\sqrt{\dfrac{5}{6}}=\dfrac{\sqrt{5}}{\sqrt{6}}=\dfrac{\sqrt{5}}{\sqrt{6}}\cdot\dfrac{\sqrt{6}}{\sqrt{6}}=\dfrac{\sqrt{5\cdot 6}}{6}=\dfrac{\sqrt{30}}{6}$

60. $\sqrt{\dfrac{7}{10}}=\dfrac{\sqrt{7}}{\sqrt{10}}=\dfrac{\sqrt{7}}{\sqrt{10}}\cdot\dfrac{\sqrt{10}}{\sqrt{10}}=\dfrac{\sqrt{7\cdot 10}}{10}=\dfrac{\sqrt{70}}{10}$

61. $\dfrac{1}{\sqrt{5x}}=\dfrac{1}{\sqrt{5x}}\cdot\dfrac{\sqrt{5x}}{\sqrt{5x}}=\dfrac{\sqrt{5x}}{5x}$

62. $\dfrac{5}{\sqrt{3y}}=\dfrac{5}{\sqrt{3y}}\cdot\dfrac{\sqrt{3y}}{\sqrt{3y}}=\dfrac{5\sqrt{3y}}{3y}$

63. $\sqrt{\dfrac{3}{x}}=\dfrac{\sqrt{3}}{\sqrt{x}}=\dfrac{\sqrt{3}}{\sqrt{x}}\cdot\dfrac{\sqrt{x}}{\sqrt{x}}=\dfrac{\sqrt{3x}}{x}$

64. $\sqrt{\dfrac{6}{y}}=\dfrac{\sqrt{6}}{\sqrt{y}}=\dfrac{\sqrt{6}}{\sqrt{y}}\cdot\dfrac{\sqrt{y}}{\sqrt{y}}=\dfrac{\sqrt{6y}}{y}$

Copyright © 2012 Pearson Education, Inc. Publishing as Prentice Hall.

65. $\frac{3}{\sqrt{5}-2}=\frac{3}{\sqrt{5}-2}\cdot\frac{\sqrt{5}+2}{\sqrt{5}+2}$
$=\frac{3\left(\sqrt{5}+2\right)}{5-4}$
$=\frac{3\left(\sqrt{5}+2\right)}{1}$
$=3\left(\sqrt{5}+2\right)$ or $3\sqrt{5}+6$

66. $\frac{8}{\sqrt{10}-3}=\frac{8}{\sqrt{10}-3}\cdot\frac{\sqrt{10}+3}{\sqrt{10}+3}$
$=\frac{8\left(\sqrt{10}+3\right)}{10-9}$
$=\frac{8\left(\sqrt{10}+3\right)}{1}$
$=8\left(\sqrt{10}+3\right)$ or $8\sqrt{10}+24$

67. $\frac{\sqrt{2}+1}{\sqrt{3}-1}=\frac{\sqrt{2}+1}{\sqrt{3}-1}\cdot\frac{\sqrt{3}+1}{\sqrt{3}+1}$
$=\frac{\sqrt{2}\sqrt{3}+1\cdot\sqrt{2}+1\cdot\sqrt{3}+1\cdot 1}{3-1}$
$=\frac{\sqrt{6}+\sqrt{2}+\sqrt{3}+1}{2}$

68. $\frac{\sqrt{3}-2}{\sqrt{5}+2}=\frac{\sqrt{3}-2}{\sqrt{5}+2}\cdot\frac{\sqrt{5}-2}{\sqrt{5}-2}$
$=\frac{\sqrt{3}\sqrt{5}-2\sqrt{3}-2\sqrt{5}+2\cdot 2}{5-4}$
$=\frac{\sqrt{15}-2\sqrt{3}-2\sqrt{5}+4}{1}$
$=\sqrt{15}-2\sqrt{3}-2\sqrt{5}+4$

69. $\frac{10}{\sqrt{x}+5}=\frac{10}{\sqrt{x}+5}\cdot\frac{\sqrt{x}-5}{\sqrt{x}-5}$
$=\frac{10\left(\sqrt{x}-5\right)}{x-25}$
$=\frac{10\sqrt{x}-50}{x-25}$

70. $\frac{8}{\sqrt{x}-1}=\frac{8}{\sqrt{x}-1}\cdot\frac{\sqrt{x}+1}{\sqrt{x}+1}=\frac{8\left(\sqrt{x}+1\right)}{x-1}=\frac{8\sqrt{x}+8}{x-1}$

71. $\sqrt{2x}=6$
$\left(\sqrt{2x}\right)^2=6^2$
$2x=36$
$x=18$

72. $\sqrt{x+3}=4$
$\left(\sqrt{x+3}\right)^2=4^2$
$x+3=16$
$x=13$

73. $\sqrt{x}+3=8$
$\sqrt{x}=5$
$\left(\sqrt{x}\right)^2=5^2$
$x=25$

74. $\sqrt{x}+8=3$
$\sqrt{x}=-5$
$\sqrt{x}$ cannot equal -5. Thus, the equation has no solution.

75. $\sqrt{2x+1}=x-7$
$\left(\sqrt{2x+1}\right)^2=(x-7)^2$
$2x+1=x^2-14x+49$
$0=x^2-16x+48$
$0=(x-12)(x-4)$
$x-12=0$ or $x-4=0$
$x=12$ $\quad$ $x=4$
$x=4$ does not check, so the solution is $x=12$.

76. $\sqrt{3x+1}=x-1$
$\left(\sqrt{3x+1}\right)^2=(x-1)^2$
$3x+1=x^2-2x+1$
$0=x^2-5x$
$0=x(x-5)$
$x=0$ or $x-5=0$
$x=5$
$x=0$ does not check, so the solution is $x=5$.

Copyright © 2012 Pearson Education, Inc. Publishing as Prentice Hall.

77.
$$\sqrt{x}+3=\sqrt{x+15}$$
$$\left(\sqrt{x}+3\right)^2=\left(\sqrt{x+15}\right)^2$$
$$x+6\sqrt{x}+9=x+15$$
$$6\sqrt{x}=6$$
$$\sqrt{x}=1$$
$$\left(\sqrt{x}\right)^2=1^2$$
$$x=1$$

78.
$$\sqrt{x-5}=\sqrt{x}-1$$
$$\left(\sqrt{x-5}\right)^2=\left(\sqrt{x}-1\right)^2$$
$$x-5=x-2\sqrt{x}+1$$
$$-6=-2\sqrt{x}$$
$$3=\sqrt{x}$$
$$3^2=\left(\sqrt{x}\right)^2$$
$$9=x$$

79.
$$a^2+b^2=c^2$$
$$5^2+b^2=9^2$$
$$25+b^2=81$$
$$b^2=56$$
$$\sqrt{b^2}=\sqrt{56}$$
$$b=2\sqrt{14}\approx 7.48$$

80.
$$a^2+b^2=c^2$$
$$6^2+9^2=c^2$$
$$36+81=c^2$$
$$117=c^2$$
$$\sqrt{117}=\sqrt{c^2}$$
$$c=3\sqrt{13}\approx 10.82$$

81. The distance between Romeo and Juliet is the length of the hypotenuse of a right triangle with legs of length 20 feet and 12 feet.
$$a^2+b^2=c^2$$
$$20^2+12^2=c^2$$
$$400+144=c^2$$
$$544=c^2$$
$$\sqrt{544}=\sqrt{c^2}$$
$$c=4\sqrt{34}\approx 23.32$$
The distance is exactly $4\sqrt{34}$ feet or approximately 23.32 feet.

82. The diagonal of a rectangle forms right triangles with the sides of the rectangle.
$$a^2+b^2=c^2$$
$$5^2+b^2=10^2$$
$$25+b^2=100$$
$$b^2=75$$
$$\sqrt{b^2}=\sqrt{75}$$
$$b=5\sqrt{3}\approx 8.66$$
The length of the rectangle is exactly $5\sqrt{3}$ inches or approximately 8.66 inches.

83.
$$r=\sqrt{\frac{S}{4\pi}}$$
$$r=\sqrt{\frac{72}{4\pi}}$$
$$r\approx 2.4$$
The radius is 2.4 inches.

84.
$$r=\sqrt{\frac{S}{4\pi}}$$
$$6=\sqrt{\frac{S}{4\pi}}$$
$$6^2=\left(\sqrt{\frac{S}{4\pi}}\right)^2$$
$$36=\frac{S}{4\pi}$$
$$144\pi=S$$
The surface area is 144π square inches.

85. $\sqrt{144}=12$, because $12^2=144$ and 12 is positive.

86. $-\sqrt[3]{64}=-4$, because $4^3=64$.

87. $\sqrt{16x^{16}}=4x^8$, because $(4x^8)^2=16x^{16}$.

88. $\sqrt{4x^{24}}=2x^{12}$, because $(2x^{12})^2=4x^{24}$.

89. $\sqrt{18x^7}=\sqrt{9x^6\cdot 2x^1}=\sqrt{9x^6}\sqrt{2x}=3x^3\sqrt{2x}$

90. $\sqrt{48y^6}=\sqrt{16y^6\cdot 3}=\sqrt{16y^6}\sqrt{3}=4y^3\sqrt{3}$

91. $\sqrt{\frac{y^4}{81}}=\frac{\sqrt{y^4}}{\sqrt{81}}=\frac{y^2}{9}$

Copyright © 2012 Pearson Education, Inc. Publishing as Prentice Hall.

92. $\sqrt{\frac{x^9}{9}} = \frac{\sqrt{x^9}}{\sqrt{9}} = \frac{\sqrt{x^8 \cdot x}}{3} = \frac{\sqrt{x^8}\sqrt{x}}{3} = \frac{x^4\sqrt{x}}{3}$

93. $\sqrt{12} + \sqrt{75} = \sqrt{4 \cdot 3} + \sqrt{25 \cdot 3}$
$= \sqrt{4}\sqrt{3} + \sqrt{25}\sqrt{3}$
$= 2\sqrt{3} + 5\sqrt{3}$
$= (2+5)\sqrt{3}$
$= 7\sqrt{3}$

94. $\sqrt{63} + \sqrt{28} - \sqrt{9} = \sqrt{9 \cdot 7} + \sqrt{4 \cdot 7} - 3$
$= \sqrt{9}\sqrt{7} + \sqrt{4}\sqrt{7} - 3$
$= 3\sqrt{7} + 2\sqrt{7} - 3$
$= (3+2)\sqrt{7} - 3$
$= 5\sqrt{7} - 3$

95. $\sqrt{\frac{3}{16}} - \sqrt{\frac{3}{4}} = \frac{\sqrt{3}}{\sqrt{16}} - \frac{\sqrt{3}}{\sqrt{4}}$
$= \frac{\sqrt{3}}{4} - \frac{\sqrt{3}}{2}$
$= \frac{\sqrt{3}}{4} - \frac{2\sqrt{3}}{4}$
$= -\frac{\sqrt{3}}{4}$

96. $\sqrt{45x^3} + x\sqrt{20x} - \sqrt{5x^3}$
$= \sqrt{9x^2 \cdot 5x} + x\sqrt{4 \cdot 5x} - \sqrt{x^2 \cdot 5x}$
$= \sqrt{9x^2}\sqrt{5x} + x\sqrt{4}\sqrt{5x} - \sqrt{x^2}\sqrt{5x}$
$= 3x\sqrt{5x} + 2x\sqrt{5x} - x\sqrt{5x}$
$= (3x + 2x - x)\sqrt{5x}$
$= 4x\sqrt{5x}$

97. $\sqrt{7} \cdot \sqrt{14} = \sqrt{7 \cdot 14}$
$= \sqrt{98}$
$= \sqrt{49 \cdot 2}$
$= \sqrt{49}\sqrt{2}$
$= 7\sqrt{2}$

98. $\sqrt{3}\left(\sqrt{9} - \sqrt{2}\right) = \sqrt{3}\left(3 - \sqrt{2}\right)$
$= 3\sqrt{3} - \sqrt{3}\sqrt{2}$
$= 3\sqrt{3} - \sqrt{6}$

99. $\left(\sqrt{2} + 4\right)\left(\sqrt{5} - 1\right) = \sqrt{2}\sqrt{5} - 1\sqrt{2} + 4\sqrt{5} - 4 \cdot 1$
$= \sqrt{10} - \sqrt{2} + 4\sqrt{5} - 4$

100. $\left(\sqrt{x} + 3\right)^2 = \left(\sqrt{x}\right)^2 + 2\left(\sqrt{x}\right)(3) + (3)^2$
$= x + 6\sqrt{x} + 9$

101. $\frac{\sqrt{120}}{\sqrt{5}} = \sqrt{\frac{120}{5}} = \sqrt{24} = \sqrt{4 \cdot 6} = 2\sqrt{6}$

102. $\frac{\sqrt{60x^9}}{\sqrt{15x^4}} = \sqrt{\frac{60x^9}{15x^7}} = \sqrt{4x^2} = 2x$

103. $\sqrt{\frac{2}{7}} = \frac{\sqrt{2}}{\sqrt{7}} = \frac{\sqrt{2}}{\sqrt{7}} \cdot \frac{\sqrt{7}}{\sqrt{7}} = \frac{\sqrt{14}}{7}$

104. $\frac{3}{\sqrt{2x}} = \frac{3}{\sqrt{2x}} \cdot \frac{\sqrt{2x}}{\sqrt{2x}} = \frac{3\sqrt{2x}}{2x}$

105. $\frac{3}{\sqrt{x} - 6} = \frac{3}{\sqrt{x} - 6} \cdot \frac{\sqrt{x} + 6}{\sqrt{x} + 6}$
$= \frac{3\left(\sqrt{x} + 6\right)}{x - 36}$
$= \frac{3\sqrt{x} + 18}{x - 36}$

106. $\frac{\sqrt{7} - 5}{\sqrt{5} + 3} = \frac{\sqrt{7} - 5}{\sqrt{5} + 3} \cdot \frac{\sqrt{5} - 3}{\sqrt{5} - 3}$
$= \frac{\sqrt{7}\sqrt{5} - 3\sqrt{7} - 5\sqrt{5} + 5 \cdot 3}{5 - 9}$
$= \frac{\sqrt{35} - 3\sqrt{7} - 5\sqrt{5} + 15}{-4}$

107. $\sqrt{4x} = 2$
$\left(\sqrt{4x}\right)^2 = 2^2$
$4x = 4$
$x = 1$

108. $\sqrt{x - 4} = 3$
$\left(\sqrt{x - 4}\right)^2 = 3^2$
$x - 4 = 9$
$x = 13$

Copyright © 2012 Pearson Education, Inc. Publishing as Prentice Hall.

109. $\sqrt{4x+8}+6=x$

$\sqrt{4x+8}=x-6$

$\left(\sqrt{4x+8}\right)^2=(x-6)^2$

$4x+8=x^2-12x+36$

$0=x^2-16x+28$

$0=(x-14)(x-2)$

$x-14=0$ or $x-2=0$

$x=14$ $\quad$ $x=2$

$x=2$ does not check, so the solution is $x=14$.

110. $\sqrt{x-8}=\sqrt{x}-2$

$\left(\sqrt{x-8}\right)^2=\left(\sqrt{x}-2\right)^2$

$x-8=x-4\sqrt{x}+4$

$-12=-4\sqrt{x}$

$3=\sqrt{x}$

$3^2=\left(\sqrt{x}\right)^2$

$9=x$

111. $a^2+b^2=c^2$

$3^2+7^2=c^2$

$9+49=c^2$

$58=c^2$

$\sqrt{58}=\sqrt{c^2}$

$c=\sqrt{58}\approx 7.62$

112. The diagonal is the hypotenuse of a right triangle with the sides of the rectangle as its legs.

$a^2+b^2=c^2$

$2^2+b^2=6^2$

$4+b^2=36$

$b^2=32$

$\sqrt{b^2}=\sqrt{32}$

$b=4\sqrt{2}\approx 5.66$

The length of the rectangle is exactly $4\sqrt{2}$ inches or approximately 5.66 inches.

Chapter 8 Test

1. $\sqrt{16}=4$, because $4^2=16$ and 4 is positive.

2. $\sqrt[3]{125}=5$, because $5^3=125$.

3. $\sqrt[4]{81}=3$, because $3^4=81$.

4. $\sqrt{\frac{9}{16}}=\frac{3}{4}$, because $\left(\frac{3}{4}\right)^2=\frac{9}{16}$.

5. $\sqrt[4]{-81}$ is not a real number since the index 4 is even and the radicand -81 is negative.

6. $\sqrt{x^{10}}=x^5$, because $(x^5)^2=x^{10}$.

7. $\sqrt{54}=\sqrt{9\cdot 6}=\sqrt{9}\sqrt{6}=3\sqrt{6}$

8. $\sqrt{92}=\sqrt{4\cdot 23}=\sqrt{4}\sqrt{23}=2\sqrt{23}$

9. $\sqrt{y^7}=\sqrt{y^6\cdot y}=\sqrt{y^6}\sqrt{y}=y^3\sqrt{y}$

10. $\sqrt{24x^8}=\sqrt{4x^8\cdot 6}=\sqrt{4x^8}\sqrt{6}=2x^4\sqrt{6}$

11. $\sqrt[3]{27}=3$

12. $\sqrt[3]{16}=\sqrt[3]{8\cdot 2}=\sqrt[3]{8}\sqrt[3]{2}=2\sqrt[3]{2}$

13. $\sqrt{\frac{5}{16}}=\frac{\sqrt{5}}{\sqrt{16}}=\frac{\sqrt{5}}{4}$

14. $\sqrt{\frac{y^3}{25}}=\frac{\sqrt{y^3}}{\sqrt{25}}=\frac{\sqrt{y^2\cdot y}}{5}=\frac{\sqrt{y^2}\sqrt{y}}{5}=\frac{y\sqrt{y}}{5}$

15. $\sqrt{13}+\sqrt{13}-4\sqrt{13}=(1+1-4)\sqrt{13}=-2\sqrt{13}$

16. $\sqrt{18}-\sqrt{75}+7\sqrt{3}-\sqrt{8}$

$=\sqrt{9\cdot 2}-\sqrt{25\cdot 3}+7\sqrt{3}-\sqrt{4\cdot 2}$

$=\sqrt{9}\sqrt{2}-\sqrt{25}\sqrt{3}+7\sqrt{3}-\sqrt{4}\sqrt{2}$

$=3\sqrt{2}-5\sqrt{3}+7\sqrt{3}-2\sqrt{2}$

$=(3-2)\sqrt{2}+(-5+7)\sqrt{3}$

$=\sqrt{2}+2\sqrt{3}$

Copyright © 2012 Pearson Education, Inc. Publishing as Prentice Hall.

17. $\sqrt{\frac{3}{4}}+\sqrt{\frac{3}{25}}=\frac{\sqrt{3}}{\sqrt{4}}+\frac{\sqrt{3}}{\sqrt{25}}$
$=\frac{\sqrt{3}}{2}+\frac{\sqrt{3}}{5}$
$=\frac{5\sqrt{3}}{10}+\frac{2\sqrt{3}}{10}$
$=\frac{5\sqrt{3}+2\sqrt{3}}{10}$
$=\frac{(5+2)\sqrt{3}}{10}$
$=\frac{7\sqrt{3}}{10}$

18. $\sqrt{7}\cdot\sqrt{14}=\sqrt{7\cdot 14}$
$=\sqrt{98}$
$=\sqrt{49\cdot 2}$
$=\sqrt{49}\sqrt{2}$
$=7\sqrt{2}$

19. $\sqrt{2}\left(\sqrt{6}-\sqrt{5}\right)=\sqrt{2}\sqrt{6}-\sqrt{2}\sqrt{5}$
$=\sqrt{12}-\sqrt{10}$
$=\sqrt{4\cdot 3}-\sqrt{10}$
$=2\sqrt{3}-\sqrt{10}$

20. $\left(\sqrt{x}+2\right)\left(\sqrt{x}-3\right)=\sqrt{x}\cdot\sqrt{x}-3\sqrt{x}+2\sqrt{x}-2\cdot 3$
$=x-3\sqrt{x}+2\sqrt{x}-6$
$=x-\sqrt{x}-6$

21. $\frac{\sqrt{50}}{\sqrt{10}}=\sqrt{\frac{50}{10}}=\sqrt{5}$

22. $\frac{\sqrt{40x^4}}{\sqrt{2x}}=\sqrt{\frac{40x^4}{2x}}$
$=\sqrt{20x^3}$
$=\sqrt{4x^2\cdot 5x}$
$=\sqrt{4x^2}\sqrt{5x}$
$=2x\sqrt{5x}$

23. $\sqrt{\frac{2}{3}}=\frac{\sqrt{2}}{\sqrt{3}}=\frac{\sqrt{2}}{\sqrt{3}}\cdot\frac{\sqrt{3}}{\sqrt{3}}=\frac{\sqrt{6}}{3}$

24. $\frac{8}{\sqrt{5y}}=\frac{8}{\sqrt{5y}}\cdot\frac{\sqrt{5y}}{\sqrt{5y}}=\frac{8\sqrt{5y}}{5y}$

25. $\frac{8}{\sqrt{6}+2}=\frac{8}{\sqrt{6}+2}\cdot\frac{\sqrt{6}-2}{\sqrt{6}-2}$
$=\frac{8\left(\sqrt{6}-2\right)}{\left(\sqrt{6}\right)^2-2^2}$
$=\frac{8\left(\sqrt{6}-2\right)}{6-4}$
$=\frac{8\left(\sqrt{6}-2\right)}{2}$
$=4\left(\sqrt{6}-2\right)$
$=4\sqrt{6}-8$

26. $\frac{1}{3-\sqrt{x}}=\frac{1}{3-\sqrt{x}}\cdot\frac{3+\sqrt{x}}{3+\sqrt{x}}=\frac{1\left(3+\sqrt{x}\right)}{3^2-\left(\sqrt{x}\right)^2}=\frac{3+\sqrt{x}}{9-x}$

27. $\sqrt{x}+8=11$
$\sqrt{x}=3$
$\left(\sqrt{x}\right)^2=3^2$
$x=9$

28. $\sqrt{3x-6}=\sqrt{x+4}$
$\left(\sqrt{3x-6}\right)^2=\left(\sqrt{x+4}\right)^2$
$3x-6=x+4$
$2x-6=4$
$2x=10$
$x=5$

29. $\sqrt{2x-2}=x-5$
$\left(\sqrt{2x-2}\right)^2=(x-5)^2$
$2x-2=x^2-10x+25$
$0=x^2-12x+27$
$0=(x-3)(x-9)$
$x-3=0$ or $x-9=0$
$x=3$ $\quad$ $x=9$
$x=3$ does not check, so the solution is $x=9$.

Copyright © 2012 Pearson Education, Inc. Publishing as Prentice Hall.

30. $a^2+b^2=c^2$
$8^2+b^2=12^2$
$64+b^2=144$
$b^2=80$
$\sqrt{b^2}=\sqrt{80}$
$b=4\sqrt{5}$
The length is $4\sqrt{5}$ inches.

31. $r=\sqrt{\frac{A}{\pi}}$
$r=\sqrt{\frac{15}{\pi}}$
$r\approx 2.19$
The radius is 2.19 meters.

Cumulative Review Chapters 1–8

1. $-2(-14)=2\cdot 14=28$

2. $9(-5.2)=-9\cdot 5.2=-46.8$

3. $-\frac{2}{3}\cdot\frac{4}{7}=-\frac{2\cdot 4}{3\cdot 7}=-\frac{8}{21}$

4. $-3\frac{3}{8}\cdot 5\frac{1}{3}=-\frac{27}{8}\cdot\frac{16}{3}$
$=-\frac{27\cdot 16}{8\cdot 3}$
$=-9\cdot 2$
$=-18$

5. $4(2x-3)+7=3x+5$
$8x-12+7=3x+5$
$8x-5=3x+5$
$5x-5=5$
$5x=10$
$x=2$

6. $6y-11+4+2y=8+15y-8y$
$8y-7=8+7y$
$y-7=8$
$y=15$

7. a. The sector for business represents 17% of the circle.

b. These sectors comprise a total of 17% + 4% or 21% of the circle.

c. 17% of 253 = 0.17(253) = 43.01
About 43 of the 253 American travelers would be traveling solely for business.

8. a. $\frac{4(-3)-(-6)}{-8+4}=\frac{-12+6}{-4}=\frac{-6}{-4}=\frac{3}{2}$

b. $\frac{3+(-3)(-2)^3}{-1-(-4)}=\frac{3+(-3)(-8)}{-1+4}$
$=\frac{3+24}{3}$
$=\frac{27}{3}$
$=9$

9. a. $1.02\times 10^5=102{,}000$

b. $7.358\times 10^{-3}=0.007358$

c. $8.4\times 10^7=84{,}000{,}000$

d. $3.007\times 10^{-5}=0.00003007$

10. a. $7{,}200{,}000=7.2\times 10^6$

b. $0.000308=3.08\times 10^{-4}$

11. $(3x+2)(2x-5)$
$=(3x)(2x)-(3x)(5)+2(2x)-2(5)$
$=6x^2-15x+4x-10$
$=6x^2-11x-10$

12. $(7x+1)^2=(7x)^2+2(7x)(1)+(1)^2$
$=49x^2+14x+1$

13. $xy+2x+3y+6=x(y+2)+3(y+2)$
$=(y+2)(x+3)$

14. $xy^2+5x-y^2-5=x(y^2+5)-(y^2+5)$
$=(y^2+5)(x-1)$

15. $3x^2+11x+6=(3x+2)(x+3)$

16. $3x^2+15x+18=3(x^2+5x+6)=3(x+2)(x+3)$

17. a. The expression is undefined for $x=3$, since the denominator, $x-3$, is 0 when $x=3$.

Copyright © 2012 Pearson Education, Inc. Publishing as Prentice Hall.

b. The expression is undefined for $x = 2$ and $x = 1$, since the denominator, $x^2 - 3x + 2 = (x-2)(x-1)$, is 0 when $x = 2$ or $x = 1$.

c. There are no values for which the expression is undefined, since the denominator, 3, is never 0.

18. $\dfrac{2x^2+7x+3}{x^2-9} = \dfrac{(2x+1)(x+3)}{(x+3)(x-3)} = \dfrac{2x+1}{x-3}$

19. $\dfrac{x^2+4x+4}{x^2+2x} = \dfrac{(x+2)(x+2)}{x(x+2)} = \dfrac{x+2}{x}$

20.
$$\begin{aligned}\frac{12x^2y^3}{5} \div \frac{3y^3}{x} &= \frac{12x^2y^3}{5} \cdot \frac{x}{3y^3} \\ &= \frac{4x^2}{5} \cdot \frac{x}{1} \\ &= \frac{4x^3}{5}\end{aligned}$$

21. a. $\dfrac{a}{4} - \dfrac{2a}{8} = \dfrac{a}{4} - \dfrac{a}{4} = 0$

b. $\dfrac{3}{10x^2} + \dfrac{7}{25x} = \dfrac{15}{50x^2} + \dfrac{14x}{50x^2} = \dfrac{15+14x}{50x^2}$

22. $y = mx + b$
$y = -2x + 4$

23.
$$\frac{4x}{x^2+x-30} + \frac{2}{x-5} = \frac{1}{x+6}$$
$$(x-5)(x+6)\left(\frac{4x}{(x-5)(x+6)} + \frac{2}{x-5}\right) = (x-5)(x+6)\left(\frac{1}{x+6}\right)$$
$$\begin{aligned}4x + 2(x+6) &= x-5 \\ 4x + 2x + 12 &= x-5 \\ 6x + 12 &= x-5 \\ 5x + 12 &= -5 \\ 5x &= -17 \\ x &= -\frac{17}{5}\end{aligned}$$

24.
$$\begin{aligned}4a^2 + 3a - 2a^2 + 7a - 5 &= (4-2)a^2 + (3+7)a - 5 \\ &= 2a^2 + 10a - 5\end{aligned}$$

Copyright © 2012 Pearson Education, Inc. Publishing as Prentice Hall.

25.

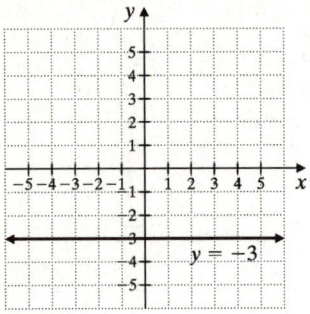

26. For $x = 0$:

$$\begin{aligned} 2x + y &= 6 \\ 2(0) + y &= 6 \\ y &= 6 \end{aligned}$$

For $y = -2$:

$$\begin{aligned} 2x + y &= 6 \\ 2x - 2 &= 6 \\ 2x &= 8 \\ x &= 4 \end{aligned}$$

For $x = 3$:

$$\begin{aligned} 2x + y &= 6 \\ 2(3) + y &= 6 \\ 6 + y &= 6 \\ y &= 0 \end{aligned}$$

x	y
0	6
4	−2
3	0

27. $y = mx + b$

$$y = \frac{1}{4}x - 3$$

28. The slope of a line perpendicular to $y = 2x + 4$ has slope $-\frac{1}{2}$.

$$\begin{aligned} y - y_1 &= m(x - x_1) \\ y - 5 &= -\frac{1}{2}(x - 1) \\ y &= -\frac{1}{2}x + \frac{1}{2} + 5 \\ y &= -\frac{1}{2}x + \frac{11}{2} \end{aligned}$$

29. $\begin{cases} 3x + 4y = 13 \\ 5x - 9y = 6 \end{cases}$

$$\begin{aligned} 15x + 20y &= 65 \\ -15x + 27y &= -18 \\ \hline 47y &= 47 \\ y &= 1 \end{aligned}$$

$$\begin{aligned} 3x + 4y &= 13 \\ 3x + 4(1) &= 13 \\ 3x &= 9 \\ x &= 3 \end{aligned}$$

The solution is (3, 1).

30. $\begin{cases} \frac{x}{2} + y = \frac{5}{6} \\ 2x - y = \frac{5}{6} \end{cases}$

$$\begin{aligned} \frac{x}{2} + y &= \frac{5}{6} \\ 2x - y &= \frac{5}{6} \\ \hline \frac{5}{2}x \quad &= \frac{10}{6} \\ x &= \frac{2}{3} \end{aligned}$$

$$\begin{aligned} \frac{x}{2} + y &= \frac{5}{6} \\ \frac{1}{2}\left(\frac{2}{3}\right) + y &= \frac{5}{6} \\ \frac{1}{3} + y &= \frac{5}{6} \\ y &= \frac{5}{6} - \frac{1}{3} \\ y &= \frac{1}{2} \end{aligned}$$

The solution is $\left(\frac{2}{3}, \frac{1}{2}\right)$.

31. Let x be Alfredo's walking speed. Then $x + 1$ is Louisa's walking speed.

$$\begin{aligned} 2x + 2(x + 1) &= 15 \\ 2x + 2x + 2 &= 15 \\ 4x + 2 &= 15 \\ 4x &= 13 \\ x &= \frac{13}{4} \text{ or } 3.25 \end{aligned}$$

$x + 1 = 3.25 + 1 = 4.25$

Alfredo's walking speed is 3.25 miles per hour, and Louisa's is 4.25 miles per hour.

Copyright © 2012 Pearson Education, Inc. Publishing as Prentice Hall.

32. Let x be the speed of the slower streetcar. Then $x + 15$ is the speed of the faster streetcar. Twelve minutes is $\frac{1}{5}$ hour.

$$\frac{1}{5}x+\frac{1}{5}(x+15)=11$$
$$\frac{1}{5}x+\frac{1}{5}x+3=11$$
$$\frac{2}{5}x+3=11$$
$$\frac{2}{5}x=8$$
$$x=20$$

$x + 15 = 20 + 15 = 35$

The streetcars are traveling at 20 miles per hour and 35 miles per hour.

33. $\sqrt[3]{1}=1$, because $1^3=1$.

34. $\sqrt{121}=11$, because $11^2=121$ and 11 is positive.

35. $\sqrt[3]{-27}=-3$, because $(-3)^3=-27$.

36. $\sqrt{\frac{1}{4}}=\frac{1}{2}$, because $\left(\frac{1}{2}\right)^2=\frac{1}{4}$.

37. $\sqrt[3]{\frac{1}{125}}=\frac{1}{5}$, because $\left(\frac{1}{5}\right)^3=\frac{1}{125}$.

38. $\sqrt{\frac{25}{144}}=\frac{5}{12}$, because $\left(\frac{5}{12}\right)^2=\frac{25}{144}$.

39. $\sqrt{54}=\sqrt{9\cdot 6}=\sqrt{9}\sqrt{6}=3\sqrt{6}$

40. $\sqrt{63}=\sqrt{9\cdot 7}=\sqrt{9}\sqrt{7}=3\sqrt{7}$

41. $\sqrt{200}=\sqrt{100\cdot 2}=\sqrt{100}\sqrt{2}=10\sqrt{2}$

42. $\sqrt{500}=\sqrt{100\cdot 5}=\sqrt{100}\sqrt{5}=10\sqrt{5}$

43.
$$\begin{aligned}7\sqrt{12}-2\sqrt{75}&=7\sqrt{4\cdot 3}-2\sqrt{25\cdot 3}\\&=7\sqrt{4}\sqrt{3}-2\sqrt{25}\sqrt{3}\\&=7\cdot 2\sqrt{3}-2\cdot 5\sqrt{3}\\&=14\sqrt{3}-10\sqrt{3}\\&=(14-10)\sqrt{3}\\&=4\sqrt{3}\end{aligned}$$

44. $\left(\sqrt{x}+5\right)\left(\sqrt{x}-5\right)=\left(\sqrt{x}\right)^2-5^2=x-25$

45.
$$\begin{aligned}&2\sqrt{x^2}-\sqrt{25x^5}+\sqrt{x^5}\\&=2x-\sqrt{25x^4\cdot x}+\sqrt{x^4\cdot x}\\&=2x-\sqrt{25x^4}\sqrt{x}+\sqrt{x^4}\sqrt{x}\\&=2x-5x^2\sqrt{x}+x^2\sqrt{x}\\&=2x+(-5x^2+x^2)\sqrt{x}\\&=2x-4x^2\sqrt{x}\end{aligned}$$

46.
$$\begin{aligned}\left(\sqrt{6}+2\right)^2&=\left(\sqrt{6}\right)^2+2\left(\sqrt{6}\right)(2)+(2)^2\\&=6+4\sqrt{6}+4\\&=10+4\sqrt{6}\end{aligned}$$

47. $\frac{2}{\sqrt{7}}=\frac{2}{\sqrt{7}}\cdot\frac{\sqrt{7}}{\sqrt{7}}=\frac{2\sqrt{7}}{7}$

48. $\frac{x+3}{\frac{1}{x}+\frac{1}{3}}=\frac{x+3}{\frac{3}{3x}+\frac{x}{3x}}=\frac{x+3}{\frac{3+x}{3x}}=(x+3)\cdot\frac{3x}{3+x}=3x$

49.
$$\begin{aligned}\sqrt{x}&=\sqrt{5x-2}\\\left(\sqrt{x}\right)^2&=\left(\sqrt{5x-2}\right)^2\\x&=5x-2\\-4x&=-2\\x&=\frac{-2}{-4}\\x&=\frac{1}{2}\end{aligned}$$

The solution is $\frac{1}{2}$.

50.
$$\begin{aligned}\sqrt{x+4}&=\sqrt{3x-1}\\\left(\sqrt{x+4}\right)^2&=\left(\sqrt{3x-1}\right)^2\\x+4&=3x-1\\-2x+4&=-1\\-2x&=-5\\x&=\frac{-5}{-2}\\x&=\frac{5}{2}\end{aligned}$$

The solution is $\frac{5}{2}$.

Copyright © 2012 Pearson Education, Inc. Publishing as Prentice Hall.

Chapter 9

Section 9.1 Practice

1. $x^2 - 25 = 0$

$(x+5)(x-5) = 0$

$x+5=0$ or $x-5=0$

$x=-5$ $\quad$ $x=5$

The solutions are –5 and 5.

2. $2x^2 - 3x = 9$

$2x^2 - 3x - 9 = 0$

$(2x+3)(x-3) = 0$

$2x+3=0$ or $x-3=0$

$2x=-3$ $\quad$ $x=3$

$x=-\frac{3}{2}$

The solutions are $-\frac{3}{2}$ and 3.

3. $x^2 - 16 = 0$

$x^2 = 16$

$x=\sqrt{16}$ or $x=-\sqrt{16}$

$x=4$ $\quad$ $x=-4$

The solutions are 4 and –4.

4. $3x^2 = 11$

$x^2 = \frac{11}{3}$

$x=\sqrt{\frac{11}{3}}$ or $x=-\sqrt{\frac{11}{3}}$

$x=\frac{\sqrt{11}\cdot\sqrt{3}}{\sqrt{3}\cdot\sqrt{3}}$ $\quad$ $x=-\frac{\sqrt{11}\cdot\sqrt{3}}{\sqrt{3}\cdot\sqrt{3}}$

$x=\frac{\sqrt{33}}{3}$ $\quad$ $x=-\frac{\sqrt{33}}{3}$

The solutions are $\frac{\sqrt{33}}{3}$ and $-\frac{\sqrt{33}}{3}$.

5. $(x-4)^2 = 49$

$x-4=\sqrt{49}$ or $x-4=-\sqrt{49}$

$x-4=7$ $\quad$ $x-4=-7$

$x=11$ $\quad$ $x=-3$

The solutions are 11 and –3.

6. $(x-5)^2 = 18$

$x-5=\sqrt{18}$ or $x-5=-\sqrt{18}$

$x-5=3\sqrt{2}$ $\quad$ $x-5=-3\sqrt{2}$

$x=5+3\sqrt{2}$ $\quad$ $x=5-3\sqrt{2}$

$x=5\pm3\sqrt{2}$

The solutions are $5\pm3\sqrt{2}$.

7. $(x+3)^2 = -5$

This equation has no real solution because the square root of –5 is not a real number.

8. $(4x+1)^2 = 15$

$4x+1=\sqrt{15}$ or $4x+1=-\sqrt{15}$

$4x=-1+\sqrt{15}$ $\quad$ $4x=-1-\sqrt{15}$

$x=\frac{-1+\sqrt{15}}{4}$ $\quad$ $x=\frac{-1-\sqrt{15}}{4}$

$x=\frac{-1\pm\sqrt{15}}{4}$

The solutions are $\frac{-1\pm\sqrt{15}}{4}$.

9. $h = 16t^2$

$650 = 16t^2$

$40.625 = t^2$

$6.4 = t$ or $-6.4 = t$

–6.4 is rejected because time cannot be negative. It takes the object 6.4 seconds to fall 650 feet.

Exercise Set 9.1

1. $k^2 - 49 = 0$

$(k+7)(k-7) = 0$

$k+7=0$ or $k-7=0$

$k=-7$ $\quad$ $k=7$

The solutions are $k = -7$ and $k = 7$.

3. $m^2 + 2m = 15$

$m^2 + 2m - 15 = 0$

$(m+5)(m-3) = 0$

$m+5=0$ or $m-3=0$

$m=-5$ $\quad$ $m=3$

The solutions are $m = -5$ and $m = 3$.

Copyright © 2012 Pearson Education, Inc. Publishing as Prentice Hall.

5. $2x^2 - 32 = 0$

$2(x^2 - 16) = 0$

$2(x+4)(x-4) = 0$

$x+4=0$ or $x-4=0$

$x=-4$ $\quad$ $x=4$

The solutions are $x = -4$ and $x = 4$.

7. $4a^2 - 36 = 0$

$4(a^2 - 9) = 0$

$4(a-3)(a+3) = 0$

$a-3=0$ or $a+3=0$

$a=3$ $\quad$ $a=-3$

The solutions are $a = 3$ and $a = -3$.

9. $x^2 + 7x = -10$

$x^2 + 7x + 10 = 0$

$(x+2)(x+5) = 0$

$x+2=0$ or $x+5=0$

$x=-2$ $\quad$ $x=-5$

The solutions are $x = -2$ and $x = -5$.

11. $x^2 = 64$

$x=\sqrt{64}$ or $x=-\sqrt{64}$

$x=8$ $\quad$ $x=-8$

The solutions are $x = \pm 8$.

13. $x^2 = 21$

$x=\sqrt{21}$ or $x=-\sqrt{21}$

The solutions are $x = \pm\sqrt{21}$.

15. $x^2 = \frac{1}{25}$

$x=\sqrt{\frac{1}{25}}$ or $x=-\sqrt{\frac{1}{25}}$

$x=\frac{1}{5}$ $\quad$ $x=-\frac{1}{5}$

The solutions are $x = \pm\frac{1}{5}$.

17. $x^2 = -4$ has no real solution because the square root of -4 is not a real number.

19. $3x^2 = 13$

$x^2 = \frac{13}{3}$

$x=\sqrt{\frac{13}{3}}$ or $x=-\sqrt{\frac{13}{3}}$

$x=\frac{\sqrt{13}}{\sqrt{3}}\cdot\frac{\sqrt{3}}{\sqrt{3}}$ $\quad$ $x=-\frac{\sqrt{13}}{\sqrt{3}}\cdot\frac{\sqrt{3}}{\sqrt{3}}$

$x=\frac{\sqrt{39}}{3}$ $\quad$ $x=-\frac{\sqrt{39}}{3}$

The solutions are $x = \pm\frac{\sqrt{39}}{3}$.

21. $7x^2 = 4$

$x^2 = \frac{4}{7}$

$x=\sqrt{\frac{4}{7}}$ or $x=-\sqrt{\frac{4}{7}}$

$x=\frac{\sqrt{4}}{\sqrt{7}}\cdot\frac{\sqrt{7}}{\sqrt{7}}$ $\quad$ $x=-\frac{\sqrt{4}}{\sqrt{7}}\cdot\frac{\sqrt{7}}{\sqrt{7}}$

$x=\frac{2\sqrt{7}}{7}$ $\quad$ $x=-\frac{2\sqrt{7}}{7}$

The solutions are $x = \pm\frac{2\sqrt{7}}{7}$.

23. $x^2 - 2 = 0$

$x^2 = 2$

$x=\sqrt{2}$ or $x=-\sqrt{2}$

The solutions are $x = \pm\sqrt{2}$.

25. $(x-5)^2 = 49$

$x-5=\sqrt{49}$ or $x-5=-\sqrt{49}$

$x-5=7$ $\quad$ $x-5=-7$

$x=12$ $\quad$ $x=-2$

The solutions are $x = -2$ and $x = 12$.

27. $(x+2)^2 = 7$

$x+2=\sqrt{7}$ or $x+2=-\sqrt{7}$

$x=-2+\sqrt{7}$ $\quad$ $x=-2-\sqrt{7}$

The solutions are $x = -2 \pm \sqrt{7}$.

Copyright © 2012 Pearson Education, Inc. Publishing as Prentice Hall.

29. $\left(m-\frac{1}{2}\right)^2=\frac{1}{4}$

$m-\frac{1}{2}=\sqrt{\frac{1}{4}}$ or $m-\frac{1}{2}=-\sqrt{\frac{1}{4}}$

$m-\frac{1}{2}=\frac{1}{2}$ $\quad$ $m-\frac{1}{2}=-\frac{1}{2}$

$m=1$ $\quad$ $m=0$

The solutions are $m=0$ and $m=1$.

31. $(p+2)^2=10$

$p+2=\sqrt{10}$ or $p+2=-\sqrt{10}$

$p=-2+\sqrt{10}$ $\quad$ $p=-2-\sqrt{10}$

The solutions are $p=-2\pm\sqrt{10}$.

33. $(3y+2)^2=100$

$3y+2=\sqrt{100}$ or $3y+2=-\sqrt{100}$

$3y+2=10$ $\quad$ $3y+2=-10$

$3y=8$ $\quad$ $3y=-12$

$y=\frac{8}{3}$ $\quad$ $y=-4$

The solutions are $y=-4$ and $y=\frac{8}{3}$.

35. $(z-4)^2=-9$ has no real solution because the square root of -9 is not a real number.

37. $(2x-11)^2=50$

$2x-11=\sqrt{50}$ or $2x-11=-\sqrt{50}$

$2x-11=5\sqrt{2}$ $\quad$ $2x-11=-5\sqrt{2}$

$2x=11+5\sqrt{2}$ $\quad$ $2x=11-5\sqrt{2}$

$x=\frac{11+5\sqrt{2}}{2}$ $\quad$ $x=\frac{11-5\sqrt{2}}{2}$

The solutions are $x=\frac{11\pm5\sqrt{2}}{2}$.

39. $(3x-7)^2=32$

$3x-7=\sqrt{32}$ or $3x-7=-\sqrt{32}$

$3x-7=4\sqrt{2}$ $\quad$ $3x-7=-4\sqrt{2}$

$3x=7+4\sqrt{2}$ $\quad$ $3x=7-4\sqrt{2}$

$x=\frac{7+4\sqrt{2}}{3}$ $\quad$ $x=\frac{7-4\sqrt{2}}{3}$

The solutions are $x=\frac{7\pm4\sqrt{2}}{3}$.

41. $x^2-29=0$

$x^2=29$

$x=\sqrt{29}$ or $x=-\sqrt{29}$

The solutions are $x=\pm\sqrt{29}$.

43. $(x+6)^2=24$

$x+6=\sqrt{24}$ or $x+6=-\sqrt{24}$

$x+6=2\sqrt{6}$ $\quad$ $x+6=-2\sqrt{6}$

$x=-6+2\sqrt{6}$ $\quad$ $x=-6-2\sqrt{6}$

The solutions are $x=-6\pm2\sqrt{6}$.

45. $\frac{1}{2}n^2=5$

$n^2=10$

$n=\sqrt{10}$ or $n=-\sqrt{10}$

The solutions are $n=\pm\sqrt{10}$.

47. $(4x-1)^2=5$

$4x-1=\sqrt{5}$ or $4x-1=-\sqrt{5}$

$4x=1+\sqrt{5}$ $\quad$ $4x=1-\sqrt{5}$

$x=\frac{1+\sqrt{5}}{4}$ $\quad$ $x=\frac{1-\sqrt{5}}{4}$

The solutions are $x=\frac{1\pm\sqrt{5}}{4}$.

49. $3z^2=36$

$z^2=12$

$z=\sqrt{12}$ or $z=-\sqrt{12}$

$z=2\sqrt{3}$ $\quad$ $z=-2\sqrt{3}$

The solutions are $z=\pm2\sqrt{3}$.

51. $(8-3x)^2-45=0$

$(8-3x)^2=45$

$8-3x=\sqrt{45}$ or $8-3x=-\sqrt{45}$

$8-3x=3\sqrt{5}$ $\quad$ $8-3x=-3\sqrt{5}$

$-3x=-8+3\sqrt{5}$ $\quad$ $-3x=-8-3\sqrt{5}$

$x=\frac{-8+3\sqrt{5}}{-3}$ $\quad$ $x=\frac{-8-3\sqrt{5}}{-3}$

The solutions are $x=\frac{-8\pm3\sqrt{5}}{-3}$.

Copyright © 2012 Pearson Education, Inc. Publishing as Prentice Hall.

53. $h = 16t^2$

$87.6 = 16t^2$

$\frac{87.6}{16} = t^2$

$5.475 = t^2$

$\sqrt{5.475} = t$ or $-\sqrt{5.475} = t$

$2.3 \approx t$ $\quad -2.3 \approx t$

Since the time of a dive is not a negative number, reject the solution –2.3. The dive lasted approximately 2.3 seconds.

55. $h = 16t^2$

$4000 = 16t^2$

$250 = t^2$

$\sqrt{250} = t$ or $-\sqrt{250} = t$

$15.8 \approx t$ $\quad -15.8 \approx t$

Since the time of a fall is not a negative number, reject the solution –15.8. It would take approximately 15.8 seconds for the object to reach the bottom of the canyon.

57. $A = \pi r^2$

$36\pi = \pi r^2$

$36 = r^2$

$\sqrt{36} = r$ or $-\sqrt{36} = r$

$6 = r$ $\quad -6 = r$

Since the radius does not have a negative length, reject the solution –6. The radius is 6 inches.

59. $A = s^2$

$20 = s^2$

$\sqrt{20} = s$ or $-\sqrt{20} = s$

$2\sqrt{5} = s$ $\quad -2\sqrt{5} = s$

Since the length of a side is not a negative number, reject the solution $-2\sqrt{5}$. The sides have length $2\sqrt{5} \approx 4.47$ inches.

61. $A = s^2$

$3039 = s^2$

$\sqrt{3039} = s$ or $-\sqrt{3039} = s$

$55.13 \approx s$ $\quad -55.13 \approx s$

Since the length of a side is not a negative number, reject the solution –55.13. The sides have length 55.13 feet.

63. $x^2 + 6x + 9 = (x)^2 + 2(x)(3) + (3)^2 = (x+3)^2$

65. $x^2 - 4x + 4 = x^2 - 2(x)(2) + (2)^2 = (x-2)^2$

67. answers may vary

69. $x^2 + 4x + 4 = 16$

$(x+2)^2 = 16$

$x + 2 = \sqrt{16}$ or $x + 2 = -\sqrt{16}$

$x + 2 = 4$ $\quad x + 2 = -4$

$x = 2$ $\quad x = -6$

The solutions are $x = 2$ and $x = -6$.

71. $x^2 = 1.78$

$x = \sqrt{1.78}$ or $x = -\sqrt{1.78}$

$x \approx \pm 1.33$

The solutions are $x = \pm 1.33$.

73. $y = 148(x+1.5)^2 + 589{,}105$

$605{,}000 = 148(x+1.5)^2 + 589{,}105$

$15{,}895 = 148(x+1.5)^2$

$\frac{15{,}895}{148} = (x+1.5)^2$

$\sqrt{\frac{15{,}895}{148}} = x + 1.5$

$-1.5 + \sqrt{\frac{15{,}895}{148}} = x$

$8.9 \approx x$

or

$-\sqrt{\frac{15{,}895}{148}} = x + 1.5$

$-1.5 - \sqrt{\frac{15{,}895}{148}} = x$

$-11.9 \approx x$

Since the time will not be a negative number in this context, reject the solution of –11.9. The solution is $x \approx 9$, so the model predicts that there were 605,000 highway bridges in 2009 (2000 + 9).

Section 9.2 Practice

1. $x^2 + 8x + 1 = 0$

$x^2 + 8x = -1$

$x^2 + 8x + 16 = -1 + 16$

$(x+4)^2 = 15$

$x + 4 = \sqrt{15}$ or $x + 4 = -\sqrt{15}$

$x = -4 + \sqrt{15}$ $\quad x = -4 - \sqrt{15}$

The solutions are $-4 \pm \sqrt{15}$.

Copyright © 2012 Pearson Education, Inc. Publishing as Prentice Hall.

2. $x^2 - 14x = -32$

$x^2 - 14x + 49 = -32 + 49$

$(x-7)^2 = 17$

$x - 7 = \sqrt{17}$ or $x - 7 = -\sqrt{17}$

$x = 7 + \sqrt{17}$ $\quad$ $x = 7 - \sqrt{17}$

The solutions are $7 \pm \sqrt{17}$.

3. $4x^2 - 16x - 9 = 0$

$x^2 - 4x - \frac{9}{4} = 0$

$x^2 - 4x = \frac{9}{4}$

$x^2 - 4x + 4 = \frac{9}{4} + 4$

$(x-2)^2 = \frac{25}{4}$

$x - 2 = \sqrt{\frac{25}{4}}$ or $x - 2 = -\sqrt{\frac{25}{4}}$

$x - 2 = \frac{5}{2}$ $\quad$ $x - 2 = -\frac{5}{2}$

$x = 2 + \frac{5}{2}$ $\quad$ $x = 2 - \frac{5}{2}$

$x = \frac{9}{2}$ $\quad$ $x = -\frac{1}{2}$

The solutions are $\frac{9}{2}$ and $-\frac{1}{2}$.

4. $2x^2 + 10x = -13$

$x^2 + 5x = -\frac{13}{2}$

$x^2 + 5x + \frac{25}{4} = -\frac{13}{2} + \frac{25}{4}$

$\left(x + \frac{5}{2}\right)^2 = -\frac{1}{4}$

There is no real solution to this equation since the square root of a negative number is not a real number.

5. $2x^2 = -6x + 5$

$x^2 = -3x + \frac{5}{2}$

$x^2 + 3x = \frac{5}{2}$

$x^2 + 3x + \frac{9}{4} = \frac{5}{2} + \frac{9}{4}$

$\left(x + \frac{3}{2}\right)^2 = \frac{19}{4}$

$x + \frac{3}{2} = \sqrt{\frac{19}{4}}$ or $x + \frac{3}{2} = -\sqrt{\frac{19}{4}}$

$x + \frac{3}{2} = \frac{\sqrt{19}}{2}$ $\quad$ $x + \frac{3}{2} = -\frac{\sqrt{19}}{2}$

$x = -\frac{3}{2} + \frac{\sqrt{19}}{2}$ $\quad$ $x = -\frac{3}{2} - \frac{\sqrt{19}}{2}$

The solutions are $\frac{-3 \pm \sqrt{19}}{2}$.

Vocabulary and Readiness Check

1. By the zero-factor property, if the product of two numbers is zero, then at least one of these two numbers must be <u>zero</u>.

2. If a is a positive number, and if $x^2 = a$, then $x = \underline{\pm\sqrt{a}}$.

3. An equation that can be written in the form $ax^2 + bx + c = 0$ where a, b, and c are real numbers and a is not zero is called a <u>quadratic equation</u>.

4. The process of solving a quadratic equation by writing it in the form $(x+a)^2 = c$ is called <u>completing the square</u>.

5. To complete the square on $x^2 + 8x$, add <u>16</u>.

6. To complete the square on $x^2 + bx$, add $\underline{\left(\frac{b}{2}\right)^2}$.

7. $\left(\frac{8}{2}\right)^2 = 4^2 = 16$; $p^2 + 8p + \underline{16}$

8. $\left(\frac{4}{2}\right)^2 = 2^2 = 2$; $p^2 + 4p + \underline{4}$

9. $\left(\frac{20}{2}\right)^2 = 10^2 = 100$; $x^2 + 20x + \underline{100}$

10. $\left(\frac{18}{2}\right)^2 = 9^2 = 81$; $x^2 + 18x + \underline{81}$

11. $\left(\frac{14}{2}\right)^2 = 7^2 = 49$; $y^2 + 14y + \underline{49}$

Copyright © 2012 Pearson Education, Inc. Publishing as Prentice Hall.

12. $\left(\frac{2}{2}\right)^2 = 1^2 = 1;\ y^2 + 2y + \underline{1}$

Exercise Set 9.2

1.
$$x^2 + 8x = -12$$
$$x^2 + 8x + \left(\frac{8}{2}\right)^2 = -12 + \left(\frac{8}{2}\right)^2$$
$$x^2 + 8x + 4^2 = -12 + 4^2$$
$$x^2 + 8x + 16 = -12 + 16$$
$$(x+4)^2 = 4$$
$$x + 4 = \sqrt{4} \quad \text{or} \quad x + 4 = -\sqrt{4}$$
$$x = -4 + 2 \qquad x = -4 - 2$$
$$x = -2 \qquad x = -6$$
The solutions are $x = -6$ and $x = -2$.

3.
$$x^2 + 2x - 7 = 0$$
$$x^2 + 2x = 7$$
$$x^2 + 2x + \left(\frac{2}{2}\right)^2 = 7 + \left(\frac{2}{2}\right)^2$$
$$x^2 + 2x + 1 = 7 + 1$$
$$(x+1)^2 = 8$$
$$x + 1 = \sqrt{8} \quad \text{or} \quad x + 1 = -\sqrt{8}$$
$$x + 1 = 2\sqrt{2} \qquad x + 1 = -2\sqrt{2}$$
$$x = -1 + 2\sqrt{2} \qquad x = -1 - 2\sqrt{2}$$
The solutions are $x = -1 \pm 2\sqrt{2}$.

5.
$$x^2 - 6x = 0$$
$$x^2 - 6x + \left(\frac{-6}{2}\right)^2 = 0 + \left(\frac{-6}{2}\right)^2$$
$$x^2 - 6x + (-3)^2 = 0 + (-3)^2$$
$$x^2 - 6x + 9 = 9$$
$$(x-3)^2 = 9$$
$$x - 3 = \sqrt{9} \quad \text{or} \quad x - 3 = -\sqrt{9}$$
$$x = 3 + 3 \qquad x = 3 - 3$$
$$x = 6 \qquad x = 0$$
The solutions are $x = 0$ and $x = 6$.

7.
$$y^2 + 5y + 4 = 0$$
$$y^2 + 5y = -4$$
$$y^2 + 5y + \left(\frac{5}{2}\right)^2 = -4 + \left(\frac{5}{2}\right)^2$$
$$\left(y + \frac{5}{2}\right)^2 = \frac{9}{4}$$
$$y + \frac{5}{2} = \sqrt{\frac{9}{4}} \quad \text{or} \quad y + \frac{5}{2} = -\sqrt{\frac{9}{4}}$$
$$y = -\frac{5}{2} + \frac{3}{2} \qquad y = -\frac{5}{2} - \frac{3}{2}$$
$$y = -\frac{2}{2} = -1 \qquad y = -\frac{8}{2} = -4$$
The solutions are $y = -1$ and $y = -4$.

9.
$$x^2 - 2x - 1 = 0$$
$$x^2 - 2x = 1$$
$$x^2 - 2x + \left(\frac{-2}{2}\right)^2 = 1 + \left(\frac{-2}{2}\right)^2$$
$$x^2 - 2x + (-1)^2 = 1 + (-1)^2$$
$$x^2 - 2x + 1 = 1 + 1$$
$$(x-1)^2 = 2$$
$$x - 1 = \sqrt{2} \quad \text{or} \quad x - 1 = -\sqrt{2}$$
$$x = 1 + \sqrt{2} \qquad x = 1 - \sqrt{2}$$
The solutions are $x = 1 \pm \sqrt{2}$.

11.
$$z^2 + 5z = 7$$
$$z^2 + 5z + \left(\frac{5}{2}\right)^2 = 7 + \left(\frac{5}{2}\right)^2$$
$$z^2 + 5z + \frac{25}{4} = 7 + \frac{25}{4}$$
$$\left(z + \frac{5}{2}\right)^2 = \frac{53}{4}$$
$$z + \frac{5}{2} = \sqrt{\frac{53}{4}} \quad \text{or} \quad z + \frac{5}{2} = -\sqrt{\frac{53}{4}}$$
$$z = -\frac{5}{2} + \frac{\sqrt{53}}{2} \qquad z = -\frac{5}{2} - \frac{\sqrt{53}}{2}$$
The solutions are $z = \frac{-5 \pm \sqrt{53}}{2}$.

Copyright © 2012 Pearson Education, Inc. Publishing as Prentice Hall.

13.
$$3x^2-6x=24$$
$$x^2-2x=8$$
$$x^2-2x+\left(\frac{-2}{2}\right)^2=8+\left(\frac{-2}{2}\right)^2$$
$$x^2-2x+(-1)^2=8+(-1)^2$$
$$x^2-2x+1=8+1$$
$$(x-1)^2=9$$
$$x-1=\sqrt{9} \quad \text{or} \quad x-1=-\sqrt{9}$$
$$x=1+3 \qquad x=1-3$$
$$x=4 \qquad x=-2$$
The solutions are $x=-2$ and $x=4$.

15.
$$5x^2+10x+6=0$$
$$x^2+2x+\frac{6}{5}=0$$
$$x^2+2x=-\frac{6}{5}$$
$$x^2+2x+\left(\frac{2}{2}\right)^2=-\frac{6}{5}+\left(\frac{2}{2}\right)^2$$
$$(x+1)^2=-\frac{1}{5}$$
This has no real solution because $\sqrt{-\frac{1}{5}}$ is not a real number.

17.
$$2x^2=6x+5$$
$$2x^2-6x=5$$
$$x^2-3x=\frac{5}{2}$$
$$x^2-3x+\left(\frac{-3}{2}\right)^2=\frac{5}{2}+\left(\frac{-3}{2}\right)^2$$
$$x^2-3x+\frac{9}{4}=\frac{5}{2}+\frac{9}{4}$$
$$\left(x-\frac{3}{2}\right)^2=\frac{10}{4}+\frac{9}{4}$$
$$\left(x-\frac{3}{2}\right)^2=\frac{19}{4}$$
$$x-\frac{3}{2}=\sqrt{\frac{19}{4}} \quad \text{or} \quad x-\frac{3}{2}=-\sqrt{\frac{19}{4}}$$
$$x=\frac{3}{2}+\frac{\sqrt{19}}{2} \qquad x=\frac{3}{2}-\frac{\sqrt{19}}{2}$$
$$x=\frac{3+\sqrt{19}}{2} \qquad x=\frac{3-\sqrt{19}}{2}$$
The solutions are $x=\frac{3\pm\sqrt{19}}{2}$.

19.
$$2y^2+8y+5=0$$
$$y^2+4y+\frac{5}{2}=0$$
$$y^2+4y=-\frac{5}{2}$$
$$y^2+4y+\left(\frac{4}{2}\right)^2=-\frac{5}{2}+\left(\frac{4}{2}\right)^2$$
$$(y+2)^2=\frac{3}{2}$$
$$y+2=\sqrt{\frac{3}{2}} \quad \text{or} \quad y+2=-\sqrt{\frac{3}{2}}$$
$$y=-2+\frac{\sqrt{6}}{2} \qquad y=-2-\frac{\sqrt{6}}{2}$$
The solutions are $y=-2\pm\frac{\sqrt{6}}{2}$ or $y=\frac{-4\pm\sqrt{6}}{2}$.

21.
$$x^2+6x-25=0$$
$$x^2+6x=25$$
$$x^2+6x+\left(\frac{6}{2}\right)^2=25+\left(\frac{6}{2}\right)^2$$
$$x^2+6x+3^2=25+3^2$$
$$x^2+6x+9=25+9$$
$$(x+3)^2=34$$
$$x+3=\sqrt{34} \quad \text{or} \quad x+3=-\sqrt{34}$$
$$x=-3+\sqrt{34} \qquad x=-3-\sqrt{34}$$
The solutions are $x=-3\pm\sqrt{34}$.

23.
$$x^2-3x-3=0$$
$$x^2-3x=3$$
$$x^2-3x+\left(-\frac{3}{2}\right)^2=3+\left(-\frac{3}{2}\right)^2$$
$$\left(x-\frac{3}{2}\right)^2=\frac{21}{4}$$

Copyright © 2012 Pearson Education, Inc. Publishing as Prentice Hall.

$$x-\frac{3}{2}=\sqrt{\frac{21}{4}} \quad \text{or} \quad x-\frac{3}{2}=-\sqrt{\frac{21}{4}}$$
$$x=\frac{3}{2}+\frac{\sqrt{21}}{2} \qquad x=\frac{3}{2}-\frac{\sqrt{21}}{2}$$

The solutions are $x=\frac{3\pm\sqrt{21}}{2}$.

25.
$$2y^2-3y+1=0$$
$$2y^2-3y=-1$$
$$y^2-\frac{3}{2}y=-\frac{1}{2}$$
$$y^2-\frac{3}{2}y+\left(\frac{-\frac{3}{2}}{2}\right)^2=-\frac{1}{2}+\left(\frac{-\frac{3}{2}}{2}\right)^2$$
$$y^2-\frac{3}{2}y+\left(-\frac{3}{4}\right)^2=-\frac{1}{2}+\left(-\frac{3}{4}\right)^2$$
$$y^2-\frac{3}{2}y+\frac{9}{16}=-\frac{1}{2}+\frac{9}{16}$$
$$\left(y-\frac{3}{4}\right)^2=\frac{1}{16}$$

$$y-\frac{3}{4}=\sqrt{\frac{1}{16}} \quad \text{or} \quad y-\frac{3}{4}=-\sqrt{\frac{1}{16}}$$
$$y=\frac{3}{4}+\frac{1}{4} \qquad y=\frac{3}{4}-\frac{1}{4}$$
$$y=1 \qquad y=\frac{1}{2}$$

The solutions are $y=\frac{1}{2}$ and $y=1$.

27.
$$x(x+3)=18$$
$$x^2+3x=18$$
$$x^2+3x+\left(\frac{3}{2}\right)^2=18+\left(\frac{3}{2}\right)^2$$
$$\left(x+\frac{3}{2}\right)^2=\frac{81}{4}$$

$$x+\frac{3}{2}=\sqrt{\frac{81}{4}} \quad \text{or} \quad x+\frac{3}{2}=-\sqrt{\frac{81}{4}}$$
$$x=-\frac{3}{2}+\frac{9}{2} \qquad x=-\frac{3}{2}-\frac{9}{2}$$
$$x=\frac{6}{2}=3 \qquad x=-\frac{12}{2}=-6$$

The solutions are $x=3$ and $x=-6$.

29.
$$3z^2+6z+4=0$$
$$3z^2+6z=-4$$
$$z^2+2z=-\frac{4}{3}$$
$$z^2+2z+\left(\frac{2}{2}\right)^2=-\frac{4}{3}+\left(\frac{2}{2}\right)^2$$
$$z^2+2z+1^2=-\frac{4}{3}+1^2$$
$$z^2+2z+1=-\frac{4}{3}+1$$
$$(z+1)^2=-\frac{1}{3}$$

The equation has no real solution, since the square root of $-\frac{1}{3}$ is not a real number.

31.
$$4x^2+16x=48$$
$$x^2+4x=12$$
$$x^2+4x+\left(\frac{4}{2}\right)^2=12+\left(\frac{4}{2}\right)^2$$
$$(x+2)^2=16$$
$$x+2=\sqrt{16} \quad \text{and} \quad x+2=-\sqrt{16}$$
$$x=-2+4 \qquad x=-2-4$$
$$x=2 \qquad x=-6$$

The solutions are $x=2$ and $x=-6$.

33.
$$\frac{3}{4}-\sqrt{\frac{25}{16}}=\frac{3}{4}-\frac{\sqrt{25}}{\sqrt{16}}$$
$$=\frac{3}{4}-\frac{5}{4}$$
$$=\frac{3-5}{4}$$
$$=\frac{-2}{4}$$
$$=-\frac{1}{2}$$

35. $\frac{1}{2}+\sqrt{\frac{9}{4}}=\frac{1}{2}+\frac{3}{2}=\frac{4}{2}=2$

37. $\frac{6+4\sqrt{5}}{2}=\frac{6}{2}+\frac{4\sqrt{5}}{2}=3+2\sqrt{5}$

39. $\frac{3-9\sqrt{2}}{6}=\frac{3}{6}-\frac{9\sqrt{2}}{6}=\frac{1}{2}-\frac{3\sqrt{2}}{2}=\frac{1-3\sqrt{2}}{2}$

41. answers may vary

Copyright © 2012 Pearson Education, Inc. Publishing as Prentice Hall.

43. a. $x^2+6x+9=11$

$(x+3)^2=11$

$x+3=\sqrt{11}$ or $x+3=-\sqrt{11}$

$x=-3+\sqrt{11}$ $\quad$ $x=-3-\sqrt{11}$

The solutions are $x=-3\pm\sqrt{11}$.

b. answers may vary

45. $x^2+kx+\left(\frac{k}{2}\right)^2$ is a perfect square trinomial. If $x^2+kx+16$ is a perfect square trinomial, then

$\left(\frac{k}{2}\right)^2=16$

$\frac{k}{2}=\sqrt{16}$ or $\frac{k}{2}=-\sqrt{16}$

$k=2\cdot 4$ $\quad$ $k=2(-4)$

$k=8$ $\quad$ $k=-8$

$x^2+kx+16$ is a perfect square trinomial when $k=8$ or $k=-8$.

47.

$y=2.5x^2+7.5x+122$

$392=2.5x^2+7.5x+122$

$270=2.5x^2+7.5x$

$108=x^2+3x$

$\left(\frac{3}{2}\right)^2+108=x^2+3x+\left(\frac{3}{2}\right)^2$

$110.25=(x+1.5)^2$

$\sqrt{110.25}=x+1.5$ or $-\sqrt{110.25}=x+1.5$

$10.5=x+1.5$ $\quad$ $-10.5=x+1.5$

$9=x$ $\quad$ $-12=x$

Since the time will not be a negative number in this context, reject the solution -12. The solution is $x=9$, so the model predicts that retail sales from online shopping will be \$392 billion in 2011 (2002 + 9).

49. The solutions are $x=-6$ and $x=-2$.

51. The solutions are $x\approx -0.68$ and $x\approx 3.68$

Section 9.3 Practice

1. $2x^2-x-5=0$

$a=2,\ b=-1,\ c=-5$

$$x=\frac{-b\pm\sqrt{b^2-4ac}}{2a}$$

$$x=\frac{-(-1)\pm\sqrt{(-1)^2-4(2)(-5)}}{2(2)}$$

$$=\frac{1\pm\sqrt{1+40}}{4}$$

$$=\frac{1\pm\sqrt{41}}{4}$$

The solutions are $\frac{1\pm\sqrt{41}}{4}$.

2. $3x^2+8x=3$

$3x^2+8x-3=0$

$a=3,\ b=8,\ c=-3$

$$x=\frac{-b\pm\sqrt{b^2-4ac}}{2a}$$

$$x=\frac{-8\pm\sqrt{8^2-4(3)(-3)}}{2(3)}$$

$$=\frac{-8\pm\sqrt{64+36}}{6}$$

$$=\frac{-8\pm\sqrt{100}}{6}$$

$$=\frac{-8\pm 10}{6}$$

$x=\frac{-8+10}{6}=\frac{1}{3}$ or $\frac{-8-10}{6}=-3$

The solutions are $\frac{1}{3}$ and -3.

3. $5x^2=2$

$5x^2-2=0$

$a=5,\ b=0,\ c=-2$

Copyright © 2012 Pearson Education, Inc. Publishing as Prentice Hall.

$$x = \frac{-b \pm \sqrt{b^2 - 4ac}}{2a}$$
$$x = \frac{-0 \pm \sqrt{0^2 - 4(5)(-2)}}{2(5)}$$
$$= \pm \frac{\sqrt{40}}{10}$$
$$= \pm \frac{2\sqrt{10}}{10}$$
$$= \pm \frac{\sqrt{10}}{5}$$

The solutions are $\pm \frac{\sqrt{10}}{5}$.

4. $x^2 = -2x - 3$

$x^2 + 2x + 3 = 0$

$a = 1, b = 2, c = 3$

$$x = \frac{-b \pm \sqrt{b^2 - 4ac}}{2a}$$
$$x = \frac{-2 \pm \sqrt{2^2 - 4(1)(3)}}{2(1)} = \frac{-2 \pm \sqrt{-8}}{2}$$

There is no real number solution because $\sqrt{-8}$ is not a real number.

5. $\frac{1}{3}x^2 - x = 1$

$\frac{1}{3}x^2 - x - 1 = 0$

$a = \frac{1}{3},\ b = -1, c = -1$

$$x = \frac{-b \pm \sqrt{b^2 - 4ac}}{2a}$$
$$x = \frac{-(-1) \pm \sqrt{(-1)^2 - 4\left(\frac{1}{3}\right)(-1)}}{2\left(\frac{1}{3}\right)}$$
$$= \frac{1 \pm \sqrt{\frac{7}{3}}}{\frac{2}{3}}$$
$$= \frac{1 \pm \frac{\sqrt{21}}{3}}{\frac{2}{3}}$$
$$= \frac{3 \pm \sqrt{21}}{2}$$

The solutions are $\frac{3 \pm \sqrt{21}}{2}$.

6. $\frac{1 + \sqrt{41}}{4} \approx 1.9$

$1 - \frac{\sqrt{41}}{4} \approx -1.4$

Vocabulary and Readiness Check

1. The quadratic formula is $\underline{x = \frac{-b \pm \sqrt{b^2 - 4ac}}{2a}}$.

2. $5x^2 - 7x + 1 = 0;\ a = \underline{5}, b = \underline{-7}, c = \underline{1}$

3. $x^2 + 3x - 7 = 0;\ a = \underline{1}, b = \underline{3}, c = \underline{-7}$

4. $x^2 - 6 = 0;\ a = \underline{1}, b = \underline{0}, c = \underline{-6}$

5. $x^2 + x - 1 = 0;\ a = \underline{1}, b = \underline{1}, c = \underline{-1}$

6. $9x^2 - 4 = 0;\ a = \underline{9}, b = \underline{0}, c = \underline{-4}$

7. $$\frac{-1 \pm \sqrt{1^2 - 4(1)(-2)}}{2(1)} = \frac{-1 \pm \sqrt{1+8}}{2}$$
$$= \frac{-1 \pm \sqrt{9}}{2}$$
$$= \frac{-1 \pm 3}{2}$$

$$\frac{-1+3}{2} = \frac{2}{2} = 1$$
$$\frac{-1-3}{2} = \frac{-4}{2} = -2$$

8. $$\frac{-(-5) \pm \sqrt{(-5)^2 - 4(2)(3)}}{2(2)} = \frac{5 \pm \sqrt{25-24}}{4}$$
$$= \frac{5 \pm \sqrt{1}}{4}$$
$$= \frac{5 \pm 1}{4}$$

$$\frac{5+1}{4} = \frac{6}{4} = \frac{3}{2}$$
$$\frac{5-1}{4} = \frac{4}{4} = 1$$

9. $$\frac{-5 \pm \sqrt{5^2 - 4(1)(2)}}{2(1)} = \frac{-5 \pm \sqrt{25-8}}{2} = \frac{-5 \pm \sqrt{17}}{2}$$

Copyright © 2012 Pearson Education, Inc. Publishing as Prentice Hall.

10. $\dfrac{-7\pm\sqrt{7^2-4(2)(1)}}{2(2)}=\dfrac{-7\pm\sqrt{49-8}}{4}=\dfrac{-7\pm\sqrt{41}}{4}$

Exercise Set 9.3

1. $x^2-3x+2=0$

$a=1,\ b=-3,\ c=2$

$$x=\frac{-b\pm\sqrt{b^2-4ac}}{2a}$$

$$x=\frac{-(-3)\pm\sqrt{(-3)^2-4(1)(2)}}{2(1)}$$

$$=\frac{3\pm\sqrt{9-8}}{2}$$

$$=\frac{3\pm\sqrt{1}}{2}$$

$$=\frac{3\pm1}{2}$$

$x=\dfrac{3+1}{2}=2$ or $x=\dfrac{3-1}{2}=1$

The solutions are $x=2$ and $x=1$.

3. $3k^2+7k+1=0$

$a=3,\ b=7,\ c=1$

$$k=\frac{-b\pm\sqrt{b^2-4ac}}{2a}$$

$$k=\frac{-7\pm\sqrt{7^2-4(3)(1)}}{2(3)}$$

$$=\frac{-7\pm\sqrt{49-12}}{6}$$

$$=\frac{-7\pm\sqrt{37}}{6}$$

The solutions are $k=\dfrac{-7\pm\sqrt{37}}{6}$.

5. $4x^2-3=0$

$4x^2+0x-3=0$

$a=4,\ b=0,\ c=-3$

$$x=\frac{-b\pm\sqrt{b^2-4ac}}{2a}$$

$$x=\frac{-0\pm\sqrt{0^2-4(4)(-3)}}{2(4)}$$

$$=\frac{0\pm\sqrt{0+48}}{8}$$

$$=\frac{\pm\sqrt{48}}{8}$$

$$=\pm\frac{4\sqrt{3}}{8}$$

$$=\pm\frac{\sqrt{3}}{2}$$

The solutions are $x=\pm\dfrac{\sqrt{3}}{2}$.

7. $5z^2-4z+3=0$

$a=5,\ b=-4,\ c=3$

$$z=\frac{-b\pm\sqrt{b^2-4ac}}{2a}$$

$$z=\frac{-(-4)\pm\sqrt{(-4)^2-4(5)(3)}}{2(5)}$$

$$=\frac{4\pm\sqrt{16-60}}{10}$$

$$=\frac{4\pm\sqrt{-44}}{10}$$

The equation has no solution, since the square root of -44 is not a real number.

9. $y^2=7y+30$

$y^2-7y-30=0$

$a=1,\ b=-7,\ c=-30$

$$y=\frac{-b\pm\sqrt{b^2-4ac}}{2a}$$

$$y=\frac{-(-7)\pm\sqrt{(-7)^2-4(1)(-30)}}{2(1)}$$

$$=\frac{7\pm\sqrt{49+120}}{2}$$

$$=\frac{7\pm\sqrt{169}}{2}$$

$$=\frac{7\pm13}{2}$$

$y=\dfrac{7+13}{2}=10$ or $y=\dfrac{7-13}{2}=-3$

The solutions are $y=10$ and $y=-3$.

Copyright © 2012 Pearson Education, Inc. Publishing as Prentice Hall.

11. $2x^2 = 10$

$2x^2 - 10 = 0$

$a = 2, b = 0, c = -10$

$$x = \frac{-b \pm \sqrt{b^2 - 4ac}}{2a}$$

$$x = \frac{-0 \pm \sqrt{0^2 - 4(2)(-10)}}{2(2)}$$

$$= \frac{\pm\sqrt{80}}{4}$$

$$= \frac{\pm 4\sqrt{5}}{4}$$

$$= \pm\sqrt{5}$$

The solutions are $x = \pm\sqrt{5}$.

13. $m^2 - 12 = m$

$m^2 - m - 12 = 0$

$a = 1, b = -1, c = -12$

$$m = \frac{-b \pm \sqrt{b^2 - 4ac}}{2a}$$

$$m = \frac{-(-1) \pm \sqrt{(-1)^2 - 4(1)(-12)}}{2(1)}$$

$$= \frac{1 \pm \sqrt{1+48}}{2}$$

$$= \frac{1 \pm \sqrt{49}}{2}$$

$$= \frac{1 \pm 7}{2}$$

$m = \frac{1+7}{2} = 4$ or $m = \frac{1-7}{2} = -3$

The solutions are $m = 4$ and $m = -3$.

15. $3 - x^2 = 4x$

$0 = x^2 + 4x - 3$

$a = 1, b = 4, c = -3$

$$x = \frac{-b \pm \sqrt{b^2 - 4ac}}{2a}$$

$$x = \frac{-4 \pm \sqrt{4^2 - 4(1)(-3)}}{2(1)}$$

$$= \frac{-4 \pm \sqrt{16+12}}{2}$$

$$= \frac{-4 \pm \sqrt{28}}{2}$$

$$= \frac{-4 \pm 2\sqrt{7}}{2}$$

$$= -2 \pm \sqrt{7}$$

The solutions are $x = -2 \pm \sqrt{7}$.

17. $6x^2 + 9x = 2$

$6x^2 + 9x - 2 = 0$

$a = 6, b = 9, c = -2$

$$x = \frac{-b \pm \sqrt{b^2 - 4ac}}{2a}$$

$$x = \frac{-9 \pm \sqrt{9^2 - 4(6)(-2)}}{2(6)}$$

$$= \frac{-9 \pm \sqrt{81+48}}{12}$$

$$= \frac{-9 \pm \sqrt{129}}{12}$$

The solutions are $x = \frac{-9 \pm \sqrt{129}}{12}$.

19. $7p^2 + 2 = 8p$

$7p^2 - 8p + 2 = 0$

$a = 7, b = -8, c = 2$

$$p = \frac{-b \pm \sqrt{b^2 - 4ac}}{2a}$$

$$p = \frac{-(-8) \pm \sqrt{(-8)^2 - 4(7)(2)}}{2(7)}$$

$$= \frac{8 \pm \sqrt{64-56}}{14}$$

$$= \frac{8 \pm \sqrt{8}}{14}$$

$$= \frac{8 \pm 2\sqrt{2}}{14}$$

$$= \frac{4 \pm \sqrt{2}}{7}$$

The solutions are $p = \frac{4 \pm \sqrt{2}}{7}$.

Copyright © 2012 Pearson Education, Inc. Publishing as Prentice Hall.

21. $x^2 - 6x + 2 = 0$

$a = 1, b = -6, c = 2$

$$x = \frac{-b \pm \sqrt{b^2 - 4ac}}{2a}$$

$$x = \frac{-(-6) \pm \sqrt{(-6)^2 - 4(1)(2)}}{2(1)}$$

$$= \frac{6 \pm \sqrt{36-8}}{2}$$

$$= \frac{6 \pm \sqrt{28}}{2}$$

$$= \frac{6 \pm 2\sqrt{7}}{2}$$

$$= \frac{2\left(3 \pm \sqrt{7}\right)}{2}$$

$$= 3 \pm \sqrt{7}$$

The solutions are $x = 3 \pm \sqrt{7}$.

23. $2x^2 - 6x + 3 = 0$

$a = 2, b = -6, c = 3$

$$x = \frac{-b \pm \sqrt{b^2 - 4ac}}{2a}$$

$$x = \frac{-(-6) \pm \sqrt{(-6)^2 - 4(2)(3)}}{2(2)}$$

$$= \frac{6 \pm \sqrt{36-24}}{4}$$

$$= \frac{6 \pm \sqrt{12}}{4}$$

$$= \frac{6 \pm 2\sqrt{3}}{4}$$

$$= \frac{3 \pm \sqrt{3}}{2}$$

The solutions are $x = \frac{3 \pm \sqrt{3}}{2}$.

25. $3x^2 = 1 - 2x$

$3x^2 + 2x - 1 = 0$

$a = 3, b = 2, c = -1$

$$x = \frac{-b \pm \sqrt{b^2 - 4ac}}{2a}$$

$$x = \frac{-2 \pm \sqrt{2^2 - 4(3)(-1)}}{2(3)}$$

$$= \frac{-2 \pm \sqrt{4+12}}{6}$$

$$= \frac{-2 \pm \sqrt{16}}{6}$$

$$= \frac{-2 \pm 4}{6}$$

$x = \frac{-2+4}{6} = \frac{1}{3}$ or $x = \frac{-2-4}{6} = -1$

The solutions are $x = \frac{1}{3}$ and $x = -1$.

27. $4y^2 = 6y + 1$

$4y^2 - 6y - 1 = 0$

$a = 4, b = -6, c = -1$

$$y = \frac{-b \pm \sqrt{b^2 - 4ac}}{2a}$$

$$y = \frac{-(-6) \pm \sqrt{(-6)^2 - 4(4)(-1)}}{2(4)}$$

$$= \frac{6 \pm \sqrt{36+16}}{8}$$

$$= \frac{6 \pm \sqrt{52}}{8}$$

$$= \frac{6 \pm 2\sqrt{13}}{8}$$

$$= \frac{3 \pm \sqrt{13}}{4}$$

The solutions are $y = \frac{3 \pm \sqrt{13}}{4}$.

29. $20y^2 = 3 - 11y$

$20y^2 + 11y - 3 = 0$

$a = 20, b = 11, c = -3$

$$y = \frac{-b \pm \sqrt{b^2 - 4ac}}{2a}$$

Copyright © 2012 Pearson Education, Inc. Publishing as Prentice Hall.

$$y=\frac{-11\pm\sqrt{11^2-4(20)(-3)}}{2(20)}$$
$$=\frac{-11\pm\sqrt{121+240}}{40}$$
$$=\frac{-11\pm\sqrt{361}}{40}$$
$$=\frac{-11\pm19}{40}$$
$$y=\frac{-11+19}{40}=\frac{1}{5} \text{ or } y=\frac{-11-19}{40}=-\frac{3}{4}$$

The solutions are $y=\frac{1}{5}$ and $y=-\frac{3}{4}$.

31. $x^2+x+2=0$

$a=1, b=1, c=2$

$$x=\frac{-b\pm\sqrt{b^2-4ac}}{2a}$$
$$x=\frac{-1\pm\sqrt{1^2-4(1)(2)}}{2(1)}=\frac{-1\pm\sqrt{1-8}}{2}=\frac{-1\pm\sqrt{-7}}{2}$$

The equation has no solution, since the square root of -7 is not a real number.

33. $$\frac{m^2}{2}=m+\frac{1}{2}$$
$$\frac{m^2}{2}-m-\frac{1}{2}=0$$
$$m^2-2m-1=0$$

$a=1, b=-2, c=-1$

$$m=\frac{-b\pm\sqrt{b^2-4ac}}{2a}$$
$$m=\frac{-(-2)\pm\sqrt{(-2)^2-4(1)(-1)}}{2(1)}$$
$$=\frac{2\pm\sqrt{4+4}}{2}$$
$$=\frac{2\pm\sqrt{8}}{2}$$
$$=\frac{2\pm2\sqrt{2}}{2}$$
$$=\frac{2\left(1\pm\sqrt{2}\right)}{2}$$
$$=1\pm\sqrt{2}$$

The solutions are $m=1\pm\sqrt{2}$.

35. $3p^2-\frac{2}{3}p+1=0$

$$9p^2-2p+3=0$$

$a=9, b=-2, c=3$

$$p=\frac{-b\pm\sqrt{b^2-4ac}}{2a}$$
$$p=\frac{-(-2)\pm\sqrt{(-2)^2-4(9)(3)}}{2(9)}$$
$$=\frac{2\pm\sqrt{4-108}}{18}$$
$$=\frac{2\pm\sqrt{-104}}{18}$$

The equation has no solution, since the square root of -104 is not a real number.

37. $$4p^2+\frac{3}{2}=-5p$$
$$4p^2+5p+\frac{3}{2}=0$$
$$8p^2+10p+3=0$$

$a=8, b=10, c=3$

$$p=\frac{-b\pm\sqrt{b^2-4ac}}{2a}$$
$$p=\frac{-10\pm\sqrt{10^2-4(8)(3)}}{2(8)}$$
$$=\frac{-10\pm\sqrt{100-96}}{16}$$
$$=\frac{-10\pm\sqrt{4}}{16}$$
$$=\frac{-10\pm2}{16}$$
$$p=\frac{-10+2}{16}=-\frac{1}{2} \text{ or } p=\frac{-10-2}{16}=-\frac{3}{4}$$

The solutions are $p=-\frac{1}{2}$ and $p=-\frac{3}{4}$.

39. $$5x^2=\frac{7}{2}x+1$$
$$5x^2-\frac{7}{2}x-1=0$$
$$10x^2-7x-2=0$$

$a=10, b=-7, c=-2$

$$x=\frac{-b\pm\sqrt{b^2-4ac}}{2a}$$

Copyright © 2012 Pearson Education, Inc. Publishing as Prentice Hall.

$$x = \frac{-(-7) \pm \sqrt{(-7)^2 - 4(10)(-2)}}{2(10)}$$
$$= \frac{7 \pm \sqrt{49+80}}{20}$$
$$= \frac{7 \pm \sqrt{129}}{20}$$

The solutions are $x = \frac{7 \pm \sqrt{129}}{20}$.

41. $x^2 - \frac{11}{2}x - \frac{1}{2} = 0$

$2x^2 - 11x - 1 = 0$

$a = 2, b = -11, c = -1$

$$x = \frac{-b \pm \sqrt{b^2 - 4ac}}{2a}$$
$$x = \frac{-(-11) \pm \sqrt{(-11)^2 - 4(2)(-1)}}{2(2)}$$
$$= \frac{11 \pm \sqrt{121+8}}{4}$$
$$= \frac{11 \pm \sqrt{129}}{4}$$

The solutions are $x = \frac{11 \pm \sqrt{129}}{4}$.

43. $5z^2 - 2z = \frac{1}{5}$

$5z^2 - 2z - \frac{1}{5} = 0$

$25z^2 - 10z - 1 = 0$

$a = 25, b = -10, c = -1$

$$z = \frac{-b \pm \sqrt{b^2 - 4ac}}{2a}$$
$$z = \frac{-(-10) \pm \sqrt{(-10)^2 - 4(25)(-1)}}{2(25)}$$
$$= \frac{10 \pm \sqrt{100+100}}{50}$$
$$= \frac{10 \pm \sqrt{200}}{50}$$
$$= \frac{10 \pm 10\sqrt{2}}{50}$$
$$= \frac{1 \pm \sqrt{2}}{5}$$

The solutions are $z = \frac{1 \pm \sqrt{2}}{5}$.

45. $3x^2 = 21$

$3x^2 - 21 = 0$

$3x^2 + 0x - 21 = 0$

$a = 3, b = 0, c = -21$

$$x = \frac{-b \pm \sqrt{b^2 - 4ac}}{2a}$$
$$x = \frac{-0 \pm \sqrt{0^2 - 4(3)(-21)}}{2(3)}$$
$$= \frac{\pm\sqrt{0+252}}{6}$$
$$= \frac{\pm\sqrt{252}}{6}$$
$$= \frac{\pm 6\sqrt{7}}{6}$$
$$= \pm\sqrt{7}$$

The solutions are $x = \pm\sqrt{7}$ or $x \approx -2.6$ and $x \approx 2.6$.

47. $x^2 + 6x + 1 = 0$

$a = 1, b = 6, c = 1$

$$x = \frac{-b \pm \sqrt{b^2 - 4ac}}{2a}$$
$$x = \frac{-6 \pm \sqrt{6^2 - 4(1)(1)}}{2(1)}$$
$$= \frac{-6 \pm \sqrt{36-4}}{2}$$
$$= \frac{-6 \pm \sqrt{32}}{2}$$
$$= \frac{-6 \pm 4\sqrt{2}}{2}$$
$$= -3 \pm 2\sqrt{2}$$

The solutions are $x = -3 \pm 2\sqrt{2}$ or $x \approx -5.8$ and $x \approx -0.2$.

Copyright © 2012 Pearson Education, Inc. Publishing as Prentice Hall.

49. $x^2 = 9x + 4$

$x^2 - 9x - 4 = 0$

$a = 1, b = -9, c = -4$

$$x = \frac{-b \pm \sqrt{b^2 - 4ac}}{2a}$$

$$x = \frac{-(-9) \pm \sqrt{(-9)^2 - 4(1)(-4)}}{2(1)}$$

$$= \frac{9 \pm \sqrt{81+16}}{2}$$

$$= \frac{9 \pm \sqrt{97}}{2}$$

The solutions are $x = \frac{9 \pm \sqrt{97}}{2}$ or $x \approx 9.4$ and $x \approx -0.4$.

51. $3x^2 - 2x - 2 = 0$

$a = 3, b = -2, c = -2$

$$x = \frac{-b \pm \sqrt{b^2 - 4ac}}{2a}$$

$$x = \frac{-(-2) \pm \sqrt{(-2)^2 - 4(3)(-2)}}{2(3)}$$

$$= \frac{2 \pm \sqrt{4+24}}{6}$$

$$= \frac{2 \pm \sqrt{28}}{6}$$

$$= \frac{2 \pm 2\sqrt{7}}{6}$$

$$= \frac{1 \pm \sqrt{7}}{3}$$

The solutions are $x = \frac{1 \pm \sqrt{7}}{3}$ or $x \approx 1.2$ and $x \approx -0.5$.

53. $y = -3$ is a horizontal line with y-intercept $(0, -3)$.

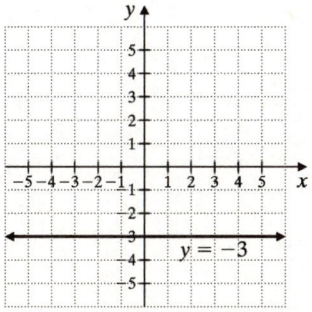

55. $y = 3x - 2$ is a line with slope of 3 and y-intercept $(0, -2)$.

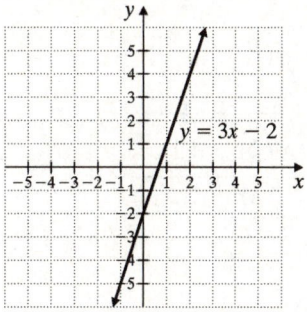

57. $5x^2 + 2 = x$

$5x^2 - x + 2 = 0$

$b = -1$, which is choice c.

59. $7y^2 = 3y$

$0 = -7y^2 + 3y + 0$

$a = -7$, which is choice b.

61. Let x be the width of the chocolate bar. Then the length is $3x - 0.6$.

Area = (length)(width)

$34.65 = (3x - 0.6)(x)$

$34.65 = 3x^2 - 0.6x$

$0 = 3x^2 - 0.6x - 34.65$

$a = 3, b = -0.6, c = -34.65$

$$x = \frac{-b \pm \sqrt{b^2 - 4ac}}{2a}$$

$$x = \frac{-(-0.6) \pm \sqrt{(-0.6)^2 - 4(3)(-34.65)}}{2(3)}$$

$$= \frac{0.6 \pm \sqrt{0.36 + 415.8}}{6}$$

$$= \frac{0.6 \pm \sqrt{416.16}}{6}$$

$$= \frac{0.6 \pm 20.4}{6}$$

$$x = \frac{0.6 + 20.4}{6} = \frac{21}{6} = 3.5$$

or

$$x = \frac{0.6 - 20.4}{6} = \frac{-19.8}{6} = -3.3$$

Since the width is not a negative number, discard the solution -3.3.

$3x - 0.6 = 3(3.5) - 0.6 = 10.5 - 0.6 = 9.9$

The width was 3.5 feet and the length was 9.9 feet.

Copyright © 2012 Pearson Education, Inc. Publishing as Prentice Hall.

63. $x^2+3\sqrt{2}x-5=0$

$a=1,\ b=3\sqrt{2},\ c=-5$

$$x=\frac{-b\pm\sqrt{b^2-4ac}}{2a}$$

$$x=\frac{-3\sqrt{2}\pm\sqrt{\left(3\sqrt{2}\right)^2-4(1)(-5)}}{2(1)}$$

$$=\frac{-3\sqrt{2}\pm\sqrt{18+20}}{2}$$

$$=\frac{-3\sqrt{2}\pm\sqrt{38}}{2}$$

The solutions are $x=\dfrac{-3\sqrt{2}\pm\sqrt{38}}{2}$.

65. answers may vary

67. $7.3^2+5.4z-1.1=0$

$a=7.3,\ b=5.4,\ c=-1.1$

$$z=\frac{-b\pm\sqrt{b^2-4ac}}{2a}$$

$$z=\frac{-5.4\pm\sqrt{(5.4)^2-4(7.3)(-1.1)}}{2(7.3)}$$

$$=\frac{-5.4\pm\sqrt{29.16+32.12}}{14.6}$$

$$=\frac{-5.4\pm\sqrt{61.28}}{14.6}$$

$$z=\frac{-5.4+\sqrt{61.28}}{14.6}\approx 0.2 \text{ or}$$

$$z=\frac{-5.4-\sqrt{61.28}}{14.6}\approx -0.9$$

The solutions are $z\approx 0.2$ and $z=-0.9$.

69. $h=-16t^2+120t+80$

$30=-16t^2+120t+80$

$0=-16t^2+120t+50$

$a=-16,\ b=120,\ c=50$

$$t=\frac{-b\pm\sqrt{b^2-4ac}}{2a}$$

$$t=\frac{-120\pm\sqrt{(120)^2-4(-16)(50)}}{2(-16)}$$

$$=\frac{-120\pm\sqrt{14,400+3200}}{-32}$$

$$=\frac{-120\pm\sqrt{17,600}}{-32}$$

$$t=\frac{-120+\sqrt{17,600}}{-32}\approx -0.4 \text{ or}$$

$$t=\frac{-120-\sqrt{17,600}}{-32}\approx 7.9$$

Since the time of the flight is not a negative umber, discard the solution –0.4. The rocket will reach the height of 30 feet approximately 7.9 second after it is launched.

71. $y=-587x^2+10,243x+26,047$

$70,423=-587x^2+10,243x+26,047$

$0=-587x^2+10,243x-44,376$

$a=-587,\ b=10,243,\ c=-44,376$

$$x=\frac{-b\pm\sqrt{b^2-4ac}}{2a}$$

$$x=\frac{-10,243\pm\sqrt{10,243^2-4(-587)(-44,376)}}{2(-587)}$$

$$=\frac{-10,243\pm\sqrt{104,919,049-104,194,848}}{-1174}$$

$$=\frac{-10,243\pm\sqrt{724,201}}{-1174}$$

$$=\frac{-10,243\pm 851}{-1174}$$

$$x=\frac{-10,243+851}{-1174}=\frac{-9392}{-1174}=8$$

$$\text{or } x=\frac{-10,243-851}{-1174}=\frac{-11,094}{-1174}\approx 9.4$$

ABIOMED's revenues from product sales will first be \$70,423 thousand in 2012 (2004 + 8).

Integrated Review

1. $5x^2-11x+2=0$

$(5x-1)(x-2)=0$

$5x-1=0$ or $x-2=0$

$5x=1$ $\quad x=2$

$x=\dfrac{1}{5}$

The solutions are $x=\dfrac{1}{5}$ and $x=2$.

Copyright © 2012 Pearson Education, Inc. Publishing as Prentice Hall.

2. $5x^2+13x-6=0$

$(5x-2)(x+3)=0$

$5x-2=0$ or $x+3=0$

$5x=2$ $\quad x=-3$

$x=\frac{2}{5}$

The solutions are $x=\frac{2}{5}$ and $x=-3$.

3. $x^2-1=2x$

$x^2-2x-1=0$

$a=1, b=-2, c=-1$

$$x=\frac{-b\pm\sqrt{b^2-4ac}}{2a}$$

$$x=\frac{-(-2)\pm\sqrt{(-2)^2-4(1)(-1)}}{2(1)}$$

$$=\frac{2\pm\sqrt{4+4}}{2}$$

$$=\frac{2\pm\sqrt{8}}{2}$$

$$=\frac{2\pm2\sqrt{2}}{2}$$

$$=1\pm\sqrt{2}$$

The solutions are $x=1\pm\sqrt{2}$.

4. $x^2+7=6x$

$x^2-6x+7=0$

$a=1, b=-6, c=7$

$$x=\frac{-b\pm\sqrt{b^2-4ac}}{2a}$$

$$x=\frac{-(-6)\pm\sqrt{(-6)^2-4(1)(7)}}{2(1)}$$

$$=\frac{6\pm\sqrt{36-28}}{2}$$

$$=\frac{6\pm\sqrt{8}}{2}$$

$$=\frac{6\pm2\sqrt{2}}{2}$$

$$=3\pm\sqrt{2}$$

The solutions are $x=3\pm\sqrt{2}$.

5. $a^2=20$

$a=\pm\sqrt{20}$

$a=\pm2\sqrt{5}$

The solutions are $a=\pm2\sqrt{5}$.

6. $a^2=72$

$a=\pm\sqrt{72}$

$a=\pm6\sqrt{2}$

The solutions are $a=\pm6\sqrt{2}$.

7. $x^2-x+4=0$

$a=1, b=-1, c=4$

$$x=\frac{-b\pm\sqrt{b^2-4ac}}{2a}$$

$$x=\frac{-(-1)\pm\sqrt{(-1)^2-4(1)(4)}}{2(1)}$$

$$=\frac{1\pm\sqrt{1-16}}{2}$$

$$=\frac{1\pm\sqrt{-15}}{2}$$

The equation has no solution, since the square root of -15 is not a real number.

8. $x^2-2x+7=0$

$a=1, b=-2, c=7$

$$x=\frac{-b\pm\sqrt{b^2-4ac}}{2a}$$

$$x=\frac{-(-2)\pm\sqrt{(-2)^2-4(1)(7)}}{2(1)}$$

$$=\frac{2\pm\sqrt{4-28}}{2}$$

$$=\frac{2\pm\sqrt{-24}}{2}$$

The equation has no solution, since the square root of -24 is not a real number.

9. $3x^2-12x+12=0$

$3(x^2-4x+4)=0$

$3(x-2)(x-2)=0$

$x-2=0$ or $x-2=0$

$x=2$ $\quad x=2$

The solution is $x=2$.

Copyright © 2012 Pearson Education, Inc. Publishing as Prentice Hall.

10. $5x^2 - 30x + 45 = 0$

$5(x^2 - 6x + 9) = 0$

$5(x-3)(x-3) = 0$

$x - 3 = 0$ or $x - 3 = 0$

$x = 3$ $\quad$ $x = 3$

The solution is $x = 3$.

11. $9 - 6p + p^2 = 0$

$(3-p)(3-p) = 0$

$3 - p = 0$ or $3 - p = 0$

$3 = p$ $\quad$ $3 = p$

The solution is $p = 3$.

12. $49 - 28p + 4p^2 = 0$

$(7-2p)(7-2p) = 0$

$7 - 2p = 0$ or $7 - 2p = 0$

$7 = 2p$ $\quad$ $7 = 2p$

$\frac{7}{2} = p$ $\quad$ $\frac{7}{2} = p$

The solution is $p = \frac{7}{2}$.

13. $4y^2 - 16 = 0$

$4(y^2 - 4) = 0$

$4(y-2)(y+2) = 0$

$y - 2 = 0$ or $y + 2 = 0$

$y = 2$ $\quad$ $y = -2$

The solutions are $y = \pm 2$.

14. $3y^2 - 27 = 0$

$3(y^2 - 9) = 0$

$3(y-3)(y+3) = 0$

$y - 3 = 0$ or $y + 3 = 0$

$y = 3$ $\quad$ $y = -3$

The solutions are $y = \pm 3$.

15. $x^2 - 3x + 2 = 0$

$(x-2)(x-1) = 0$

$x - 2 = 0$ or $x - 1 = 0$

$x = 2$ $\quad$ $x = 1$

The solutions are $x = 1$ and $x = 2$.

16. $x^2 + 7x + 12 = 0$

$(x+3)(x+4) = 0$

$x + 3 = 0$ or $x + 4 = 0$

$x = -3$ $\quad$ $x = -4$

The solutions are $x = -3$ and $x = -4$.

17. $(2z+5)^2 = 25$

$2z + 5 = \sqrt{25}$ or $2z + 5 = -\sqrt{25}$

$2z + 5 = 5$ $\quad$ $2z + 5 = -5$

$2z = 0$ $\quad$ $2z = -10$

$z = 0$ $\quad$ $z = -5$

The solutions are $z = 0$ and $z = -5$.

18. $(3z-4)^2 = 16$

$3z - 4 = \sqrt{16}$ or $3z - 4 = -\sqrt{16}$

$3z - 4 = 4$ $\quad$ $3z - 4 = -4$

$3z = 8$ $\quad$ $3z = 0$

$z = \frac{8}{3}$ $\quad$ $z = 0$

The solutions are $z = \frac{8}{3}$ and $z = 0$.

19. $30x = 25x^2 + 2$

$0 = 25x^2 - 30x + 2$

$a = 25,\ b = -30,\ c = 2$

$$x = \frac{-b \pm \sqrt{b^2 - 4ac}}{2a}$$

$$x = \frac{-(-30) \pm \sqrt{(-30)^2 - 4(25)(2)}}{2(25)}$$

$$= \frac{30 \pm \sqrt{900 - 200}}{50}$$

$$= \frac{30 \pm \sqrt{700}}{50}$$

$$= \frac{30 \pm 10\sqrt{7}}{50}$$

$$= \frac{3 \pm \sqrt{7}}{5}$$

The solutions are $x = \frac{3 \pm \sqrt{7}}{5}$.

20. $12x = 4x^2 + 4$

$0 = 4x^2 - 12x + 4$

$0 = 4(x^2 - 3x + 1)$

$a = 1,\ b = -3,\ c = 1$

$$x = \frac{-b \pm \sqrt{b^2 - 4ac}}{2a}$$

Copyright © 2012 Pearson Education, Inc. Publishing as Prentice Hall.

$$x = \frac{-(-3) \pm \sqrt{(-3)^2 - 4(1)(1)}}{2(1)}$$
$$= \frac{3 \pm \sqrt{9-4}}{2}$$
$$= \frac{3 \pm \sqrt{5}}{2}$$

The solutions are $x = \frac{3 \pm \sqrt{5}}{2}$.

21. $\frac{2}{3}m^2 - \frac{1}{3}m - 1 = 0$

$2m^2 - m - 3 = 0$

$(2m-3)(m+1) = 0$

$2m - 3 = 0$ or $m + 1 = 0$

$2m = 3$ $\quad m = -1$

$m = \frac{3}{2}$

The solutions are $m = \frac{3}{2}$ and $m = -1$.

22. $\frac{5}{8}m^2 + m - \frac{1}{2} = 0$

$5m^2 + 8m - 4 = 0$

$(5m-2)(m+2) = 0$

$5m - 2 = 0$ or $m + 2 = 0$

$5m = 2$ $\quad m = -2$

$m = \frac{2}{5}$

The solutions are $m = \frac{2}{5}$ and $m = -2$.

23. $x^2 - \frac{1}{2}x - \frac{1}{5} = 0$

$10x^2 - 5x - 2 = 0$

$a = 10, b = -5, c = -2$

$$x = \frac{-b \pm \sqrt{b^2 - 4ac}}{2a}$$
$$x = \frac{-(-5) \pm \sqrt{(-5)^2 - 4(10)(-2)}}{2(10)}$$
$$= \frac{5 \pm \sqrt{25+80}}{20}$$
$$= \frac{5 \pm \sqrt{105}}{20}$$

The solutions are $x = \frac{5 \pm \sqrt{105}}{20}$.

24. $x^2 + \frac{1}{2}x - \frac{1}{8} = 0$

$8x^2 + 4x - 1 = 0$

$a = 8, b = 4, c = -1$

$$x = \frac{-b \pm \sqrt{b^2 - 4ac}}{2a}$$
$$x = \frac{-4 \pm \sqrt{4^2 - 4(8)(-1)}}{2(8)}$$
$$= \frac{-4 \pm \sqrt{16+32}}{16}$$
$$= \frac{-4 \pm \sqrt{48}}{16}$$
$$= \frac{-4 \pm 4\sqrt{3}}{16}$$
$$= \frac{-1 \pm \sqrt{3}}{4}$$

The solutions are $\frac{-1 \pm \sqrt{3}}{4}$.

25. $4x^2 - 27x + 35 = 0$

$(4x-7)(x-5) = 0$

$4x - 7 = 0$ or $x - 5 = 0$

$4x = 7$ $\quad x = 5$

$x = \frac{7}{4}$

The solutions are $x = \frac{7}{4}$ and $x = 5$.

26. $9x^2 - 16x + 7 = 0$

$(9x-7)(x-1) = 0$

$9x - 7 = 0$ or $x - 1 = 0$

$9x = 7$ $\quad x = 1$

$x = \frac{7}{9}$

The solutions are $x = \frac{7}{9}$ and $x = 1$.

27. $(7-5x)^2 = 18$

$7 - 5x = \sqrt{18}$

$-5x = -7 + 3\sqrt{2}$

$$x = \frac{-7+3\sqrt{2}}{-5} = \frac{7-3\sqrt{2}}{5}$$

or

Copyright © 2012 Pearson Education, Inc. Publishing as Prentice Hall.

$7-5x=-\sqrt{18}$

$-5x=-7-3\sqrt{2}$

$x=\dfrac{-7-3\sqrt{2}}{-5}=\dfrac{7+3\sqrt{2}}{5}$

The solutions are $x=\dfrac{-7\pm3\sqrt{2}}{-5}$ or $\dfrac{7\pm3\sqrt{2}}{5}$.

28. $(5-4x)^2=75$

$5-4x=\sqrt{75}$

$-4x=-5+5\sqrt{3}$

$x=\dfrac{-5+5\sqrt{3}}{-4}=\dfrac{5-5\sqrt{3}}{4}$

or

$5-4x=-\sqrt{75}$

$-4x=-5-5\sqrt{3}$

$x=\dfrac{-5-5\sqrt{3}}{-4}=\dfrac{5+5\sqrt{3}}{4}$

The solutions are $x=\dfrac{-5\pm5\sqrt{3}}{-4}$ or $\dfrac{5\pm5\sqrt{3}}{4}$.

29. $3z^2-7z=12$

$3z^2-7z-12=0$

$a=3,\ b=-7,\ c=-12$

$z=\dfrac{-b\pm\sqrt{b^2-4ac}}{2a}$

$z=\dfrac{-(-7)\pm\sqrt{(-7)^2-4(3)(-12)}}{2(3)}$

$=\dfrac{7\pm\sqrt{49+144}}{6}$

$=\dfrac{7\pm\sqrt{193}}{6}$

The solutions are $z=\dfrac{7\pm\sqrt{193}}{6}$.

30. $6z^2+7z=6$

$6z^2+7z-6=0$

$a=6,\ b=7,\ c=-6$

$z=\dfrac{-b\pm\sqrt{b^2-4ac}}{2a}$

$z=\dfrac{-7\pm\sqrt{(7)^2-4(6)(-6)}}{2(6)}$

$=\dfrac{-7\pm\sqrt{49+144}}{12}$

$=\dfrac{-7\pm\sqrt{193}}{12}$

The solutions are $z=\dfrac{-7\pm\sqrt{193}}{12}$.

31. $x=x^2-110$

$0=x^2-x-110$

$0=(x-11)(x+10)$

$x-11=0$ or $x+10=0$

$x=11$ $\quad$ $x=-10$

The solutions are $x=11$ and $x=-10$.

32. $x=56-x^2$

$0=56-x-x^2$

$0=(8+x)(7-x)$

$8+x=0$ or $7-x=0$

$x=-8$ $\quad$ $x=7$

The solutions are $x=7$ and $x=-8$.

33. $\dfrac{3}{4}x^2-\dfrac{5}{2}x-2=0$

$3x^2-10x-8=0$

$(3x+2)(x-4)=0$

$3x+2=0$ or $x-4=0$

$3x=-2$ $\quad$ $x=4$

$x=-\dfrac{2}{3}$

The solutions are $x=-\dfrac{2}{3}$ and $x=4$.

34. $x^2-\dfrac{6}{5}x-\dfrac{8}{5}=0$

$5x^2-6x-8=0$

$(5x+4)(x-2)=0$

$5x+4=0$ or $x-2=0$

$5x=-4$ $\quad$ $x=2$

$x=-\dfrac{4}{5}$

The solutions are $x=-\dfrac{4}{5}$ and $x=2$.

Copyright © 2012 Pearson Education, Inc. Publishing as Prentice Hall.

35. $x^2 - 0.6x + 0.05 = 0$
$(x-0.5)(x-0.1) = 0$
$x - 0.5 = 0$ or $x - 0.1 = 0$
$x = 0.5$ $x = 0.1$
The solutions are $x = 0.5$ and $x = 0.1$.

36. $x^2 - 0.1x - 0.06 = 0$
$(x-0.3)(x+0.2) = 0$
$x - 0.3 = 0$ or $x + 0.2 = 0$
$x = 0.3$ $x = -0.2$
The solutions are $x = 0.3$ and $x = -0.2$.

37. $10x^2 - 11x + 2 = 0$
$a = 10, b = -11, c = 2$

$$x = \frac{-b \pm \sqrt{b^2 - 4ac}}{2a}$$
$$x = \frac{-(-11) \pm \sqrt{(-11)^2 - 4(10)(2)}}{2(10)}$$
$$= \frac{11 \pm \sqrt{121 - 80}}{20}$$
$$= \frac{11 \pm \sqrt{41}}{20}$$

The solutions are $x = \frac{11 \pm \sqrt{41}}{20}$.

38. $20x^2 - 11x + 1 = 0$
$a = 20, b = -11, c = 1$

$$x = \frac{-b \pm \sqrt{b^2 - 4ac}}{2a}$$
$$x = \frac{-(-11) \pm \sqrt{(-11)^2 - 4(20)(1)}}{2(20)}$$
$$= \frac{11 \pm \sqrt{121 - 80}}{40}$$
$$= \frac{11 \pm \sqrt{41}}{40}$$

The solutions are $x = \frac{11 \pm \sqrt{41}}{40}$.

39. $\frac{1}{2}z^2 - 2z + \frac{3}{4} = 0$
$2z^2 - 8z + 3 = 0$
$a = 2, b = -8, c = 3$

$$z = \frac{-b \pm \sqrt{b^2 - 4ac}}{2a}$$
$$z = \frac{-(-8) \pm \sqrt{(-8)^2 - 4(2)(3)}}{2(2)}$$
$$= \frac{8 \pm \sqrt{64 - 24}}{4}$$
$$= \frac{8 \pm \sqrt{40}}{4}$$
$$= \frac{8 \pm 2\sqrt{10}}{4}$$
$$= \frac{4 \pm \sqrt{10}}{2}$$

The solutions are $z = \frac{4 \pm \sqrt{10}}{2}$.

40. $\frac{1}{5}z^2 - \frac{1}{2}z - 2 = 0$
$2z^2 - 5z - 20 = 0$
$a = 2, b = -5, c = -20$

$$z = \frac{-b \pm \sqrt{b^2 - 4ac}}{2a}$$
$$z = \frac{-(-5) \pm \sqrt{(-5)^2 - 4(2)(-20)}}{2(2)}$$
$$= \frac{5 \pm \sqrt{25 + 160}}{4}$$
$$= \frac{5 \pm \sqrt{185}}{4}$$

The solutions are $z = \frac{5 \pm \sqrt{185}}{4}$.

41. answers may vary

Section 9.4 Practice

1. $y = -3x^2$

x	y
-2	-12
-1	-3
0	0
1	-3
2	-12

Copyright © 2012 Pearson Education, Inc. Publishing as Prentice Hall.

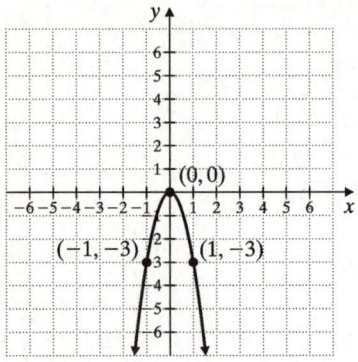

2. $y = x^2 - 9$

Let $x = 0$.

$y = 0^2 - 9 = -9$

The y-intercept is $(0, -9)$.

Let $y = 0$.

$0 = x^2 - 9$

$0 = (x-3)(x+3)$

$x - 3 = 0$ or $x + 3 = 0$

$x = 3$ $\quad$ $x = -3$

The x-intercepts are $(3, 0)$ and $(-3, 0)$.

x	y
−3	0
−2	−5
−1	−8
0	−9
1	−8
2	−5
3	0

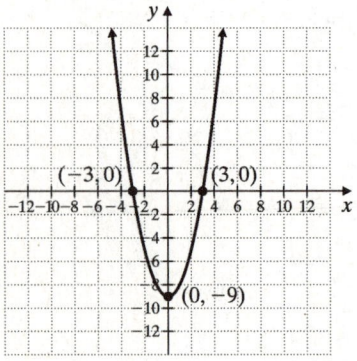

3. $y = x^2 - 2x - 3$

$a = 1, b = -2, c = -3$

$x = \dfrac{-b}{2a} = \dfrac{-(-2)}{2(1)} = \dfrac{2}{2} = 1$

$y = 1^2 - 2(1) - 3 = 1 - 2 - 3 = -4$

The vertex is $(1, -4)$.

Let $x = 0$.

$y = 0^2 - 2(0) - 3 = -3$

The y-intercept is $(0, -3)$.

Let $y = 0$.

$0 = x^2 - 2x - 3$

$0 = (x-3)(x+1)$

$x - 3 = 0$ or $x + 1 = 0$

$x = 3$ $\quad$ $x = -1$

The x-intercepts are $(3, 0)$ and $(-1, 0)$.

x	y
−1	0
0	−3
1	−4
2	−3
3	0

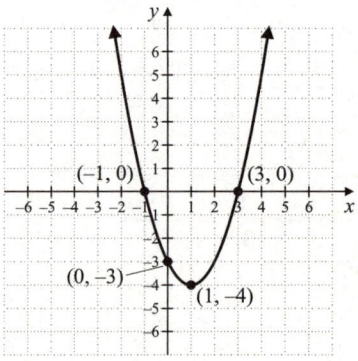

4. $y = x^2 - 4x + 1$

$a = 1, b = -4, c = 1$

$x = \dfrac{-b}{2a} = \dfrac{-(-4)}{2(1)} = \dfrac{4}{2} = 2$

$y = 2^2 - 4(2) + 1 = 4 - 8 + 1 = -3$

vertex $= (2, -3)$

$x = \dfrac{-b \pm \sqrt{b^2 - 4ac}}{2a}$

Copyright © 2012 Pearson Education, Inc. Publishing as Prentice Hall.

$$x = \frac{-(-4) \pm \sqrt{(-4)^2 - 4(1)(1)}}{2(1)}$$
$$= \frac{4 \pm \sqrt{16-4}}{2}$$
$$= \frac{4 \pm \sqrt{12}}{2}$$
$$= \frac{4 \pm 2\sqrt{3}}{2}$$
$$= 2 \pm \sqrt{3}$$
$$= 3.7 \text{ or } 0.3$$

x	y
2	−3
$2+\sqrt{3} \approx 3.7$	0
$2-\sqrt{3} \approx 0.3$	0
0	1
4	1

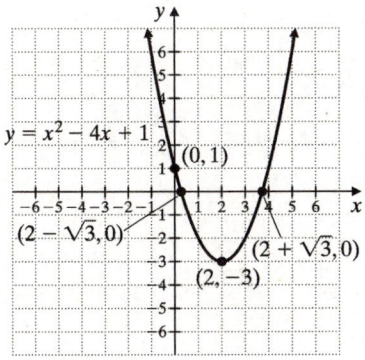

Calculator Explorations

1. $x^2 - 7x - 3 = 0$ $\quad x \approx -0.41, 7.41$
2. $2x^2 - 11x - 1 = 0$ $\quad x \approx -0.09, 5.59$
3. $-1.7x^2 + 5.6x - 3.7 = 0$ $\quad x \approx 0.91, 2.38$
4. $-5.8x^2 + 2.3x - 3.9 = 0$ $\quad$ No real solution
5. $5.8x^2 - 2.6x - 1.9 = 0$ $\quad x \approx -0.39, 0.84$
6. $7.5x^2 - 3.7x - 1.1 = 0$ $\quad x \approx -0.21, 0.70$

Exercise Set 9.4

1.

x	$y = 2x^2$
−2	$2(-2)^2 = 2(4) = 8$
−1	$2(-1)^2 = 2(1) = 2$
0	$2(0)^2 = 2(0) = 0$
1	$2(1)^2 = 2(1) = 2$
2	$2(2)^2 = 2(4) = 8$

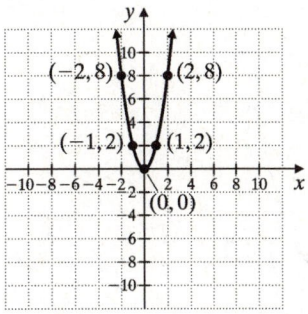

3.

x	$y = -x^2$
−2	$-(-2)^2 = -(4) = -4$
−1	$-(-1)^2 = -(1) = -1$
0	$-(0)^2 = -(0) = 0$
1	$-(1)^2 = -(1) = -1$
2	$-(2)^2 = -(4) = -4$

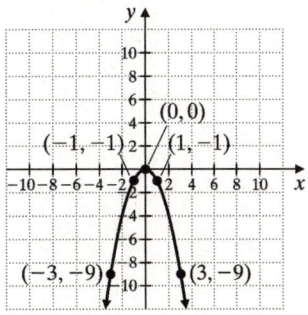

Copyright © 2012 Pearson Education, Inc. Publishing as Prentice Hall.

5. $y = x^2 - 1$

$y = 0$: $0 = x^2 - 1$

$1 = x^2$

$\pm\sqrt{1} = x$

$\pm 1 = x$

x-intercepts: $(-1, 0)$, $(1, 0)$

$x = 0$: $y = 0^2 - 1 = -1$

y-intercept: $(0, -1)$

$y = x^2 + 0x - 1$

$a = 1, b = 0, c = 1$

$\frac{-b}{2a} = \frac{-0}{2(1)} = \frac{0}{2} = 0$

$x = 0$: $y = 0^2 - 1 = -1$

The vertex is $(0, -1)$.

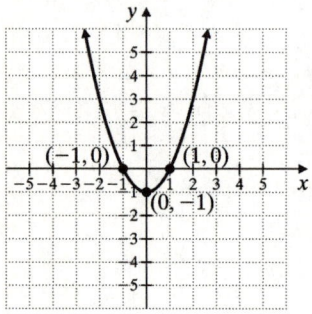

7. $y = x^2 + 4$

$y = 0$: $0 = x^2 + 4$

$-4 = x^2$

$\pm\sqrt{-4} = x$

x-intercepts: none

$x = 0$: $y = 0^2 + 4 = 4$

y-intercept: $(0, 4)$

$y = x^2 + 0x + 4$

$a = 1, b = 0, c = 4$

$\frac{-b}{2a} = \frac{-0}{2(1)} = \frac{0}{2} = 0$

$x = 0$: $y = 0^2 + 4 = 4$

The vertex is $(0, 4)$.

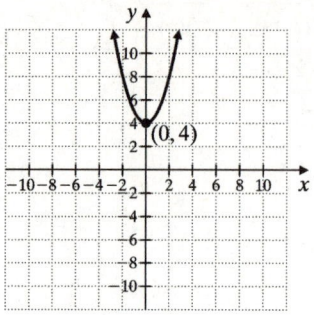

9. $y = -x^2 + 4x - 4$

$y = 0$: $0 = -x^2 + 4x - 4$

$0 = x^2 - 4x + 4$

$0 = (x - 2)^2$

$0 = x - 2$

$2 = x$

The x-intercept is $(2, 0)$.

$x = 0$: $y = -0^2 + 4(0) - 4 = -4$

The y-intercept is $(0, -4)$.

$y = -x^2 + 4x - 4$

$a = -1, b = 4, c = -4$

$\frac{-b}{2a} = \frac{-4}{2(-1)} = \frac{-4}{-2} = 2$

$x = 2$: $y = -(2)^2 + 4(2) - 4 = -4 + 8 - 4 = 0$

The vertex is $(2, 0)$.

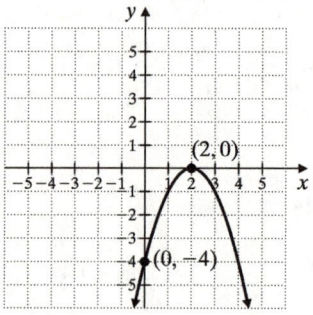

11. $y = x^2 + 5x + 4$

$y = 0$: $0 = x^2 + 5x + 4$

$0 = (x + 4)(x + 1)$

$0 = x + 4$ or $0 = x + 1$

$-4 = x$ $\quad$ $-1 = x$

x-intercepts: $(-4, 0)$ and $(-1, 0)$

$x = 0$: $y = 0^2 + 5(0) + 4 = 4$

y-intercept: $(0, 4)$

$y = x^2 + 5x + 4$

$a = 1, b = 5, c = 4$

Copyright © 2012 Pearson Education, Inc. Publishing as Prentice Hall.

$$\frac{-b}{2a}=\frac{-5}{2(1)}=\frac{-5}{2}$$

$$x=\frac{-5}{2}:\ y=\left(-\frac{5}{2}\right)^2+5\left(-\frac{5}{2}\right)+4$$
$$=\frac{25}{4}-\frac{25}{2}+4$$
$$=\frac{25}{4}-\frac{50}{4}+\frac{16}{4}$$
$$=-\frac{9}{4}$$

The vertex is $\left(-\frac{5}{2},-\frac{9}{4}\right)$ or $\left(-2\frac{1}{2},-2\frac{1}{4}\right)$.

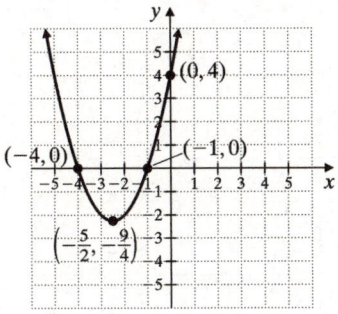

13. $y=x^2-4x+5$

$y=0$: $0=x^2-4x+5$

$a=1,\ b=-4,\ c=5$

$$y=\frac{-b\pm\sqrt{b^2-4ac}}{2a}$$
$$y=\frac{-(-4)\pm\sqrt{(-4)^2-4(1)(5)}}{2(1)}$$
$$=\frac{4\pm\sqrt{16-20}}{2}$$
$$=\frac{4\pm\sqrt{-4}}{2}$$

There are no x-intercepts, since $\sqrt{-4}$ is not a real number.

$x=0$: $y=0^2-4(0)+5=5$

y-intercept: (0, 5)

$y=x^2-4x+5$

$a=1,\ b=-4,\ c=5$

$$\frac{-b}{2a}=\frac{-(-4)}{2(1)}=\frac{4}{2}=2$$

$x=2$: $y=2^2-4(2)+5=4-8+5=1$

vertex: (2, 1)

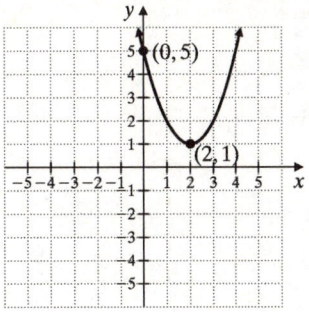

15. $y=2-x^2$

$y=0$: $0=2-x^2$

$x^2=2$

$x=\pm\sqrt{2}$

x-intercepts: $\left(\sqrt{2},0\right)$ and $\left(-\sqrt{2},0\right)$

$x=0$: $y=2-0^2=2$

y-intercept: (0, 2)

$y=2-x^2$

$a=-1,\ b=0,\ c=2$

$$\frac{-b}{2a}=\frac{-0}{2(-1)}=\frac{0}{-2}=0$$

$x=0$: $y=2-0^2=2$

vertex: (0, 2)

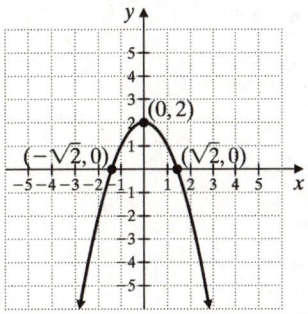

17. $y=\frac{1}{3}x^2$

$y=0$: $0=\frac{1}{3}x^2$

$0=x^2$

$0=x$

x-intercept: (0, 0)

The y-intercept is also (0, 0).

$y=\frac{1}{3}x^2+0x+0$

$a=\frac{1}{3},\ b=0,\ c=0$

Copyright © 2012 Pearson Education, Inc. Publishing as Prentice Hall.

$$\frac{-b}{2a} = \frac{-0}{2\left(\frac{1}{3}\right)} = \frac{0}{\frac{2}{3}} = 0$$

$x = 0$: $y = \frac{1}{3}(0)^2 = \frac{1}{3}(0) = 0$

vertex: (0, 0)

For additional points, use $x = \pm 3$.

$x = -3$: $y = \frac{1}{3}(-3)^2 = \frac{1}{3}(9) = 3$

$x = 3$: $y = \frac{1}{3}(3)^2 = \frac{1}{3}(9) = 3$

(−3, 3) and (3, 3) are also on the graph.

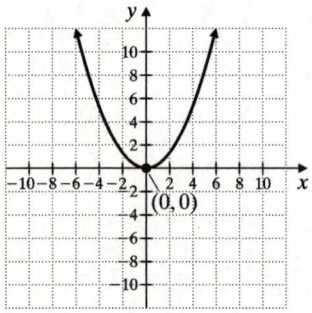

19. $y = x^2 + 6x$

$y = 0$: $0 = x^2 + 6x$

$0 = x(x+6)$

$x = 0$ or $x + 6 = 0$

$x = -6$

x-intercepts: (0, 0) and (−6, 0)

$x = 0$: $y = 0^2 + 6(0) = 0$

y-intercept: (0, 0)

$y = x^2 + 6x + 0$

$a = 1, b = 6, c = 0$

$$\frac{-b}{2a} = \frac{-6}{2(1)} = \frac{-6}{2} = -3$$

$x = -3$: $y = (-3)^2 + 6(-3) = 9 - 18 = -9$

vertex: (−3, −9)

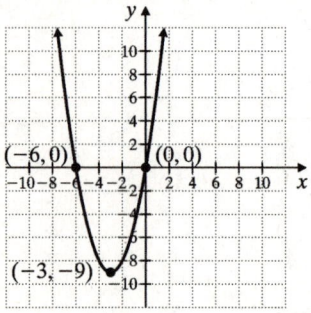

21. $y = x^2 + 2x - 8$

$y = 0$: $0 = x^2 + 2x - 8$

$0 = (x+4)(x-2)$

$0 = x + 4$ or $0 = x - 2$

$-4 = x$ $\quad$ $2 = x$

x-intercepts: (−4, 0), (2, 0)

$x = 0$: $y = 0^2 + 2(0) - 8 = -8$

y-intercept: (0, −8)

$y = x^2 + 2x - 8$

$a = 1, b = 2, c = -8$

$$\frac{-b}{2a} = \frac{-2}{2(1)} = \frac{-2}{2} = -1$$

$x = -1$: $y = (-1)^2 + 2(-1) - 8 = 1 - 2 - 8 = -9$

vertex: (−1, −9)

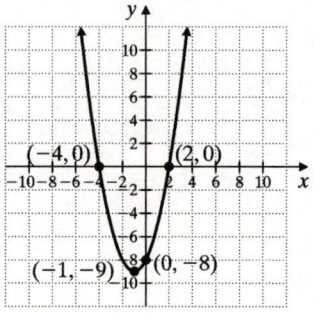

23. $y = -\frac{1}{2}x^2$

$y = 0$: $0 = -\frac{1}{2}x^2$

$0 = x^2$

$0 = x$

x-intercept: (0, 0)

The y-intercept is also (0, 0).

$y = -\frac{1}{2}x^2 + 0x + 0$

$a = -\frac{1}{2}$, $b = 0, c = 0$

$$\frac{-b}{2a} = \frac{-0}{2\left(-\frac{1}{2}\right)} = \frac{0}{-1} = 0$$

$x = 0$: $y = -\frac{1}{2}(0)^2 = -\frac{1}{2}(0) = 0$

vertex: (0, 0)

Copyright © 2012 Pearson Education, Inc. Publishing as Prentice Hall.

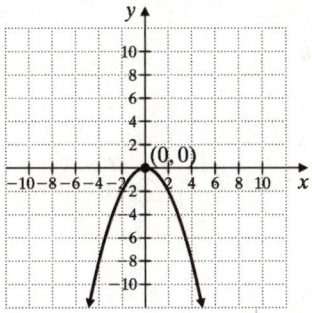

25. $y = 2x^2 - 11x + 5$

$y = 0$: $0 = 2x^2 - 11x + 5$

$0 = (2x-1)(x-5)$

$0 = 2x - 1$ or $0 = x - 5$

$1 = 2x$ $\quad 5 = x$

$\frac{1}{2} = x$

x-intercepts: $\left(\frac{1}{2}, 0\right)$, $(5, 0)$

$x = 0$: $y = 2(0)^2 - 11(0) + 5 = 5$

y-intercept: $(0, 5)$

$y = 2x^2 - 11x + 5$

$a = 2, b = -11, c = 5$

$\frac{-b}{2a} = \frac{-(-11)}{2(2)} = \frac{11}{4}$

$$x = \frac{11}{4}: \quad y = 2\left(\frac{11}{4}\right)^2 - 11\left(\frac{11}{4}\right) + 5$$
$$= 2\left(\frac{121}{16}\right) - \frac{121}{4} + 5$$
$$= \frac{121}{8} - \frac{121}{4} + 5$$
$$= \frac{121}{8} - \frac{242}{8} + \frac{40}{8}$$
$$= -\frac{81}{8}$$

vertex: $\left(\frac{11}{4}, -\frac{81}{8}\right)$

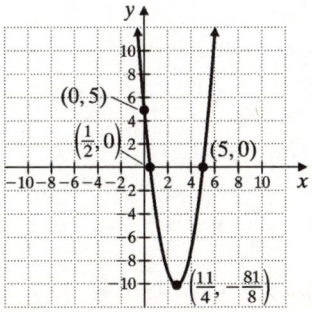

27. $y = -x^2 + 4x - 3$

$y = 0$: $0 = -x^2 + 4x - 3$

$0 = x^2 - 4x + 3$

$0 = (x-3)(x-1)$

$x - 3 = 0$ or $x - 1 = 0$

$x = 3$ $\quad x = 1$

x-intercepts: $(3, 0)$ and $(1, 0)$

$x = 0$: $y = -0^2 + 4(0) - 3 = -3$

y-intercept: $(0, -3)$

$y = -x^2 + 4x - 3$

$a = -1, b = 4, c = -3$

$\frac{-b}{2a} = \frac{-4}{2(-1)} = \frac{-4}{-2} = 2$

$x = 2$: $y = -(2)^2 + 4(2) - 3 = -4 + 8 - 3 = 1$

vertex: $(2, 1)$

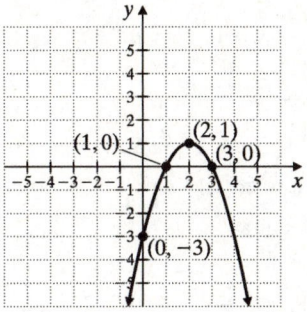

29. $\frac{\frac{1}{7}}{\frac{2}{5}} = \frac{1}{7} \div \frac{2}{5} = \frac{1}{7} \cdot \frac{5}{2} = \frac{5}{14}$

31. $\frac{\frac{1}{x}}{\frac{2}{x^2}} = \frac{1}{x} \div \frac{2}{x^2} = \frac{1}{x} \cdot \frac{x^2}{2} = \frac{x}{2}$

33. $\frac{2x}{1 - \frac{1}{x}} = \frac{x(2x)}{x\left(1 - \frac{1}{x}\right)} = \frac{2x^2}{x-1}$

35. $\frac{\frac{a-b}{2b}}{\frac{b-a}{8b^2}} = \frac{a-b}{2b} \div \frac{b-a}{8b^2} = \frac{a-b}{2b} \cdot \frac{8b^2}{b-a} = -4b$

37. **a.** The maximum height appears to be about 256 feet.

b. The fireball appears to reach its maximum height when $t = 4$ seconds.

c. The fireball appears to return to the ground when $t = 8$ seconds.

Copyright © 2012 Pearson Education, Inc. Publishing as Prentice Hall.

39. With $a > 0$, the parabola opens upward. An upward-opening parabola that crosses the x-axis twice is graph A.

41. With $a < 0$, the parabola opens downward. A downward-opening parabola that does not touch or cross the x-axis (no x-intercept) is graph D.

43. With $a > 0$, the parabola opens upward. An upward-opening parabola that crosses the x-axis once is graph F.

Chapter 9 Vocabulary Check

1. If $x^2 = a$, then $x = \sqrt{a}$ or $x = -\sqrt{a}$. This property is called the square root property.

2. The graph of $y = x^2$ is called a parabola.

3. The formula $\frac{-b}{2a}$ where $y = ax^2 + bx + c$ is called the vertex formula.

4. The process of solving a quadratic equation by writing it in the form $(x+a)^2 = c$ is called completing the square.

5. The formula $x = \frac{-b \pm \sqrt{b^2 - 4ac}}{2a}$ is called the quadratic formula.

6. The lowest point on a parabola that opens upward is called the vertex.

7. The zero-factor property states that if the product of two numbers is zero, then at least one of the two numbers is zero.

Chapter 9 Review

1. $x^2 - 121 = 0$
$x^2 = 121$
$x = \pm\sqrt{121}$
$x = \pm 11$
The solutions are $x = \pm 11$.

2. $y^2 - 100 = 0$
$y^2 = 100$
$y = \pm\sqrt{100}$
$y = \pm 10$
The solutions are $y = \pm 10$.

3. $3m^2 - 5m = 2$
$3m^2 - 5m - 2 = 0$
$(3m+1)(m-2) = 0$
$3m + 1 = 0$ or $m - 2 = 0$
$3m = -1$ $\quad$ $m = 2$
$m = -\frac{1}{3}$
The solutions are $m = -\frac{1}{3}$ and $m = 2$.

4. $7m^2 + 2m = 5$
$7m^2 + 2m - 5 = 0$
$(7m-5)(m+1) = 0$
$7m - 5 = 0$ or $m + 1 = 0$
$7m = 5$ $\quad$ $m = -1$
$m = \frac{5}{7}$
The solutions are $m = \frac{5}{7}$ and $m = -1$.

5. $x^2 = 36$
$x = \pm\sqrt{36}$
$x = \pm 6$
The solutions are $x = \pm 6$.

6. $x^2 = 81$
$x = \pm\sqrt{81}$
$x = \pm 9$
The solutions are $x = \pm 9$.

7. $k^2 = 50$
$k = \pm\sqrt{50}$
$k = \pm 5\sqrt{2}$
The solutions are $k = \pm 5\sqrt{2}$.

8. $k^2 = 45$
$k = \pm\sqrt{45}$
$k = \pm 3\sqrt{5}$
The solutions are $k = \pm 3\sqrt{5}$.

9. $(x-11)^2 = 49$
$x - 11 = \pm\sqrt{49}$
$x - 11 = \pm 7$
$x = 11 + 7$ or $x = 11 - 7$
$x = 18$ $\quad$ $x = 4$
The solutions are $x = 18$ and $x = 4$.

Copyright © 2012 Pearson Education, Inc. Publishing as Prentice Hall.

10. $(x+3)^2 = 100$

$x+3 = \pm\sqrt{100}$

$x+3 = \pm 10$

$x = -3+10$ or $x = -3-10$

$x = 7$ $\quad$ $x = -13$

The solutions are $x = 7$ and $x = -13$.

11. $(4p+5)^2 = 41$

$4p+5 = \pm\sqrt{41}$

$4p = -5 \pm \sqrt{41}$

$p = \dfrac{-5 \pm \sqrt{41}}{4}$

The solutions are $p = \dfrac{-5 \pm \sqrt{41}}{4}$.

12. $(3p+7)^2 = 37$

$3p+7 = \pm\sqrt{37}$

$3p = -7 \pm \sqrt{37}$

$p = \dfrac{-7 \pm \sqrt{37}}{3}$

The solutions are $p = \dfrac{-7 \pm \sqrt{37}}{3}$.

13. $h = 16t^2$

$100 = 16t^2$

$6.25 = t^2$

$\pm 2.5 = t$

Reject –2.5 because time is not negative. It will take Kara 2.5 seconds to hit the water.

14. $5 \text{ miles} \cdot \dfrac{5280 \text{ ft}}{1 \text{ mile}} = 26{,}400 \text{ feet}$

$y = 16t^2$

$26{,}400 = 16t^2$

$1650 = t^2$

$\pm 40.6 = t$

Reject –40.6 because time is not negative. A 5-mile freefall will take 40.6 seconds.

15. $x^2 - 9x = -8$

$x^2 - 9x + \left(-\dfrac{9}{2}\right)^2 = -8 + \left(-\dfrac{9}{2}\right)^2$

$\left(x - \dfrac{9}{2}\right)^2 = -8 + \dfrac{81}{4}$

$\left(x - \dfrac{9}{2}\right)^2 = \dfrac{49}{4}$

$x - \dfrac{9}{2} = \pm\sqrt{\dfrac{49}{4}}$

$x - \dfrac{9}{2} = \pm\dfrac{7}{2}$

$x = \dfrac{9}{2} + \dfrac{7}{2} = \dfrac{16}{2} = 8$ or $x = \dfrac{9}{2} - \dfrac{7}{2} = \dfrac{2}{2} = 1$

The solutions are $x = 8$ and $x = 1$.

16. $x^2 + 8x = 20$

$x^2 + 8x + \left(\dfrac{8}{2}\right)^2 = 20 + \left(\dfrac{8}{2}\right)^2$

$(x+4)^2 = 36$

$x+4 = \pm\sqrt{36}$

$x+4 = \pm 6$

$x = -4 + 6 = 2$ or $x = -4 - 6 = -10$

The solutions are $x = 7$ and $x = -10$.

17. $x^2 + 4x = 1$

$x^2 + 4x + \left(\dfrac{4}{2}\right)^2 = 1 + \left(\dfrac{4}{2}\right)^2$

$(x+2)^2 = 5$

$x+2 = \pm\sqrt{5}$

$x = -2 \pm \sqrt{5}$

The solutions are $x = -2 \pm \sqrt{5}$.

18. $x^2 - 8x = 3$

$x^2 - 8x + \left(-\dfrac{8}{2}\right)^2 = 3 + \left(-\dfrac{8}{2}\right)^2$

$(x-4)^2 = 19$

$x-4 = \pm\sqrt{19}$

$x = 4 \pm \sqrt{19}$

The solutions are $x = 4 \pm \sqrt{19}$.

Copyright © 2012 Pearson Education, Inc. Publishing as Prentice Hall.

19.
$$x^2-6x+7=0$$
$$x^2-6x=-7$$
$$x^2-6x+\left(-\frac{6}{2}\right)^2=-7+\left(-\frac{6}{2}\right)^2$$
$$(x-3)^2=2$$
$$x-3=\pm\sqrt{2}$$
$$x=3\pm\sqrt{2}$$
The solutions are $3\pm\sqrt{2}$.

20.
$$x^2+6x+7=0$$
$$x^2+6x=-7$$
$$x^2+6x+\left(\frac{6}{2}\right)^2=-7+\left(\frac{6}{2}\right)^2$$
$$(x+3)^2=2$$
$$x+3=\pm\sqrt{2}$$
$$x=-3\pm\sqrt{2}$$
The solutions are $-3\pm\sqrt{2}$.

21.
$$2y^2+y-1=0$$
$$2y^2+y=1$$
$$y^2+\frac{1}{2}y=\frac{1}{2}$$
$$y^2+\frac{1}{2}y+\left(\frac{1}{4}\right)^2=\frac{1}{2}+\left(\frac{1}{4}\right)^2$$
$$\left(y+\frac{1}{4}\right)^2=\frac{9}{16}$$
$$y+\frac{1}{4}=\pm\sqrt{\frac{9}{16}}$$
$$y+\frac{1}{4}=\pm\frac{3}{4}$$
$$y=-\frac{1}{4}+\frac{3}{4}=\frac{2}{4}=\frac{1}{2} \quad \text{or}$$
$$y=-\frac{1}{4}-\frac{3}{4}=-\frac{4}{4}=-1$$
The solutions are $y=\frac{1}{2}$ and $y=-1$.

22.
$$4y^2+3y-1=0$$
$$4y^2+3y=1$$
$$y^2+\frac{3}{4}y=\frac{1}{4}$$
$$y^2+\frac{3}{4}y+\left(\frac{3}{8}\right)^2=\frac{1}{4}+\left(\frac{3}{8}\right)^2$$
$$\left(y+\frac{3}{8}\right)^2=\frac{25}{64}$$
$$y+\frac{3}{8}=\pm\sqrt{\frac{25}{64}}$$
$$y+\frac{3}{8}=\pm\frac{5}{8}$$
$$y=-\frac{3}{8}+\frac{5}{8}=\frac{2}{8}=\frac{1}{4} \text{ or } y=-\frac{3}{8}-\frac{5}{8}=-\frac{8}{8}=-1$$
The solutions are $y=\frac{1}{4}$ and $y=-1$.

23. $9x^2+30x+25=0$
$a=9,\ b=30,\ c=25$
$$x=\frac{-b\pm\sqrt{b^2-4ac}}{2a}$$
$$x=\frac{-30\pm\sqrt{30^2-4(9)(25)}}{2(9)}$$
$$=\frac{-30\pm\sqrt{900-900}}{18}$$
$$=\frac{-30\pm\sqrt{0}}{18}$$
$$=\frac{-30}{18}$$
$$=-\frac{5}{3}$$
The solution is $x=-\frac{5}{3}$.

24. $16x^2-72x+81=0$
$a=16,\ b=-72,\ c=81$
$$x=\frac{-b\pm\sqrt{b^2-4ac}}{2a}$$

Copyright © 2012 Pearson Education, Inc. Publishing as Prentice Hall.

$$x = \frac{-(-72) \pm \sqrt{(-72)^2 - 4(16)(81)}}{2(16)}$$
$$= \frac{72 \pm \sqrt{5184 - 5184}}{32}$$
$$= \frac{72 \pm \sqrt{0}}{32}$$
$$= \frac{72}{32}$$
$$= \frac{9}{4}$$

The solution is $x = \frac{9}{4}$.

25. $7x^2 = 35$

$7x^2 - 35 = 0$

$a = 7, b = 0, c = -35$

$$x = \frac{-b \pm \sqrt{b^2 - 4ac}}{2a}$$
$$x = \frac{-0 \pm \sqrt{0^2 - 4(7)(-35)}}{2(7)}$$
$$= \frac{\pm\sqrt{980}}{14}$$
$$= \frac{\pm 14\sqrt{5}}{14}$$
$$= \pm\sqrt{5}$$

The solutions are $x = \pm\sqrt{5}$.

26. $11x^2 = 33$

$11x^2 - 33 = 0$

$a = 11, b = 0, c = -33$

$$x = \frac{-b \pm \sqrt{b^2 - 4ac}}{2a}$$
$$x = \frac{-0 \pm \sqrt{0^2 - 4(11)(-33)}}{2(11)}$$
$$= \frac{\pm\sqrt{1452}}{22}$$
$$= \frac{\pm 22\sqrt{3}}{22}$$
$$= \pm\sqrt{3}$$

The solutions are $x = \pm\sqrt{3}$.

27. $x^2 - 10x + 7 = 0$

$a = 1, b = -10, c = 7$

$$x = \frac{-b \pm \sqrt{b^2 - 4ac}}{2a}$$
$$x = \frac{-(-10) \pm \sqrt{(-10)^2 - 4(1)(7)}}{2(1)}$$
$$= \frac{10 \pm \sqrt{100 - 28}}{2}$$
$$= \frac{10 \pm \sqrt{72}}{2}$$
$$= \frac{10 \pm 6\sqrt{2}}{2}$$
$$= 5 \pm 3\sqrt{2}$$

The solutions are $x = 5 \pm 3\sqrt{2}$.

28. $x^2 + 4x - 7 = 0$

$a = 1, b = 4, c = -7$

$$x = \frac{-b \pm \sqrt{b^2 - 4ac}}{2a}$$
$$x = \frac{-4 \pm \sqrt{4^2 - 4(1)(-7)}}{2(1)}$$
$$= \frac{-4 \pm \sqrt{16 + 28}}{2}$$
$$= \frac{-4 \pm \sqrt{44}}{2}$$
$$= \frac{-4 \pm 2\sqrt{11}}{2}$$
$$= -2 \pm \sqrt{11}$$

The solutions are $x = -2 \pm \sqrt{11}$.

29. $3x^2 + x - 1 = 0$

$a = 3, b = 1, c = -1$

$$x = \frac{-b \pm \sqrt{b^2 - 4ac}}{2a}$$
$$x = \frac{-1 \pm \sqrt{(1)^2 - 4(3)(-1)}}{2(3)}$$
$$= \frac{-1 \pm \sqrt{1 + 12}}{6}$$
$$= \frac{-1 \pm \sqrt{13}}{6}$$

The solutions are $x = \frac{-1 \pm \sqrt{13}}{6}$.

Copyright © 2012 Pearson Education, Inc. Publishing as Prentice Hall.

30. $x^2+3x-1=0$

$a=1,\ b=3,\ c=-1$

$$x=\frac{-b\pm\sqrt{b^2-4ac}}{2a}$$

$$x=\frac{-3\pm\sqrt{3^2-4(1)(-1)}}{2(1)}$$

$$=\frac{-3\pm\sqrt{9+4}}{2}$$

$$=\frac{-3\pm\sqrt{13}}{2}$$

The solutions are $x=\dfrac{-3\pm\sqrt{13}}{2}$.

31. $2x^2+x+5=0$

$a=2,\ b=1,\ c=5$

$$x=\frac{-b\pm\sqrt{b^2-4ac}}{2a}$$

$$x=\frac{-1\pm\sqrt{1^2-4(2)(5)}}{2(2)}$$

$$=\frac{-1\pm\sqrt{1-40}}{4}$$

$$=\frac{-1\pm\sqrt{-39}}{4}$$

The equation has no solution, since the square root of -39 is not a real number.

32. $7x^2-3x+1=0$

$a=7,\ b=-3,\ c=1$

$$x=\frac{-b\pm\sqrt{b^2-4ac}}{2a}$$

$$x=\frac{-(-3)\pm\sqrt{(-3)^2-4(7)(1)}}{2(7)}$$

$$=\frac{3\pm\sqrt{9-28}}{14}$$

$$=\frac{3\pm\sqrt{-19}}{14}$$

The equation has no solution since the square root of -19 is not a real number.

33. $x=\dfrac{-1+\sqrt{13}}{6}\approx 0.4$ or $x=\dfrac{-1-\sqrt{13}}{6}\approx -0.8$

34. $x=\dfrac{-3+\sqrt{13}}{2}\approx 0.3$ or $x=\dfrac{-3-\sqrt{13}}{2}\approx -3.3$

35. $y=8x^2+109x+700$

$2590=8x^2+109x+700$

$0=8x^2+109x-1890$

$a=8,\ b=109,\ c=-1890$

$$x=\frac{-b\pm\sqrt{b^2-4ac}}{2a}$$

$$x=\frac{-109\pm\sqrt{109^2-4(8)(-1890)}}{2(8)}$$

$$=\frac{-109\pm\sqrt{11,881+60,480}}{16}$$

$$=\frac{-109\pm\sqrt{72,361}}{16}$$

$$=\frac{-109\pm 269}{16}$$

$$x=\frac{-109+269}{16}=\frac{160}{16}=10$$

or

$$x=\frac{-109-269}{16}=\frac{-378}{16}=-23.625$$

Since time will not be a negative number in this context, discard the solution -23.625. The solution is $x=10$, so the model predicts that the price of platinum will be \$2590 per ounce in 2013 (2003 + 10).

36. $y=26x^2+115x+493$

$3077=26x^2+115x+493$

$0=26x^2+115x-2584$

$a=26,\ b=115,\ c=-2584$

$$x=\frac{-b\pm\sqrt{b^2-4ac}}{2a}$$

$$x=\frac{-115\pm\sqrt{115^2-4(26)(-2584)}}{2(26)}$$

$$=\frac{-115\pm\sqrt{13,225+268,736}}{52}$$

$$=\frac{-115\pm\sqrt{281,961}}{52}$$

$$=\frac{-115\pm 531}{52}$$

$$x=\frac{-115+531}{52}=\frac{416}{52}=8$$

or

$$x=\frac{-115-531}{52}=\frac{-646}{52}\approx -12.4$$

Since time will not be a negative number in this

Copyright © 2012 Pearson Education, Inc. Publishing as Prentice Hall.

context, discard the solution −12.4. The solution is $x = 8$, so the model predicts that the price of silver will be 3077 cents per ounce in 2011 (2003 + 8).

37. $y = 5x^2$

$y = 0$: $0 = 5x^2$

$0 = x^2$

$0 = x$

x-intercept: (0, 0)

The y-intercept is also (0, 0).

x	$y = 5x^2$
−2	$5(-2)^2 = 5(4) = 20$
−1	$5(-1)^2 = 5(1) = 5$
0	$5(0)^2 = 5(0) = 0$
1	$5(1)^2 = 5(1) = 5$
2	$5(2)^2 = 5(4) = 20$

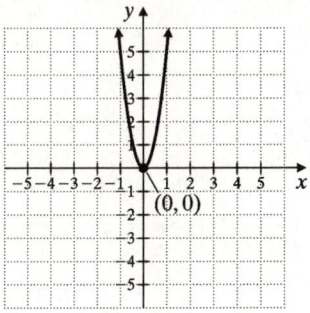

38. $y = -\frac{1}{2}x^2$

$y = 0$: $0 = -\frac{1}{2}x^2$

$0 = x^2$

$0 = x$

x-intercept: (0, 0)

The y-intercept is also (0, 0).

x	$y = -\frac{1}{2}x^2$
−2	$-\frac{1}{2}(-2)^2 = -\frac{1}{2}(4) = -2$
−1	$-\frac{1}{2}(-1)^2 = -\frac{1}{2}(1) = -\frac{1}{2}$
0	$-\frac{1}{2}(0)^2 = -\frac{1}{2}(0) = 0$
1	$-\frac{1}{2}(1)^2 = -\frac{1}{2}(1) = -\frac{1}{2}$
2	$-\frac{1}{2}(2)^2 = -\frac{1}{2}(4) = -2$

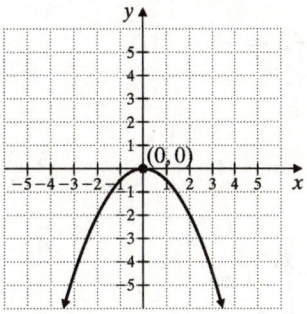

39. $y = x^2 - 25$

$y = 0$: $0 = x^2 - 25$

$25 = x^2$

$\pm\sqrt{25} = x$

$\pm 5 = x$

x-intercepts: (5, 0) and (−5, 0)

$x = 0$: $y = 0^2 - 25 = -25$

y-intercept: (0, −25)

$y = x^2 + 0x - 25$

$a = 1, b = 0, c = -25$

$\frac{-b}{2a} = \frac{-0}{2(1)} = \frac{0}{2} = 0$

$x = 0$: $y = 0^2 - 25 = -25$

vertex: (0, −25)

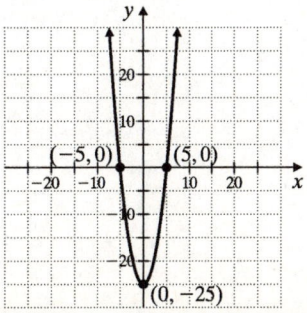

Copyright © 2012 Pearson Education, Inc. Publishing as Prentice Hall.

40. $y = x^2 - 36$

$y = 0$: $0 = x^2 - 36$

$36 = x^2$

$\pm\sqrt{36} = x$

$\pm 6 = x$

x-intercepts: (6, 0) and (–6, 0)

$x = 0$: $y = 0^2 - 36 = -36$

y-intercept: (0, –36)

$y = x^2 + 0x - 36$

$a = 1, b = 0, c = -36$

$\frac{-b}{2a} = \frac{-0}{2(1)} = \frac{0}{2} = 0$

$x = 0$: $y = 0^2 - 36 = -36$

vertex: (0, –36)

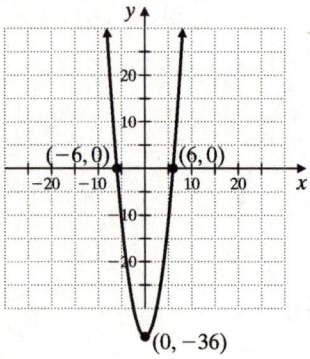

41. $y = x^2 + 3$

$y = 0$: $0 = x^2 + 3$

$-3 = x^2$

$\pm\sqrt{-3} = x$

x-intercepts: none

$x = 0$: $y = 0^2 + 3 = 3$

y-intercept: (0, 3)

$y = x^2 + 0x + 3$

$a = 1, b = 0, c = 3$

$\frac{-b}{2a} = \frac{-0}{2(1)} = \frac{0}{2} = 0$

$x = 0$: $y = 0^2 + 3 = 3$

vertex: (0, 3)

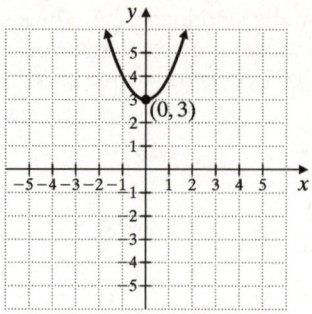

42. $y = x^2 + 8$

$y = 0$: $0 = x^2 + 8$

$-8 = x^2$

$\pm\sqrt{-8} = x$

x-intercepts: none

$x = 0$: $y = 0^2 + 8 = 8$

y-intercept: (0, 8)

$y = x^2 + 0x + 8$

$a = 1, b = 0, c = 8$

$\frac{-b}{2a} = \frac{0}{2(1)} = \frac{0}{2} = 0$

$x = 0$: $y = 0^2 + 8 = 8$

vertex: (0, 8)

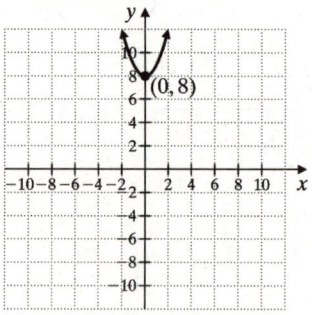

43. $y = -4x^2 + 8$

$y = 0$: $0 = -4x^2 + 8$

$0 = -4(x^2 - 2)$

$0 = x^2 - 2$

$2 = x^2$

$\pm\sqrt{2} = x$

x-intercepts: $\left(\sqrt{2}, 0\right)$ and $\left(-\sqrt{2}, 0\right)$

$x = 0$: $y = -4(0)^2 + 8 = 8$

y-intercept: (0, 8)

$y = -4x^2 + 0x + 8$

$a = -4, b = 0, c = 8$

Copyright © 2012 Pearson Education, Inc. Publishing as Prentice Hall.

$\frac{-b}{2a} = \frac{-0}{2(-4)} = \frac{0}{-8} = 0$

$x = 0$: $y = -4(0)^2 + 8 = 8$

vertex: (0, 8)

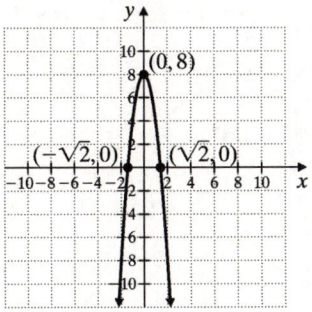

44. $y = -3x^2 + 9$

$y = 0$: $0 = -3x^2 + 9$

$0 = -3(x^2 - 3)$

$0 = x^2 - 3$

$3 = x^2$

$\pm\sqrt{3} = x$

x-intercepts: $\left(\sqrt{3}, 0\right)$ and $\left(-\sqrt{3}, 0\right)$

$x = 0$: $y = -3(0)^2 + 9 = 9$

y-intercept: (0, 9)

$y = -3x^2 + 0x + 9$

$a = -3$, $b = 0$, $c = 8$

$\frac{-b}{2a} = \frac{-0}{2(-3)} = \frac{0}{-6} = 0$

$x = 0$: $y = -3(0)^2 + 9 = 9$

vertex: (0, 9)

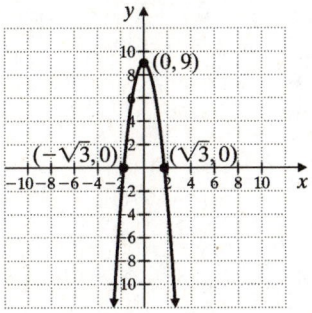

45. $y = x^2 + 3x - 10$

$y = 0$: $0 = x^2 + 3x - 10$

$0 = (x+5)(x-2)$

$0 = x + 5$ or $0 = x - 2$

$-5 = x$ $\quad$ $2 = x$

x-intercepts: (–5, 0) and (2, 0)

$x = 0$: $y = 0^2 + 3(0) - 10 = -10$

y-intercept: (0, –10)

$y = x^2 + 3x - 10$

$a = 1$, $b = 3$, $c = -10$

$\frac{-b}{2a} = \frac{-3}{2(1)} = -\frac{3}{2}$

$x = -\frac{3}{2}$: $y = \left(-\frac{3}{2}\right)^2 + 3\left(-\frac{3}{2}\right) - 10$

$= \frac{9}{4} - \frac{9}{2} - 10$

$= \frac{9}{4} - \frac{18}{4} - \frac{40}{4}$

$= -\frac{49}{4}$

vertex: $\left(-\frac{3}{2}, -\frac{49}{4}\right)$

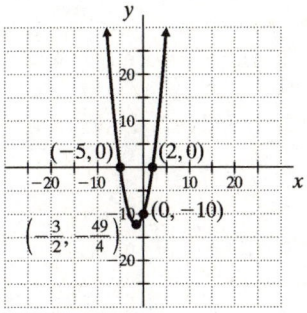

46. $y = x^2 + 3x - 4$

$y = 0$: $0 = x^2 + 3x - 4$

$0 = (x+4)(x-1)$

$0 = x + 4$ or $0 = x - 1$

$-4 = x$ $\quad$ $1 = x$

x-intercepts: (–4, 0) and (1, 0)

$x = 0$: $y = 0^2 + 3(0) - 4 = -4$

y-intercept: (0, –4)

$y = x^2 + 3x - 4$

$a = 1$, $b = 3$, $c = -4$

$\frac{-b}{2a} = \frac{-3}{2(1)} = \frac{-3}{2}$

Copyright © 2012 Pearson Education, Inc. Publishing as Prentice Hall.

$$x=-\frac{3}{2}:\ y=\left(-\frac{3}{2}\right)^2+3\left(-\frac{3}{2}\right)-4$$
$$=\frac{9}{4}-\frac{9}{2}-4$$
$$=\frac{9}{4}-\frac{18}{4}-\frac{16}{4}$$
$$=-\frac{25}{4}$$

vertex: $\left(-\frac{3}{2},-\frac{25}{4}\right)$

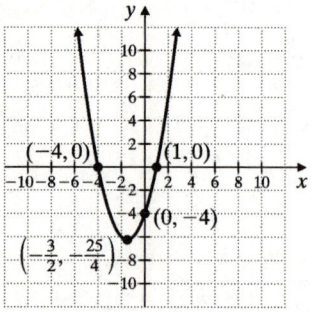

47. $y=-x^2-5x-6$

$y=0$: $0=-x^2-5x-6$
$$0=x^2+5x+6$$
$$0=(x+3)(x+2)$$
$$0=x+3 \quad \text{or} \quad 0=x+2$$
$$-3=x \qquad\qquad -2=x$$

x-intercepts: $(-3, 0)$ and $(-2, 0)$

$x=0$: $y=-0^2-5(0)-6=-6$

y-intercept: $(0, -6)$

$y=-x^2-5x-6$

$a=-1,\ b=-5,\ c=-6$

$$\frac{-b}{2a}=\frac{-(-5)}{2(-1)}=\frac{5}{-2}=-\frac{5}{2}$$

$$x=-\frac{5}{2}:\ y=-\left(-\frac{5}{2}\right)^2-5\left(-\frac{5}{2}\right)-6$$
$$y=-\frac{25}{4}+\frac{25}{2}-6$$
$$y=-\frac{25}{4}+\frac{50}{4}-\frac{24}{4}$$
$$y=\frac{1}{4}$$

vertex: $\left(-\frac{5}{2},\frac{1}{4}\right)$

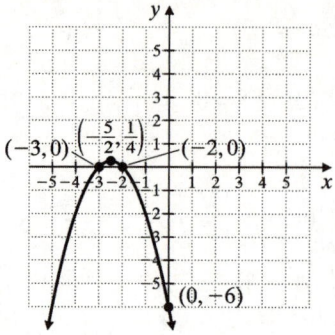

48. $y=3x^2-x-2$

$y=0$: $0=3x^2-x-2$
$$0=(3x+2)(x-1)$$
$$0=3x+2 \quad \text{or} \quad 0=x-1$$
$$-2=3x \qquad\qquad 1=x$$
$$-\frac{2}{3}=x$$

x-intercepts: $\left(-\frac{2}{3}, 0\right)$ and $(1, 0)$

$x=0$: $y=3(0)^2-0-2=-2$

y-intercept: $(0, -2)$

$y=3x^2-x-2$

$a=3,\ b=-1,\ c=-2$

$$\frac{-b}{2a}=\frac{-(-1)}{2(3)}=\frac{1}{6}$$

$$x=\frac{1}{6}:\ y=3\left(\frac{1}{6}\right)^2-\frac{1}{6}-2$$
$$y=\frac{3}{36}-\frac{1}{6}-2$$
$$y=\frac{3}{36}-\frac{6}{36}-\frac{72}{36}$$
$$y=-\frac{75}{36}=-\frac{25}{12}$$

vertex: $\left(\frac{1}{6},-\frac{25}{12}\right)$

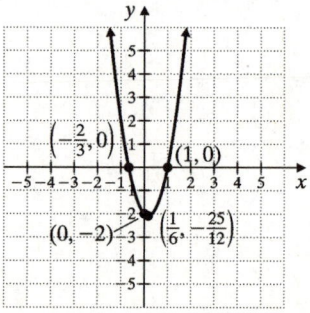

Copyright © 2012 Pearson Education, Inc. Publishing as Prentice Hall.

49. $y = 2x^2 - 11x - 6$

$y = 0$: $0 = 2x^2 - 11x - 6$

$0 = (2x+1)(x-6)$

$0 = 2x+1$ or $0 = x-6$

$-1 = 2x$ $\quad 6 = x$

$-\frac{1}{2} = x$

x-intercepts: $\left(-\frac{1}{2}, 0\right)$ and $(6, 0)$

$x = 0$: $y = 2(0)^2 - 11(0) - 6 = -6$

y-intercept: $(0, -6)$

$y = 2x^2 - 11x - 6$

$a = 2, b = -11, c = -6$

$\frac{-b}{2a} = \frac{-(-11)}{2(2)} = \frac{11}{4}$

$x = \frac{11}{4}$: $y = 2\left(\frac{11}{4}\right)^2 - 11\left(\frac{11}{4}\right) - 6$

$y = \frac{242}{16} - \frac{121}{4} - 6$

$y = \frac{242}{16} - \frac{484}{16} - \frac{96}{16}$

$y = -\frac{338}{16} = -\frac{169}{8}$

vertex: $\left(\frac{11}{4}, -\frac{169}{8}\right)$

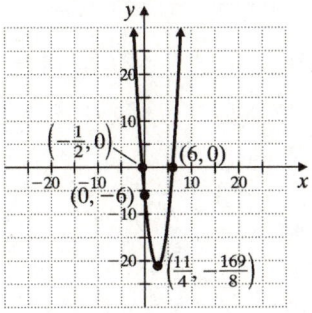

50. $y = -x^2 + 4x + 8$

$y = 0$: $0 = -x^2 + 4x + 8$

$a = -1, b = 4, c = 8$

$x = \frac{-b \pm \sqrt{b^2 - 4ac}}{2a}$

$x = \frac{-4 \pm \sqrt{4^2 - 4(-1)(8)}}{2(-1)}$

$= \frac{-4 \pm \sqrt{16+32}}{2}$

$= \frac{-4 \pm \sqrt{48}}{-2}$

$= \frac{-4 \pm 4\sqrt{3}}{-2}$

$= 2 \pm 2\sqrt{3}$

x-intercepts: $\left(2+2\sqrt{3}, 0\right)$ and $\left(2-2\sqrt{3}, 0\right)$

$x = 0$: $y = -0^2 + 4(0) + 8 = 8$

y-intercept: $(0, 8)$

$y = -x^2 + 4x + 8$

$a = -1, b = 4, c = 8$

$\frac{-b}{2a} = \frac{-4}{2(-1)} = \frac{-4}{-2} = 2$

$x = 2$: $y = -(2)^2 + 4(2) + 8 = -4 + 8 + 8 = 12$

vertex: $(2, 12)$

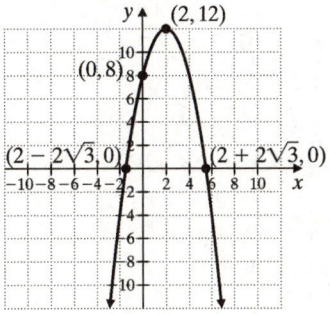

51. $y = 2x^2$ matches with graph A.

52. $y = -x^2$ matches with graph D.

53. $y = x^2 + 4x + 4$ matches with graph B.

54. $y = x^2 + 5x + 4$ matches with graph C.

55. The graph crosses the x-axis once so there is one real solution.

56. The graph crosses the x-axis twice so there are two real solutions.

57. The graph doesn't cross the x-axis so there are no real solutions.

58. The graph crosses the x-axis twice so there are two real solutions.

Copyright © 2012 Pearson Education, Inc. Publishing as Prentice Hall.

59. $x^2 = 49$

$x = \pm\sqrt{49}$

$x = \pm 7$

The solutions are $x = \pm 7$.

60. $y^2 = 75$

$y = \pm\sqrt{75}$

$y = \pm 5\sqrt{3}$

The solutions are $y = \pm 5\sqrt{3}$.

61. $(x-7)^2 = 64$

$x - 7 = \pm\sqrt{64}$

$x - 7 = \pm 8$

$x - 7 = 8$ or $x - 7 = -8$

$x = 15$ $\quad$ $x = -1$

The solutions are $x = 15$ and $x = -1$.

62.

$$x^2 + 4x = 6$$

$$x^2 + 4x + \left(\frac{4}{2}\right)^2 = 6 + \left(\frac{4}{2}\right)^2$$

$$(x+2)^2 = 10$$

$$x + 2 = \pm\sqrt{10}$$

$$x = -2 \pm \sqrt{10}$$

The solutions are $x = -2 \pm \sqrt{10}$.

63.

$$3x^2 + x = 2$$

$$x^2 + \frac{1}{3}x = \frac{2}{3}$$

$$x^2 + \frac{1}{3}x + \left(\frac{1}{6}\right)^2 = \frac{2}{3} + \left(\frac{1}{6}\right)^2$$

$$\left(x + \frac{1}{6}\right)^2 = \frac{25}{36}$$

$$x + \frac{1}{6} = \pm\sqrt{\frac{25}{36}}$$

$$x + \frac{1}{6} = \pm\frac{5}{6}$$

$$x = -\frac{1}{6} \pm \frac{5}{6}$$

$$x = -\frac{1}{6} + \frac{5}{6} = \frac{4}{6} = \frac{2}{3} \quad \text{or} \quad x = -\frac{1}{6} - \frac{5}{6} = -\frac{6}{6} = -1$$

The solutions are $x = \frac{2}{3}$ and $x = -1$.

64.

$$4x^2 - x - 2 = 0$$

$$x^2 - \frac{1}{4}x - \frac{1}{2} = 0$$

$$x^2 - \frac{1}{4}x = \frac{1}{2}$$

$$x^2 - \frac{1}{4}x + \left(-\frac{1}{8}\right)^2 = \frac{1}{2} + \left(-\frac{1}{8}\right)^2$$

$$\left(x - \frac{1}{8}\right)^2 = \frac{33}{64}$$

$$x - \frac{1}{8} = \pm\sqrt{\frac{33}{64}}$$

$$x - \frac{1}{8} = \pm\frac{\sqrt{33}}{8}$$

$$x = \frac{1}{8} \pm \frac{\sqrt{33}}{8}$$

$$x = \frac{1 \pm \sqrt{33}}{8}$$

The solutions are $x = \frac{1 \pm \sqrt{33}}{8}$.

65. $4x^2 - 3x - 2 = 0$

$a = 4,\ b = -3,\ c = -2$

$$x = \frac{-b \pm \sqrt{b^2 - 4ac}}{2a}$$

$$x = \frac{-(-3) \pm \sqrt{(-3)^2 - 4(4)(-2)}}{2(4)}$$

$$= \frac{3 \pm \sqrt{9 + 32}}{8}$$

$$= \frac{3 \pm \sqrt{41}}{8}$$

The solutions are $x = \frac{3 \pm \sqrt{41}}{8}$.

66. $5x^2 + x - 2 = 0$

$a = 5,\ b = 1,\ c = -2$

$$x = \frac{-b \pm \sqrt{b^2 - 4ac}}{2a}$$

Copyright © 2012 Pearson Education, Inc. Publishing as Prentice Hall.

$$x = \frac{-1 \pm \sqrt{1^2 - 4(5)(-2)}}{2(5)}$$
$$= \frac{-1 \pm \sqrt{1+40}}{10}$$
$$= \frac{-1 \pm \sqrt{41}}{10}$$

The solutions are $x = \frac{-1 \pm \sqrt{41}}{10}$.

67. $4x^2 + 12x + 9 = 0$

$a = 4,\ b = 12,\ c = 9$

$$x = \frac{-b \pm \sqrt{b^2 - 4ac}}{2a}$$
$$x = \frac{-12 \pm \sqrt{12^2 - 4(4)(9)}}{2(4)}$$
$$= \frac{-12 \pm \sqrt{144 - 144}}{8}$$
$$= \frac{-12 \pm \sqrt{0}}{8}$$
$$= -\frac{12}{8}$$
$$= -\frac{3}{2}$$

The solution is $x = -\frac{3}{2}$.

68. $2x^2 + x + 4 = 0$

$a = 2,\ b = 1,\ c = 4$

$$x = \frac{-b \pm \sqrt{b^2 - 4ac}}{2a}$$
$$x = \frac{-1 \pm \sqrt{1^2 - 4(2)(4)}}{2(2)}$$
$$= \frac{-1 \pm \sqrt{1-32}}{4}$$
$$= \frac{-1 \pm \sqrt{-31}}{4}$$

The equation has no solution, since the square root of -31 is not a real number.

69. $y = 4 - x^2$

$y = 0$: $0 = 4 - x^2$

$x^2 = 4$

$x = \pm\sqrt{4}$

$x = \pm 2$

x-intercepts: (2, 0) and (−2, 0)

$x = 0$: $y = 4 - 0^2 = 4$

y-intercept: (0, 4)

$y = 4 + 0x - x^2$

$a = -1,\ b = 0,\ c = 4$

$$\frac{-b}{2a} = \frac{-0}{2(-1)} = \frac{0}{-2} = 0$$

$x = 0$: $y = 4 - 0^2 = 4$

vertex: (0, 4)

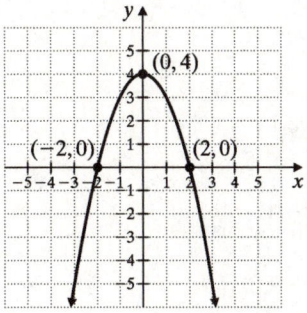

70. $y = x^2 + 4$

$y = 0$: $0 = x^2 + 4$

$-4 = x^2$

$\pm\sqrt{-4} = x$

x-intercepts: none

$x = 0$: $y = 0^2 + 4 = 4$

y-intercept: (0, 4)

$y = x^2 + 0x + 4$

$a = 1,\ b = 0,\ c = 4$

$$\frac{-b}{2a} = \frac{-0}{2(1)} = \frac{0}{2} = 0$$

$x = 0$: $y = 0^2 + 4 = 4$

vertex: (0, 4)

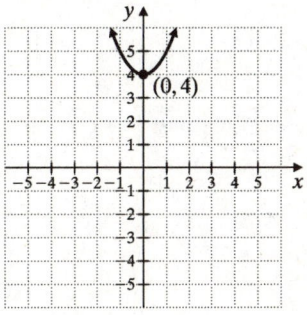

71. $y = x^2 + 6x + 8$

$y = 0$: $0 = x^2 + 6x + 8$

$0 = (x+4)(x+2)$

Copyright © 2012 Pearson Education, Inc. Publishing as Prentice Hall.

$x+4=0$ or $x+2=0$

$x=-4$ $x=-2$

x-intercepts: $(-4, 0)$ and $(-2, 0)$

$x=0$: $y=0^2+6(0)+8=8$

y-intercept: $(0, 8)$

$y=x^2+6x+8$

$a=1, b=6, c=8$

$\frac{-b}{2a}=\frac{-6}{2(1)}=\frac{-6}{2}=-3$

$x=-3$: $y=(-3)^2+6(-3)+8=9-18+8=-1$

vertex: $(-3, -1)$

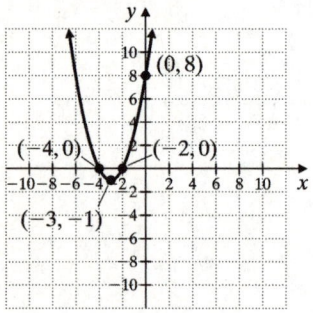

72. $y=x^2-2x-4$

$y=0$: $0=x^2-2x-4$

$a=1, b=-2, c=-4$

$$x=\frac{-b\pm\sqrt{b^2-4ac}}{2a}$$

$$x=\frac{-(-2)\pm\sqrt{(-2)^2-4(1)(-4)}}{2(1)}$$

$$=\frac{2\pm\sqrt{4+16}}{2}$$

$$=\frac{2\pm\sqrt{20}}{2}$$

$$=\frac{2\pm2\sqrt{5}}{2}$$

$$=1\pm\sqrt{5}$$

x-intercepts: $\left(1+\sqrt{5}, 0\right)$ and $\left(1-\sqrt{5}, 0\right)$

$x=0$: $y=0^2-2(0)-4=-4$

x-intercept: $(0, -4)$

$y=x^2-2x-4$

$a=1, b=-2, c=-4$

$\frac{-b}{2a}=\frac{-(-2)}{2(1)}=\frac{2}{2}=1$

$x=1$: $y=1^2-2(1)-4=1-2-4=-5$

vertex: $(1, -5)$

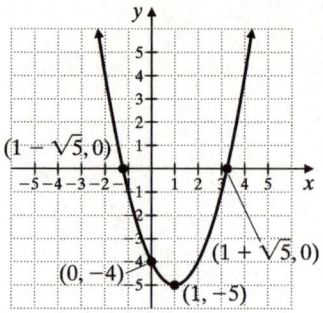

Chapter 9 Test

1. $x^2-400=0$

$(x+20)(x-20)=0$

$x+20=0$ or $x-20=0$

$x=-20$ $x=20$

The solutions are $x=\pm20$.

2. $2x^2-11x=21$

$2x^2-11x-21=0$

$(2x+3)(x-7)=0$

$2x+3=0$ or $x-7=0$

$2x=-3$ $x=7$

$x=-\frac{3}{2}$

The solutions are $x=-\frac{3}{2}$ and $x=7$.

3. $5k^2=80$

$k^2=16$

$k=\pm\sqrt{16}$

$k=\pm4$

The solutions are $k=\pm4$.

4. $(3m-5)^2=8$

$3m-5=\pm\sqrt{8}$

$3m-5=\pm2\sqrt{2}$

$3m=5\pm2\sqrt{2}$

$m=\frac{5\pm2\sqrt{2}}{3}$

The solutions are $m=\frac{5\pm2\sqrt{2}}{3}$.

Copyright © 2012 Pearson Education, Inc. Publishing as Prentice Hall.

5. $x^2 - 26x + 160 = 0$

$$x^2 - 26x = -160$$

$$x^2 - 26x + \left(\frac{-26}{2}\right)^2 = -160 + \left(\frac{-26}{2}\right)^2$$

$$x^2 - 26x + (-13)^2 = -160 + (-13)^2$$

$$x^2 - 26x + 169 = -160 + 169$$

$$(x-13)^2 = 9$$

$x - 13 = \sqrt{9}$ or $x - 13 = -\sqrt{9}$

$x - 13 = 3$ $\quad$ $x - 13 = -3$

$x = 16$ $\quad$ $x = 10$

The solutions are $x = 10$ and $x = 16$.

6. $3x^2 + 12x - 4 = 0$

$$x^2 + 4x - \frac{4}{3} = 0$$

$$x^2 + 4x = \frac{4}{3}$$

$$x^2 + 4x + \left(\frac{4}{2}\right)^2 = \frac{4}{3} + \left(\frac{4}{2}\right)^2$$

$$x^2 + 4x + (2)^2 = \frac{4}{3} + 2^2$$

$$(x+2)^2 = \frac{16}{3}$$

$$x + 2 = \pm\sqrt{\frac{16}{3}}$$

$$x + 2 = \pm\sqrt{\frac{16}{3}} \cdot \sqrt{\frac{3}{3}}$$

$$x + 2 = \pm\frac{4\sqrt{3}}{3}$$

$$x = -2 \pm \frac{4\sqrt{3}}{3}$$

The solutions are $x = -2 \pm \frac{4\sqrt{3}}{3}$.

7. $x^2 - 3x - 10 = 0$

$a = 1, b = -3, c = -10$

$$x = \frac{-b \pm \sqrt{b^2 - 4ac}}{2a}$$

$$x = \frac{-(-3) \pm \sqrt{(-3)^2 - 4(1)(-10)}}{2(1)}$$

$$= \frac{3 \pm \sqrt{9 + 40}}{2}$$

$$= \frac{3 \pm \sqrt{49}}{2}$$

$$= \frac{3 \pm 7}{2}$$

$x = \frac{3+7}{2} = \frac{10}{2} = 5$ or $x = \frac{3-7}{2} = \frac{-4}{2} = -2$

The solutions are $x = 5$ and $x = -2$.

8. $p^2 - \frac{5}{3}p - \frac{1}{3} = 0$

$a = 1,\ b = -\frac{5}{3},\ c = -\frac{1}{3}$

$$p = \frac{-b \pm \sqrt{b^2 - 4ac}}{2a}$$

$$p = \frac{-\left(-\frac{5}{3}\right) \pm \sqrt{\left(-\frac{5}{3}\right)^2 - 4(1)\left(-\frac{1}{3}\right)}}{2(1)}$$

$$= \frac{\frac{5}{3} \pm \sqrt{\frac{25}{9} + \frac{4}{3}}}{2}$$

$$= \frac{\frac{5}{3} \pm \sqrt{\frac{25}{9} + \frac{12}{9}}}{2}$$

$$= \frac{\frac{5}{3} \pm \sqrt{\frac{37}{9}}}{2}$$

$$= \frac{\frac{5}{3} \pm \frac{\sqrt{37}}{3}}{2}$$

$$= \frac{5 \pm \sqrt{37}}{6}$$

The solutions are $p = \frac{5 \pm \sqrt{37}}{6}$.

9. $(3x-5)(x+2) = -6$

$$3x^2 + 6x - 5x - 10 = -6$$

$$3x^2 + x - 10 = -6$$

$$3x^2 + x - 4 = 0$$

$a = 3, b = 1, c = -4$

$$x = \frac{-b \pm \sqrt{b^2 - 4ac}}{2a}$$

Copyright © 2012 Pearson Education, Inc. Publishing as Prentice Hall.

$$x = \frac{-1 \pm \sqrt{1^2 - 4(3)(-4)}}{2(3)}$$
$$= \frac{-1 \pm \sqrt{1+48}}{6}$$
$$= \frac{-1 \pm \sqrt{49}}{6}$$
$$= \frac{-1 \pm 7}{6}$$
$$x = \frac{-1+7}{6} = 1 \text{ or } x = \frac{-1-7}{6} = -\frac{4}{3}$$

The solutions are $x = 1$ and $x = -\frac{4}{3}$.

10. $(3x-1)^2 = 16$
$$3x - 1 = \pm\sqrt{16}$$
$$3x - 1 = \pm 4$$
$$3x = 1 \pm 4$$
$$x = \frac{1 \pm 4}{3}$$
$$x = \frac{1+4}{3} = \frac{5}{3} \quad \text{or} \quad x = \frac{1-4}{3} = \frac{-3}{3} = -1$$

The solutions are $x = \frac{5}{3}$ and $x = -1$.

11. $3x^2 - 7x - 2 = 0$
$a = 3, b = -7, c = -2$
$$x = \frac{-b \pm \sqrt{b^2 - 4ac}}{2a}$$
$$x = \frac{-(-7) \pm \sqrt{(-7)^2 - 4(3)(-2)}}{2(3)}$$
$$= \frac{7 \pm \sqrt{49+24}}{6}$$
$$= \frac{7 \pm \sqrt{73}}{6}$$

The solutions are $x = \frac{7 \pm \sqrt{73}}{6}$.

12. $x^2 - 4x - 5 = 0$
$(x-5)(x+1) = 0$
$x - 5 = 0$ or $x + 1 = 0$
$x = 5$ $\quad$ $x = -1$
The solutions are $x = 5$ and $x = -1$.

13. $3x^2 - 7x + 2 = 0$
$a = 3, b = -7, c = 2$
$$x = \frac{-b \pm \sqrt{b^2 - 4ac}}{2a}$$
$$x = \frac{-(-7) \pm \sqrt{(-7)^2 - 4(3)(2)}}{2(3)}$$
$$= \frac{7 \pm \sqrt{49-24}}{6}$$
$$= \frac{7 \pm \sqrt{25}}{6}$$
$$= \frac{7 \pm 5}{6}$$
$$x = \frac{7+5}{6} = 2 \text{ or } x = \frac{7-5}{6} = \frac{1}{3}$$

The solutions are $x = \frac{1}{3}$ and $x = 2$.

14. $2x^2 - 6x + 1 = 0$
$a = 2, b = -6, c = 1$
$$x = \frac{-b \pm \sqrt{b^2 - 4ac}}{2a}$$
$$x = \frac{-(-6) \pm \sqrt{(-6)^2 - 4(2)(1)}}{2(2)}$$
$$= \frac{6 \pm \sqrt{36-8}}{4}$$
$$= \frac{6 \pm \sqrt{28}}{4}$$
$$= \frac{6 \pm 2\sqrt{7}}{4}$$
$$= \frac{3 \pm \sqrt{7}}{2}$$

The solutions are $x = \frac{3 \pm \sqrt{7}}{2}$.

15. Let x = the length of the base.
$4x$ = height
$$A = \frac{1}{2}bh$$
$$18 = \frac{1}{2}(x)(4x)$$
$$36 = 4x^2$$
$$9 = x^2$$
$$\pm\sqrt{9} = x$$
$$\pm 3 = x$$
$4x = 4(3) = 12$
The base is 3 feet. The height is 12 feet.

Copyright © 2012 Pearson Education, Inc. Publishing as Prentice Hall.

16. $y = -5x^2$

$y = 0$: $0 = -5x^2$

$0 = x^2$

$0 = x$

x-intercept: (0, 0)

The y-intercept is also (0, 0).

$y = -5x^2 + 0x + 0$

$a = -5,\ b = 0,\ c = 0$

$\frac{-b}{2a} = \frac{-0}{2(-5)} = \frac{0}{-10} = 0$

$x = 0$: $y = -5(0)^2 = 0$

vertex: (0, 0)

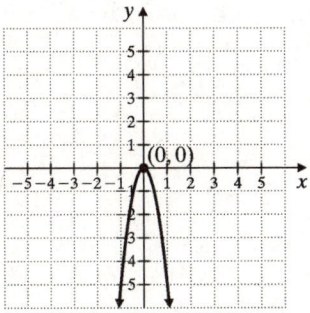

17. $y = x^2 - 4$

$y = 0$: $0 = x^2 - 4$

$0 = (x+2)(x-2)$

$0 = x + 2$ or $0 = x - 2$

$-2 = x$ $\quad$ $2 = x$

x-intercepts: (−2, 0), (2, 0)

$x = 0$: $y = 0^2 - 4 = -4$

y-intercept: (0, −4)

$y = x^2 + 0x - 4$

$a = 1,\ b = 0,\ c = -4$

$\frac{-b}{2a} = \frac{-0}{2(1)} = \frac{0}{2} = 0$

$x = 0$: $y = 0^2 - 4 = -4$

vertex: (0, −4)

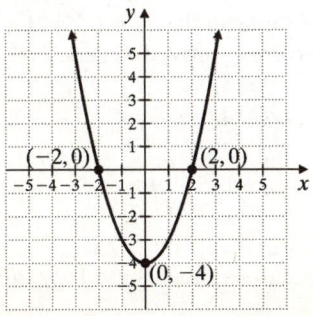

18. $y = x^2 - 7x + 10$

$y = 0$: $0 = x^2 - 7x + 10$

$0 = (x-5)(x-2)$

$x - 5 = 0$ or $x - 2 = 0$

$x = 5$ $\quad$ $x = 2$

x-intercepts: (5, 0) and (2, 0)

$x = 0$: $y = 0^2 - 7(0) + 10 = 10$

y-intercept: (0, 10)

$y = x^2 - 7x + 10$

$a = 1,\ b = -7,\ c = 10$

$\frac{-b}{2a} = \frac{-(-7)}{2(1)} = \frac{7}{2}$

$x = \frac{7}{2}$: $y = \left(\frac{7}{2}\right)^2 - 7\left(\frac{7}{2}\right) + 10$

$= \frac{49}{4} - \frac{49}{2} + 10$

$= \frac{49}{4} - \frac{98}{4} + \frac{40}{4}$

$= -\frac{9}{4}$

vertex: $\left(\frac{7}{2}, -\frac{9}{4}\right)$

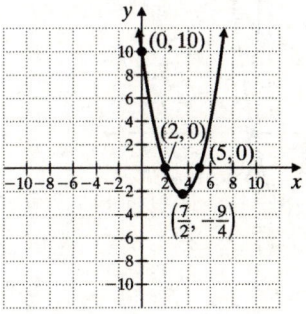

19. $y = 2x^2 + 4x - 1$

$y = 0$: $0 = 2x^2 + 4x - 1$

$a = 2,\ b = 4,\ c = -1$

$x = \frac{-b \pm \sqrt{b^2 - 4ac}}{2a}$

Copyright © 2012 Pearson Education, Inc. Publishing as Prentice Hall.

$$x = \frac{-4 \pm \sqrt{4^2 - 4(2)(-1)}}{2(2)}$$
$$= \frac{-4 \pm \sqrt{16+8}}{4}$$
$$= \frac{-4 \pm \sqrt{24}}{4}$$
$$= \frac{-4 \pm 2\sqrt{6}}{4}$$
$$= \frac{-2 \pm \sqrt{6}}{2}$$

x-intercepts: $\left(\frac{-2-\sqrt{6}}{2}, 0\right)$ and $\left(\frac{-2+\sqrt{6}}{2}, 0\right)$

$x = 0$: $y = 2(0)^2 + 4(0) - 1 = -1$

y-intercept: $(0, -1)$

$y = 2x^2 + 4x - 1$

$a = 2, b = 4, c = -1$

$\frac{-b}{2a} = \frac{-4}{2(2)} = \frac{-4}{4} = -1$

$x = -1$: $y = 2(-1)^2 + 4(-1) - 1 = 2 - 4 - 1 = -3$

vertex: $(-1, -3)$

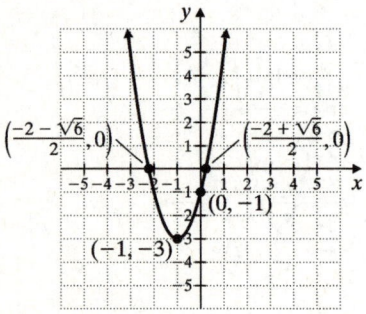

20. $d = \frac{n^2 - 3n}{2}$

$9 = \frac{n^2 - 3n}{2}$

$18 = n^2 - 3n$

$0 = n^2 - 3n - 18$

$0 = (n-6)(n+3)$

$n - 6 = 0$ or $n + 3 = 0$

$n = 6$ $\quad$ $n = -3$

A 6-sided polygon has 9 diagonals.

21. $h = 16t^2$

$120.75 = 16t^2$

$\frac{120.75}{16} = t^2$

$7.546875 = t^2$

$\sqrt{7.546875} = t$ or $-\sqrt{7.546875} = t$

$2.7 \approx t$ $\quad$ $-2.7 \approx t$

Since the time of the dive is not a negative number, discard the solution $t \approx -2.7$. The dive took approximately 2.7 seconds.

Cumulative Review Chapters 1–9

1. $y + 0.6 = -1.0$

$y = -1.6$

2. $8x - 14 = 6x - 20$

$2x = -6$

$x = -3$

3. $8(2 - t) = -5t$

$16 - 8t = -5t$

$16 = 3t$

$\frac{16}{3} = t$

4. $2(x + 7) = 5(2x - 3)$

$2x + 14 = 10x - 15$

$29 = 8x$

$\frac{29}{8} = x$

5. Let x be the number of Republican representatives. Then $x + 78$ was the number of Democratic representatives.

$x + x + 78 = 434$

$2x + 78 = 434$

$2x = 356$

$x = 178$

$x + 78 = 178 + 78 = 256$

There were 256 Democratic representatives and 178 Republican representatives.

6. Let n be the first integer. Then $n + 1$ and $n + 2$ are the next two consecutive integers.

$n + (n+1) + (n+2) = 438$

$3n + 3 = 438$

$3n = 435$

$n = 145$

$n + 1 = 146$

$n + 2 = 147$

The three consecutive integers are 145, 146, and 147.

Copyright © 2012 Pearson Education, Inc. Publishing as Prentice Hall.

7. $3^0 = 1$

8. $\left(\dfrac{-6x}{y^3}\right)^3 = \dfrac{(-6x)^3}{(y^3)^3} = \dfrac{(-6)^3 x^3}{y^{3\cdot 3}} = -\dfrac{216x^3}{y^9}$

9. $(5x^3y^2)^0 = 1$

10. $\dfrac{a^2b^7}{(2b^2)^5} = \dfrac{a^2b^7}{32b^{10}} = \dfrac{a^2}{32b^3}$

11. $-4^0 = -1$

12. $\dfrac{(3y)^2}{y^2} = \dfrac{9y^2}{y^2} = 9$

13. $(3y+2)^2 = (3y)^2 + 2(2)(3y) + 2^2$
$= 9y^2 + 12y + 4$

14. $(x^2+5)(y-1) = x^2y - x^2 + 5y - 5$

15. $\dfrac{x^2+7x+12}{x+3} = x+4$

$$\begin{array}{r} x+4 \\ x+3\overline{\big)\,x^2+7x+12} \\ \underline{x^2+3x} \\ 4x+12 \\ \underline{4x+12} \\ 0 \end{array}$$

16. $2 + 8.1a + a - 6 = 8.1a + a + 2 - 6$
$= (8.1+1) + (2-6)$
$= 9.1a - 4$

17. $r^2 - r - 42 = (r-7)(r+6)$

18. a. $\dfrac{x-y}{7-x} = \dfrac{-4-7}{7-(-4)} = \dfrac{-11}{11} = -1$

b. $x^2 + 2y = (-4)^2 + 2(7) = 16 + 14 = 30$

19. $10x^2 - 13xy - 3y^2 = (2x-3y)(5x+y)$

20. $\dfrac{1}{x+2} + \dfrac{7}{x-1} = \dfrac{1(x-1)}{(x+2)(x-1)} + \dfrac{7(x+2)}{(x-1)(x+2)}$
$= \dfrac{x-1+7x+14}{(x+2)(x-1)}$
$= \dfrac{8x+13}{(x+2)(x-1)}$

21. $8x^2 - 14x + 5 = 8x^2 - 4x - 10x + 5$
$= 4x(2x-1) - 5(2x-1)$
$= (2x-1)(4x-5)$

22. $\dfrac{x^2+7x}{5x} \cdot \dfrac{10x+25}{x^2-49} = \dfrac{x(x+7)}{5x} \cdot \dfrac{5(2x+5)}{(x-7)(x+7)}$
$= \dfrac{2x+5}{x-7}$

23. a. $4x^3 - 49x = x(4x^2 - 49) = x(2x-7)(2x+7)$

b. $162x^4 - 2 = 2(81x^4 - 1)$
$= 2(9x^2-1)(9x^2+1)$
$= 2(9x^2+1)(3x-1)(3x+1)$

24. $\dfrac{2x+7}{3} = \dfrac{x-6}{2}$
$6\left(\dfrac{2x+7}{3}\right) = 6\left(\dfrac{x-6}{2}\right)$
$2(2x+7) = 3(x-6)$
$4x + 14 = 3x - 18$
$x = -32$

25. $(5x-1)(2x^2+15x+18) = 0$
$(5x-1)(2x+3)(x+6) = 0$

$5x-1=0$	or	$2x+3=0$	or	$x+6=0$
$5x=1$		$2x=-3$		$x=-6$
$x=\dfrac{1}{5}$		$x=-\dfrac{3}{2}$		

The solutions are $x = \dfrac{1}{5}$, $x = -\dfrac{3}{2}$, and $x = -6$.

26. a. $4x - 3 + 7 - 5x = -x + 4$

b. $-6y + 3y - 8 + 8y = 5y - 8$

c. $7 + 10.1a - a - 11 = 10.1a - a + 7 - 11$
$= 9.1a - 4$

d. $2x^2 - 2x = 2x^2 - 2x$

Copyright © 2012 Pearson Education, Inc. Publishing as Prentice Hall.

27. $\dfrac{x^2+8x+7}{x^2-4x-5}=\dfrac{(x+7)(x+1)}{(x-5)(x+1)}=\dfrac{x+7}{x-5}$

28.
$$2x^2+5x=7$$
$$2x^2+5x-7=0$$
$$(2x+7)(x-1)=0$$
$$2x+7=0 \quad \text{or} \quad x-1=0$$
$$2x=-7 \qquad x=1$$
$$x=-\frac{7}{2}$$

The solutions are $x=-\dfrac{7}{2}$ and $x=1$.

29.
$$\frac{n}{6}-\frac{5}{3}=\frac{n}{2}$$
$$6\left(\frac{n}{6}-\frac{5}{3}\right)=6\left(\frac{n}{2}\right)$$
$$n-10=3n$$
$$-10=2n$$
$$-5=n$$

The number is -5.

30.
$$d=\sqrt{(x_2-x_1)^2+(y_2-y_1)^2}$$
$$d=\sqrt{(2-(-7))^2+(5-4)^2}$$
$$d=\sqrt{(9)^2+(1)^2}$$
$$d=\sqrt{81+1}$$
$$d=\sqrt{82}$$

The distance is $\sqrt{82}$ units.

31.

	x	$y=3x$
a.	-1	-3
b.	0	0
c.	-3	-9

32. a. x-intercept: (4, 0)
y-intercept: (0, 1)

b. x-intercepts: (–2, 0), (0, 0), (3, 0)
y-intercept: (0, 0)

33. a.
$$y=-\frac{1}{5}x+1$$
$$2x+10y=3$$
$$10y=-2x+3$$
$$y=\frac{-2}{10}x+\frac{3}{10}$$
$$y=-\frac{1}{5}x+\frac{3}{10}$$

These two equations have the same slope, therefore they are parallel.

b.
$$x+y=3 \Rightarrow y=-x+3$$
$$-x+y=4 \Rightarrow y=x+4$$

These two equations have slopes whose product is -1, therefore they are perpendicular.

c.
$$3x+y=5 \Rightarrow y=-3x+5$$
$$2x+3y=6 \Rightarrow 3y=-2x+6$$
$$\Rightarrow y=-\frac{2}{3}x+2$$

These two equations have different slopes (and their product is $\neq -1$), so they are neither parallel nor perpendicular.

34.
$$y=3x+7$$
$$x+3y=-15 \Rightarrow 3y=-x-15$$
$$y=-\frac{x}{3}-5$$

These two equations have slopes whose product is -1; therefore these lines are perpendicular.

35. a. {(–1, 1), (2, 3), (7, 3), (8, 6)}
This is a function because each x-value is assigned to only one y-value.

b. {(0, –2), (1, 5), (0, 3), (7, 7)}
This is not a function because the x-value 0 is paired with two different y-values.

36. a. $\sqrt{80}+\sqrt{20}=4\sqrt{5}+2\sqrt{5}=6\sqrt{5}$

b.
$$2\sqrt{98}-2\sqrt{18}=2\cdot 7\sqrt{2}-2\cdot 3\sqrt{2}$$
$$=14\sqrt{2}-6\sqrt{2}$$
$$=8\sqrt{2}$$

c. $\sqrt{32}+\sqrt{121}-\sqrt{12}=4\sqrt{2}+11-2\sqrt{3}$

Copyright © 2012 Pearson Education, Inc. Publishing as Prentice Hall.

37. $\begin{cases} 2x+y=10 \\ x=y+2 \end{cases}$

Substitute $y + 2$ for x in the first equation.

$$\begin{aligned} 2(y+2)+y&=10 \\ 2y+4+y&=10 \\ 3y&=6 \\ y&=2 \end{aligned}$$

Solve for x.

$x = y + 2 = 2 + 2 = 4$

The solution for this system is (4, 2).

38. $\begin{cases} 5x+y=3 \\ y=-5x \end{cases}$

Substitute $-5x$ for y is the first equation. Solve for x.

$$\begin{aligned} 5x-5x&=3 \\ 0&=3 \quad \text{False} \end{aligned}$$

The system has no solution.

39. $\begin{cases} 2x-y=7 \\ 8x-4y=1 \end{cases}$

Solve the first equation for y.

$$\begin{aligned} 2x-y&=7 \\ 2x-7&=y \end{aligned}$$

Substitute $2x - 7$ for y in the second equation and solve for x.

$$\begin{aligned} 8x-4(2x-7)&=1 \\ 8x-8x+28&=1 \\ 28&=1 \quad \text{False} \end{aligned}$$

The system has no solution.

40. $\begin{cases} -2x+y=7 \\ 6x-3y=-21 \Rightarrow -2x+y=7 \end{cases}$

The two equations are the same. Therefore, there is an infinite number of solutions.

41. $\sqrt{36}=6$ because $6^2=36$.

42. $\sqrt{\dfrac{4}{25}}=\dfrac{2}{5}$ because $\left(\dfrac{2}{5}\right)^2=\dfrac{4}{25}$.

43. $\sqrt{\dfrac{9}{100}}=\dfrac{3}{10}$ because $\left(\dfrac{3}{10}\right)^2=\dfrac{9}{100}$.

44. $\sqrt{\dfrac{16}{121}}=\dfrac{4}{11}$ because $\left(\dfrac{4}{11}\right)^2=\dfrac{16}{121}$.

45.
$$\begin{aligned} \frac{2}{1+\sqrt{3}}&=\frac{2}{1+\sqrt{3}}\cdot\frac{1-\sqrt{3}}{1-\sqrt{3}} \\ &=\frac{2-2\sqrt{3}}{1-3} \\ &=\frac{2-2\sqrt{3}}{-2} \\ &=-1+\sqrt{3} \end{aligned}$$

46. $\dfrac{5}{\sqrt{8}}=\dfrac{5}{2\sqrt{2}}\cdot\dfrac{\sqrt{2}}{\sqrt{2}}=\dfrac{5\sqrt{2}}{4}$

47.
$$\begin{aligned} (x-3)^2&=16 \\ x-3&=\pm\sqrt{16} \\ x-3&=\pm 4 \\ x&=3\pm 4 \end{aligned}$$

$x = 3 + 4 = 7$ or $x = 3 - 4 = -1$

The solutions are $x = 7$ and $x = -1$.

48.
$$\begin{aligned} 3(x-4)^2&=9 \\ (x-4)^2&=3 \\ x-4&=\pm\sqrt{3} \\ x&=4\pm\sqrt{3} \end{aligned}$$

The solutions are $x=4\pm\sqrt{3}$.

49.
$$\begin{aligned} \frac{1}{2}x^2-x&=2 \\ \frac{1}{2}x^2-x-2&=0 \end{aligned}$$

$a=\dfrac{1}{2},\ b=-1,\ c=-2$

$$\begin{aligned} x&=\frac{-b\pm\sqrt{b^2-4ac}}{2a} \\ x&=\frac{-(-1)\pm\sqrt{(-1)^2-4\left(\frac{1}{2}\right)(-2)}}{2\left(\frac{1}{2}\right)} \\ &=\frac{1\pm\sqrt{1+4}}{1} \\ &=1\pm\sqrt{5} \end{aligned}$$

The solutions are $x=1\pm\sqrt{5}$.

Copyright © 2012 Pearson Education, Inc. Publishing as Prentice Hall.

50. $x^2 + 4x = 8$

$x^2 + 4x - 8 = 0$

$a = 1, b = 4, c = -8$

$$x = \frac{-b \pm \sqrt{b^2 - 4ac}}{2a}$$

$$x = \frac{-4 \pm \sqrt{4^2 - 4(1)(-8)}}{2(1)}$$

$$= \frac{-4 \pm \sqrt{16 + 32}}{2}$$

$$= \frac{-4 \pm \sqrt{48}}{2}$$

$$= \frac{-4 \pm 4\sqrt{3}}{2}$$

$$= -2 \pm 2\sqrt{3}$$

The solutions are $x = -2 \pm 2\sqrt{3}$.

Copyright © 2012 Pearson Education, Inc. Publishing as Prentice Hall.

Appendices

Appendix A Exercise Set

1. $a^3+27=a^3+3^3$
$=(a+3)[a^2-(a)(3)+3^2]$
$=(a+3)(a^2-3a+9)$

3. $8a^3+1=(2a)^3+1^3$
$=(2a+1)[(2a)^2-(2a)(1)+1^2]$
$=(2a+1)(4a^2-2a+1)$

5. $5k^3+40=5(k^3+8)$
$=5(k^3+2^3)$
$=5(k+2)[k^2-(k)(2)+2^2]$
$=5(k+2)(k^2-2k+4)$

7. $x^3y^3-64=(xy)^3-4^3$
$=(xy-4)[(xy)^2+(xy)(4)+4^2]$
$=(xy-4)(x^2y^2+4xy+16)$

9. $x^3+125=x^3+5^3$
$=(x+5)[x^2-(x)(5)+5^2]$
$=(x+5)(x^2-5x+25)$

11. $24x^4-81xy^3$
$=3x(8x^3-27y^3)$
$=3x[(2x)^3-(3y)^3]$
$=3x(2x-3y)[(2x)^2+(2x)(3y)+(3y)^2]$
$=3x(2x-3y)(4x^2+6xy+9y^2)$

13. $27-t^3=3^3-t^3$
$=(3-t)[3^2+(3)(t)+t^2]$
$=(3-t)(9+3t+t^2)$

15. $8r^3-64=8(r^3-8)$
$=8(r^3-2^3)$
$=8(r-2)[r^2+(r)(2)+2^2]$
$=8(r-2)(r^2+2r+4)$

17. $t^3-343=t^3-7^3$
$=(t-7)[t^2+(t)(7)+7^2]$
$=(t-7)(t^2+7t+49)$

19. $s^3-64t^3=s^3-(4t)^3$
$=(s-4t)[s^2+(s)(4t)+(4t)^2]$
$=(s-4t)(s^2+4st+16t^2)$

Appendix B Exercise Set

1. mean $=\dfrac{21+28+16+42+38}{5}=\dfrac{145}{5}=29$
16, 21, <u>28</u>, 38, 42
median = 28
no mode

3. mean $=\dfrac{7.6+8.2+8.2+9.6+5.7+9.1}{6}$
$=\dfrac{48.4}{6}$
≈ 8.1
5.7, 7.6, <u>8.2, 8.2</u>, 9.1, 9.6
median $=\dfrac{8.2+8.2}{2}=8.2$
mode: 8.2

5. mean
$=\dfrac{0.2+0.3+0.5+0.6+0.6+0.9+0.2+0.7+1.1}{9}$
$=\dfrac{5.1}{9}$
≈ 0.6
0.2, 0.2, 0.3, 0.5, <u>0.6</u>, 0.6, 0.7, 0.9, 1.1
median = 0.6
mode: 0.2 and 0.6

7. mean
$=\dfrac{231+543+601+293+588+109+334+268}{8}$
$=\dfrac{2967}{8}$
≈ 370.9
109, 231, 268, <u>293, 334</u>, 543, 588, 601
median $=\dfrac{293+334}{2}=313.5$
no mode

Copyright © 2012 Pearson Education, Inc. Publishing as Prentice Hall.

9. $\dfrac{1454+1250+1136+1127+1107}{5}=\dfrac{6074}{5}$
$=1214.8$

The mean height of the five tallest buildings is 1214.8 feet.

11. 1002, 1023, 1046, <u>1107, 1127</u>, 1136, 1250, 1454

$\text{median}=\dfrac{1107+1127}{2}=1117$

The median height of the eight tallest buildings is 1117 feet.

13. $\dfrac{7.8+6.9+7.5+4.7+6.9+7.0}{6}=\dfrac{40.8}{6}=6.8$

The mean time was 6.8 seconds.

15. The mode time is 6.9 seconds.

17. 74, 77, <u>85, 86</u>, 91, 95

$\text{median}=\dfrac{85+86}{2}=85.5$

The median test score was 85.5.

19. $\dfrac{78+80+66+\cdots+72}{15}=\dfrac{1095}{15}=73$

The mean pulse rate is 73.

21. The values 70 and 71 both occur twice while other values only occur once, so the mode of the pulse rates is 70 and 71.

23. There were 9 rates lower than the mean of 73.

25. Since the mode is 21, the value 21 must occur at least twice in the set of numbers. Since there are an odd number of numbers in the set, the median is the middle number. That is, the median of 20 is one of the numbers in the set.
The missing numbers are 21, 21, and 20.

Appendix C Exercise Set

1. The set of negative integers from -10 to -5 in roster form is $\{-9, -8, -7, -6\}$.

3. The set of the days of the week starting with the letter T in roster form is {Tuesday, Thursday}.

5. The set of whole numbers in roster form is $\{0, 1, 2, 3, 4, \ldots\}$.

7. There are no integers between 1 and 2, so the answer is { } or $\varnothing$.

9. Since 3 is a listed element of the set $\{1, 3, 5, 7, 9\}$, the statement $3 \in \{1, 3, 5, 7, 9\}$ is true.

11. $\{3\} \subseteq \{1, 3, 5, 7, 9\}$ since every element of the left-hand set is also an element of the right-hand set.

13. $\{a, e, i, o, u\} \subseteq \{a, e, i, o, u\}$ since every element of the left-hand set is also an element of the right-hand set.

15. {May} $\subseteq$ the set of the days of the week is false since the element in the left-hand set is not an element of the right-hand set. Note that the days of the week are Sunday, Monday, Tuesday, Wednesday, Thursday, Friday, and Saturday.

17. Since 9 is not an even number, it is not an element of the set $\{x|x \text{ is an even number}\}$. Thus the statement $9 \notin \{x|x \text{ is an even number}\}$ is true.

19. $\{a\} \not\subseteq$ the set of vowels is false since the element in the left-hand set is a vowel.

21. $A \cup B = \{1, 2, 3, 4, 5, 6\} \cup \{2, 4, 6\}$
$= \{1, 2, 3, 4, 5, 6\}$

23. $A \cap B = \{1, 2, 3, 4, 5, 6\} \cap \{2, 4, 6\}$
$= \{2, 4, 6\}$

25. $C \cup D = \{1, 3, 5\} \cup \{7\} = \{1, 3, 5, 7\}$

27. $B \cap D = \{2, 4, 6\} \cap \{7\} = \varnothing$

29. Since every element of $B = \{2, 4, 6\}$ is also an element of $A = \{1, 2, 3, 4, 5, 6\}$, the statement $B \subseteq A$ is true.

31. Since the element of $D = \{7\}$ is not an element of $C = \{1, 3, 5\}$, the statement $D \subseteq C$ is false.

33. Since 2 is an element of $A = \{1, 2, 3, 4, 5, 6\}$, but not of $C = \{1, 3, 5\}$, the statement $A \subseteq C$ is false.

35. Since not every element of $B = \{2, 4, 6\}$ is also an element of $C = \{1, 3, 5\}$, the statement $B \not\subseteq C$ is true.

37. Since the empty set is a subset of every set, the statement $\varnothing \subseteq A$ is true.

Copyright © 2012 Pearson Education, Inc. Publishing as Prentice Hall.

39. $A \cup D = \{1, 2, 3, 4, 5, 6\} \cup \{7\}$
$= \{1, 2, 3, 4, 5, 6, 7\}$
The statement is true.

41. Since the union of a set with the empty set is the original set, the statement that $\{a, b, c\} \cup \{\ \}$ is $\{a, b, c\}$ is true.

Appendix D Exercise Set

1. $90° - 19° = 71°$
The complement of a 19° angle is a 71° angle.

3. $90° - 70.8° = 19.2°$
The complement of a 70.8° angle is a 19.2° angle.

5. $90° - 11\frac{1}{4}° = 78\frac{3}{4}°$
The complement of an $11\frac{1}{4}°$ angle is a $78\frac{3}{4}°$ angle.

7. $180° - 150° = 30°$
The supplement of a 150° angle is a 30° angle.

9. $180° - 30.2° = 149.8°$
The supplement of a 30.2° angle is a 149.8° angle.

11. $180° - 79\frac{1}{2}° = 100\frac{1}{2}°$
The supplement of a $79\frac{1}{2}°$ angle is a $100\frac{1}{2}°$ angle.

13. $\angle 1$ and the angle marked 110° are vertical angles, so $m\angle 1 = 110°$.
$\angle 2$ and the angle marked 110° are supplementary angles, so
$m\angle 2 = 180° - 110° = 70°$.
$\angle 3$ and the angle marked 110° are supplementary angles, so
$m\angle 3 = 180° - 110° = 70°$.
$\angle 4$ and $\angle 3$ are alternate interior angles, so
$m\angle 4 = m\angle 3 = 70°$.
$\angle 5$ and the angle marked 110° are alternate interior angles, so $m\angle 5 = 110°$.
$\angle 6$ and $\angle 5$ are supplementary angles, so
$m\angle 6 = 180° - m\angle 5 = 180° - 110° = 70°$.
$\angle 7$ and the angle marked 110° are corresponding angles, so $m\angle 7 = 110°$.

15. $180° - 11° - 79° = 90°$
The third angle measures 90°.

17. $180° - 25° - 65° = 90°$
The third angle measures 90°.

19. $180° - 30° - 60° = 90°$
The third angle measures 90°.

21. Since the triangle is a right triangle, one angle measures 90°.
$180° - 45° - 90° = 45°$
The other two angles of the triangle measure 45° and 90°.

23. Since the triangle is a right triangle, one angle measures 90°.
$180° - 17° - 90° = 73°$
The other two angles of the triangle measure 90° and 73°.

25. Since the triangle is a right triangle, one angle measures 90°.
$180° - 39\frac{3}{4}° - 90° = 50\frac{1}{4}°$
The other two angles of the triangle measure $50\frac{1}{4}°$ and 90°.

27.
$$\frac{12}{4} = \frac{18}{x}$$
$$12x = 72$$
$$x = \frac{72}{12}$$
$$x = 6$$

29.
$$\frac{6}{9} = \frac{3}{x}$$
$$6x = 27$$
$$x = \frac{27}{6}$$
$$x = 4.5$$

Copyright © 2012 Pearson Education, Inc. Publishing as Prentice Hall.

31. $a^2 + b^2 = c^2$

$6^2 + 8^2 = c^2$

$36 + 64 = c^2$

$100 = c^2$

Since c represents a length, we assume that c is positive. Since $c^2 = 100$, $c = 10$. The hypotenuse of the right triangle has length 10.

33. $a^2 + b^2 = c^2$

$a^2 + 5^2 = 13^2$

$a^2 + 25 = 169$

$a^2 = 144$

Since a represents a length, we assume that a is positive. Since $a^2 = 144$, $a = 12$. The other leg of the right triangle has length 12.

Copyright © 2012 Pearson Education, Inc. Publishing as Prentice Hall.

Practice Final Exam

1. $-3^4 = -(3\cdot 3\cdot 3\cdot 3) = -81$

2. $4^{-3} = \dfrac{1}{4^3} = \dfrac{1}{64}$

3.
$$\begin{aligned}6[5+2(3-8)-3] &= 6[5+2(3+(-8))-3]\\ &= 6[5+2(-5)-3]\\ &= 6[5+(-10)+(-3)]\\ &= 6[-5+(-3)]\\ &= 6[-8]\\ &= -48\end{aligned}$$

4.
$$\begin{array}{r} 5x^3 + x^2 + 5x - 2 \\ -(8x^3 - 4x^2 + x - 7) \\ \hline \end{array} \qquad \begin{array}{r} 5x^3 + x^2 + 5x - 2 \\ -8x^3 + 4x^2 - x + 7 \\ \hline -3x^3 + 5x^2 + 4x + 5 \end{array}$$

5.
$$\begin{aligned}(4x-2)^2 &= (4x)^2 - 2(4x)(2) + 2^2\\ &= 16x^2 - 16x + 4\end{aligned}$$

6.
$$\begin{aligned}&(3x+7)(x^2+5x+2)\\ &= 3x(x^2+5x+2) + 7(x^2+5x+2)\\ &= 3x(x^2) + 3x(5x) + 3x(2) + 7(x^2) + 7(5x) + 7(2)\\ &= 3x^3 + 15x^2 + 6x + 7x^2 + 35x + 14\\ &= 3x^3 + 22x^2 + 41x + 14\end{aligned}$$

7.
$$\begin{aligned}6t^2 - t - 5 &= 6t^2 + 5t - 6t - 5\\ &= t(6t+5) - 1(6t+5)\\ &= (6t+5)(t-1)\end{aligned}$$

8.
$$\begin{aligned}3x^3 - 21x^2 + 30x &= 3x(x^2 - 7x + 10)\\ &= 3x(x-5)(x-2)\end{aligned}$$

9.
$$\begin{aligned}180 - 5x^2 &= 5(36 - x^2)\\ &= 5(6^2 - x^2)\\ &= 5(6-x)(6+x)\end{aligned}$$

10.
$$\begin{aligned}3a^2 + 3ab - 7a - 7b &= 3a(a+b) - 7(a+b)\\ &= (a+b)(3a-7)\end{aligned}$$

11.
$$\begin{aligned}x - x^5 &= x(1 - x^4)\\ &= x[1^2 - (x^2)^2]\\ &= x(1+x^2)(1-x^2)\\ &= x(1+x^2)(1+x)(1-x)\end{aligned}$$

12.
$$\begin{aligned}\left(\frac{4x^2y^3}{x^3y^{-4}}\right)^2 &= \frac{4^2x^{2\cdot 2}y^{3\cdot 2}}{x^{3\cdot 2}y^{-4\cdot 2}}\\ &= \frac{16x^4y^6}{x^6y^{-8}}\\ &= 16x^{4-6}y^{6-(-8)}\\ &= 16x^{-2}y^{14}\\ &= \frac{16y^{14}}{x^2}\end{aligned}$$

13.
$$\frac{5-\frac{1}{y^2}}{\frac{1}{y}+\frac{2}{y^2}} = \frac{y^2\left(5-\frac{1}{y^2}\right)}{y^2\left(\frac{1}{y}+\frac{2}{y^2}\right)} = \frac{5y^2-1}{y+2}$$

14.
$$\begin{aligned}&\frac{x^2-13x+42}{x^2+10x+21} \div \frac{x^2-4}{x^2+x-6}\\ &= \frac{x^2-13x+42}{x^2+10x+21}\cdot\frac{x^2+x-6}{x^2-4}\\ &= \frac{(x-6)(x-7)}{(x+3)(x+7)}\cdot\frac{(x+3)(x-2)}{(x+2)(x-2)}\\ &= \frac{(x-6)(x-7)}{(x+7)(x+2)}\end{aligned}$$

15.
$$\begin{aligned}&\frac{5a}{a^2-a-6} - \frac{2}{a-3}\\ &= \frac{5a}{(a-3)(a+2)} - \frac{2}{a-3}\\ &= \frac{5a}{(a-3)(a+2)} - \frac{2(a+2)}{(a-3)(a+2)}\\ &= \frac{5a-2(a+2)}{(a-3)(a+2)}\\ &= \frac{5a-2a-4}{(a-3)(a+2)}\\ &= \frac{3a-4}{(a-3)(a+2)}\end{aligned}$$

Copyright © 2012 Pearson Education, Inc. Publishing as Prentice Hall.

16.
$$\begin{aligned} 4(n-5) &= -(4-2n) \\ 4n-20 &= -4+2n \\ 4n-20-2n &= -4+2n-2n \\ 2n-20 &= -4 \\ 2n-20+20 &= -4+20 \\ 2n &= 16 \\ \frac{2n}{2} &= \frac{16}{2} \\ n &= 8 \end{aligned}$$

17.
$$\begin{aligned} x(x+6) &= 7 \\ x^2+6x &= 7 \\ x^2+6x-7 &= 0 \\ (x+7)(x-1) &= 0 \end{aligned}$$
$x+7=0$ or $x-1=0$
$x=-7$ $\quad$ $x=1$

The solutions are –7 and 1.

18.
$$\begin{aligned} 3x-5 &\geq 7x+3 \\ 3x-5-3x &\geq 7x+3-3x \\ -5 &\geq 4x+3 \\ -5-3 &\geq 4x+3-3 \\ -8 &\geq 4x \\ \frac{-8}{4} &\geq \frac{4x}{4} \\ -2 &\geq x \end{aligned}$$
$\{x|x \leq -2\}$

19. $2x^2-6x+1=0$

$a=2, b=-6, c=1$
$$\begin{aligned} x &= \frac{-b \pm \sqrt{b^2-4ac}}{2a} \\ x &= \frac{-(-6) \pm \sqrt{(-6)^2-4(2)(1)}}{2(2)} \\ &= \frac{6 \pm \sqrt{36-8}}{4} \\ &= \frac{6 \pm \sqrt{28}}{4} \\ &= \frac{6 \pm 2\sqrt{7}}{4} \\ &= \frac{3 \pm \sqrt{7}}{2} \end{aligned}$$

The solutions are $x = \frac{3 \pm \sqrt{7}}{2}$.

20. The LCD is $3 \cdot 5 \cdot y = 15y$.
$$\begin{aligned} \frac{4}{y} - \frac{5}{3} &= -\frac{1}{5} \\ 15y\left(\frac{4}{y} - \frac{5}{3}\right) &= 15y\left(-\frac{1}{5}\right) \\ 60-25y &= -3y \\ 60 &= 22y \\ \frac{60}{22} &= y \\ \frac{30}{11} &= y \end{aligned}$$

The solution is $\frac{30}{11}$.

21.
$$\begin{aligned} \frac{5}{y+1} &= \frac{4}{y+2} \\ 5(y+2) &= 4(y+1) \\ 5y+10 &= 4y+4 \\ y &= -6 \end{aligned}$$
The solution is –6.

22. The LCD is $2(a-3)$.
$$\begin{aligned} \frac{a}{a-3} &= \frac{3}{a-3} - \frac{3}{2} \\ 2(a-3)\left(\frac{a}{a-3}\right) &= 2(a-3)\left(\frac{3}{a-3} - \frac{3}{2}\right) \\ 2a &= 6-3(a-3) \\ 2a &= 6-3a+9 \\ 5a &= 15 \\ a &= 3 \end{aligned}$$
Since $a = 3$ causes the denominator $a - 3$ to be 0, the equation has no solution.

23.
$$\begin{aligned} \sqrt{2x-2} &= x-5 \\ \left(\sqrt{2x-2}\right)^2 &= (x-5)^2 \\ 2x-2 &= x^2-10x+25 \\ 0 &= x^2-12x+27 \\ 0 &= (x-3)(x-9) \end{aligned}$$
$x-3=0$ or $x-9=0$
$x=3$ $\quad$ $x=9$

$x = 3$ does not check, so the solution is $x = 9$.

24. $5x-7y=10$

$y=0$: $5x-7(0)=10$
$$\begin{aligned} 5x-0 &= 10 \\ 5x &= 10 \\ x &= 2 \end{aligned}$$
x-intercept: (2, 0)

Copyright © 2012 Pearson Education, Inc. Publishing as Prentice Hall.

$x = 0$: $5(0) - 7y = 10$
$0 - 7y = 10$
$-7y = 10$
$y = -\frac{10}{7}$

y-intercept: $\left(0, -\frac{10}{7}\right)$

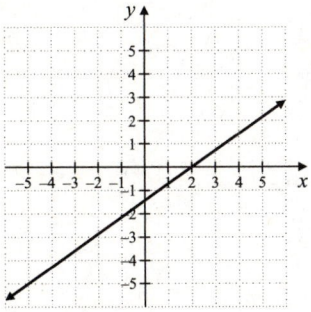

25. The graph of $y = -1$ is a horizontal line with a y-intercept of $(0, -1)$.

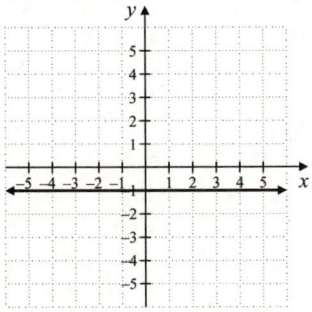

26. Graph the boundary line, $y = -4x$, with a solid line.
Test $(1, 1)$: $y \geq -4x$
$1 \geq -4(1)$
$1 \geq -4$ True
Shade the half-plane containing $(1, 1)$.

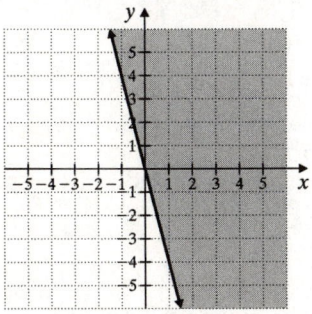

27. $m = \frac{y_2 - y_1}{x_2 - x_1} = \frac{2-(-5)}{-1-6} = \frac{7}{-7} = -1$

28. $y = mx + b$
$-3x + y = 5$
$y = 3x + 5$
$m = 3$

29. $m = \frac{y_2 - y_1}{x_2 - x_1} = \frac{3-(-5)}{1-2} = \frac{3+5}{-1} = -8$
$y - y_1 = m(x - x_1)$
$y - (-5) = -8(x - 2)$
$y + 5 = -8x + 16$
$y = -8x + 11$
$8x + y = 11$

30. $m = \frac{1}{8}$; $b = 12$
$y = \frac{1}{8}x + 12$
$8y = x + 8(12)$
$8y = x + 96$
$x - 8y = -96$

31. $\begin{cases} 3x - 2y = -14 \\ y = x + 5 \end{cases}$
Substitute $x + 5$ for y in the first equation and solve for x.
$3x - 2y = -14$
$3x - 2(x + 5) = -14$
$3x - 2x - 10 = -14$
$x - 10 = -14$
$x = -4$
$y = x + 5 = -4 + 5 = 1$
The solution of the system is $(-4, 1)$.

32. $\begin{cases} 4x - 6y = 7 \\ -2x + 3y = 0 \end{cases}$
Multiply the second equation by 2.
$\begin{cases} 4x - 6y = 7 \\ 2(-2x + 3y) = 2(0) \end{cases} \rightarrow \begin{cases} 4x - 6y = 7 \\ -4x + 6y = 0 \end{cases}$
Add the equations to eliminate x.
$$\begin{array}{r} 4x - 6y = 7 \\ -4x + 6y = 0 \\ \hline 0 = 7 \end{array}$$
Since the statement $0 = 7$ is false, the system has no solution.

Copyright © 2012 Pearson Education, Inc. Publishing as Prentice Hall.

33. $f(x) = x^3 - x$

a. $f(-1) = (-1)^3 - (-1) = -1 + 1 = 0$

b. $f(0) = 0^3 - 0 = 0 - 0 = 0$

c. $f(4) = 4^3 - 4 = 64 - 4 = 60$

34. No vertical line will intersect the graph more than once, so the graph is the graph of a function.
From the graph, the domain is all real numbers and the range is $\{2\}$.

35. $\sqrt{16} = 4$, because $4^2 = 16$ and 4 is positive.

36. $\sqrt[3]{125} = 5$, because $5^3 = 125$.

37. $\sqrt{\frac{9}{16}} = \frac{3}{4}$, because $\left(\frac{3}{4}\right)^2 = \frac{9}{16}$.

38. $\sqrt{54} = \sqrt{9 \cdot 6} = \sqrt{9}\sqrt{6} = 3\sqrt{6}$

39. $\sqrt{24x^8} = \sqrt{4x^8 \cdot 6} = \sqrt{4x^8}\sqrt{6} = 2x^4\sqrt{6}$

40.
$$\begin{aligned}
&\sqrt{18} - \sqrt{75} + 7\sqrt{3} - \sqrt{8} \\
&= \sqrt{9 \cdot 2} - \sqrt{25 \cdot 3} + 7\sqrt{3} - \sqrt{4 \cdot 2} \\
&= \sqrt{9}\sqrt{2} - \sqrt{25}\sqrt{3} + 7\sqrt{3} - \sqrt{4}\sqrt{2} \\
&= 3\sqrt{2} - 5\sqrt{3} + 7\sqrt{3} - 2\sqrt{2} \\
&= (3-2)\sqrt{2} + (-5+7)\sqrt{3} \\
&= \sqrt{2} + 2\sqrt{3}
\end{aligned}$$

41.
$$\begin{aligned}
\frac{\sqrt{40x^4}}{\sqrt{2x}} &= \sqrt{\frac{40x^4}{2x}} \\
&= \sqrt{20x^3} \\
&= \sqrt{4x^2 \cdot 5x} \\
&= \sqrt{4x^2}\sqrt{5x} \\
&= 2x\sqrt{5x}
\end{aligned}$$

42.
$$\begin{aligned}
\sqrt{2}\left(\sqrt{6} - \sqrt{5}\right) &= \sqrt{2}\sqrt{6} - \sqrt{2}\sqrt{5} \\
&= \sqrt{12} - \sqrt{10} \\
&= \sqrt{4 \cdot 3} - \sqrt{10} \\
&= 2\sqrt{3} - \sqrt{10}
\end{aligned}$$

43. $\frac{8}{\sqrt{5y}} = \frac{8}{\sqrt{5y}} \cdot \frac{\sqrt{5y}}{\sqrt{5y}} = \frac{8\sqrt{5y}}{5y}$

44.
$$\begin{aligned}
\frac{8}{\sqrt{6}+2} &= \frac{8}{\sqrt{6}+2} \cdot \frac{\sqrt{6}-2}{\sqrt{6}-2} \\
&= \frac{8\left(\sqrt{6}-2\right)}{\left(\sqrt{6}\right)^2 - 2^2} \\
&= \frac{8\left(\sqrt{6}-2\right)}{6-4} \\
&= \frac{8\left(\sqrt{6}-2\right)}{2} \\
&= 4\left(\sqrt{6}-2\right) \\
&= 4\sqrt{6} - 8
\end{aligned}$$

45. Let n be the number.
$$\begin{aligned}
n + 5\left(\frac{1}{n}\right) &= 6 \\
n + \frac{5}{n} &= 6 \\
n\left(n + \frac{5}{n}\right) &= n(6) \\
n^2 + 5 &= 6n \\
n^2 - 6n + 5 &= 0 \\
(n-5)(n-1) &= 0 \\
n - 5 = 0 \quad &\text{or} \quad n - 1 = 0 \\
n = 5 \quad & \qquad\quad n = 1
\end{aligned}$$
The number is 1 or 5.

46. Let x = area code 1, then $2x$ = area code 2.
$$\begin{aligned}
x + 2x &= 1203 \\
3x &= 1203 \\
\frac{3x}{3} &= \frac{1203}{3} \\
x &= 401
\end{aligned}$$
$2x = 2(401) = 802$
The area codes are 401 and 802.

47. Let x be speed of one hiker and y be the speed of the other hiker.
$$\begin{cases} 4x + 4y = 36 \\ y = 2x \end{cases}$$
Substitute $2x$ for y in the first equation; then solve for x.

Copyright © 2012 Pearson Education, Inc. Publishing as Prentice Hall.

$$4x+4y=36$$
$$4x+4(2x)=36$$
$$4x+8x=36$$
$$12x=36$$
$$x=3$$

Now solve for y.
$y = 2x = 2(3) = 6$
One hiker hikes at 3 mph and the other hikes at 6 mph.

48. Let x = number of cc of 12% solution and y = number of cc of 16% solution.

$$\begin{cases} x+80=y \\ 0.12x+0.22(80)=0.16y \end{cases}$$

Substitute $x + 80$ for y in the second equation; then solve for x.

$$0.12x+17.6=0.16y$$
$$0.12x+17.6=0.16(x+80)$$
$$0.12x+17.6=0.16x+12.8$$
$$4.8=0.04x$$
$$120=x$$

120 cc of the 12% saline is needed to make the 16% solution.

Copyright © 2012 Pearson Education, Inc. Publishing as Prentice Hall.